SOLAR MAGNETIC FIELDS

ASTROPHYSICS AND SPACE SCIENCE LIBRARY

VOLUME 189

SOLAR MAGNETIC FIELDS

POLARIZED RADIATION DIAGNOSTICS

by

JAN OLOF STENFLO

Institute of Astronomy,
Swiss Federal Institute of Technology,
Zurich, Switzerland

KLUWER ACADEMIC PUBLISHERS

DORDRECHT / BOSTON / LONDON

A C.I.P. Catalogue record for this book is available from the Library of Congress

ISBN 0-7923-2793-4

Published by Kluwer Academic Publishers,
P.O. Box 17, 3300 AA Dordrecht, The Netherlands.

Kluwer Academic Publishers incorporates
the publishing programmes of
D. Reidel, Martinus Nijhoff, Dr W. Junk and MTP Press.

Sold and distributed in the U.S.A. and Canada
by Kluwer Academic Publishers,
101 Philip Drive, Norwell, MA 02061, U.S.A.

In all other countries, sold and distributed
by Kluwer Academic Publishers Group,
P.O. Box 322, 3300 AH Dordrecht, The Netherlands.

Printed on acid-free paper

Printed in the Netherlands

To Joyce,
Erik and Martin

Cover: Magnetic fields are present all over the solar surface in a highly fragmented or intermittent form. They can be mapped by recording the Zeeman-effect polarization in selected lines of the solar spectrum. The picture shows a portion of the quiet sun at the center of the solar disk. Blue and red indicate fields directed towards or away from us, against a green background with vanishing line-of-sight component. This magnetogram is a color version of Fig. 1.12, courtesy J.W. Harvey, National Solar Observatory / Kitt Peak.

CONTENTS

PREFACE

Magnetic fields are responsible for much of the variability and structuring of cosmical objects, from active galactic nuclei and accretion disks to stellar flares and activity cycles. Yet it is only for the sun and the earth's magnetosphere that we have the possibility to resolve and study in detail the physics of the interaction between magnetic fields and cosmic plasma. Being a typical star the sun offers us a unique astrophysical laboratory, in which an otherwise inaccessible parameter domain of plasma physics and magnetohydrodynamics can be explored. It is thus not surprising that research about the sun has increasingly focused on the sun's magnetic field. This area of solar physics has seen an explosive growth during the past few decades.

Our knowledge about the sun and stars derives almost exclusively from analysis of their electromagnetic radiation. In the case of cosmic magnetic fields the most direct information is obtained from analysis of the *polarization* of the emitted radiation. It is for this reason that the present monograph has the subtitle "Polarized Radiation Diagnostics", since our aim is to develop a solid foundation on which our knowledge about the magnetic field can be based.

Although information on the coronal magnetic fields can also be gained from the polarization of the radio continuum, it is the polarization of the Fraunhofer lines that allows us to systematically explore and map solar magnetic fields all over the solar disk with high spatial resolution. This spectral-line polarization is caused by the Zeeman effect as well as by coherent scattering processes, which are modified in the presence of a magnetic field (the Hanle effect). It is the understanding of the processes by which the line polarization is generated that allows us to diagnose the physical conditions in the magnetic flux elements on the sun and on other non-degenerate stars.

The now "classical" treatises of the physics of stellar atmospheres by Unsöld (1955) and Mihalas (1978) have provided us with a diagnostic foundation for the analysis of unpolarized stellar spectra and the derivation of unmagnetized model stellar atmospheres from observational data. In contrast the theory of line formation in magnetic fields has been developed in bits and pieces over the past decades, but a unifying monograph has been missing. Different authors have generally addressed only certain aspects of the theory while using different formalisms and theoretical frameworks. Time is therefore overdue for a textbook that tries to expound the basic theory in some depth in a comprehensive way, to provide a diagnostic foundation for investigations of solar and stellar magnetic fields.

The present monograph may serve as an introduction to solar and stellar magnetism for astronomers and physicists at the graduate or advanced undergraduate level. Since no similar systematic treatment of this area exists, it is hoped that this text may also become a resource book for more senior scientists with a general interest in cosmic magnetic fields.

The structure of the book is as follows: A fairly extensive overview of our knowledge about the sun's magnetic field in Chapter 1 provides the background for the ensuing systematic development of the foundation for the diagnostics of solar magnetic fields. The theory of line formation is developed within a classical framework from first principles (Maxwell's equations) in Chapters 2–5. While Chapter 2 presents the basic mathematical tools to deal with polarization physics, Chapter 3 gives a general treatment of the interaction between matter and radiation that is used in Chapters 4 and 5 to develop the theory of polarized radiative transfer in magnetic fields. First we treat the simpler special case when radiative scattering is ignored in Chapter 4, while radiative scattering in a magnetic field, including coherence and the Hanle effect as well as a preliminary treatment of partial redistribution, are the topics dealt with in Chapter 5.

The classical theory is extended in Chapter 6 in a phenomenological way to multilevel atoms and anomalous Zeeman splitting patterns, with transition probabilities and the equations of statistical equilibrium. This gives us a non-LTE formalism for line formation in magnetic fields, which however does not allow us to calculate certain quantum phenomena, like the polarizability coefficients in Rayleigh and Raman scattering, or general interference effects between polarized atomic levels. To establish a more general theory we need to quantize not only the atomic system but also the radiation field (since basic processes like spontaneous emission cannot be understood in quantum mechanics without a zero-point energy of the vacuum). The quantum field theory of polarized radiation and matter is introduced in Chapter 7 and used to develop a very general radiative-transfer theory in Chapter 8. The physical understanding of the often obstruse formalism of quantum field theory is greatly aided by making frequent comparisons with the corresponding classical theory of Chapters 2–5, to make the relation between classical and quantum radiation physics more transparent.

While Chapter 8 provides a basis for our physical insight, one rarely needs to use the theory in its general form for practical applications. Instead one deals with commonly occurring special cases, or uses the general theory to derive algebraic expressions in terms of the quantum numbers for certain coefficients that appear in the non-LTE radiative transfer formalism. For most practical purposes one may ignore the possible coherences in the initial state of scattering processes. A convenient phase-matrix formalism is developed in Chapter 9 for this case, including explicit algebraic expressions in terms of the quantum numbers for all the possible cases of excited m- and J-state interferences that may occur in Rayleigh and Raman scattering, illustrated by a solar observation showing such a phenomenon. This scattering theory is extended in Chapter 10 to include the effects of collisional destruction of the atomic alignment and orientation, which allows us to make a comprehensive formulation of the theory of partial redistribution of polarized radiation in a magnetic field. This theory is needed as a diagnostic tool to explore the properties of spatially unresolved turbulent magnetic fields on the sun via the Hanle effect, as reviewed in Chapter 10.

Before the book turns to the practical applications, various ways of solving the polarized radiative-transfer problem are presented in Chapter 11, including analytical and numerical solutions and Stokes inversions. The fundamental problem in applying these methods lies in the circumstance that the sun's magnetic field is so inhomogeneous that the best observations do not spatially resolve the basic flux elements or "flux tubes". Chapter 12 is devoted to this problem area, how to extract information on solar and stellar magnetic fields although they are spatially unresolved. This is the main chapter that directly deals with the interpretation of actual observations and which presents the empirical results on the derived magnetic and thermodynamic structure of the basic magnetic building blocks of the sun, the flux tubes. Chapter 13 finally reviews the observational tools and techniques to measure the full state of polarization needed to determine vector magnetic fields.

SI units are consistently used throughout the book, with the exception that numerical *results* on field strengths are given in G (Gauss), although the SI unit T (Tesla, $= 10^4$ G) is to be used in the formulae. In addition, we give wavelengths in Å instead of nm. As most previous literature on radiation physics has been in terms of the cgs system, and since the various conversion factors combine in a non-trivial way in the different radiation parameters, it has been of value to develop the theory from first principles to be sure to get the various expressions right in terms of the SI system.

In preparing this book I have greatly benefitted from discussions and support from my colleagues at the Institute of Astronomy at ETH Zurich, in particular from Christoph Keller, Sami Solanki, and Susi Weber. Among colleagues elsewhere the clarifying scientific discussions with Egidio Landi Degl'Innocenti and Marianne Faurobert-Scholl deserve special mention. Illustrations for the book have kindly been provided by Jack Harvey, Christoph Keller, Bruce Lites, Bill Livingston, Gordon Newkirk, Peter Povel, Bogdan Rompolt, Dick Shine, Yutaka Uchida, Hal Zirin, and the NSO/Sacramento Peak Observatory. Other figures have been reproduced with kind permission of Kluwer, ESO, American Astronomical Society, Macmillan Magazines Ltd, Oxford University Press, Plenum Press, and The Institute of Physics. The most important support has however been that of my family. I thank my wife, Joyce, and our two sons, Erik and Martin, for their understanding, patience, and encouragement in this seemingly endless undertaking.

Institute of Astronomy
ETH Zurich

January 1994 *Jan Olof Stenflo*

CHAPTER 1

SOLAR MAGNETISM — AN OVERVIEW

1.1. Introduction

The sun is a unique object in astrophysics, since it is the only star on which the physical processes can be resolved in some detail. By observing the sun we know that stellar atmospheres are highly structured, governed by an intricate interplay between magnetic forces, turbulent motions, and heating processes. These interactions manifest themselves in various signatures of stellar activity, like dark spots, bright faculae, flare eruptions, X-ray emissions, coronal transients, wind streams, etc. On a global scale they form a complex pattern that has a cyclic evolution. The sun serves as a prototype for stellar activity.

The sun can also be seen as a unique plasma physics laboratory, in which a domain of physics not accessible to experiments in earth laboratories can be explored. If we apply similarity transformations of plasma physics (Alfvén and Fälthammar, 1963) to reduce the solar dimensions down to those of terrestrial laboratories, then the scaled values of the magnetic field strengths and electric current densities become so large that they are orders of magnitude beyond reach of MHD simulations on earth. The sun continually performs new experiments in its laboratory. By remote sensing of these experiments we can explore an otherwise inaccessible domain of physics. This physics is needed for an understanding of the origins and mechanisms of magnetic activity in cosmical objects.

The sun is further the energy source for our planetary system, and all life on earth depends on it. Interplanetary space is filled with the tenuous and hot, streaming medium of the supersonically expanding solar corona, which forms the heliosphere that reaches far beyond the orbit of Pluto before being braked as it encounters the interstellar medium. The "climate" of the heliosphere fluctuates violently in response to the magnetic activity on the sun, but the thin terrestrial atmosphere and the magnetosphere shield us from the bombardment of high-energy particles, UV radiation, etc.

In spite of this shielding, even the terrestrial climate that we experience on the earth's surface gets modulated by the sun's magnetic activity. The total radiative energy received from the sun by the earth is found to vary in phase with the 11 yr activity cycle. On shorter time scales it shows larger fluctuations that are correlated with the appearance of dark sunspots and bright faculae. Thus the energy that is fed into the "climate engine" is modulated and perturbed by the magnetic processes on the sun. These solar effects need to be identified when trying to determine the residual, man-made disturbances that are caused by our human activity.

The source of the solar irradiance variations is to be found in the accumulated effect of highly localized heating and cooling processes in regions with concentrated magnetic fields. An understanding of the thermodynamics of these flux elements, which in the following will be referred to as flux tubes, is thus of central importance to solar-terrestrial physics.

The key physical parameter behind these various forms of solar activity is the magnetic field. The only way that it can be directly diagnosed through remote sensing is by observations of the polarization of the solar electromagnetic radiation. Although qualitative information on the presence of magnetic fields and the shapes of the field lines can be obtained by observing the shapes of the emission structures with filters at various wavelengths, *quantitative* information requires polarization measurements. The present monograph will focus on the diagnostic foundations for remote sensing of the sun's magnetic field, including a systematic treatment of the physics of spectral line formation in the presence of magnetic fields. These foundations are general in nature and can be applied to a variety of other astrophysical contexts as well.

At the core of polarized radiation diagnostics of the magnetic field is the Zeeman effect, which causes a splitting of the atomic energy levels due to a precession of the atom around the magnetic field. The split components of the spectral lines become polarized depending on the magnitude and direction of the magnetic field vector. By measuring the full state of polarization (including circular and linear polarization) the magnetic field vector can thus be retrieved, at least in principle.

The inversion of polarimetric data to determine the magnetic field depends on the process of line formation (involving polarized radiative transfer) in an inhomogeneous, magnetized solar atmosphere, in which the physical parameters vary strongly along the line of sight. The problem is greatly complicated by the circumstance that the magnetic field is highly intermittent, with the basic building blocks of the magnetic field structure being smaller than what can currently be spatially resolved on the sun. Various fairly sophisticated indirect techniques have been devised to overcome this obstacle and obtain information on the intrinsic physical parameters independent of any spatial smearing. In parallel with these developments the resolution limit has been pushed towards ever smaller scales by various advances in observational techniques, to allow direct imaging of the partially resolved magnetic field pattern.

A different class of polarization effects, due to coherent scattering, can also be used for magnetic field diagnostics. Like the blue sky is polarized due to Rayleigh scattering at molecules, certain lines in the solar spectrum become polarized when they are partly formed by coherent scattering. In a magnetic field the precession of the atoms induces a modification, called the *Hanle effect*, of this scattering polarization. The Hanle and Zeeman effect complement each other, since the Hanle effect is sensitive to a different magnetic field regime (weak fields and fields of mixed polarity) as compared with the ordinary Zeeman effect. The present monograph also tries to provide a diagnostic foundation for the use of the Hanle effect by including it in a unified treatment of the interaction between the radiation and the magnetic field.

Overviews of various aspects of solar magnetic fields can be found in the proceedings of recent conferences, in particular the IAU Symposium in Kiev (Stenflo,

Fig. 1.1. Evidence that the sun is a magnetized sphere. This photograph of the white-light solar corona was obtained during the eclipse of 12 November 1966 in Bolivia. The overexposed image of Venus is seen to the left. Courtesy G. Newkirk, Jr., High Altitude Observatory, Boulder, Colorado.

1990) and the IAU Colloquium in Helsinki (Tuominen et al., 1991). The most recent extensive reviews on the properties and diagnostics of small scale solar magnetic fields are those of Schüssler (1987), Zwaan (1987), Stenflo (1989), and Solanki (1993). These reviews contain abundant references to previous work. For a general introduction to solar physics the book "The Sun" by Stix (1989) is highly recommended. A theoretical treatise on "Solar Magnetohydrodynamics" has been given by Priest (1982).

1.2. Emission Patterns and Field Topology

From a glance at the solar corona during a total eclipse (cf. Fig. 1.1) it is evident that the sun is a magnetized sphere. This conclusion led Hale to look for and find Zeeman splittings of spectral lines in sunpots, which constituted the first detection of magnetic fields on the sun. The dynamics of the tenuous, hot coronal plasma is largely dominated by magnetic forces, since the kinetic energy density

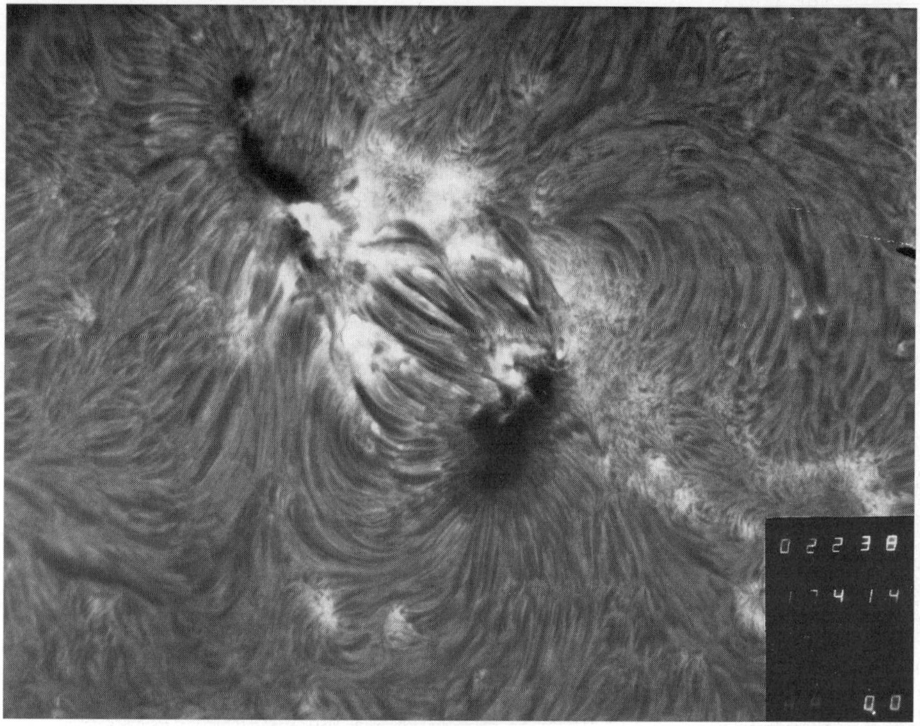

Fig. 1.2. Visualization of the magnetic field lines by the dark Hα fibrils connecting two sunspots of opposite magnetic polarities in a newly emerged active region. The photograph was made on 23 February 1981. Courtesy H. Zirin, Big Bear Solar Observatory, California.

of the plasma is much smaller than the magnetic energy density. Under such conditions the magnetic field is said to be nearly force-free, which means that the non-magnetic forces are small in comparison with the magnetic pressure and tension forces in the force balance equation. The emission structures (which for the white-light corona represent regions of enhanced electron density) are governed by the shapes of the magnetic field lines, since the field channels the mass and energy flows. In Fig. 1.1 the orientation of the magnetic axis, which here also coincides with the rotational axis, is evident: it is inclined from upper left to lower right in the figure.

Magnetic field lines get visualized in a way that is reminiscent of the iron filings above a bar magnet when the sun is viewed in the light of the hydrogen Hα line, as shown by Fig. 1.2. While the dark fibrils outline the horizontal component of the magnetic field, the bright so-called plages are areas of predominantly vertical magnetic fields (Veeder and Zirin, 1970). The core of the Hα line is formed in the upper chromosphere, which like the corona is magnetically controlled and largely "force-free". The field lines are however anchored in the denser layers of the solar body, where the conditions are not force-free. There the field lines are instead passively carried around by the plasma, controlled by ordinary, non-magnetic gas dynamics.

The pattern of magnetic fields on the sun evolves with the 11 yr activity cycle, but the different magnetic features vary over much shorter time scales, depending on the size of the structures. New magnetic flux is continually being produced inside the sun. Bipolar magnetic regions are formed when this flux emerges at the solar surface. Larger flux concentrations inhibit convective energy transport from the sun's interior and are seen as darker structures, called sunspots. Since new magnetic flux can only emerge in bipolar form (due to the absence of magnetic monopoles), sunspots usually appear in pairs of opposite magnetic polarities. Figure 1.2 is an example of a recently emerged sunspot group with fibrils, being proxies of the magnetic field, connecting the spots of opposite polarities.

Fig. 1.3. Visualization of the 3-D magnetic field structure above the solar limb, here in the form of an arcade of Hα loops, which developed immediately above and after a solar flare. Courtesy B. Rompolt, Astronomical Observatory, Wrocław, Poland.

The dark Hα fibrils actually represent 3-D structures seen in projection against the solar disk. The height dimension becomes visible if we observe the structures at the solar limb with a coronagraph, while blocking the intense radiation from the solar disk. An example of a system of magnetic loops that developed above the limb region where a solar flare had just occured is shown in Fig. 1.3. Hα movies of such nested loop systems show streaming motions of the emission structures along the arches.

The bipolar magnetic regions where sunspots and flaring activity occur are called "active regions". Magnetic fields are however in no ways limited to active regions but are present all over the sun. In the "quiet regions" the field gets organized in a cell pattern, the so-called *supergranulation*, which has its origin in convective motions at intermediate depths in the solar convection zone (which occupies the outer part of the solar body, down to a depth of 30 % of the solar radius). The horizontal pressure gradients across the convection cells propagate to the visible surface layers and drive horizontal motions there, forming the supergranular velocity cells that have typical diameters of 30 Mm (megameters). As the magnetic field lines are "frozen-in" due to the high electrical conductivity of the solar plasma, they are carried by the horizontal flows to the cell boundaries, where the magnetic flux gets concentrated. This is visualized by Fig. 1.4, a photograph of a region near the solar limb taken in the wing of the Hα line.

In Fig. 1.4 the short, dark, fibril-like structures mark the location of the magnetic fields. At the limb these absorption features would appear as the bright *spicules*, plasma jets shooting out with speeds of 10 km s^{-1} or more from the magnetic-field concentrations. At the foot-points of the "dark spicules" in Fig. 1.4 one can notice tiny, bright emission points, marking the foot-points of the corresponding magnetic-field concentrations with enhanced heating. These bright emission points occur along the boundaries of the supergranular cells and form what is called the *network*, which can be seen in spectral lines formed in the upper photosphere, chromosphere, as well as in the chromosphere-corona transition region. The structure of the emission network gets coarser the higher up we go in the solar atmosphere, suggesting that the cross section of the structures diverges with height.

We expect that the cross section of the magnetic flux elements should also rapidly expand with height, due to the almost exponential decrease of the external gas pressure and the transition from a line-tied to a force-free atmosphere. Thus a magnetic field that is highly intermittent in the photosphere should be volume-filling in the corona. This conclusion may seem to be contradicted by Figs. 1.2–1.4, which show that the fibrils and loops have an emission or absorption cross section that does not vary much with height. It is however important to remember that what we are seeing is not the magnetic field itself but only proxies outlining the shape of the field. These proxies do not provide information on the flux densities. Only certain field lines get visualized as absorption or emission features, depending on complex wave and heating processes, with fluctuating, "hidden" boundary conditions determining the localized mass and energy input in the photosphere. To find out how the field lines expand we have to measure the field itself and not rely on proxies.

In Hα radiation coronal magnetic fields can be visualized as in Fig. 1.3 only for condensations that are much denser and cooler (with temperatures of about 10^4 K) than the surrounding corona (10^6 K). The hot coronal plasma radiates primarily in the soft X-ray wavelength range. The 3-D coronal magnetic fields therefore get beautifully visualized in projection against the solar disk in photographs with soft X-ray telescopes, like the one in Fig. 1.5, obtained with the Japanese Yohkoh

Fig. 1.4. Visualization of the supergranulation cell pattern on the "quiet" sun near the solar limb, as seen in the wing of the Hα line. The magnetic flux elements get concentrated at the cell boundaries, where the bright points forming the emission network are located, and from which dark plasma jets (spicules) are ejected. An active region is seen in the lower portion of the figure. Courtesy NSO/Sacramento Peak, AURA, Inc.

satellite (Ogawara et al., 1991). Since the heat capacity of the low-density coronal material is so low, small energy fluctuations or mass injections can have large temperature and density effects, resulting in highly structured, high-contrast coronal X-ray emission. The shapes of the emission structures are governed by magnetic fields as was the case for the white-light corona (cf. Fig. 1.1).

The X-ray pictures allow us to see not only the shapes of the field lines but also the *topological connections* between regions of opposite magnetic polarities. In principle all magnetic regions should be topologically connected via field lines

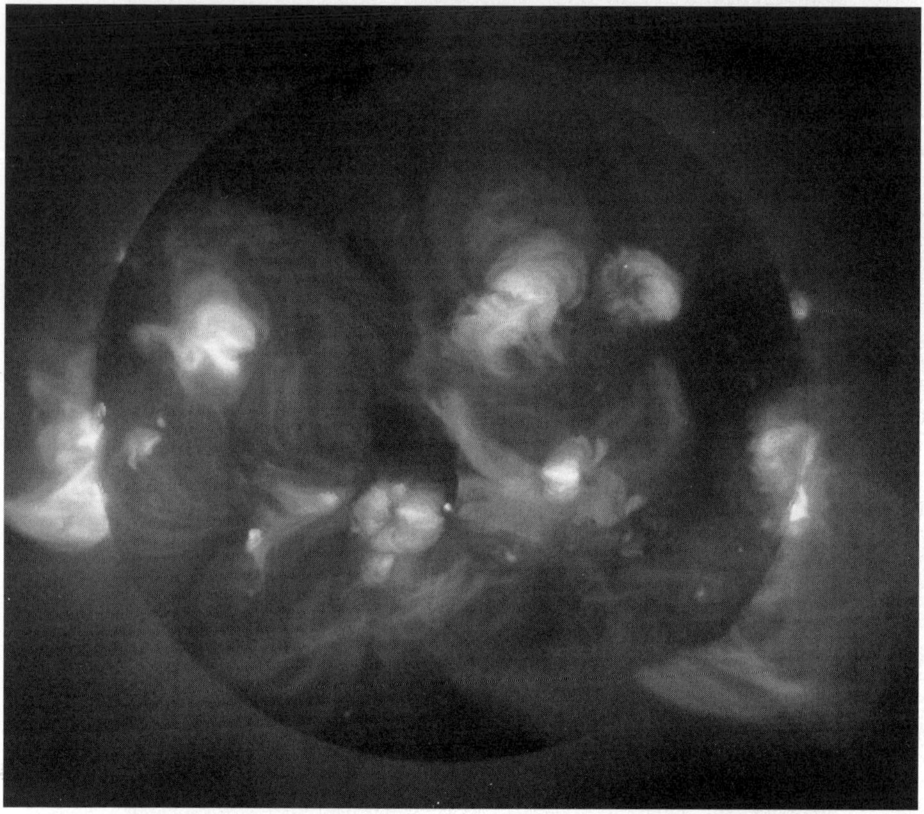

Fig. 1.5. Visualization of the global topology of the sun's magnetic field by the X-ray emission from the million-degree hot coronal plasma. The photograph, obtained with the Japanese Yohkoh satellite on 10 December 1991, shows intense emission from the closed magnetic loop structures, whereas the regions of open, diverging field lines connected to the heliosphere (the solar wind region) appear dark, as "coronal holes". Courtesy Y. Uchida, University of Tokyo.

to another region of opposite polarity, due to the divergence-free condition of Maxwell's equations (absence of magnetic monopoles). If however the two footpoints of the field line are widely separated, the field line can reach to such heights in the corona, where the magnetic tension forces can no longer withstand the pressure forces that drive the solar wind. Due to the high thermal conductivity of the coronal plasma the outer corona maintains such a high temperature that it cannot remain in hydrostatic equilibrium with gravitation but has to expand (leading to the solar wind), unless it can be contained by magnetic tension forces. Such containment works for magnetic loops in the lower corona, but high loops may be pushed out by the coronal gas pressure gradients. The field lines then get convected out through the heliosphere until the solar wind is braked when it encounters the interstellar medium, at a distance from the sun a few times larger

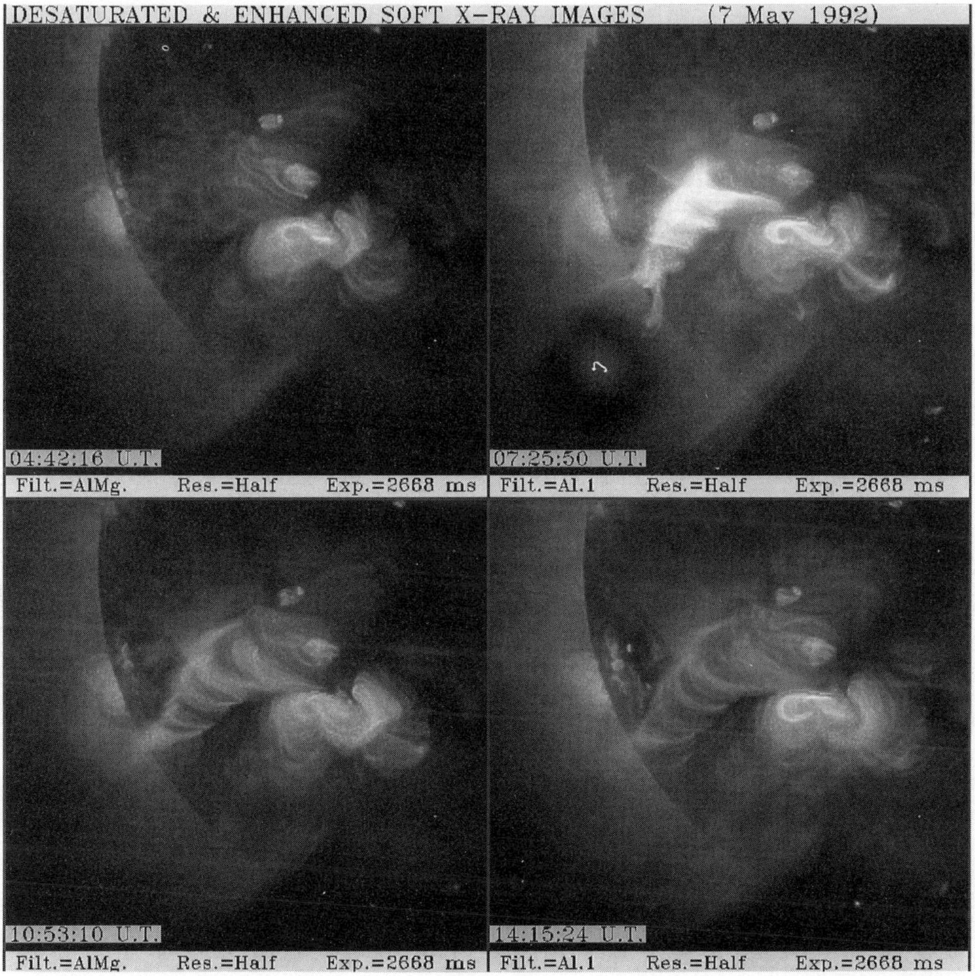

Fig. 1.6. Explosive reorganization of the coronal magnetic field with the formation of an arcade of magnetic loops. This sequence of X-ray images with a time step of about 3 hr was obtained with the Yohkoh satellite on 7 May 1992. Courtesy Y. Uchida, University of Tokyo.

than that of Pluto. Only at these outskirts of the solar system do the field lines "close" to topologically connect with an opposite-polarity region on the sun. As the closure takes place at such a large distance, the corresponding field lines can be considered as "open" in the context of the thermodynamics of the corona.

We can thus distinguish between two main topological classes of coronal magnetic fields: open and closed. In the open regions the field lines extend through interplanetary space, and the coronal plasma is allowed to freely expand. The closed loop regions on the other hand can be regarded as "magnetic bottles", in which heat and mass can get trapped and accumulate. They are therefore found to have higher densities and temperatures, resulting in much higher X-ray emission as compared with the open regions.

The regions of open magnetic fields are called *coronal holes* (since they look dark in X-ray radiation). They may occur at all latitudes but are most prominent over the polar caps during periods of minimum solar activity, since then the sun's magnetic field has a global, dipole-type structure with diverging field lines of predominantly one polarity over each pole (cf. the shape of the field in Fig. 1.1). The closed-field regions can occur in different sizes. The largest and most prominent ones in the X-ray pictures are the intense loops over active regions with sunspots. The loop systems often become unstable and explode into what is called a *coronal transient*. Smaller closed loops are seen as *X-ray bright points*. They appear at all latitudes and are not limited to the zones where normal active regions with sunspots occur. They coincide with small bipolar magnetic regions and have short life times, typically 8–12 hr.

The coronal magnetic field is in a highly dynamic state and is subject to various instabilities. Figure 1.6, also obtained with the Yohkoh satellite, gives an example of an explosive magnetic reconfiguration of the corona accompanied by the formation of an arcade of magnetic loops. The sequence that is shown spans about 10 hr (Uchida, 1993).

1.3. Zeeman Effect on the Sun

Quantitative measurements of solar magnetic fields are made possible by the Zeeman effect, which was discovered in the laboratory in 1896 (Zeeman, 1897a,b,c) and was first seen on the sun by Hale (1908) in sunspots. Figure 1.7 illustrates the Zeeman effect in a sunspot. Due to the precession of the atoms in a magnetic field and the resulting mixing of the precession frequency with the atomic transition frequency, spectral lines get split up in separate components if there is a magnetic field in the atmospheric layers where the respective spectral lines are formed. The splitting pattern depends on the quantum numbers of the transition and on the strength and direction of the magnetic field. Certain transitions, like that of the Fe I 5250.22 Å line, represent so-called normal Zeeman triplets, with one unshifted π component and two σ components shifted to either side of the π component. Other transitions usually have "anomalous" splitting patterns with several π and σ components.

While the amount of wavelength shift is proportional to the field strength and the Landé factor of the line, the polarization and the relative strengths of the components are determined by the field direction. For a field directed along the line of sight (longitudinal Zeeman effect) the σ components have right- and left-handed circular polarization, while for a field perpendicular to the line of sight (transverse Zeeman effect) the σ and π components have mutually orthogonal linear polarizations, parallel and perpendicular to the direction of the field (assuming an absorption line — for an emission line it is the π component that is polarized parallel to the magnetic field). When the field has an arbitrary direction the Zeeman line components are elliptically polarized.

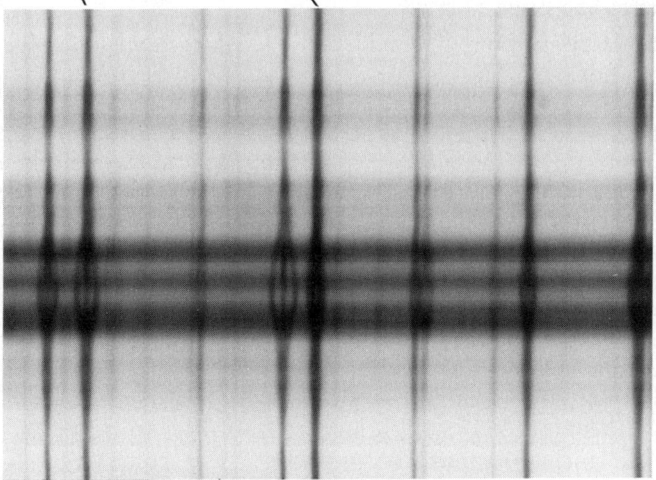

Fig. 1.7. Zeeman splitting in a sunspot, which appears as the dark horizontal band due to its reduced continuum radiation. The maximum splitting corresponds to a field strength of 4.1 kG. This region of the spectrum has been extensively used for the diagnostics of small-scale magnetic fields, as explained in Chapter 12. A tracing of the corresponding circularly polarized line profiles is shown in Fig. 12.1 for a plage region. Courtesy W.C. Livingston, NSO/Kitt Peak, Tucson.

The spectral region shown in Fig. 1.7 has played an important role for the diagnostics of small-scale magnetic fields, as we will indicate below and in greater detail in Chapter 12. The slit of the spectrograph cuts across a sunspot (appearing as a dark band in the continuous spectrum), which exhibits an unusually large Zeeman splitting, corresponding to a field-strength amplitude of 4.1 kG. The spatial resolution along the slit is however rather low. An example of a spectrogram with high spatial resolution is given in Fig. 1.8. The slit-jaw white-light images to the left show how the vertical slit cuts across the sunspot penumbra in the upper picture, across an umbra with many bright "umbral dots" in the lower picture. The spectrograms to the right show the broad, Zeeman-split Fe I 6302 Å line surrounded by two narrow telluric lines (formed in the earth's atmosphere), which may serve as wavelength references. We notice the considerable amount of spatial fine structure of the Zeeman splitting, which indicates large magnetic field gradients and therefore, as a consequence of Maxwell's equations, the presence of strong electric currents.

It is almost only in sunspots that we can see and directly measure the Zeeman splitting in unpolarized radiation, as in Figs. 1.7 and 1.8. Although sunspots represent the regions on the sun with the largest flux concentrations, magnetic flux is present all over the sun, as evidenced by the emission patterns shown in the preceding section (e.g. Fig. 1.4). Instead of the splitting the polarization produced by the Zeeman effect is a much more useful observable that allows us to detect

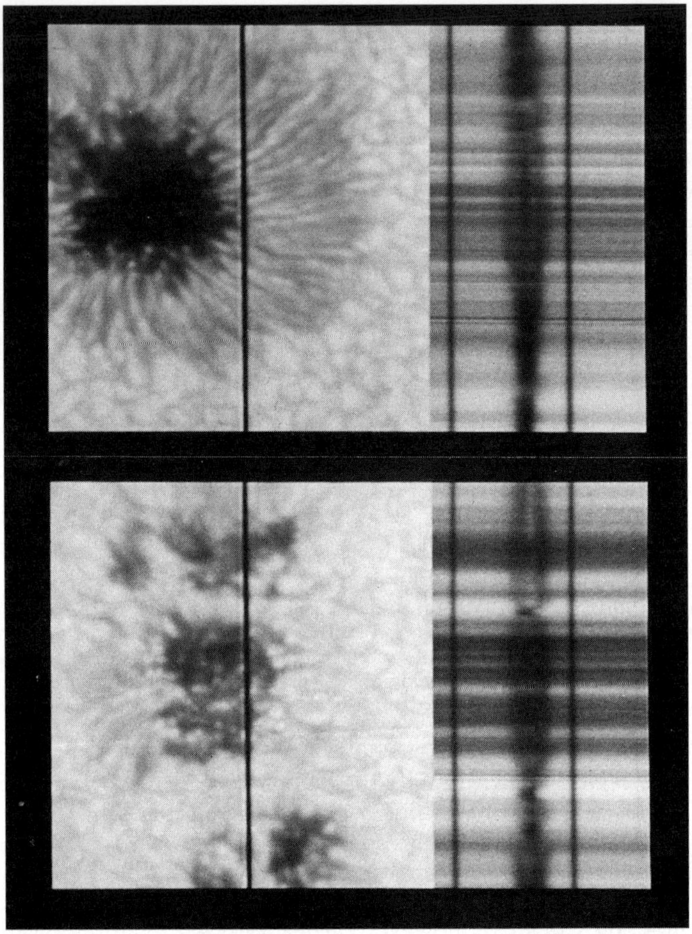

Fig. 1.8. Zeeman splitting in sunspots, recorded with high spatial resolution at the Swedish La Palma observatory. The white-light slit-jaw images to the left show the location of the spectrograph slit (dark, vertical strip). The corresponding spectra are displayed to the right, with the Zeeman-split solar Fe I 6302 Å line surrounded by two narrow and straight telluric lines. Courtesy B.W. Lites, High Altitude Observatory, Boulder, Colorado.

and quantitatively determine magnetic fields outside sunspots, anywhere on the surface of the sun. The most complete observational information is of course obtained if we can record the full state of polarization (circular *and* linear) across the spectral lines with high spectral resolution (to give us polarized line profiles).

Since the Zeeman splitting varies with λ^2, while the line width is approximately proportional to the wavelength λ, the Zeeman splitting becomes more complete when we go to longer wavelengths. It is therefore advantageous to work in the infrared, which has been increasingly done in recent years in the spectral regions around 1.6 and 12 μm (cf. Chapter 12). With few exceptions the lines in the visible are formed in the photosphere, between the level of continuum formation and the temperature minimum region, over a height range of about 500 km. By choosing lines of different elements, line strengths, and excitation potentials, one may sample different photospheric heights. This fairly restricted range can be extended somewhat by the infrared lines, since the 1.6 μm region, representing the opacity minimum of the solar spectrum, is formed at the "bottom" of the photosphere, while the Mg I 12 μm lines are formed in the low chromosphere, and the He I 10830 Å line in the middle chromosphere.

Spectral lines that are formed still higher, in the chromosphere-corona transition region or in the solar corona, occur in the vacuum ultraviolet or at still shorter wavelengths. Due to the wavelength dependence of the Zeeman splitting the polarization effects produced will be tiny and hard to measure, although the Hanle effect (see below) offers additional diagnostic possibilities. Radio observations can also provide certain limited information on the coronal magnetic fields (cf. Stenflo, 1978a; Klein, 1992). So far, however, the main method to derive quantitative information on the 3-D coronal magnetic fields has been to extrapolate the measured photospheric field into the corona and try to match the shapes of the extrapolated field lines with the shapes of the corresponding emission structures. Agreement between these shapes may be taken as evidence that the idealizations used for the extrapolation are valid approximations.

The extrapolation procedure consists of using the observed photospheric field as an inner boundary condition to solve Maxwell's equations for a force-free magnetic field outside the boundary. The simplest and most used special case is that of a current-free vacuum field, but constant-α force-free fields have also been employed, whereby α (which is proportional to the ratio between the electric current density and the field strength) is usually treated as a free parameter that is adjusted to fit the shapes of the emission structures. In reality there is no reason why α should be spatially homogeneous. Furthermore the conditions in the photosphere, where the fields are measured, are certainly far from being force free. It is therefore not physically correct to use the observed fields directly as boundary conditions for force-free solutions. Due to all these complications (cf. Stenflo, 1989, for further arguments) one should use extrapolated coronal fields with great caution.

As we cannot depend much on secondary information like emission proxies or extrapolations, accurate polarimetric diagnostics is indispensible. Remote sensing by solar polarimetry is however full of complications of different kinds. As most magnetic fluxes involved are small, the polarization degrees to be measured are generally small fractions of one percent, except when strong-field flux tubes are spatially resolved. Oblique reflections in the telescope may corrupt the incoming polarization information, and seeing effects of the earth's atmosphere may produce spurious polarization features. These instrumental problems will be addressed in Chapter 13.

One major diagnostic obstacle is due to the circumstance that the solar magnetic flux is extremely fragmented or intermittent, with fine structures on scales beyond the reach of the spatial resolution capabilities of existing instruments. To overcome the resolution limit and determine the intrinsic properties of the unresolved magnetic field, indirect methods that make use of observations with good *spectral* resolution have been developed (cf. Chapter 12). At the same time there is an ongoing quest for improving the spatial resolution to resolve more of the flux fragments or "flux tubes", since it is being increasingly realized that these smallest building blocks of solar magnetism hold the key to a unified understanding of solar activity. However, high spatial resolution also has to be combined with relatively high time resolution, since the smaller the solar structures are, the shorter their evolutionary time scale is. When we have to combine high spatial, spectral, and time resolution with high polarimetric accuracy, we quickly run out of photons, even with the largest available solar telescopes. All solar observations therefore involve some trade-off between the various resolutions or accuracies, depending on the objectives of the particular observing program. High-resolution solar polarimetry can be advanced by the construction of larger telescopes.

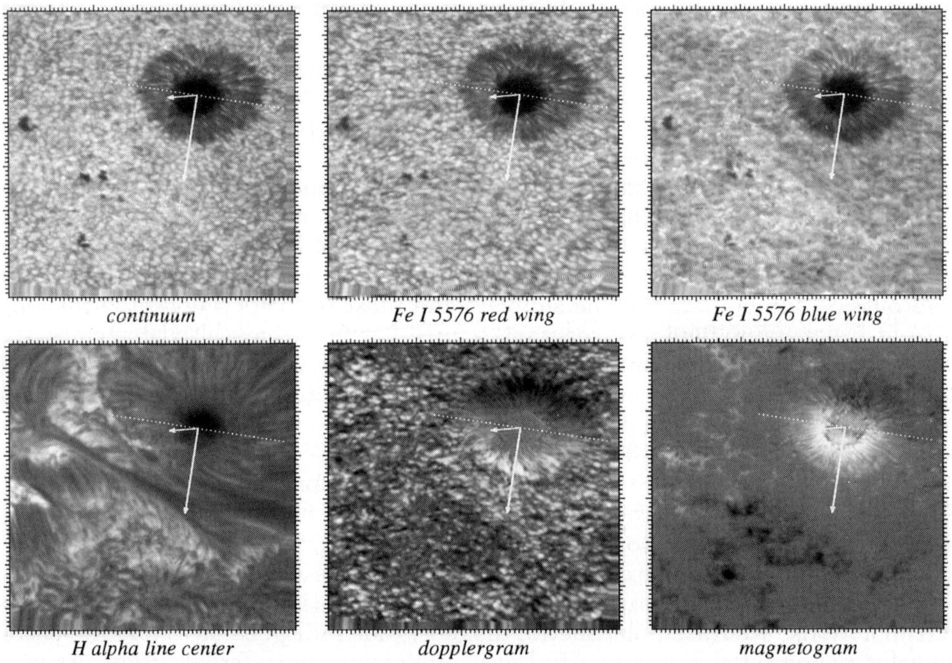

Fig. 1.9. Different representations of an active region, as observed with a narrow-band filter. The recordings were made with the Swedish La Palma telescope and the Lockheed universal filter system on 22 June 1990. The dopplergram maps the wavelength shifts of the Fe I 5576 Å line, while the magnetogram represents the degree of circular polarization in the blue wing of the Fe I 6302 Å line. While the long arrow points towards the center of the solar disk, the short arrow points in the direction of heliographic north. The dotted line is perpendicular to the long arrow and has a dot spacing of one arcsec, the same as the tick-mark spacing. Courtesy R. Shine, Lockheed Palo Alto Research Laboratories, California.

To explore the role of magnetic fields for the thermodynamics and activity of the solar atmosphere it is common to combine Zeeman-effect diagnostics with unpolarized diagnostics of radiation that responds to other physical parameters of the atmosphere. An example of a set of such correlated, complementary observations is given in Fig. 1.9, which shows an active region dominated by the sunspot in the upper right portion of the field of view. The different, almost simultaneous frames of the same region of the solar disk have been obtained with the passband of the narrow-band filter positioned at different wavelengths. The dopplergram represents a map of the wavelength displacements of the unpolarized Fe I 5576 Å line (which has a Landé factor of zero) determined from filter images at four different positions within this line. Since these displacements are induced by the Doppler effect, they are proportional to the line-of-sight component of the photospheric velocities. The magnetogram is a map of the degree of circular polarization in the blue wing of the Fe I 6302 Å line (which was shown in the spectrograms of Fig. 1.8). This Zeeman-induced polarization is in a first approximation proportional to the line-of-sight component of the magnetic flux through the spatial resolution element. The brighter and darker areas represent the opposite magnetic polarities against a grey background of apparently zero magnetic flux.

1.4. Global Evolution of the Magnetic Field

A monochromatic circular-polarization map of the full solar disk can be built up by scanning the solar image across the spectrograph entrance slit while recording the circular polarization within a narrow spectral window. If this window is in the wing of a Zeeman-sensitive spectral line, such a circular-polarization map is usually called a magnetogram, since it indicates the distribution and morphology of the line-of-sight component of the magnetic flux across the solar disk. In Fig. 1.9 we saw an example of a magnetogram obtained with a narrow-band filter instead of a spectrograph. Figure 1.10 shows an example of a spectrograph-based full-disk magnetogram. It provides a nice overview of the large-scale flux pattern on the sun at a time of high solar activity. Heliographic north (as defined by the sun's rotational axis) is at top, east is to the left. Bright areas represent regions where the magnetic field is directed towards the observer, dark regions where it is directed away from the observer.

Although the pattern in Fig. 1.10 appears quite complex, we may identify a number of characteristic features: young and old bipolar magnetic regions, latitude zones of activity, background fields, the magnetic network. The dominating bipolar magnetic regions are found in two bands at intermediate latitudes, one in the northern, the other in the southern hemisphere. When a new bipolar region with sunspots first emerges from the sun's interior, it is compact and has its polarities oriented in the approximate east-west direction. Over a time scale of months the region decays, whereby the flux gets dispersed over a larger area. The flux becomes highly intermittent, so that it consists of tiny flux fragments that mix with the intermittent background fields. The flux elements are carried around in a random-walk process by the granular and supergranular motions. They are swept to the boundaries of the supergranular cells and thereby form a network pattern with a typical cell size of 30 Mm. The individual photospheric flux elements that make up this network have a typical size expected to be about 100 km only, i.e., their areas are about 10^5 times smaller than the area of a supergranular cell. This network

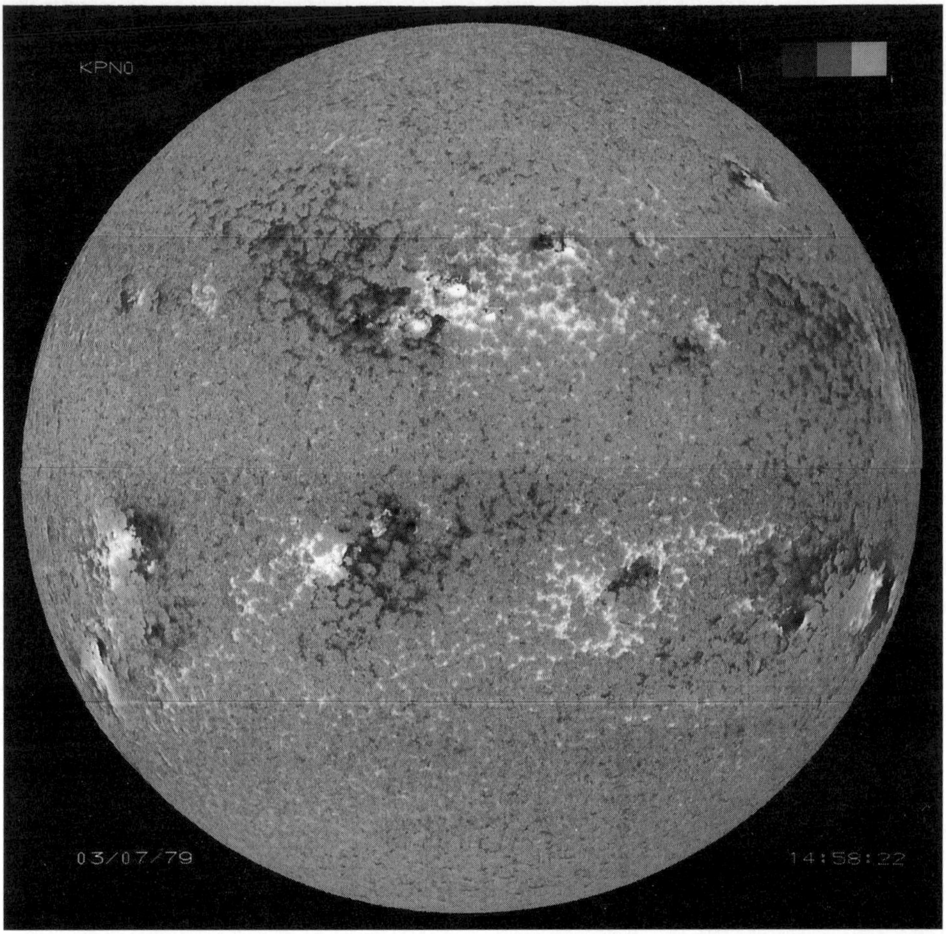

Fig. 1.10. Full-disk solar magnetogram recorded on 7 March 1979 at the National Solar Observatory (Kitt Peak). North is at top, east to the left. Bright (dark) areas represent magnetic fields directed towards (away from) us. Courtesy J.W. Harvey, NSO/Kitt Peak, Tucson.

pattern may be discerned in Fig. 1.10, although the individual flux elements are far smaller than the spatial resolution of about 2 Mm. Since the heating of the upper atmosphere is mediated by these flux tubes, the emission network that was discussed in Sect. 1.2 is cospatial with the magnetic network.

The newly born, compact bipolar magnetic regions with sunspots always obey *Hale's polarity law*, which states that the east-west polarity orientation is opposite in the north and south hemispheres, and that the orientation gets reversed when a new 11 yr activity cycle starts. As a result of this rule we find in Fig. 1.10 that in the northern hemisphere the bipolar orientation from left to right (E to W) is $-$ to $+$ (dark to bright), while the orientation is opposite in the southern hemisphere. In the case of older, partially decayed magnetic regions the law is not always evident, since the spread-out fluxes merge with the remnants of other bipolar regions and with the "background" field.

During the course of the 11 yr cycle the zones of emergence of bipolar magnetic flux migrate towards the equator. A new cycle starts when sunspot groups of reversed polarity orientation first appear (at higher latitudes). The background magnetic field develops a global pattern, which during sunspot activity minimum is approximately that of a global N-S dipole, with flux of predominantly one polarity over the north polar cap and the opposite polarity over the south pole. The polarity of a polar cap is the same as the polarity of the eastern, following (with respect to the solar rotation) portion of the bipolar magnetic regions in the same hemisphere during the preceding sunspot maximum. To come back to the same polarity orientations one has to go through two 11 yr activity cycles. The length of the *magnetic activity cycle* is therefore 22 yr.

Since the sun rotates with a period of about 27 days, one may use a series of daily full-disk magnetograms to extract information at all latitudes and longitudes and construct *synoptic maps* that show the distribution of the line-of-sight magnetic flux as a function of sin(latitude) and longitude over the whole solar surface. As the sun does not rotate like a solid body but does it differentially, faster at lower than at higher latitudes, the longitude system, which rotates rigidly, has to be adopted by convention to refer to some fixed rate. Usually the Carrington system is used, which is based on a rotation period of 27.2753 days (corresponding to the sunspot rotation rate at a latitude of about 16°).

Figure 1.11 gives an example of synoptic maps of two consecutive Carrington rotation periods. In this figure the gray-scale representation is opposite to that of Fig. 1.10: brighter areas represent negative polarity (fields directed away from us), darker areas positive polarity. The south pole is at the bottom, the north pole at the top of the diagrams. The longitudes increase from left to right. The regularities caused by Hale's polarity law are clearly seen. The most striking demonstration provided by Fig. 1.11 is however how much the flux pattern may change from one month to the next. Thus the lower synoptic chart shows a number of intense bipolar magnetic regions that are not present in the upper chart and thus must have emerged from the inside of the sun during the intervening time.

The global evolution of the sun's magnetic field can also be explored by projecting the flux pattern of the synoptic charts (after the longitudinal field has been converted to a radial field) on orthogonal functions. The spherical harmonics are the most natural choice of basis functions for the spherical surface of the sun (Stenflo and Vogel, 1986; Stenflo and Güdel, 1988). Eigenvalue problems in spherical geometry, like the time-independent Schrödinger equation for atomic systems, the equations describing the oscillations of stars, or the dynamo equations for the generation and maintenance of stellar magnetic fields, use spherical harmonics to construct the eigenfunctions that characterize the various resonances of the system. The spherical harmonic coeffients c_ℓ^m are classified by the "quantum numbers" ℓ and m, the harmonic degree and order.

Synoptic observations of the sun's magnetic field are available since 1959, when they were initiated by H.W. Babcock at the Mount Wilson Observatory. Since 1975 synoptic magnetic field data are also available from the Kitt Peak Observatory.

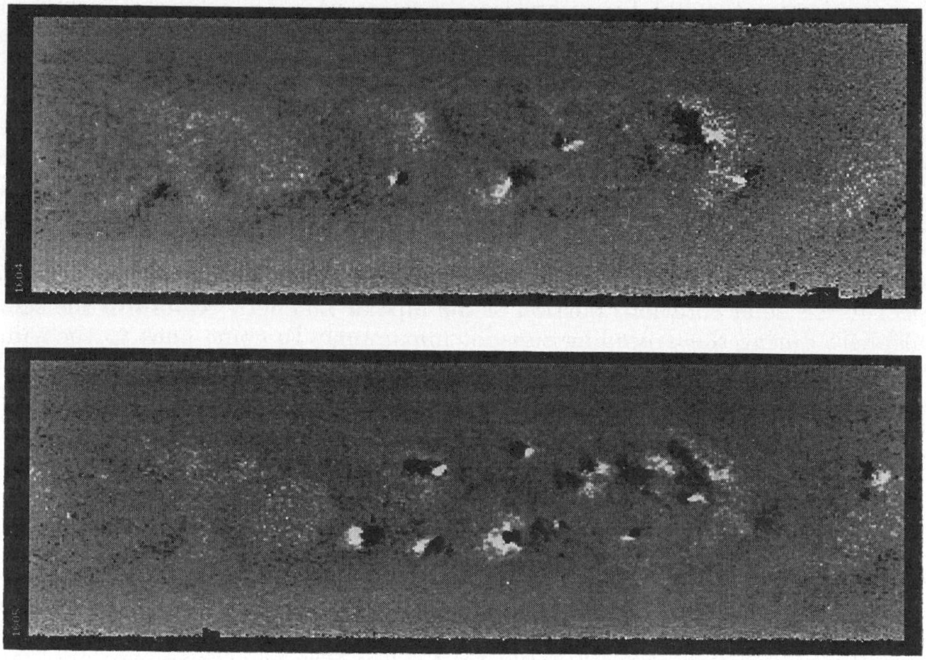

Fig. 1.11. Synoptic magnetic maps of the sun for two consecutive rotation periods (Carrington rotations 1604 and 1605, corresponding to 26 July – 22 August and 22 August – 18 September 1973). The vertical scale represents sin(latitude), from the south pole (bottom) to the north pole (top), while the longitudes run from 0° (left) to 359° (right). Bright (dark) areas represent magnetic fields directed away from (towards) us. Courtesy J.W. Harvey, NSO/Kitt Peak, Tucson.

Time series analysis of the harmonic coefficients has uncovered the hidden modal pattern of the global field evolution (Stenflo and Vogel, 1986; Stenflo and Güdel, 1988; Stenflo, 1988). The power spectra show that the axisymmetric modes (with $m = 0$) strictly obey a parity selection rule. Modes of odd parity (odd values of ℓ), which represent patterns that are anti-symmetric with respect to reflections in the equatorial plane, are dominated by the 22 yr resonance, with less than 10 % of the power at the second harmonic at frequency $(11\,\mathrm{yr})^{-1}$. In contrast the modes of even parity (symmetric with respect to reflections in the equatorial plane) show no trace of the 22 yr cycle, but instead exhibit weaker power spectrum peaks scattered over a frequency range corresponding to periods on the order of 2 yr. The non-axisymmetric modes (with $m \neq 0$) are dominated by the 22 yr resonance, but modes of odd and even parity (odd and even values of $\ell - m$) behave the same way. A number of low-amplitude, discrete modes of higher frequencies are also found.

Explanations of the pattern of solar magnetic fields and of solar activity are generally sought within the framework of dynamo theory. The magnetic fields observed in planets, stars, and galaxies are believed to be produced by dynamo processes, whereby a weak, initial "seed" field gets amplified and acquires a large-scale organization through the interaction between magnetic fields and turbulence in a rotating, electrically conducting medium. Rotation induces Coriolis forces, which make the turbulence "cyclonic" and break the left-right symmetry (Parker, 1955; Zeldovich et al., 1983).

In this framework the sun represents the special case of an oscillating dynamo that can be observed in great detail. The turbulence in the convection zone makes the rotation differential (with radial and latitudinal gradients in the angular velocity) through transport of angular momentum, while rotation makes the turbulence cyclonic. Differential rotation winds up an initially poloidal field (which has no azimuthal components) and thereby builds up a toroidal field. Cyclonic turbulence induces electric currents along the field and thereby regenerates locally a poloidal field component from the toroidal one. Turbulent transport spreads this new poloidal field, which has a polarity opposite to that of the previous poloidal field, which therefore gets cancelled and replaced. This scenario leads to an oscillation between a poloidal and a toroidal magnetic field, both of which reverse their directions during the course of the oscillatory cycle, which can be identified with the observed 22 yr magnetic cycle.

Our limited understanding of convection and turbulence is the main obstacle to making good quantitative models of the above scenario. Turbulence shreds, amplifies, and transports the field. It is the detailed properties of the turbulence that determine what the dynamo will look like. Additional governing factors are buoyancy and viscous drag forces, which depend on the fragmentation and degree of intermittency of the field that the turbulence has caused. No adequate theory for turbulent magnetoconvection in the gravitationally stratified interiors of stars exists, only highly idealized parametrized models or coarse-grained numerical simulations. When the model is described in terms of many free parameters, as is the case for the kinematic $\alpha - \omega$ dynamos (Steenbeck and Krause, 1969; Krause and Rädler, 1980), one can fit the cycle length and the migration of the activity latitude belts (the "butterfly diagram") by suitably adjusting the free parameters. When numerical simulations are done without free parameters (e.g. Gilman, 1983), the observed properties of the sun cannot be reproduced, most likely because of the vastly insufficient spatial resolution of the computer grids used, since the turbulent spectrum extends down to very small scales.

Magnetic flux tubes are highly buoyant in the convection zone and should rise to the surface in a time scale that appears to be too short with respect to the needs for field amplification by dynamo processes. Long-term flux storage only seems possible at the bottom of the convection zone or in the convectively stable layer immediately below (Spiegel and Weiss, 1980). It is therefore now generally believed that much of the dynamo operates near this boundary between the radiative core and the convective zone of the sun.

It is being increasingly realized that magnetic flux emerges from the sun's interior at all spatial scales, and that the active regions only represent the large-scale portion of the size spectrum. The binning done in synoptic charts like those of Fig. 1.11 effectively filters out the smaller, bipolar regions (due to cancelling of nearby opposite polarities within a bin). We therefore only see the large-scale features that are left and might be led to believe that this is everything that there is. The spatial filtering is less severe in full-disk Kitt Peak magnetograms like that of Fig. 1.10. Careful analysis of such magnetograms has led to the identification of bipolar magnetic regions at intermediate scales, so-called *ephemeral active regions* (Harvey and Martin, 1973; Harvey et al., 1975; Martin and Harvey, 1979). At still smaller scales numerous mixed-polarity magnetic regions, called *intranetwork fields*, may be seen all over the sun in high-resolution magnetograms if the polarimetric sensitivity is sufficiently high.

Estimates of the amount of magnetic flux that emerges to the surface at the various scales give about 10^{20} Mx/day in the form of active regions (AR), 10^{22} Mx/day as ephemeral regions (ER), and 10^{24} Mx/day as intranetwork fields (IN) (Zirin, 1987). While the flux emergence rates are in the proportions 1:100:10,000 for AR:ER:IN, the typical sizes are in the approximate proportions 25:5:1 (or 75:15:3 arcsec). Although the magnetic flux seen on the sun at any given time is thus dominated by the smallest scales and has a short "turn-over" or replenishment time scale, this is no guarantee that the small-scale flux emergence possesses sufficient large-scale or non-random properties (like active longitudes, obedience of Hale's polarity law, and systematic tilts of the bipolar magnetic axes) to also be a major contributor to the *large-scale* magnetic pattern. It is known that the orientations of ER statistically prefer the directions dictated by Hale's polarity law, although their latitude distribution is considerably wider than that of ordinary AR. Still wider is the latitude distribution of IN, but it is not known whether or not the orientations of the IN bipoles are entirely random. A tiny, non-random component would however suffice for the IN emergence to be the dominating source of the large-scale pattern, since the IN flux emergence rate is 10^4 times larger than that of AR (cf. Stenflo, 1992).

Indirect evidence that the large-scale flux pattern is indeed replenished over a short time scale, on the order of weeks rather than 11 yr, and that it is thus produced by the accumulated effect of many small-scale emergence processes, comes from cross- and auto-correlation analysis of the large-scale magnetic field pattern. Such analysis has revealed the coexistence of two distinctly different differential rotation laws that govern the rotational phase velocity of the magnetic pattern. If the pattern were not rapidly regenerated, the smoothing transport processes on the solar surface would make its phase velocity quasi-rigid on scales larger than the supergranulation scale (Sheeley et al., 1987), in contradiction with the steep differential rotation law found from cross-correlation analysis at a resolution of about one arcmin (Snodgrass, 1983). More detailed arguments have been given in Stenflo (1992). A recent theoretical treatment by Petrovay and Szakály (1993) of the origin of the intranetwork fields inside the sun also reaches the conclusion that the replenishment time scale for the large-scale fields is on the order of weeks.

The relations between the large- and small-scale processes, between global convection and small-scale turbulent amplification, between boundary-layer and distributed dynamos, etc., are still subject to intense debate. Various critical viewpoints can be found in the proceedings of the Solar Cycle Workshop (Harvey, 1992) and in Durney et al. (1993).

1.5. Small-scale Magnetic Fields on the Quiet Sun

From the previous section we have seen that there is a coexistence of large- and small-scale magnetic patterns on the sun. Although the magnetic flux exhibits a global organization, it is also highly fragmented or intermittent. As the spatial resolution has been improved, ever more magnetic structures not observable with lower resolution have been revealed. An example is given by the high-resolution magnetogram of a quiet-sun region at disk center in Fig. 1.12. The size of the field of view is $230 \times 220\,\mathrm{arcsec}^2$, i.e., the linear dimension is about 10 % of the solar diameter. We see how the quiet sun is everywhere filled with intermittent flux fragments of mixed polarities (giving rise to the "salt and pepper" appearance of the magnetogram), invisible in coarser magnetograms like that of Fig. 1.10.

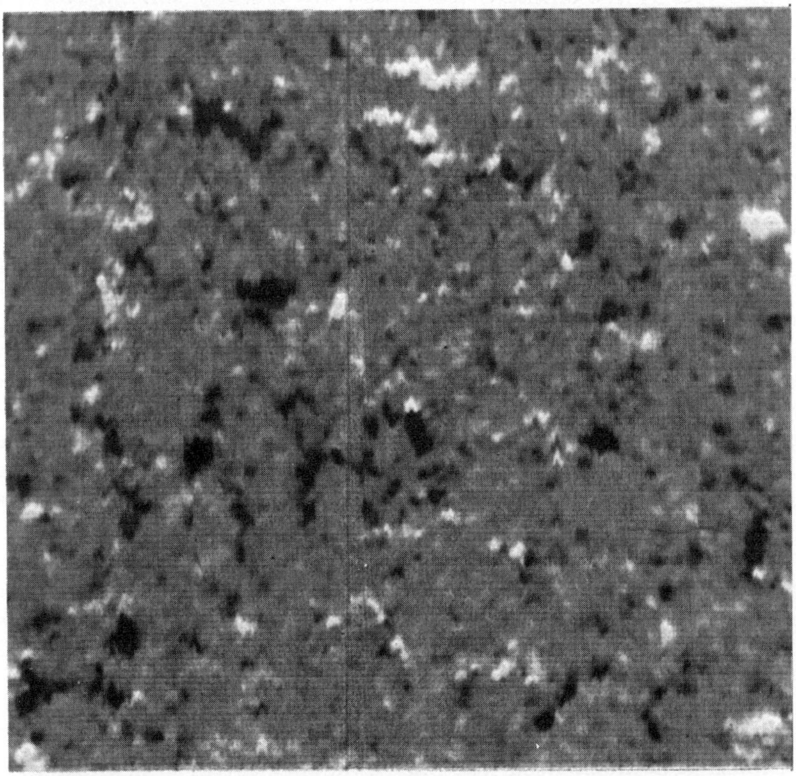

Fig. 1.12. High-resolution magnetogram of a quiet-sun region at disk center. As before the brighter and darker areas represent magnetic flux of opposite polarities. The field of view is $230 \times 220\,\mathrm{arcsec}^2$. Courtesy J.W. Harvey, NSO/Kitt Peak, Tucson.

Most of what we see in Fig. 1.12 are mixed-polarity *network fields* preferentially located at the supergranular cell boundaries, but there are also indications of apparently weak, mixed-polarity fluxes inside the supergranular cells, so-called *intranetwork fields.* The flux fragments on the quiet sun that contribute to the background magnetic pattern in full-disk magnetograms like that of Fig. 1.10 represents flux that does not mix the polarities on such small scales and therefore survives some spatial averaging. As these fluxes forming the photospheric network provide the main contribution to the polarization signals in observations of intermediate resolution, they are the ones that have been extensively explored, while fairly little is known about the properties of the intranetwork fields apart from their morphology and evolution (cf. Martin, 1990). At the smallest scales we expect the photosphere to be filled with a mixed-polarity *turbulent field* that has a size spectrum extending far beyond the present resolution limit. The existence and some properties of this "hidden" field may be inferred by indirect methods (cf. Stenflo, 1982, 1987, 1989; Faurobert-Scholl, 1993), as outlined in Chapter 10. The results of these indirect methods agree with recently derived theoretical properties (Petrovay and Szakály, 1993). The portion of the turbulent spectrum at scales below about 100 km would be practically invisible even in magnetograms with infinite spatial resolution, since in this "microturbulent regime" the turbulent elements are optically thin with cancellation of the polarization contributions from the opposite polarities along the line of sight over the region of spectral line formation. The indirect diagnostic methods of Chapter 10 however make use of different physical effects and are therefore unaffected by these line-of-sight cancellations.

The key to a unified understanding of solar (and stellar) activity is to be found in the physics of the basic building blocks of the magnetic field. As these building blocks have remained beyond the attainable spatial resolution, an arsenal of indirect diagnostic techniques have been developed to overcome the resolution limit and derive the statistical, physical properties of the spatially unresolved magnetic field. This situation is not uncommon in physics. One can deduce the internal structure of atoms without "resolving" them. Similarly, detailed stellar physics is developed without spatially resolving the stars.

The basic building blocks of the network field are the so-called magnetic "flux tubes", which like the turbulent field are generally spatially unresolved. Chapter 12 provides a detailed overview of the methods to diagnose the magnetic field and thermodynamics of flux tubes as a function of height in the atmosphere. With these indirect techniques the missing spatial information is compensated for by complementary information in the spectral domain. An early method that has played a key role for our understanding of the small-scale nature of the field has been the so-called *line-ratio technique* (Stenflo, 1973), with which the Zeeman-effect polarization is recorded simultaneously in two or more carefully selected spectral lines. The most used line pair to deduce intrinsic field strengths of the spatially unresolved flux elements has been Fe I 5247.06 and 5250.22 Å, displayed before in Fig. 1.7. These two lines belong to the same multiplet, have nearly the same line strength and excitation potential, but differ in their Landé factors (2.0 and 3.0, respectively). For weak magnetic fields, when the Zeeman splitting is small in comparison with the line width, the circular-polarization amplitudes of the two lines are in proportion to their Landé factors. When the splitting gets comparable to the line width, however, the polarization signal saturates and does so more in the line with the larger Landé factor than in the other line. By measuring this differential Zeeman saturation the intrinsic field strength can be determined, regardless of how small the magnetic filling factor is (the relative area of the resolution element that is occupied by the contributing flux).

The surprising result obtained when the line-ratio technique was first applied was that the magnetic field in the quiet-sun photospheric network is everywhere strong, 1–2 kG, and that the field strength appears to be almost independent of the magnetic filling factor. Thus similar values of the intrinsic field strength are found in the quiet sun as in active-region plages, where the amount of flux (and thus the filling factor) is larger by an order of magnitude or more (Stenflo, 1973; Frazier and Stenflo, 1978; Stenflo and Harvey, 1985). This result led to the concept of magnetic flux tubes with unique or similar physical properties as the basic building blocks of both the quiet network and of active-region plages. The main difference between quiet and active regions (outside sunspots) is then primarily in the number density and size distribution of the flux tubes.

With the advent of the Fourier Transform Spectrometer (FTS) as a polarimeter it became possible to simultaneously record the spectrally fully resolved polarized line profiles of not just a few, but of hundreds of spectral lines, with high signal-to-noise ratio in the degree of polarization (Stenflo et al., 1984). This new wealth of diagnostic information has been used not only to confirm the previous line-ratio results, but to extend them with diagnostics of the internal thermodynamic flux tube structure and with information on the height variation of the physical parameters. This has led to the rapid growth of a model building "industry" for the construction of empirical flux tube models of ever increasing levels of sophistication and physical realism (cf. Solanki, 1987, 1993).

One of the results of this modeling concerns the thermal structure: although the field strengths of flux tubes in the quiet network and active-region plages are not much different, the network flux tubes are found to be significantly hotter than the ones in active regions. The general thermal behavior is however the same: cooling of the lower levels, heating of the upper levels (relative to the flux tube surroundings).

In recent years the near infrared region around the opacity minimum of the solar spectrum at 1.6 μm has opened up new diagnostic possibilities for exploring the spatially unresolved fields (Harvey, 1977; Stenflo et al., 1987b; Zayer et al., 1989; Rüedi et al., 1992; Rabin, 1992a,b). In particular, the much larger Zeeman splitting in the infrared provides information on the *distribution* of field strengths within the spatial resolution element and along the line of sight, the effects of which are too small to be spectrally resolved at visible wavelengths. Furthermore, the 1.6 μm region is formed deeper than the visible region and therefore helps to extend the height range that can be covered by the various diagnostic methods. For details we refer to Chapter 12.

As we cannot tell from a given filling factor whether it is due to one or many flux tubes within the spatial resolution element, we cannot derive flux tube *sizes* from the spectral diagnostics, only set upper limits (which however are quite small). Comparison with counts of the number of bright points in the emission network helps us to get decent estimates of the actual sizes. Thus typical diameters on the order of one or a few hundred km are to be expected. Recently it was for the first time possible to directly spatially resolve a flux tube with a diameter of 200 km by the technique of speckle polarimetry (Keller, 1992), which allows the reconstruction of a polarimetric image that is free from seeing distortions and smearing (cf. Chapter 13 for details).

From a theoretical point of view 100 km is a natural length scale in the photosphere, since it represents the approximate horizontal mean free path of the visible photons. As we will see in the next section, flux tubes may spontaneously collapse if their interiors are radiatively insulated from their surroundings. This situation occurs for flux tubes thicker than the photon mean free path.

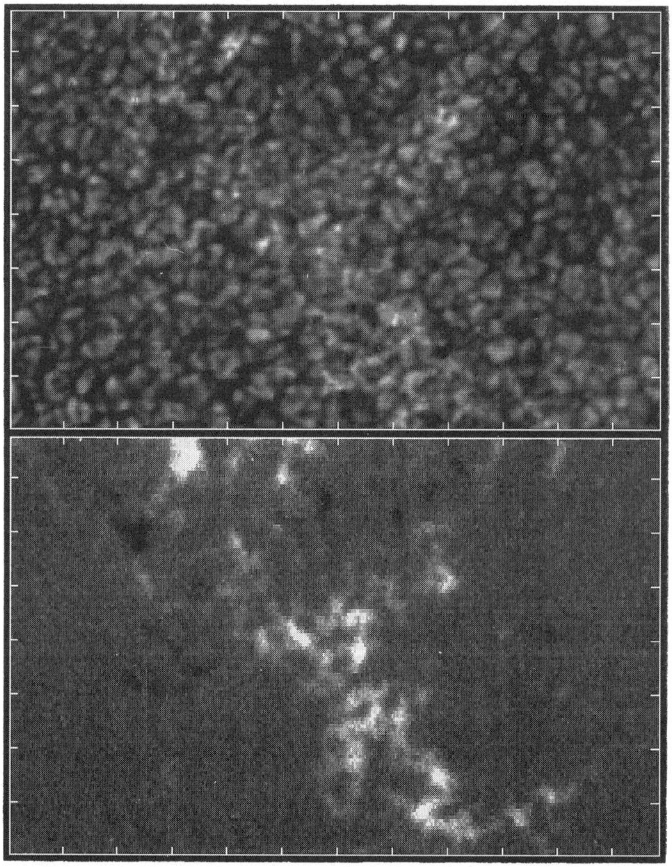

Fig. 1.13. High-resolution magnetogram (lower frame), in the form of a spatial map of the degree of circular polarization in the blue wing of the Fe I 5250.22 Å line. The tick-mark separation is 5 arcsec and the field of view 59×38 arcsec^2, so the area covered is about twenty times smaller than the field of view of Fig. 1.12. The recording was made at the Swedish La Palma observatory on 16 May 1993 with a narrow-band filter and the ZIMPOL I (**Z**urich **Im**aging Stokes **Pol**arimeter) polarimetric imaging system (Keller et al., 1992) in a quiet region at μ (cosine of the heliocentric angle) $= 0.6$. The upper frame is the simultaneously recorded intensity image showing the pattern of solar granulation.

Another length scale of similar magnitude is the pressure-density scale height, which is about 150 km in the photosphere. It governs the height divergence of the field in the small-scale flux tubes and therefore also the curvature of the field lines, which is a key factor for the stability of the flux tubes against fluting. Resolved flux tubes of different spatial scales appear to decay by fragmentation. The rate of flux decrease is $\mathrm{d}\Phi/\mathrm{d}t \approx -10^{15}\,\mathrm{Mx\,s^{-1}}$ (Gokhale and Zwaan, 1972; Zwaan, 1987; Martínez-Pillet et al., 1993). As the field strength does not vary greatly with the size of the flux tubes (Stenflo, 1977), an approximately constant flux decay rate implies a fragmentation time scale $t \sim d^2$ (where d is the flux tube diameter), proportional to the cross section area of the flux tubes (Stenflo, 1976a). This "law" cannot be extrapolated to indefinitely small sizes, since different, partly counteracting processes will come into play when approaching length scales of 100 km (about 0.1 arcsec), like turbulent flux coalescence followed by flux collapse and stabilizing vorticity formation. We need telescopes that can resolve these scales to obtain information on morphology and evolution.

Polarimetric imaging is now possible with 2-D CCD (charge-coupled device) detectors in combination with electrooptical modulation of the full state of polarization (circular and linear polarization). This allows efficient mapping of vector magnetic fields with high polarimetric accuracy and can be combined with techniques like speckle polarimetry and phase diversity (cf. Chapter 13) to remove the distorting effects of the earth's atmosphere. An example of a magnetogram obtained with such an imaging system, called ZIMPOL (**Z**urich **Im**aging Stokes **Pol**arimeter), is shown in Fig. 1.13. The magnetogram covers an area of the quiet sun that is smaller by about a factor of twenty as compared with that of Fig. 1.12. Still most of what we see in this high-resolution magnetogram is due to network fields, while the intranetwork and turbulent fields remain elusive at this level of polarimetric sensitivity.

Although mapping of the circular polarization with a narrow-band filter like in Fig. 1.13 is most efficient for morphological and evolutionary studies, quantitative determinations of the physical parameters requires additional spectral information, which may be provided by a spectrograph. The polarimetric images of Fig. 1.14 in the focal plane of a spectrograph were obtained on 19 June 1992 with the ASP (**A**dvanced **S**tokes **P**olarimeter) of the High Altitude Observatory at the National Solar Observatory/Sacramento Peak. This instrument, which has been described by Lites et al. (1991), uses a rotating wave plate as a polarization modulator. The spectrograph slit cuts across a plage area in an active region at μ (cosine of the angle between the line of sight and the vertical direction) = 0.91. The 1-D field of view is 83 arcsec. The central portion of the covered 3.0 Å wavelength range is the same as the range in Fig. 1.8. The two strong solar lines shown are due to Fe I 6301.5 and 6302.5 Å of iron multiplet no. 816. Their Landé factors are 1.67 and 2.50, respectively. The 6302.5 Å line is surrounded by two narrow, telluric lines of molecular oxygen.

The four simultaneous spectrograms in Fig. 1.14 represent the four Stokes parameters I (unpolarized intensity), Q (linear polarization at position angle 0°), U (linear polarization at 45°), and V (right-handed circular polarization). These four parameters form the *Stokes vector* $\boldsymbol{I}$, which gives a complete description of

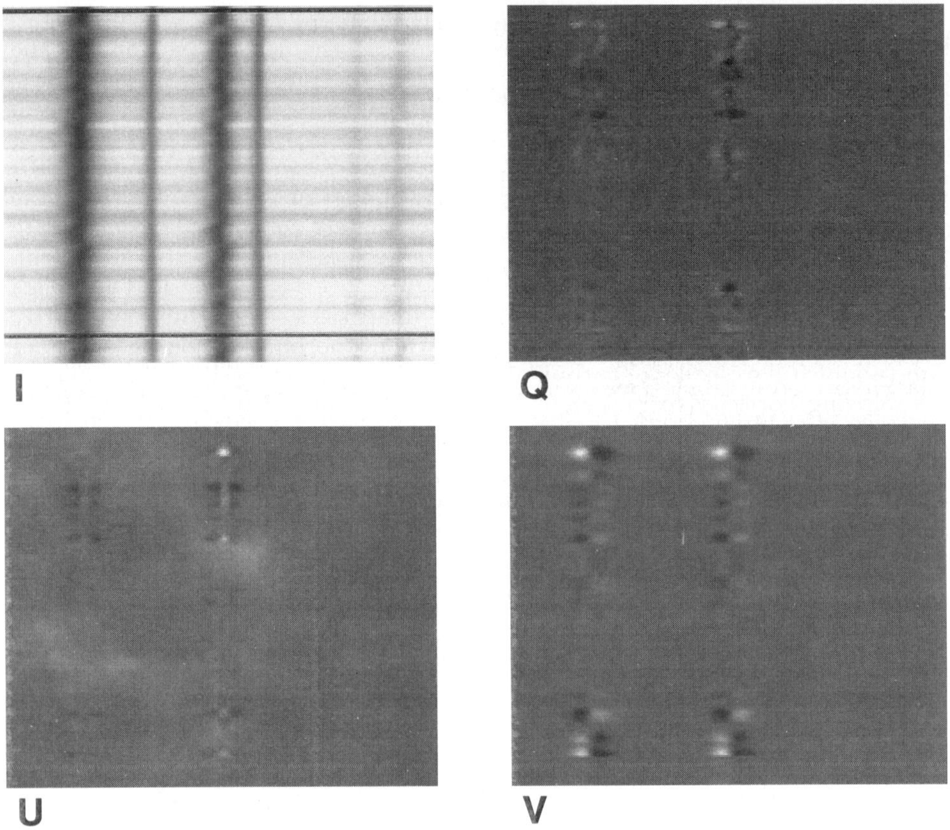

Fig. 1.14. High-resolution 2-D polarimetric spectral images of the four Stokes parameters I, Q, U, V, obtained in a plage region at $\mu = 0.91$ on 19 June 1992 with the Advanced Stokes Polarimeter of HAO/NSO. The 1-D field of view is 83 arcsec. The 3.0 Å wavelength range includes the 816 Fe I 6301.5 (left) and 6302.5 Å (right) solar lines as well as two telluric lines. Courtesy B.W. Lites, High Altitude Observatory, Boulder, Colorado.

the polarization state, as we will see in Chapter 2. Such Stokes line profiles allow the *vector* magnetic field (magnitude *and* direction of the field) to be determined.

1.6. Summary of Observed Properties and Theoretical Concepts

Quantitative information on the physical parameters in the magnetic field elements can be derived from polarimetric data by applying the theory of polarized radiative transfer in an inhomogeneous magnetic field. This is the reason why this theory occupies a major portion (Chapters 2–11) of the present monograph. Although it constitutes the diagnostic foundation for our knowledge about the sun's magnetic field, it has not previously been presented in a systematic way as attempted here. Each spectral line responds in a highly non-linear way to perturbations of the solar atmosphere, but since this response is different for different spectral lines, the information may be untangled and the physical parameters may be deduced from the observed polarization. With such an inversion approach a suitable set of polarimetric observables have to be carefully selected, so that the values of the free parameters characterizing the flux tube model used to compute the synthetic observables can be determined by an iterative least-squares fit. In this way a set of empirical flux tube models of increasing sophistication may be derived, and they may be improved upon as more powerful observational constraints become available, allowing more physical realism to be added to the models.

The models should be self-consistent and satisfy the MHD equations with horizontal and vertical pressure balance. There is thus an interplay between MHD and polarization physics. The problem is however not yet tractable in its full generality, since for a theoretically complete model one would need to couple the MHD equations to an energy equation that includes not only radiative and turbulent energy transport but also wave heating processes, which have not yet been formulated for realistic 3-D geometries of diverging flux tubes.

Construction of empirical flux tube models have therefore so far proceeded without an energy equation, leaving the temperature structure with its height variation as a free parameter to be fitted from the observations together with the magnetic field structure. The resulting kG flux tubes are found to have a low plasma β, the ratio between gas and magnetic pressure, of about 0.3. They are also found to be cooler than their surroundings in the lower, hotter in the higher layers of the atmosphere (cf. Chapter 12).

Theoretical modelling with a simplified energy equation containing idealized convective energy transport and 2-D radiative transfer shows that cooling results from magnetic inhibition of convective transport, while radiation from the hot bottom and walls of the flux tube contributes to the heating of the higher layers. Radiative heating may however need to be supplemented by wave heating to fit the observational data.

Observations of asymmetries in the polarized line profiles have shown that the flux tubes are surrounded by a convection cell with a downflow immediately outside the flux tube walls. The low-pressure regions where the flux tubes are located are reminiscent of the drain in a bathtub. The vortex flow that may develop in the downdraft region helps to stabilize the flux tube against breaking apart by the fluting instability (Schüssler, 1984). While larger flux tubes are stabilized by buoyancy forces (Meyer et al., 1977), smaller flux tubes need the help of whirl flows to survive as individual entities (Bünte et al., 1993a,c; Bünte, 1993).

Inside the sun convective motions sweep the frozen-in magnetic field to the boundaries of the convection cells. This process is continued by smaller-scale turbulence, making the field increasingly intermittent before it emerges at the surface. The granular and supergranular flows continue this flux concentration process, with the supergranular network as a byproduct. The dynamic pressure of the surface flows however only corresponds to a magnetic pressure of fields with a strength of about 100 G. *Convective expulsion* is thus insufficient to produce and contain the observed kG fields. Only the gas pressure can match the magnetic pressure of such fields.

To explain how amplification to kG values can be achieved the mechanism of *convective collapse* has been introduced (Parker, 1978; Spruit, 1979; Spruit and Zweibel, 1979; Unno and Ando, 1979). Magnetic fields inhibit convective energy transport since the plasma cannot move freely perpendicular to the magnetic field lines. The resulting reduction of heat transport from the solar interior leads to cooling of the lower layers of the flux tubes, as observed. Radiative exchange processes on the other hand try to smooth out any lateral temperature inhomogeneities. If however the horizontal optical thickness of the magnetic field region (flux tube) is larger than unity (i.e., if the photon mean free path is smaller than the flux tube diameter), then the interior of the flux tube is partially (within the radiative exchange time scale) thermally insulated from the surrounding atmosphere. The thermal insulation implies that any mass flow along the field lines will be approximately adiabatic, while the stratification of the external atmosphere, at depths where convection occurs, is superadiabatic (which means that the outwards temperature decrease is steeper than the corresponding adiabatic decrease, the condition for convective instability).

Let us now consider a perturbation in the form of a weak downdraft along the magnetic field lines in a region that is initially in thermal equilibrium with its surroundings. If the flow is thermally insulated from its environment (which is the case for thick flux regions), it will set up an adiabatic temperature gradient inside the flux tube. Since the outside temperature gradient is steeper, the flow induces a temperature deficit with respect to the surroundings, accompanied by a pressure deficit. To restore horizontal pressure balance the magnetic flux region has to be compressed, which enhances the downdraft, leading to further compression, etc., until a partially evacuated, strong-field flux tube has been formed. The *adiabatic cooling* of the flux region by the downdraft thus leads to a spontaneous collapse, until the magnetic pressure of the amplified field has become sufficiently strong to brake the contraction and stabilize the flux tube against further collapse.

According to this scenario we expect that flux with field strengths below a certain threshold will be unstable with respect to such convective collapse, while sufficiently strong fields will be stable, something that has been confirmed by linear stability analysis (Spruit and Zweibel, 1979). The stabilizing field needs to have a magnetic pressure comparable to the gas pressure of the surroundings, which implies field strengths in the solar photosphere of 1–2 kG. We thus expect the photospheric flux of optically thick elements to end up in such a strong-field state. This is in agreement with observations, which show that more than 90 % of the total flux that is seen by magnetographs is in strong-field, kG form (Howard and Stenflo, 1972; Frazier and Stenflo, 1972; cf. Chapter 12). An actual collapse process has however not yet been observed, since it occurs on a small spatial scale as well as on a time scale, on the order of 5–15 min, that is short in comparison with the expected life time of flux in a concentrated state.

Almost nothing is known from observations about the further fate of the concentrated fields. Their life times with respect to fragmentation by the fluting instability depends on maintaining vortex flows around them. At scales below about one hundred km the convective collapse mechanism will no longer be effective due to radiative exchange. Flux on such scales may therefore have intrinsically weaker field strengths and might then not be able to withstand the tangling effect of the turbulent motions. Smaller flux fragments may coalesce to larger structures, for which collapse becomes possible, etc. The circumstance that more than 90 % of the spatially resolved flux in magnetograms is found to have its sources in strong, spatially unresolved kG flux tubes indicates that the flux fragmentation process does not go on indefinitely but is at some small scale (probably around 100 km) statistically balanced by concentration and coalescence.

It has been described in Sect. 1.4 above how new, small-scale magnetic flux emerges from the solar interior at a tremendous rate, on a time scale that is four orders of magnitude shorter than the length of the activity cycle in the case of the intranetwork fields. Flux must then also be removed from the surface at a similar rate, but little is known how this is done, since the basic processes occur on small scales. Frequent cases of flux cancellation, when opposite-polarity magnetic elements "collide", are observed (Martin, 1990). The cancellation events may be due to either submergence (retraction of the flux loops below the surface) or annihilation of the fluxes by ohmic diffusion with field-line reconnection. A theoretical scenario for the uprooting of flux from the sun has been described by Spruit et al. (1987). Turbulent eddies may bring small sections of a horizontal, subsurface flux rope to the surface, forming a mixed-polarity field. The topology of this connected "sea serpent" is changed by reconnection in the upper portions of the emerged sections, such that a number of disconnected, small, closed loops are formed. These loops decay into heat when turbulence has reduced their length scales down to the ohmic diffusion scale.

Magneto-convection is the underlying process that is involved in most aspects of stellar activity: the dynamo mechanism, flux amplification, transport, destruction, intermittency, etc. Still we are far from a good theoretical understanding of magneto-convection and turbulence in the gravitationally stratified stars. 3-D numerical simulations of magneto-convection in the outer layers of the sun (Nordlund, 1983, 1986; Nordlund and Stein, 1990) have been successful in reproducing the observed properties of solar granulation, but the computational grid has been too coarse to properly describe such processes as convective collapse, fluting instability, ohmic decay of current sheets, etc. More efficient numerical codes with grid refinement procedures and more powerful computers will help advance our understanding of the basic physical processes. The problem is however very complex, since so many length scales are involved, down to the ohmic diffusion scale (on the order of 1 km in the photosphere).

The theory thus has to be guided by observations, but what we have so far seen on the solar surface only represents resolved, large-scale signatures of magneto-convection. Increased spatial resolution, which in the coming decade should make the crucial length scale of the photon mean free path accessible to observation, will significantly enhance our understanding of the physics involved. Even with such improved resolution, however, there will always remain an important small-scale tail of the magneto-turbulent spectrum beyond the attained resolution limit. Furthermore, when we enter the optically thin regime below a scale of about 100 km, the visibility of the structures in a magnetogram will decrease. Although unresolved or invisible, the statistical properties of this tail can still be explored by indirect methods, which will thus always remain indispensible as a complement to the direct resolution methods. In Chapter 10 we try to establish a diagnostic foundation for explorations of the unresolved tail of the magneto-turbulent spectrum. Some important diagnostic tools depend on the theories of coherent radiative scattering, the Hanle effect, and atomic collisions, complex areas that are interwoven with the Zeeman effect and non-LTE line-formation physics. The presentation in Chapter 10 builds on the systematic development of the overall theory in Chapters 2–9.

CHAPTER 2

THEORY OF POLARIZED RADIATION

2.1. Introduction

The framework of classical physics will be used in the following four chapters to develop the theory of polarized radiative transfer in stellar atmospheres permeated by magnetic fields. Scattering and the Hanle effect will also be covered by the classical treatment. This allows us to establish the direct correspondence between the classical and quantum worlds, when the quantum field theory is developed in Chapters 7 and 8. The prior development of the classical theory is needed for a good understanding of the physical contents of the often abstruse formalism of quantum field theory.

In the classical treatment the link between the macroscopic and microscopic properties of the medium is most conveniently described in terms of the macroscopic complex refractive index n, which is determined by the collective effect of the oscillating electric dipoles in the individual atoms. The atomic dipole moment is induced by the electric field of the electromagnetic radiation and is modified by the ambient magnetic field. The presence of the dipoles determines the dispersion relation for the electromagnetic waves, and this dispersion relation can be parametrized in terms of the refractive index n.

The polarization of the electromagnetic waves can be treated using several different powerful tools, in particular the Jones vector, the coherency matrix, or the Stokes vector. Within these formalisms, the effect of the medium is generally described by non-commuting matrix operators, leading to matrix representations of the radiative transfer problem.

The primary physical effects, derived from first principles within a classical framework, are most naturally expressed in terms of Jones vectors. The Jones formalism is however unable to describe a statistical ensemble of uncorrelated photons, which is needed to relate the theory to the world of observations. The statistical properties of the radiation field can be described by the coherency matrix and Stokes vector formalisms, which are equivalent to each other, and which can be derived from the Jones vectors. The Stokes formalism is directly related to the effect of measuring devices and is therefore the natural one to use for the interpretation of observations. The coherency matrix, which contains the equivalent polarization information, is the most natural representation for the radiation field in the quantum field theory, since it relates most directly to the creation and annihilation operators used there.

A fourth way of describing polarized radiation is in terms of complex spherical vectors. It is needed when evaluating the interaction of the electromagnetic waves with the oscillating dipoles in a magnetic field, since the coupling between the dipole vector components caused by the $\boldsymbol{v} \times \boldsymbol{B}$ force of the magnetic field can be removed when transforming from a Cartesian to a spherical vector system. This formalism is most naturally introduced in the next chapter, where the interaction between radiation and matter is treated.

2.2. Maxwell's Equations

The standard set of equations for the electromagnetic field can be written (cf. Jackson, 1975)

$$\begin{aligned}\nabla \cdot \boldsymbol{D} &= \rho_c\,, \\ \nabla \cdot \boldsymbol{B} &= 0\,, \\ \nabla \times \boldsymbol{E} &= -\frac{\partial \boldsymbol{B}}{\partial t}\,, \\ \nabla \times \boldsymbol{H} &= \boldsymbol{j} + \frac{\partial \boldsymbol{D}}{\partial t}\,.\end{aligned} \tag{2.1}$$

The first of these four equations expresses Coulomb's law after Gauss' theorem has been applied (ρ_c is the charge density), while the second expresses the absence of magnetic monopoles. The third equation represents Faraday's law, the fourth Ampère's law. The displacement current $\partial \boldsymbol{D}/\partial t$ is necessary to ensure that the continuity equation for the charge and current densities,

$$\frac{\partial \rho_c}{\partial t} + \nabla \cdot \boldsymbol{j} = 0\,, \tag{2.2}$$

is satisfied.

In addition we need two equations relating $\boldsymbol{D}$ and $\boldsymbol{B}$ to $\boldsymbol{E}$ and $\boldsymbol{H}$. We will only make use of the equations valid for vacuum, i.e.,

$$\begin{aligned}\boldsymbol{D} &= \epsilon_0 \boldsymbol{E}\,, \\ \boldsymbol{B} &= \mu_0 \boldsymbol{H},\end{aligned} \tag{2.3}$$

which contain the universal constants

$$\begin{aligned}\epsilon_0 &= 8.854 \times 10^{-12} \text{ farad m}^{-1}\,, \\ \mu_0 &= 4\pi \times 10^{-7} \text{ henry m}^{-1}\,,\end{aligned}$$

representing the permittivity and permeability of the vacuum.

Often in electromagnetic theory the macroscopic effects of a medium are described in terms of a dielectric constant ϵ and a magnetic permeability μ in Eq. (2.3). Such concepts however represent a phenomenological element of the theory, which hides the underlying physics and has led to physically inconsistent treatments in the past. We will therefore entirely refrain from introducing any ϵ or μ at all, to allow us to work from a rigorous version of Maxwell's theory. All the effects of matter will enter exclusively via the charge density ρ_c and the current density $\boldsymbol{j}$.

2.3. The Poynting Flux

An energy equation containing an expression for the radiative energy flux (the Poynting flux) can be obtained by multiplying in Eq. (2.1) Faraday's law by $-\boldsymbol{H}$, Ampère's law by $\boldsymbol{E}$, and adding them. This gives

$$-\boldsymbol{H}\cdot(\nabla\times\boldsymbol{E})+\boldsymbol{E}\cdot(\nabla\times\boldsymbol{H})=\boldsymbol{j}\cdot\boldsymbol{E}+\boldsymbol{E}\cdot\frac{\partial\boldsymbol{D}}{\partial t}+\boldsymbol{H}\cdot\frac{\partial\boldsymbol{B}}{\partial t}\,. \tag{2.4}$$

Since the left hand side can be written as $-\nabla\cdot(\boldsymbol{E}\times\boldsymbol{H})$, we obtain

$$\frac{\partial}{\partial t}\left(\frac{\epsilon_0}{2}\boldsymbol{E}^2+\frac{1}{2\mu_0}\boldsymbol{B}^2\right)+\nabla\cdot(\boldsymbol{E}\times\boldsymbol{H})=-\boldsymbol{j}\cdot\boldsymbol{E}\,. \tag{2.5}$$

This represents a continuity equation for the energy. The first term on the left hand side is the time variation of the energy density of the electromagnetic field, the second term is the divergence of the energy flux $\boldsymbol{E}\times\boldsymbol{H}$, while the right hand side describes the energy losses due to Joule heating by the electric currents.

If we are dealing with complex vectors $\boldsymbol{E}$ and $\boldsymbol{H}$, e.g. plane waves of the type $\boldsymbol{E}=|\boldsymbol{E}|\exp(-i\omega t+\boldsymbol{k}\cdot\boldsymbol{r})$, then a similar derivation gives

$$\frac{\partial}{\partial t}\left(\frac{\epsilon_0}{2}|\boldsymbol{E}|^2+\frac{1}{2\mu_0}|\boldsymbol{B}|^2\right)+\mathrm{Re}\,\nabla\cdot(\boldsymbol{E}\times\boldsymbol{H}^*)=-\mathrm{Re}\,\boldsymbol{j}\cdot\boldsymbol{E}^*\,. \tag{2.6}$$

Since here the energy term is formed from the *amplitudes* $|\boldsymbol{E}|$ and $|\boldsymbol{B}|$ of the oscillations, the time averaged energy terms for a harmonically oscillating field are obtained if we multiply Eq. (2.6) by $\frac{1}{2}$ (since $\langle\sin^2\rangle=\frac{1}{2}$). The average radiative flux is thus $\mathrm{Re}\,\boldsymbol{E}\times\boldsymbol{H}^*/2$.

2.4. The Electromagnetic Wave Equation

For the treatment of radiation problems it is convenient to reformulate Maxwell's equations in terms of a vector potential $\boldsymbol{A}$ and a scalar potential Φ, defined by

$$\begin{aligned}\boldsymbol{B}&=\nabla\times\boldsymbol{A}\,,\\ \boldsymbol{E}&=-\nabla\Phi-\frac{\partial\boldsymbol{A}}{\partial t}\,.\end{aligned} \tag{2.7}$$

These definitions assure that $\boldsymbol{B}$ is divergence free, and that $\boldsymbol{E}$ satisfies Faraday's law.

The remaining two Maxwell equations relating $\boldsymbol{D}$ to the charge density ρ_c and $\boldsymbol{H}$ to the current density $\boldsymbol{j}$ can then be written (since the spatial and temporal derivatives commute)

$$\begin{aligned}\nabla^2\Phi+\frac{\partial}{\partial t}(\nabla\cdot\boldsymbol{A})&=-\rho_c/\epsilon_0\,,\\ \nabla\times(\nabla\times\boldsymbol{A})&=\mu_0\boldsymbol{j}-\epsilon_0\mu_0\left(\frac{\partial}{\partial t}\nabla\Phi+\frac{\partial^2\boldsymbol{A}}{\partial t^2}\right).\end{aligned} \tag{2.8}$$

Making use of the vector identity

$$\nabla \times (\nabla \times \boldsymbol{A}) = \nabla (\nabla \cdot \boldsymbol{A}) - \nabla^2 \boldsymbol{A} \,, \tag{2.9}$$

and defining

$$c = (\epsilon_0 \mu_0)^{-1/2} \,, \tag{2.10}$$

the second of Eqs. (2.8) can be written

$$\nabla^2 \boldsymbol{A} - \nabla (\nabla \cdot \boldsymbol{A}) - \frac{1}{c^2} \left(\frac{\partial}{\partial t} \nabla \Phi + \frac{\partial^2 \boldsymbol{A}}{\partial t^2} \right) = -\mu_0 \boldsymbol{j} \,. \tag{2.11}$$

c represents the velocity of the electromagnetic wave in vacuum (speed of light).

The potentials are not uniquely determined by the above equations, but one has the freedom to choose a gauge, the effect of which vanishes when applying the differential operators to derive the values of the physical fields $\boldsymbol{E}$ and $\boldsymbol{B}$. For radiation problems the natural gauge to choose is the so-called Coulomb or radiation gauge defined by

$$\nabla \cdot \boldsymbol{A} = 0 \,. \tag{2.12}$$

With this gauge condition Eqs. (2.8) and (2.11) become

$$\begin{aligned} \nabla^2 \Phi &= -\rho_c / \epsilon_0 \,, \\ \nabla^2 \boldsymbol{A} - \frac{1}{c^2} \frac{\partial^2 \boldsymbol{A}}{\partial t^2} &= -\mu_0 \boldsymbol{j} + \frac{1}{c^2} \frac{\partial}{\partial t} \nabla \Phi \,. \end{aligned} \tag{2.13}$$

The first of these two equations is a Poisson equation for the scalar potential Φ, implying an *instantaneous* response of Φ to fluctuations in the charge density, without any wave propagation. This apparent non-local behaviour of Φ is allowed, since Φ is not the observed physical quantity, and the corresponding relevant physical field $\boldsymbol{E}$ always has local behaviour (disturbances propagate with the speed of light). The second equation on the other hand is a wave equation for the vector potential $\boldsymbol{A}$, with the sources on the right hand side. These sources also contain a Φ-term, which however does not give any contributions to the $\boldsymbol{A}$ waves, for the following reason.

Let us consider the solution to the Poisson equation,

$$\Phi(\boldsymbol{r}) = \frac{1}{4\pi\epsilon_0} \int \frac{\rho_c(\boldsymbol{r}')\, \mathrm{d}V'}{|\boldsymbol{r} - \boldsymbol{r}'|} \,. \tag{2.14}$$

At distances large as compared with the size of the charge region $\Phi(\boldsymbol{r})$ becomes proportional to $1/r$ with asymptotically vanishing deviations from spherical symmetry. Time variations in the charge density may therefore, via the $\partial \nabla \Phi / \partial t$ term, contribute to $\boldsymbol{A}$ components parallel to $\nabla \Phi$, i.e., parallel to $\boldsymbol{r}$. These varying components can however never lead to $\boldsymbol{A}$ waves propagating from or to the charge region, since the waves can only be transversal and thus perpendicular to $\boldsymbol{r}$.

The transverse nature of the waves is a direct consequence of the gauge condition (2.12). If we introduce plane waves

$$\boldsymbol{A}(t,\boldsymbol{r}) = \boldsymbol{A}_0(\omega)\, e^{-i\omega t + i\boldsymbol{k}\cdot\boldsymbol{r}}\,, \tag{2.15}$$

the gauge condition implies

$$\boldsymbol{k}\cdot\boldsymbol{A} = 0\,, \tag{2.16}$$

i.e., the waves must be transverse.

The scalar potential Φ thus decouples from the wave propagation problem, and the electromagnetic waves become described by the single equation

$$\nabla^2\boldsymbol{A} - \frac{1}{c^2}\frac{\partial^2\boldsymbol{A}}{\partial t^2} = -\mu_0\boldsymbol{j}_t\,. \tag{2.17}$$

Only the transverse component of $\boldsymbol{j}$ (marked by index t), i.e., the component of the current density that is perpendicular to the propagation direction given by the wave vector $\boldsymbol{k}$, can serve as a source for the waves. The electric field $\boldsymbol{E}$ of the waves is obtained via Eq. (2.7) as

$$\boldsymbol{E} = -\frac{\partial\boldsymbol{A}}{\partial t}\,, \tag{2.18}$$

since Φ does not contribute to the wave.

From Eq. (2.7) also follows that $\boldsymbol{B}$ is perpendicular to both $\boldsymbol{E}$ and $\boldsymbol{k}$. The oscillations of $\boldsymbol{E}$ and $\boldsymbol{B}$ are in phase with each other, and lie in a plane perpendicular to the propagation direction.

2.5. Dipole Moment and Refractive Index

Plane wave solutions of the type (2.15) to the wave equation (2.17) imply

$$\left(-k^2 + \frac{\omega^2}{c^2}\right)\boldsymbol{A}_0 e^{-i\omega t + i\boldsymbol{k}\cdot\boldsymbol{r}} = -\mu_0\boldsymbol{j}_t\,. \tag{2.19}$$

If the currents $\boldsymbol{j}$ are carried by moving electrons, each electron i contributes to the current density by the amount

$$\boldsymbol{j}_i = -e\dot{\boldsymbol{r}}_i\,. \tag{2.20}$$

For the radiation problems that we are interested in here we let the motion of the electrons represent oscillations around an average position $\boldsymbol{r}_{0i}$ (that of the atomic nucleus). The position vector of electron no. i can then be written

$$\boldsymbol{r}_i = \boldsymbol{r}_{0i} + \boldsymbol{r}'_i\,. \tag{2.21}$$

As we may disregard the motion $\dot{\boldsymbol{r}}_{0i}$ of the average position (because of its much larger mass the velocity of the atomic nucleus is much smaller than that of the electrons), we have

$$\dot{\boldsymbol{r}}_i = \dot{\boldsymbol{r}}'_i\,. \tag{2.22}$$

If we denote the electric dipole moment per unit volume by $\boldsymbol{d}$, its time derivative is according to Eq. (2.22)

$$\dot{\boldsymbol{d}} = -\sum_i e\dot{\boldsymbol{r}}_i\,, \tag{2.23}$$

i.e., it equals the total current density $\boldsymbol{j}$.

Let us now *define* a refractive index n as embodying the dispersion relation between the wave number k and angular frequency ω through

$$k \equiv n\omega/c\,. \tag{2.24}$$

Then we obtain from Eqs. (2.10), (2.19), (2.20), and (2.23)

$$\dot{\boldsymbol{d}}_t = \epsilon_0\omega^2\,(n^2-1)\,\boldsymbol{A}_0 e^{-i\omega t+i\boldsymbol{k}\cdot\boldsymbol{r}}\,, \tag{2.25}$$

where index t denotes the transverse component. This constitutes a relation between the wave amplitude $\boldsymbol{A}_0$ and the amplitude of the electric dipole moment per volume element of the medium, expressed in terms of the complex refractive index n.

The dipole moment density can be determined from the radiation-matter interaction term in the Hamiltonian of the system (see next chapter). In this way a second relation between $\boldsymbol{d}$ and $\boldsymbol{A}_0$ gets established, which when combined with Eq. (2.25) allows a solution for the refractive index n to be found.

The physical interpretation of the complex refractive index is straightforward. From Eqs. (2.15) and (2.24) it is clear that the imaginary part corresponds to damping or absorption of the wave, while the real part corresponds to an oscillating phase factor determining the phase velocity of the wave and thus the ordinary refraction effects. The dispersion effects of the medium get large near the angular frequencies of the atomic resonances.

2.6. Representations of Polarized Light

As we have seen above, light represents transverse vibrations of the electric and magnetic vectors, both of which are derivable from the transverse vibrations of the vector potential $\boldsymbol{A}$. As $\boldsymbol{E}$ and $\boldsymbol{B}$ have a fixed relation relative to each other (they are always in phase and perpendicular to each other), it is sufficient to describe light in terms of one of them, and normally the choice is $\boldsymbol{E}$.

The best way to picture the state of polarization is to describe the oscillations *in a fixed plane as seen from the direction of the observer.* It is clear what we mean by linear plarization, but the definition of the sign of the circular polarization requires some care. The convention that is generally used, and which defines the sign unambiguously, is that when the electric vector rotates *clockwise* in a fixed plane as seen from the observer, we speak of *right-handed* circular polarization, when the sense of rotation is counterclockwise we have left-handed circular polarization. We will stick to this terminology. Other definitions in terms of the handedness of the spiral pattern traced out in space by the electric vectors tend to lead to considerable confusion.

A complete description of polarized light needs four parameters. Various powerful tools allowing us to deal efficiently with all aspects of polarization will be introduced in the next subsections. For further details we refer to the monographs by Shurcliff (1962), Robson (1974), and Collett (1993). In all these formalisms the physical contents can be visualized by considering the effect of inserting four different, idealized filters, F_k, $k = 0, 1, 2, 3$, in the light beam. Their properties are illustrated in Fig. 2.1. Filter F_0 simply represents empty space, without any absorbing effects, while F_1 and F_2 transmit linear polarization with the electric vector at position angles 0 and 45°, respectively, and F_3 transmits right-handed circular polarization. Since F_1, F_2, and F_3 block the orthogonal polarization state, the intensity of an unpolarized beam is reduced by a factor of two.

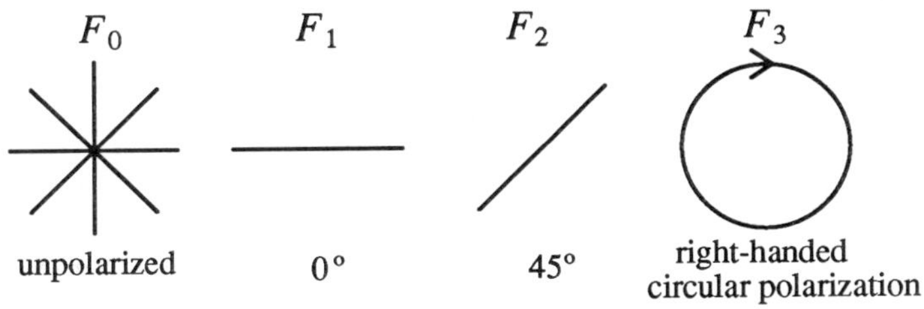

Fig. 2.1. Symbolic properties of the four idealized filters described in the text.

A detector records the light intensity I_k transmitted by each filter. These four intensity readings uniquely determine the full state of polarization of the incident beam.

2.6.1. Jones Calculus

Let us introduce a set of orthogonal basis vectors $\boldsymbol{e}_1$ and $\boldsymbol{e}_2$ in a plane perpendicular to the propagation direction. At any fixed point in space the electric vector $\boldsymbol{E}$ can then be decomposed as

$$\boldsymbol{E} = \mathrm{Re}\,(E_1\boldsymbol{e}_1 + E_2\boldsymbol{e}_2)\,, \tag{2.26}$$

where

$$E_k = E_{0k}e^{-i\omega t}\,, \quad k = 1, 2, \tag{2.27}$$

and E_{0k} are complex amplitudes. Since both amplitudes and phases are involved, the two complex numbers represent four parameters characterizing the light. The oscillating phase factor $e^{-i\omega t}$ can be ignored for polarization problems, since it is the same for both components of the electric vector, and it disappears when forming observable quantities (involving products with complex conjugate components, cf. the following subsections).

The Jones vector $\boldsymbol{J}$ is simply defined as

$$\boldsymbol{J} = \begin{pmatrix} E_1 \\ E_2 \end{pmatrix}. \tag{2.28}$$

The interaction with a medium can be described by a matrix $\boldsymbol{w}$ operating on $\boldsymbol{J}$:

$$\boldsymbol{J}' = \boldsymbol{w}\boldsymbol{J}\,. \tag{2.29}$$

The Jones matrices for the four filters of Fig. (2.1) are (cf. Shurcliff, 1966)

$$\begin{aligned} \boldsymbol{w}_0 &= \begin{pmatrix} 1 & 0 \\ 0 & 1 \end{pmatrix}, \qquad & \boldsymbol{w}_1 &= \begin{pmatrix} 1 & 0 \\ 0 & 0 \end{pmatrix}, \\ \boldsymbol{w}_2 &= \frac{1}{2}\begin{pmatrix} 1 & 1 \\ 1 & 1 \end{pmatrix}, \qquad & \boldsymbol{w}_3 &= \frac{1}{2}\begin{pmatrix} 1 & i \\ -i & 1 \end{pmatrix}. \end{aligned} \tag{2.30}$$

To derive $\boldsymbol{w}_1$ and $\boldsymbol{w}_2$ for the linearly polarizing filters, one needs to use a linear polarization basis $\boldsymbol{e}_1$ and $\boldsymbol{e}_2$, while for the derivation of $\boldsymbol{w}_3$ a circular polarization basis is the appropriate one. (Note that filter F_3 does not represent a $\lambda/4$ plate + linear polarizer, as is normally used in practical instruments to detect circular polarization, since the light that emerges would then be linearly instead of circularly polarized. Another $\lambda/4$ plate has to be added to make the transmitted light have the same right-handed circular polarization as the incident beam, in order to represent the function of F_3.)

It is useful to express a 2×2 matrix in terms of its components with respect to an orthogonal basis (analogous to the decomposition of a vector in its components). Such an orthogonal basis is provided by the Pauli spin matrices

$$\begin{aligned} \boldsymbol{\sigma}_0 &= \begin{pmatrix} 1 & 0 \\ 0 & 1 \end{pmatrix}, \qquad & \boldsymbol{\sigma}_1 &= \begin{pmatrix} 1 & 0 \\ 0 & -1 \end{pmatrix}, \\ \boldsymbol{\sigma}_2 &= \begin{pmatrix} 0 & 1 \\ 1 & 0 \end{pmatrix}, \qquad & \boldsymbol{\sigma}_3 &= \begin{pmatrix} 0 & i \\ -i & 0 \end{pmatrix}. \end{aligned} \tag{2.31}$$

The Jones matrices of the four filters can now be expressed in the conveniently compact form

$$\boldsymbol{w}_k = \tfrac{1}{2}\,(\boldsymbol{\sigma}_0 + \boldsymbol{\sigma}_k)\,, \qquad k = 0, 1, 2, 3. \tag{2.32}$$

The coherency matrix and Stokes formalisms will allow us to obtain a deeper physical understanding of these various 2×2 matrices.

The Jones vectors provide the most direct representation of the polarization of electromagnetic waves. They therefore emerge naturally from the basic theory of the interaction between matter and radiation. As however each Jones vector always represents 100 % polarization (since each photon is 100 % polarized), the Jones formalism is unable to describe partially polarized light. Partial polarization arises due to the *incoherent* superposition of mutually uncorrelated photons with different polarization states in a statistical ensemble of photons. The photons are mutually uncorrelated since the atomic processes by which they are created are stochastically independent of each other. To treat statistical ensembles we need to use the coherency matrix of the radiation field, which however can be directly derived from the Jones vectors.

2.6.2. Coherency Matrix Formalism

The 2×2 coherency matrix $\boldsymbol{D}$ of the radiation field is directly obtained from the Jones vector by

$$\boldsymbol{D} = \boldsymbol{J}\boldsymbol{J}^{\dagger} , \tag{2.33}$$

where $\boldsymbol{J}^{\dagger}$ denotes the adjoint of $\boldsymbol{J}$ (transposition and complex conjugation of $\boldsymbol{J}$).

The averaging of $\boldsymbol{D}$ over a statistical ensemble of photons represents the case of *incoherent* superposition of the photon states, since the phase factors $e^{-i\omega t}$ of the different photons disappear when forming the bilinear products of the Jones vector in Eq. (2.33). If we instead do the averaging of the Jones vectors of the individual photons *before* forming the product (2.33), this corresponds to a *coherent* superposition of the states. Since the different photons in a stellar atmosphere arise from stochastically independent atomic processes, the incoherent superposition is the appropriate one to use.

The intensity I of a light beam is proportional to $|\boldsymbol{E}_0|^2$, the square of the amplitude of the electric vector. As the constant of proportionality is unimportant for the description of the polarization state, it is convenient in the context of polarization theory to choose it to be unity. Thus we define

$$I = |\boldsymbol{E}_{01}|^2 + |\boldsymbol{E}_{02}|^2 , \tag{2.34}$$

which implies

$$I = \mathrm{Tr}\,\boldsymbol{D} , \tag{2.35}$$

where Tr means the trace (sum over the diagonal matrix elements).

If the Jones matrix of the medium is $\boldsymbol{w}$, then the coherency matrix transforms according to

$$\boldsymbol{D}' = \boldsymbol{w}\boldsymbol{D}\boldsymbol{w}^{\dagger} , \tag{2.36}$$

which is readily seen from Eqs. (2.29) and (2.33).

The 2×2 coherency matrix can also be represented in the form of a 4-dimensional vector $\boldsymbol{D}_v$, defined as

$$\boldsymbol{D}_v = \begin{pmatrix} D_{11} \\ D_{12} \\ D_{21} \\ D_{22} \end{pmatrix} , \tag{2.37}$$

where D_{ij} are the components of $\boldsymbol{D}$. This vector is transformed by a medium according to

$$\boldsymbol{D}_v' = \boldsymbol{W}\boldsymbol{D}_v , \tag{2.38}$$

where

$$\boldsymbol{W} = \boldsymbol{w} \otimes \boldsymbol{w}^* = \begin{pmatrix} w_{11}w_{11}^* & w_{11}w_{12}^* & w_{12}w_{11}^* & w_{12}w_{12}^* \\ w_{11}w_{21}^* & w_{11}w_{22}^* & w_{12}w_{21}^* & w_{12}w_{22}^* \\ w_{21}w_{11}^* & w_{21}w_{12}^* & w_{22}w_{11}^* & w_{22}w_{12}^* \\ w_{21}w_{21}^* & w_{21}w_{22}^* & w_{22}w_{21}^* & w_{22}w_{22}^* \end{pmatrix} \tag{2.39}$$

(cf. Robson, 1974). The symbols $\otimes$ and $*$ denote tensor product and complex conjugation, respectively.

Using Eqs. (2.35) and (2.36), we can now write down the intensities I_k ($k = 0,1,2,3$) measured behind the four filters F_k of Fig. 2.1, which were represented by the Jones matrices $\boldsymbol{w}_k$. They are

$$I_k = \mathrm{Tr}\,(\boldsymbol{w}_k \boldsymbol{D} \boldsymbol{w}_k^\dagger)\,. \tag{2.40}$$

Inserting the formula (2.32) for $\boldsymbol{w}_k$, we get an expression that subsequently can be reduced to the simple form

$$I_k = \tfrac{1}{2}\,[I + \mathrm{Tr}(\boldsymbol{\sigma}_k \boldsymbol{D})]\,, \quad k = 0,1,2,3, \tag{2.41}$$

by noting that $\boldsymbol{\sigma}_k^\dagger = \boldsymbol{\sigma}_k$, $\boldsymbol{D}^\dagger = \boldsymbol{D}$ (as a consequence of the definition (2.33)), $\boldsymbol{\sigma}_0 \boldsymbol{D} = \boldsymbol{D}$, $\boldsymbol{\sigma}_k^2 = \boldsymbol{\sigma}_0$, and $\mathrm{Tr}\,(\boldsymbol{\sigma D}) = \mathrm{Tr}\,(\boldsymbol{D\sigma})$.

2.6.3. STOKES PARAMETERS

The Stokes parameters S_k can be defined operationally in terms of the intensity measurements I_k with the four filters of Fig. 2.1 as

$$S_k = 2I_k - I_0\,. \tag{2.42}$$

S_0 thus represents the ordinary intensity, S_1 and S_2 the amount of linear polarization along position angles 0 and 45°, and S_3 the amount of right-handed circular polarization.

Using expression (2.41) in Eq. (2.42), we obtain the relation between the Stokes parameters and the coherency matrix:

$$S_k = \mathrm{Tr}(\boldsymbol{\sigma}_k \boldsymbol{D})\,. \tag{2.43}$$

The Stokes parameters, which are often denoted I, Q, U, and V instead of S_k, form a 4-vector

$$\boldsymbol{S} = \begin{pmatrix} S_0 \\ S_1 \\ S_2 \\ S_3 \end{pmatrix} \equiv \begin{pmatrix} I \\ Q \\ U \\ V \end{pmatrix}, \tag{2.44}$$

which in a medium transforms according to

$$\boldsymbol{S}' = \boldsymbol{M}\boldsymbol{S}\,. \tag{2.45}$$

$\boldsymbol{M}$ is the 4×4 Mueller matrix. The formalism for calculating the effect of a medium or of an optical train on the Stokes vector is called Mueller calculus.

The Stokes and coherency matrix formalisms are equivalent to each other and give a complete description of the state of polarization of a beam of electromagnetic radiation. The inverse of Eq. (2.43) can be written

$$\boldsymbol{D} = \tfrac{1}{2} \sum_k S_k \boldsymbol{\sigma}_k\,, \tag{2.46}$$

i.e., the coherency matrix can be expanded in terms of the basis matrices $\boldsymbol{\sigma}_k$, with the Stokes parameters divided by two as the expansion coefficients. Explicitly, in terms of a single matrix, Eq. (2.46) becomes

$$\boldsymbol{D} = \frac{1}{2} \begin{pmatrix} I+Q & U+iV \\ U-iV & I-Q \end{pmatrix} . \tag{2.47}$$

The transformation matrix $\boldsymbol{W}$ for the vector version $\boldsymbol{D}_v$ of the coherency matrix, as defined by Eqs. (2.37)–(2.39), can be transformed into a Mueller matrix through

$$\boldsymbol{M} = \boldsymbol{T}\boldsymbol{W}\boldsymbol{T}^{-1} , \tag{2.48}$$

where

$$\begin{aligned} \boldsymbol{T} &= \begin{pmatrix} 1 & 0 & 0 & 1 \\ 1 & 0 & 0 & -1 \\ 0 & 1 & 1 & 0 \\ 0 & -i & i & 0 \end{pmatrix} , \\ \boldsymbol{T}^{-1} &= \frac{1}{2} \begin{pmatrix} 1 & 1 & 0 & 0 \\ 0 & 0 & 1 & i \\ 0 & 0 & 1 & -i \\ 1 & -1 & 0 & 0 \end{pmatrix} . \end{aligned} \tag{2.49}$$

CHAPTER 3

INTERACTION OF MATTER WITH RADIATION

3.1. Introduction

In the preceding chapter it was shown that the radiation field interacts with matter via the current density term in the electromagnetic wave equation (2.17). Thereby a fluctuating electric dipole moment given by Eq. (2.25) is produced. Three fundamental processes are part of this interaction: (1) Absorption of the radiation when the dipole oscillations are excited. (2) Scattering of the radiation as the excited dipoles reradiate the absorbed energy. (3) Damping and exponential decay of the oscillating dipoles, as they lose energy through reradiation.

To develop the theory in a logical and unified fashion we will start with a brief treatment of the radiation from the excited dipoles, which will serve as a foundation for the more complete treatment of scattering and the Hanle effect in Chapter 5. This will also allow us as a next step to calculate the radiative reaction force leading to an expression for the damping and exponential decay, which is subsequently used as an added perturbation term in the equation governing the dynamics of the dipoles. This equation is derived from the interaction term of the Hamiltonian of the system in the presence of an external magnetic field. Its oscillatory solutions give us explicit expressions for the complex refractive index, which describes the absorbing and refractive properties of the medium, and which will later be used to describe the scattering matrix and the Hanle effect. A condensed presentation of this theory has been given in Stenflo (1991b).

3.2. Dipole Radiation

We start by considering the radiation from the medium. It can be obtained from the solution of the wave equation (2.17) for the vector potential $\boldsymbol{A}$,

$$\boldsymbol{A}(\boldsymbol{r},t)=\frac{\mu_0}{4\pi}\int\frac{\boldsymbol{j}_t(\boldsymbol{r}',t')}{|\boldsymbol{r}-\boldsymbol{r}'|}\,\mathrm{d}V' \tag{3.1}$$

(Biot-Savart's law), if we only consider the contribution to $\boldsymbol{A}$ from currents $\boldsymbol{j}$ within a localized region.

$$t'=t-|\boldsymbol{r}-\boldsymbol{r}'|/v \tag{3.2}$$

is the retarded time, where $v=\omega/k$ is the speed of light in the medium.

As shown in Sect. 2.5, the currents that are the sources of the vector potential $\boldsymbol{A}$ are due to oscillating electric dipoles. The oscillating component of the electric dipole moment $\boldsymbol{d}$ per unit volume (which is the only component of $\boldsymbol{d}$ that is of relevance for the radiation problem) can according to Eq. (2.25) be written

$$\boldsymbol{d}_t(\boldsymbol{r}',t') = \boldsymbol{d}_{0t}(\boldsymbol{r}')\, e^{-i\omega t'} . \tag{3.3}$$

It then follows from Eq. (2.20) that

$$\boldsymbol{j}_t(\boldsymbol{r}',t') = -i\omega \boldsymbol{d}_{0t}(\boldsymbol{r}')\, e^{-i\omega t'} . \tag{3.4}$$

As according to Eq. (3.2)

$$e^{-i\omega t'} = e^{-i\omega t} e^{ik|\boldsymbol{r}-\boldsymbol{r}'|} , \tag{3.5}$$

Eq. (3.1) becomes

$$\boldsymbol{A}(\boldsymbol{r},t) = -\frac{i\mu_0\omega}{4\pi}\, e^{-i\omega t} \int \frac{\boldsymbol{d}_{0t}(\boldsymbol{r}')\, e^{ik|\boldsymbol{r}-\boldsymbol{r}'|}}{|\boldsymbol{r}-\boldsymbol{r}'|}\, \mathrm{d}V' . \tag{3.6}$$

For our purposes only the so-called "wave zone" needs to be considered, where the distances r are large as compared with the size of the source region ($r \gg r'$). There the waves become aymptotically plane, and $|\boldsymbol{r}-\boldsymbol{r}'| \approx r - \boldsymbol{r}' \cdot \boldsymbol{e}_r$ and $1/|\boldsymbol{r} - \boldsymbol{r}'| \approx 1/r$. A further simplification is obtained by the "dipole approximation", which uses the assumption that the dimension of the dipoles (the size of the atoms) is small in comparison with the wavelength of the radiation, which is the case for the spectrum above the X-ray range. This approximation implies that

$$e^{ik\boldsymbol{r}'\cdot\boldsymbol{e}_r} \approx 1 . \tag{3.7}$$

If we let the vector potential $\boldsymbol{A}$ represent the wave-zone field due to dipoles within a region of unit volume, then the approximations we have introduced allow us to rewrite Eq. (3.6) in the form

$$\boldsymbol{A}(\boldsymbol{r},t) = -\frac{i\mu_0\omega}{4\pi}\, \frac{e^{-i\omega t+ikr}}{r}\, \boldsymbol{d}_{0t} . \tag{3.8}$$

This general expression can be used to find the component E_α of the vector $\boldsymbol{E}$ along some direction $\boldsymbol{e}_\alpha$ (which has to lie in a plane perpendicular to the propagation direction given by $\boldsymbol{e}_r$). The component of the full dipole moment amplitude $\boldsymbol{d}_0$ that is parallel to this direction is obtained by scalar multiplication with $\boldsymbol{e}_\alpha$. Accordingly we obtain from Eqs. (3.8) and (2.18)

$$E_\alpha = \frac{\mu_0\omega^2}{4\pi}\, \frac{e^{-i\omega t+ikr}}{r}\, \boldsymbol{d}_0 \cdot \boldsymbol{e}_\alpha . \tag{3.9}$$

This is a key expression for describing the polarization of the scattered radiation and the Hanle effect, as will be seen later.

The corresponding component of the magnetic vector $\boldsymbol{H}$ is obtained according to Eqs. (2.7) and (2.3) from $\nabla \times \boldsymbol{A}/\mu_0$. Thus its amplitude is $k/(\mu_0\omega)$ times that of the $\boldsymbol{E}$ component, and $\boldsymbol{E}$ and $\boldsymbol{H}$ oscillate in phase with each other. As the average radiative energy flux according to Eq. (2.6) is $\mathrm{Re}\,\boldsymbol{E} \times \boldsymbol{H}^*/2$, and as it oscillates as a $\cos^2$ function, the *instantaneous* Poynting flux $\boldsymbol{P}$ is

$$\boldsymbol{P} = \mathrm{Re}\,\boldsymbol{E} \times \boldsymbol{H}^* \cos^2(\omega t - kr) . \tag{3.10}$$

If we write $\boldsymbol{E} = |E|\boldsymbol{e}_E$, we obtain $|E|$ from Eq. (3.9) if we replace $\boldsymbol{e}_\alpha$ by $\boldsymbol{e}_E$. Using the scaling between $\boldsymbol{E}$ and $\boldsymbol{H}$ we thus obtain

$$|\boldsymbol{P}| = \frac{\mu_0 k\,\omega^3}{16\pi^2}\,\frac{1}{r^2}\,|\boldsymbol{d}_0\cdot\boldsymbol{e}_E|^2\cos^2(\omega t - kr)\,. \tag{3.11}$$

The loss of energy $\mathcal{E}$ per unit time and unit volume from the radiating dipoles is obtained by integrating (3.11) over all directions:

$$\frac{\mathrm{d}\mathcal{E}}{\mathrm{d}t} = -\int_0^{2\pi}\mathrm{d}\varphi\int_0^{\pi}\mathrm{d}\theta\,r^2\sin\theta\,|\boldsymbol{P}|\,. \tag{3.12}$$

Let us choose the coordinate system such that the z axis (the polar axis) is parallel to the oscillating dipole moment vector $\boldsymbol{d}_0$. Radiation in a direction at an angle θ (the colatitude) with the z axis then has its transverse $\boldsymbol{E}$ vector along a direction that makes an angle $\pi/2-\theta$ with the z axis. This means that $\boldsymbol{d}_0\cdot\boldsymbol{e}_E = d_0\sin\theta$ for radiation at colatitude θ. Then Eqs. (3.11) and (3.12) give

$$\frac{\mathrm{d}\mathcal{E}}{\mathrm{d}t} = -\frac{\mu_0 k\,\omega^3 d_0^2}{8\pi}\cos^2(\omega t - kr)\int_0^{\pi}\sin^3\theta\,\mathrm{d}\theta\,, \tag{3.13}$$

or

$$\frac{\mathrm{d}\mathcal{E}}{\mathrm{d}t} = -\frac{\mu_0}{6\pi}k\,\omega^3 d_0^2\cos^2(\omega t - kr)\,. \tag{3.14}$$

3.3. Radiative Reaction Force

The radiative energy loss can be described in terms of the work done on the oscillating dipole by a radiative reaction force F_R. This type of description is needed to obtain the complete equation for the dynamics of a classical oscillator, to be derived in the next section.

The relation between the energy loss rate and the reaction force is

$$\int\frac{\mathrm{d}\mathcal{E}}{\mathrm{d}t} = \int F_R\,\mathrm{d}z = \int F_R\,\dot{z}\,\mathrm{d}t\,. \tag{3.15}$$

The only time-dependent factor on the left-hand side is $\cos^2(\omega t - kr)$ according to Eq. (3.14). Partial integration of this factor over an interval of length t_1 gives (the phase term kr has been omitted, since it can be eliminated by choosing the zero point of the time scale for fixed r)

$$\int_0^{t_1}\cos^2\omega t\,\mathrm{d}t = \frac{1}{\omega}\sin\omega t\cos\omega t\Big|_0^{t_1} + \int_0^{t_1}\sin^2\omega\,\mathrm{d}t\,. \tag{3.16}$$

While the second term on the right-hand side grows linearly with t_1, approaching $t_1/2$, the first term always remains smaller than $1/\omega$. Thus for $t_1 \gg 2\pi/\omega$ (the oscillation period), the first term on the right-hand side can be disregarded. This means that the $\cos^2$ under the integral sign may be replaced by $\sin^2$ or shifted in phase by any amount, which proves useful for the derivation of the radiative reaction force below. As the radiation loss is very slow (the damping time is typically seven orders of magnitude larger than the oscillation period), the radiative reaction force represents a tiny perturbation, and it is fully adequate to treat it by using time smoothing over many oscillation periods (and thereby being able to exchange $\cos^2$ and $\sin^2$).

The harmonic oscillation along the z axis can be described by

$$\begin{aligned} z &= z_0 \cos \omega t\,, \\ \dot{z} &= -\omega z_0 \sin \omega t \end{aligned}\,, \tag{3.17}$$

where z_0 is the oscillation amplitude, and $d_0 = e z_0$. Using Eqs. (3.14), (3.16) and (3.17), Eq. (3.15) becomes

$$-\frac{\mu_0}{6\pi} k\,\omega^3\, e^2 z_0^2 \int \sin^2 \omega t\, \mathrm{d}t = -\omega z_0 \int F_R \sin \omega t\, \mathrm{d}t\,. \tag{3.18}$$

Applying Eq. (3.17) again on the left-hand side of Eq. (3.18), we get

$$\frac{\mu_0}{6\pi}\, e^2 k\,\omega \int \dot{z} \sin \omega t\, \mathrm{d}t = -\int F_R \sin \omega t\, \mathrm{d}t\,. \tag{3.19}$$

From this follows that

$$F_R = -\frac{\mu_0\, e^2}{6\pi} k\,\omega \dot{z}\,. \tag{3.20}$$

The usual form for the radiative reaction force is

$$F_R = -m\gamma\dot{z}\,, \tag{3.21}$$

where m is the mass of the oscillating charge (the electron), and γ is the *damping constant.* If we replace the wave number k by ω/c (valid for propagation in vacuum), Eq. (3.20) gives us the damping constant:

$$\gamma = \frac{\mu_0\, e^2 \omega^2}{6\pi m c}\,. \tag{3.22}$$

3.4. Classical Description of the Electromagnetic Interaction of an Electron in a Central Potential

For a conservative system, the Hamiltonian for an electron in an electromagnetic field is

$$H = \frac{1}{2m}(\boldsymbol{p} + e\boldsymbol{A})^2 + V(r)\,, \tag{3.23}$$

where the canonical momentum is

$$\boldsymbol{p} = m\boldsymbol{v} - e\boldsymbol{A}\,, \tag{3.24}$$

and where we let V represent a Coulomb central potential that has its source in an atomic nucleus of charge $+Ze$:

$$V(r) = -\frac{Z\,e^2}{4\pi\varepsilon_0 r}\,. \tag{3.25}$$

The equation of motion for the electron is obtained from one of Hamilton's equations (here expressed in vector notation):

$$\dot{\boldsymbol{p}} = -\nabla H\,. \tag{3.26}$$

The contribution to this equation from the kinetic energy term in Eq. (3.23) only comes from the cross terms (containing $\boldsymbol{p}\cdot\boldsymbol{A}$) describing the interaction between radiation and matter, since for the weak radiation fields that we are always dealing with in stellar atmospheres, the $\boldsymbol{A}^2$ term (describing the radiative self-interaction) can safely be disregarded. We thus obtain

$$m\dot{\boldsymbol{v}} - e\frac{\partial \boldsymbol{A}}{\partial t} - e\,(\boldsymbol{v}\cdot\nabla)\boldsymbol{A} = -\frac{e}{m}\nabla(\boldsymbol{p}\cdot\boldsymbol{A}) - \nabla V\,. \tag{3.27}$$

As according to Eq. (3.7)

$$\boldsymbol{v}\times\boldsymbol{B} = \boldsymbol{v}\times(\nabla\times\boldsymbol{A}) = \nabla(\boldsymbol{v}\cdot\boldsymbol{A}) - (\boldsymbol{v}\cdot\nabla)\boldsymbol{A} \tag{3.28}$$

and $\boldsymbol{A}^2$ terms can be neglected, we obtain from Eqs. (3.27), (3.28), (3.24), and (2.18)

$$m\dot{\boldsymbol{v}} + e\boldsymbol{E} + e\,(\boldsymbol{v}\times\boldsymbol{B}) + \nabla V = 0\,. \tag{3.29}$$

We see from this derivation that half of the $\boldsymbol{v}\times\boldsymbol{B}$ term gets its contribution from $\dot{\boldsymbol{p}}$ (the $-(\boldsymbol{v}\cdot\nabla)\boldsymbol{A}$ term), the other half from ∇H (the $\nabla(\boldsymbol{v}\cdot\boldsymbol{A})$ term).

In the case of non-relativistic velocities ($v \ll c$), which applies to the bound electron orbits except for elements of high atomic number, the magnetic field $\boldsymbol{B}$ of the radiation does not contribute significantly to the $\boldsymbol{v}\times\boldsymbol{B}$ term as compared with the $\boldsymbol{E}$ term. Therefore, while $\boldsymbol{E}$ is due exclusively to the radiation field, $\boldsymbol{B}$ in Eq. (3.29) can be regarded as being due exclusively to an external magnetic field.

Let us denote the relative position vector of the electron by $\boldsymbol{x}$ (identical to the previously in Sect. 2.5 used notation $\boldsymbol{r}'_i$, which however is inconvenient to use in the present context). Then the potential gradient can be written

$$\nabla V = \frac{Z\,e^2}{4\pi\varepsilon_0}\frac{\boldsymbol{x}}{r^3}\,. \tag{3.30}$$

We will limit our classical treatment to the case of circular electron orbits (r constant). In this case the potential gradient produces a simple harmonic oscillator, since it can be expressed as

$$\nabla V = m\omega_0^2\boldsymbol{x}\,, \tag{3.31}$$

where

$$\omega_0^2 = \frac{Z\,e^2}{4\pi\varepsilon_0 m}\frac{1}{r^3}\,. \tag{3.32}$$

This relation expresses Kepler's third law for the electron.

The momentum equation (3.29) was derived for a conservative system without taking into account the energy loss due to the radiation from the oscillating electron. As the back-reaction force $m\gamma\boldsymbol{v}$ is about 10^7 times smaller than the ∇V term (since at optical frequencies $\gamma \approx 10^8$ while $\omega_0 \approx 10^{15}$ s^{-1}), it may be added to the momentum equation as a small perturbation. The complete equation for the classical oscillator is thus

$$\ddot{\boldsymbol{x}} + \frac{e}{m}(\dot{\boldsymbol{x}}\times\boldsymbol{B}) + \gamma\dot{\boldsymbol{x}} + \omega_0^2\boldsymbol{x} = -\frac{e}{m}\boldsymbol{E}\,, \tag{3.33}$$

where the driving term from the external electromagnetic field has been placed on the right-hand side. This is the equation that we need to solve to obtain the dipole moment of the medium, which then allows us to find the refractive index from Eq. (2.25).

Due to the $\dot{\boldsymbol{x}}\times\boldsymbol{B}$ term the three component equations of (3.33) are coupled to each other if Cartesian coordinates are used, which greatly complicates the problem. If we however make a transformation from a Cartesian coordinate basis to a basis of complex spherical vectors, the component equations decouple from each other, and we can regard the system as consisting of three independent, one-dimensional harmonic oscillators. The solution of this problem is then straightforward. The important formalism of spherical vectors will therefore be introduced below, but after we have first made a quantum reinterpretation of the oscillator frequency ω_0.

3.5. Quantum Reinterpretation of the Oscillator Frequency

The above classical theory fails to describe the radiative interaction of real atoms, which are characterized by discrete energy levels and by an elevated ground level that cannot decay. The resonant frequency ω_0 of a radiative interaction is determined by the energy difference between the levels ($\omega_0 = \Delta E/\hbar$). It is not directly related to the electron orbital frequency that is used in the classical treatment.

In the asymptotic limit of high quantum numbers n, however, when we have a transition for which the quantum number changes by one step ($\Delta n = 1$), the radiative frequency becomes equal to the orbital frequency. This asymptotic agreement between the quantum and classical pictures is an expression of the correspondence principle.

In the general case (when the quantum number n is not very large) there is no way to derive the transition frequencies from a classical model. Nevertheless our classical oscillator equation (3.33) turns out to generate the same results for absorption and scattering of polarized radiation as the quantum theory does, even for small quantum numbers, provided that ω_0 in Eq. (3.33) is interpreted as representing the transition frequency $\Delta E/\hbar$ rather than the orbital frequency. In all what follows it will be assumed that this reinterpretation has been made.

3.6. Spherical Vectors and the Decoupling of the Component Equations

The complex spherical unit vectors $\boldsymbol{e}_q$, $q = 0, \pm 1$, can be defined in terms of the Cartesian linear unit vectors $\boldsymbol{e}_x$, $\boldsymbol{e}_y$, and $\boldsymbol{e}_z$ as

$$\begin{aligned} \boldsymbol{e}_0 &= \boldsymbol{e}_z \,, \\ \boldsymbol{e}_\pm &= \mp(\boldsymbol{e}_x \pm i\boldsymbol{e}_y)/\sqrt{2} \,, \end{aligned} \tag{3.34}$$

where we have used the more compact notation $\boldsymbol{e}_\pm$ instead of $\boldsymbol{e}_{\pm 1}$ (cf. Shore and Menzel, 1968).

Now let the Cartesian components of the vector $\boldsymbol{E}$ be E_x, E_y, and E_z, while the corresponding spherical vector components are denoted E_q, $q = 0, \pm 1$. We define these spherical vector components such that their relations to the Cartesian components are formally the same as the relations between the corresponding unit vectors:

$$\begin{aligned} E_0 &= E_z \,, \\ E_\pm &= \mp(E_x \pm iE_y)/\sqrt{2} \,. \end{aligned} \tag{3.35}$$

Then, with these definitions, the spherical vector decomposition of the real, linear vector $\boldsymbol{E}$ can be done in the following forms:

$$\boldsymbol{E} = \sum_q E_q^* \boldsymbol{e}_q = \sum_q E_q \, \boldsymbol{e}_q^* = \sum_q (-1)^q E_q \boldsymbol{e}_{-q} \,. \tag{3.36}$$

A scalar product becomes

$$\boldsymbol{a} \cdot \boldsymbol{b} = \sum_q a_q b_q^* = \sum_q (-1)^q a_q b_{-q} \,. \tag{3.37}$$

The fundamental role of the spherical vectors is seen if we choose a coordinate system such that the z axis is along the direction of the magnetic field $\boldsymbol{B}$, and express the momentum equation (3.33) in terms of the spherical vector components. The troublesome $\boldsymbol{v} \times \boldsymbol{B}$ term then becomes

$$\boldsymbol{v} \times \boldsymbol{B} = -iB \sum_q q v_q \boldsymbol{e}_q^* \,, \tag{3.38}$$

where $B = |\boldsymbol{B}|$.

Eq. (3.38) shows that the different components of $\boldsymbol{v}$ and $\boldsymbol{B}$ are no longer coupled to each other (different q values do not mix in the products), in contrast to the Cartesian case. This means that the momentum vector equation (3.33) can be expressed as three *independent, scalar* equations,

$$\hat{C}_q x_q = -\frac{e}{m} E_q \,, \quad q = 0, \pm 1, \tag{3.39}$$

where the operator $\hat{C}_q$ is

$$\hat{C}_q = \frac{\mathrm{d}^2}{\mathrm{d}t^2} - \left(qi\,\frac{eB}{m} - \gamma \right) \frac{\mathrm{d}}{\mathrm{d}t} + \omega_0^2 \,. \tag{3.40}$$

These three equations describe three independent, damped harmonic oscillators, which have different oscillation frequencies because of the q-dependent term in Eq. (3.40). In the limit of vanishing magnetic field B, the three frequencies coincide.

As the phases of the driving electric field components E_q are the same (see next section), there will be phase relations between the three oscillators. It is these phase relations that are the reason for the polarization of the scattered radiation. The evolution of the phase relations depends both on the frequency differences (i.e., on B) and on the damping (lifetime of the excited state). The polarization effects arising from this fascinating interplay are collectively called the Hanle effect.

3.7. Oscillatory Solutions

With our previous plane-wave expression

$$\boldsymbol{A} = \boldsymbol{A_0}\, e^{-i\omega t + i\boldsymbol{k}\cdot\boldsymbol{r}} \tag{2.15}$$

used to derive the relation (2.25) between the dipole moment and the refractive index of the medium, it follows from Eq. (2.18) that E_q in Eq. (3.39) can be written as

$$E_q = i\omega A_{0q}\, e^{-i\omega t + i\boldsymbol{k}\cdot\boldsymbol{r}} \,. \tag{3.41}$$

Oscillatory solutions to Eq. (3.39) may be found by inserting

$$x_q = x_{0q}\, e^{-i\omega t} \,, \tag{3.42}$$

which immediately gives us the solution

$$x_{0q} = \frac{-ie\omega A_{0q}\, e^{i\boldsymbol{k}\cdot\boldsymbol{r}}/m}{\omega_0^2 - \omega^2 - 2q\omega_L\omega - i\gamma\omega} \,, \tag{3.43}$$

where we have introduced the Larmor frequency

$$\omega_L = \frac{e}{2m} B \,. \tag{3.44}$$

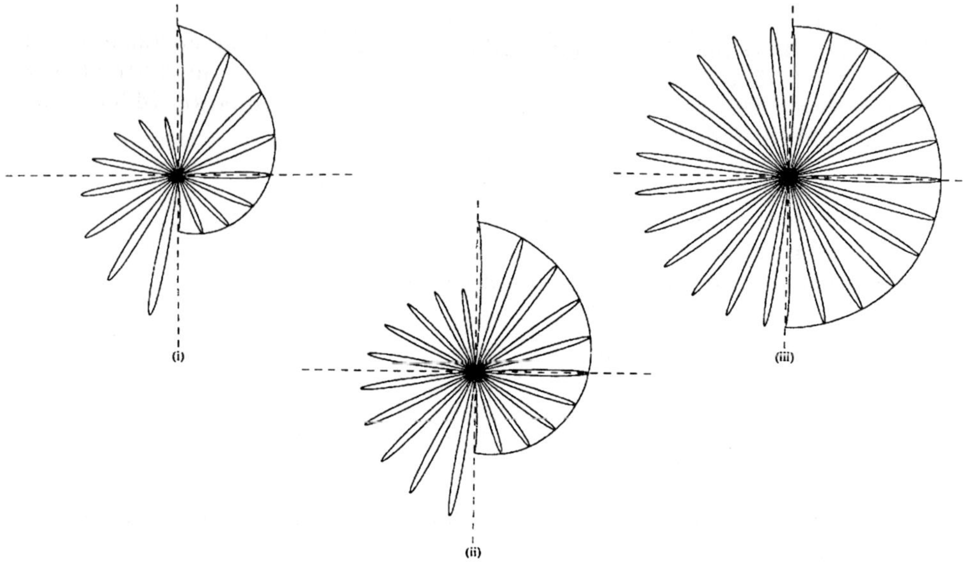

Fig. 3.1. Rosette motion of a damped classical oscillator in a magnetic field that is directed along the line of sight. The value of ω_L/γ increases from the left to the right diagram in the figure. From Corney (1977).

The solution for an oscillator precessing in a plane perpendicular to the direction of the magnetic field is illustrated in Fig. 3.1.

3.8. Solutions in Terms of a Complex Refractive Index

As

$$\dot{x}_q = -i\omega x_{0q}\, e^{-i\omega t} \tag{3.45}$$

and

$$\dot{d}_q = -eN\dot{x}_q\,, \tag{3.46}$$

where N is the number density of oscillators, Eq. (3.43) gives us

$$\dot{d}_q = \frac{N\, e^2\omega^2 A_{0q}\, e^{-i\omega t + i\boldsymbol{k}\cdot\boldsymbol{r}}/m}{\omega_0^2 - \omega^2 - 2q\omega_L\omega - i\gamma\omega}\,. \tag{3.47}$$

This can be compared with our previous expression (2.25) in terms of a refractive index. If we by n_q denote the refractive index induced by the vector component q of the vibrations, (2.25) can be written

$$\dot{d}_q = \epsilon_0\, \omega^2\, (n_q^2 - 1) A_{0q}\, e^{-i\omega t + i\boldsymbol{k}\cdot\boldsymbol{r}}\,. \tag{3.48}$$

Identification of Eqs. (3.47) and (3.48) gives the refractive index:

$$n_q^2 - 1 = \frac{\omega_A^2}{\omega_0^2 - \omega^2 - 2q\omega_L\omega - i\gamma\omega}\,, \tag{3.49}$$

where

$$\omega_A = \sqrt{\frac{e^2\,N}{\epsilon_0\,m}} \tag{3.50}$$

would be identical to the plasma frequency ω_p if N were equal to N_e, the electron number density. In our case N is the number density of classical oscillators.

As $|n_q - 1| \ll 1$ and $|\omega_0 - \omega| \ll \omega_0$, Eq. (3.49) can be simplified to

$$n_q - 1 \approx \frac{\omega_A^2}{4\omega_0}\,\frac{1}{\omega_0 - \omega - q\omega_L - i\gamma/2}\,, \tag{3.51}$$

or

$$n_q - 1 \approx \frac{\omega_A^2}{4\omega_0}\,\frac{\omega_0 - \omega - q\omega_L + i\gamma/2}{(\omega_0 - \omega - q\omega_L)^2 + (\gamma/2)^2}\,. \tag{3.52}$$

Let us now introduce dimensionless units by dividing all frequencies by a frequency width $\Delta\omega_D$, which we will later choose to be the Doppler width. If we further introduce the notations

$$\begin{aligned} v_q &= (\omega_0 - \omega - q\omega_L)/\Delta\omega_D\,, \\ a &= \gamma/(2\Delta\omega_D)\,, \end{aligned} \tag{3.53}$$

the refractive index may be expressed as

$$n_q = 1 + \frac{k_N}{\pi}\,\frac{v_q + ia}{v_q^2 + a^2}\,, \tag{3.54}$$

where

$$k_N = \frac{\pi\omega_A^2}{4\omega_0\,\Delta\omega_D}\,. \tag{3.55}$$

The reason for the explicit use of the factor $1/\pi$ is that it makes the dispersion function area normalized, since $\int_{-\infty}^{+\infty} \mathrm{d}v_q\,(a/\pi)/(v_q^2 + a^2) = 1$, which will be utilized in the next chapter when convolving with a distribution of Doppler velocities.

The real part of n_q represents anomalous dispersion, while the imaginary part (a Lorentz or dispersion profile) represents absorption. Another way of expressing Eq. (3.54) is

$$n_q - 1 = \frac{ik_N}{\pi a}\,\cos\Theta_q\; e^{i\Theta_q}\,, \tag{3.56}$$

where the "phase angle" Θ_q is determined by

$$\tan\Theta_q = -v_q/a\,. \tag{3.57}$$

When $|v_q| \gg a$, $\Theta_q \approx \pm\pi/2$, and dispersive effects dominate, although with vanishing amplitude. When on the other hand $|v_q| \ll a$, $\Theta_q \approx 0$, and the absorption effects dominate. Formally similar expressions will be derived in our treatment of the Hanle effect in Chapter 5 (cf. Eqs. (5.53) and (5.54)), where the concept of a phase angle will prove useful for describing the polarization effects.

3.9. Absorption and Dispersion of the Electromagnetic Wave

Let us now consider an electromagnetic wave propagating in a given direction $\boldsymbol{s}$. The spherical vector components of the electric field of the wave can then according to Eqs. (3.41) and (2.24) be associated with the refractive index n_q via

$$E_q \sim e^{-i\omega t + i n_q \omega s/c}\,, \tag{3.58}$$

where s is the spatial coordinate along the propagation direction. The interactions with the medium of the three E_q vector components are mutually independent, each being described by its own refractive index n_q. This mutual independence is a consequence of the decoupling of the oscillator component equations when complex spherical vectors are used.

From Eq. (3.58) it follows that the differential change in E_q caused by the medium is given by the scalar equation

$$\frac{\mathrm{d}E_q}{\mathrm{d}s} = i\frac{\omega}{c} n_q E_q\,, \tag{3.59}$$

which describes the absorption and dispersion effects in a magnetic field. This simple equation, together with the expression for n_q, e.g. as given by (3.54), contains all the physics needed to derive the transfer equation for polarized radiation (except for the scattering processes, which however can be dealt with in a similar manner, cf. Sect. 3.12 below). The remaining derivation involves only geometrical considerations and transformations between different coordinate systems.

As soon as another polarization basis is used instead of the spherical vectors, Eq. (3.59) no longer remains scalar, but has to involve matrix multiplication. The normal polarization basis to use when representing the polarized radiation in terms of Jones vectors (cf. Sect. 2.6.1) is a basis of real, linear unit vectors, in terms of which the electric vector $\boldsymbol{E}$ has components E_1 and E_2. These complex components form the Jones vector as in (2.28). We will now derive from (3.59) how the Jones vector is affected by the medium.

We start by expanding $\boldsymbol{E}$ in the linear polarization basis:

$$\boldsymbol{E} = \sum_{\alpha=1}^{2} E_\alpha \boldsymbol{e}_\alpha\,. \tag{3.60}$$

The linear unit vectors can on the other hand be expanded in complex spherical unit vectors $\boldsymbol{e}_q$:

$$\boldsymbol{e}_\alpha = \sum_q \varepsilon_q^\alpha \boldsymbol{e}_q^* = \sum_q \varepsilon_q^{\alpha *} \boldsymbol{e}_q\,, \tag{3.61}$$

where ε_q^α are the spherical vector components of the linear unit vector $\boldsymbol{e}_\alpha$ (cf. Eq. (3.36)).

The linear vector component E_α can be expressed in terms of the spherical vector components E_q through

$$E_\alpha = \boldsymbol{E} \cdot \boldsymbol{e}_\alpha = \sum_q E_q \boldsymbol{e}_q^* \cdot \boldsymbol{e}_\alpha \,. \tag{3.62}$$

Since

$$\boldsymbol{e}_q^* \boldsymbol{e}_{q'} = \delta_{qq'} \,, \tag{3.63}$$

we obtain from Eq. (3.61)

$$\boldsymbol{e}_q^* \cdot \boldsymbol{e}_\alpha = \varepsilon_q^{\alpha *} \,. \tag{3.64}$$

Thus

$$E_\alpha = \sum_q \varepsilon_q^{\alpha *} E_q \,. \tag{3.65}$$

The derivation of the inverse relation proceeds similarly:

$$E_q = \boldsymbol{E} \cdot \boldsymbol{e}_q = \sum_\alpha E_\alpha \boldsymbol{e}_\alpha \cdot \boldsymbol{e}_q \,, \tag{3.66}$$

which results in

$$E_q = \sum_\alpha \varepsilon_q^\alpha E_\alpha \,. \tag{3.67}$$

Using Eqs. (3.65) and (3.67) with (3.59), we obtain

$$\frac{\mathrm{d}E_\alpha}{\mathrm{d}s} = i\frac{\omega}{c} \sum_q n_q \varepsilon_q^{\alpha *} E_q = i\frac{\omega}{c} \sum_{\alpha'} \sum_q n_q \varepsilon_q^{\alpha *} \varepsilon_q^{\alpha'} E_{\alpha'} \,, \tag{3.68}$$

which can be expressed as matrix multiplication:

$$\frac{\mathrm{d}}{\mathrm{d}s} \begin{pmatrix} E_1 \\ E_2 \end{pmatrix} = i\frac{\omega}{c} \begin{pmatrix} C_{11} & C_{12} \\ C_{21} & C_{22} \end{pmatrix} \begin{pmatrix} E_1 \\ E_2 \end{pmatrix} , \tag{3.69}$$

where

$$C_{\alpha\alpha'} = \sum_q n_q \varepsilon_q^{\alpha *} \varepsilon_q^{\alpha'} \,. \tag{3.70}$$

The physics is here cleanly separated in the different factors: While $\varepsilon_q^{\alpha *} \varepsilon_q^{\alpha'}$ contains all the information on the geometry but nothing on absorption and dispersion, the opposite is the case for n_q.

3.10. Evaluation of the Spherical Vector Components

If we decompose the linear polarization unit vector $\boldsymbol{e}_\alpha$ of Fig. 3.2 in its Cartesian components, and then use the defining Eq. (3.35) to convert to the spherical vector components ε_q^α of Eq. (3.61), we easily find

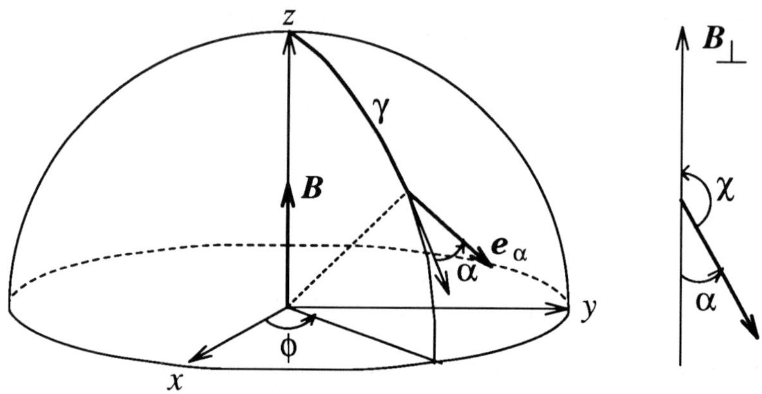

Fig. 3.2. Description of the coordinate system used in the text. $\boldsymbol{B}$ is the magnetic field vector, $\boldsymbol{B}_\perp$ its projection on a plane perpendicular to the propagation direction. γ is the angle between the beam and the magnetic field, χ the angle of $\boldsymbol{B}_\perp$ with respect to the polarization vector $\boldsymbol{e}_\alpha$.

$$\begin{aligned} \varepsilon^\alpha_\pm &= \mp \frac{1}{\sqrt{2}}(\cos\gamma\cos\alpha \pm i\sin\alpha)\,e^{\pm i\phi}\,, \\ \varepsilon^\alpha_0 &= -\sin\gamma\cos\alpha\,. \end{aligned} \tag{3.71}$$

For the absorption-dispersion problems that we are considering here, the azimuth phase factor $e^{\pm i\phi}$ is of no consequence, since it vanishes when forming the product $\varepsilon^{\alpha *}_q \varepsilon^{\alpha'}_q$ in the matrix elements $C_{\alpha\alpha'}$. Therefore we are free to choose any value of ϕ, e.g. zero.

Cancelling of the phase in the product of the two ε factors occurs because $\varepsilon^{\alpha *}_q$ and $\varepsilon^{\alpha'}_q$ refer to the same coordinate system, with the same ϕ. This is not the case for scattering problems, since the angles of the incident (exciting) and emitted (scattered) rays are generally different. We will return to the scattering problem in Sect. 3.12.

Let us now choose a linear polarization basis $\boldsymbol{e}_1$ and $\boldsymbol{e}_2$, such that the projection $\boldsymbol{B}_\perp$ of the magnetic-field vector makes an angle χ (in the counter-clockwise direction) with $\boldsymbol{e}_1$, $\chi - \pi/2$ with $\boldsymbol{e}_2$. Thus, as seen from Fig. 3.2, $\alpha = \pi - \chi$ for $\boldsymbol{e}_1$, $\pi - (\chi - \pi/2)$ for $\boldsymbol{e}_2$. If we further make use of our freedom to choose $\phi = 0$, we obtain from Eq. (3.71)

$$\begin{aligned} \varepsilon^1_0 &= \sin\gamma\cos\chi\,, \\ \varepsilon^2_0 &= \sin\gamma\sin\chi\,, \\ \varepsilon^1_\pm &= \pm(\cos\gamma\cos\chi \mp i\sin\chi)/\sqrt{2}\,, \\ \varepsilon^2_\pm &= \pm(\cos\gamma\sin\chi \pm i\cos\chi)/\sqrt{2}\,. \end{aligned} \tag{3.72}$$

For the special case that $\chi = 0$ this system reduces to

$$\begin{aligned} \varepsilon_0^1 &= \sin\gamma\,, \\ \varepsilon_0^2 &= 0\,, \\ \varepsilon_\pm^1 &= \pm\cos\gamma/\sqrt{2}\,, \\ \varepsilon_\pm^2 &= i/\sqrt{2}\,. \end{aligned} \tag{3.73}$$

3.11. Expansion of the Absorption-Dispersion Matrix in Terms of the Pauli Matrices

With our expressions (3.70) and (3.72) it is now straightforward to derive the components of the absorption-dispersion matrix for arbitrary values of the magnetic-field azimuth χ. It is convenient to introduce the following notation:

$$\begin{aligned} n_\Delta &\equiv \frac{1}{2}\left(n_0 - \frac{n_+ + n_-}{2}\right)\,, \\ n^0 &\equiv n_I \equiv n_\Delta \sin^2\gamma + \frac{1}{2}(n_+ + n_-)\,, \\ n^1 &\equiv n_Q \equiv n_\Delta \sin^2\gamma \cos 2\chi\,, \\ n^2 &\equiv n_U \equiv n_\Delta \sin^2\gamma \sin 2\chi\,, \\ n^3 &\equiv n_V \equiv \tfrac{1}{2}(n_+ - n_-)\cos\gamma\,. \end{aligned} \tag{3.74}$$

The absorption-dispersion matrix can then be written as

$$C = \begin{pmatrix} n_I + n_Q & n_U + i n_V \\ n_U - i n_V & n_I - n_Q \end{pmatrix}\,. \tag{3.75}$$

The index notation I, Q, U, and V has been introduced because of the correspondence with the Stokes parameters, which were defined in the preceding chapter (cf. Eq. (2.44)), and which will be introduced into the radiative transfer problem in the next chapter.

Eq. (3.75) is similar to the relation (2.47) between the coherency matrix and the Stokes parameters. As with the coherency matrix in Eq. (2.46), we can expand C in terms of the orthogonal basis of the Pauli spin matrices, which were defined in Eq. (2.31). Thus

$$C = n_I \boldsymbol{\sigma}_0 + n_Q \boldsymbol{\sigma}_1 + n_U \boldsymbol{\sigma}_2 + n_V \boldsymbol{\sigma}_3\,, \tag{3.76}$$

or, using the alternative notation of Eq. (3.74),

$$C = \sum_{k=0}^{3} n^k \boldsymbol{\sigma}_k\,. \tag{3.77}$$

In the analogous expansion (2.46) of the coherency matrix, the expansion coefficients were half the Stokes parameters S_k. This analogy suggests (as will be explicitly verified later) that the four complex refractive indices $n_{I,Q,U,V}$ really describe the absorption-dispersion effects on each of the four Stokes parameters I, Q, U, and V.

Let us here again recall that the imaginary part of $\boldsymbol{C}$ represents the absorption effects, the real part the anomalous dispersion (refraction) effects.

3.12. Jones Matrix for Radiative Scattering in a Magnetic Field

The scattering process consists of two physical components: (1) Excitation of an oscillating dipole moment $\boldsymbol{d}$ in the medium by the incident radiation. (2) Radiation from the dipoles.

The relative strengths of the three different spherical vector components d_q of the dipole moment, i.e., the *atomic polarization*, depends on the polarization and direction of the incident radiation relative to the magnetic field vector. To obtain the strength and polarization of the radiation that is scattered in a given direction we need to project the d_q components on a plane perpendicular to the scattering direction, since it is only these projected components that can contribute to the scattered wave. Accordingly the polarization of the scattered beam will depend on the relative geometry of the incident and scattered beams and the magnetic field, as well as on the polarization of the incident beam.

For harmonically oscillating dipoles as described by Eqs. (3.3) or (3.42), we have

$$\dot{\boldsymbol{d}} = -i\omega \boldsymbol{d}_0 \, e^{-i\omega t} \,, \tag{3.78}$$

where $\boldsymbol{d}_0$ is the oscillation amplitude of the dipole moment per unit volume. The electric field of the *scattered* radiation has been given earlier in Eq. (3.9). Using Eq. (3.78) and index α for the scattered linear vector components, Eq. (3.9) becomes

$$E_\alpha = \frac{i\mu_0\omega}{4\pi} \frac{e^{ikr}}{r} \dot{\boldsymbol{d}} \cdot \boldsymbol{e}_\alpha \,. \tag{3.79}$$

To calculate E_α of the scattered radiation we need the expression for $\dot{\boldsymbol{d}}$, which is determined by the exciting, incident radiation. If we mark all incident radiation components by a prime, Eq. (3.41) becomes

$$E'_q = i\omega A'_{0q} \, e^{-i\omega t + i\boldsymbol{k}\cdot\boldsymbol{r}} \,. \tag{3.80}$$

This allows us to write Eq. (3.48) as

$$\dot{d}_q = -2i \, \epsilon_0 \, \omega \, E'_q \, (n_q - 1) \,. \tag{3.81}$$

As the expansion of $\dot{\boldsymbol{d}}$ in terms of the spherical vectors is

$$\dot{\boldsymbol{d}} = \sum_q \dot{d}_q \boldsymbol{e}^*_q \tag{3.82}$$

(cf. Eq. (3.36)), we obtain by inserting Eq. (3.82) in Eq. (3.79) and using Eqs. (3.81) and (3.64)

$$E_\alpha = \frac{\omega^2}{2\pi c^2}\,\frac{e^{ikr}}{r}\sum_q (n_q - 1)\,\varepsilon_q^{\alpha *}\,E'_q\,. \tag{3.83}$$

If we further apply our previous expansion (3.67) of a spherical vector component in terms of linear vector components for E'_q in Eq. (3.83), we get

$$E_\alpha = \frac{\omega^2}{2\pi c^2}\,\frac{e^{ikr}}{r}\sum_\beta w_{\alpha\beta} E'_\beta\,, \tag{3.84}$$

where

$$w_{\alpha\beta} = \sum_q (n_q - 1)\,\varepsilon_q^{\alpha *}\varepsilon_q^\beta \tag{3.85}$$

are the components of the Jones scattering matrix $\boldsymbol{w}$. This matrix determines the relation between the incident and scattered electrical vectors, both of which are given in terms of a real, linear polarization basis.

$w_{\alpha\beta}$ looks formally similar to the absorption-dispersion matrix $C_{\alpha\alpha'}$ of Eq. (3.70). It is however important to note the fundamental difference that the two ε factors in $w_{\alpha\beta}$ refer to two *different* systems of basis vectors, in contrast to the ε factors in $C_{\alpha\alpha'}$. Therefore the geometry enters in a much more complicated way in the scattering case, although everything is implicitly contained in the formally simple expression (3.85). ε_q^β is the geometric factor for the excitation of the dipole moment, while $\varepsilon_q^{\alpha *}$ describes the projection of the dipole components on a plane perpendicular to the scattering direction.

To derive explicit expressions for the components of $\boldsymbol{w}$ in their simplest form, we choose the special coordinate system that is shown in Fig. 3.3, in which the two systems of basis vectors are shown. The polar axis has been chosen along the direction of the magnetic field vector $\boldsymbol{B}$, and the directions of the incident and scattered rays are given by their colatitudes θ' and θ, and their azimuths ϕ' and ϕ. The basis vectors $\boldsymbol{e}_1$ and $\boldsymbol{e}'_1$ are parallel to the respective meridional planes (i.e., $\alpha = 0$ and $\chi = \pi$ in Fig. 3.2).

With these definitions, and using the notation $\mu = \cos\theta$, $\mu' = \cos\theta'$, we obtain from Eq. (3.71) for ε_q^α

$$\begin{aligned} \varepsilon_0^1 &= -\sin\theta\,,\\ \varepsilon_0^2 &= 0\,,\\ \varepsilon_\pm^1 &= \mp\mu\, e^{\pm i\phi}/\sqrt{2}\,,\\ \varepsilon_\pm^2 &= -i\, e^{\pm i\phi}/\sqrt{2}\,, \end{aligned} \tag{3.86}$$

with the same expressions for ε_q^β, if all the angular quantities in Eq. (3.86) are replaced by the corresponding primed quantities.

Let us further define

$$b_q \equiv (n_q - 1)\, e^{-iq(\phi-\phi')}\,, \tag{3.87}$$

which makes the explicit expressions for the four components (3.85) of the Jones scattering matrix particularly simple:

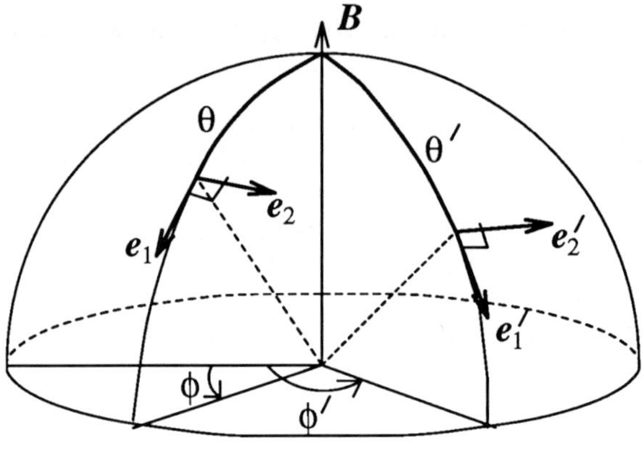

Fig. 3.3. Coordinate system and basis vectors used to derive the Jones scattering matrix. Primed quantities refer to the incident beam, unprimed to the scattered beam.

$$
\begin{aligned}
w_{11} &= b_0 \sin\theta \sin\theta' + \frac{1}{2}\mu\mu'(b_+ + b_-)\,, \\
w_{12} &= \frac{i\mu}{2}(b_+ - b_-)\,, \\
w_{21} &= -\frac{i\mu'}{2}(b_+ - b_-)\,, \\
w_{22} &= \frac{1}{2}(b_+ + b_-)\,.
\end{aligned}
\tag{3.88}
$$

For the special case that $\theta = \theta' = \gamma$ and $\phi = \phi'$ (forward scattering), we get $\boldsymbol{w} = \boldsymbol{C} - \boldsymbol{\sigma}_0$, as expected ($\boldsymbol{C}$ is the Jones matrix (3.70) for differential absorption-dispersion).

The above expressions (3.88) for the matrix components give a complete account of the polarization effects for a *single* scattering process. They contain all the information (for dipole scattering transitions) about the *Hanle effect*, with non-magnetic scattering polarization as a special case (when $\boldsymbol{B} = 0$).

The Hanle effect arises due to the coherent superposition of b_+ and b_- (or $-b_-$), with the well-defined phase factor contained in $b_\pm$. As the frequency components of b_+ and b_- are shifted by $2q\omega_L$ relative to each other, the coherent superposition of them depends on the magnetic field. In Chapter 5 we will see more explicitly how these magnetic-field effects manifest themselves in the polarization of the observed radiation. An overview of the Hanle effect with its applications to various areas of physics can be found in Moruzzi and Strumia (1991).

3.13. Modifications due to Quantum Mechanics

The classical theory that we have developed, with the previously mentioned quantum reinterpretation of the oscillator frequency ω_0 in Eq. (3.33) as being given by $\Delta E/\hbar$, is fully able to account for all the polarization effects in the case of a normal Zeeman triplet with an unsplit lower level and a Landé factor and oscillator strength of unity. It can easily be extended phenomenologically to describe most other atomic systems of practical interest. The correspondence between the classical and quantum theories will be demonstrated in Chapters 6–8.

The Landé factors (which depend on the electron spin) may be accounted for if we replace $-q\omega_L$ in the expression for the refractive index n_q (e.g. in Eq. (3.51)) by $\Delta\omega_H$, where

$$\Delta\omega_H = (g_\ell M_\ell - g_u M_u)\,\omega_L\,. \tag{3.89}$$

g_ℓ and g_u are the Landé factors of the lower and upper levels, while M_ℓ and M_u are the corresponding magnetic quantum numbers.

The inclusion of anomalous Zeeman splitting patterns is straightforward, since it only entails the weighted superposition of the various split components, as shown explicitly in Chapter 6, where Eq. (3.89) will be derived.

The oscillator strength f can simply be attached as a multiplicative factor to the expressions for the refractive index n_q, i.e., we replace N, the number density of oscillating dipoles, by $N_\ell f_{\ell u}$, where N_ℓ refers to the number density of the absorbing atoms in the lower state, while the oscillator strength $f_{\ell u}$ connects the lower and upper levels in an absorption transition.

The radiative damping is determined by the inverse lifetime of the excited level, and may differ from the classical damping rate. The lifetime is further limited by collisions. In scattering problems collisions affect the way in which the scattered photons get redistributed in frequency, as we will see in Chapter 10.

Collisions have another effect as well, which is to scramble the relative phases between the spherical vector components of the oscillator (i.e., between the states with different magnetic quantum number), thereby destroying the atomic polarization. The scattering atom will then "lose its memory" of the way in which the excitation occurred, which means that the scattered radiation will become isotropic and unpolarized. If we know the depolarizing collision rate, we can describe the scattered radiation as the incoherent superposition of two components: (1) Isotropic, unpolarized radiation due to the fraction of scattering processes that have suffered a collision while the atom was in its excited state. (2) Coherently scattered radiation as derived from the theory in the absence of depolarizing collisions.

For arbitrary atomic transitions the scattered radiation (in the absence of depolarizing collisions) does not generally behave like dipole radiation. In a fashion similar to our handling of the depolarizing collisions it is however always possible to mathematically treat the linear polarization of the scattered radiation in terms of two components, one that is describable as pure dipole radiation, the other as isotropic, unpolarized radiation. The linear polarization of the scattered light can then be accounted for in terms of an incoherent superposition of these two contributions, as will be shown in Chapters 5, 8, and 9. Each of these contributions may be described classically, but their relative weights are determined by the J (total angular momentum) quantum numbers of the upper and lower levels of the atomic transition.

When the lower atomic level is magnetically split, the population of the different M states may be affected by the polarization of the radiation field and by the transition rates between the polarized atomic levels in a way that may only be treated quantum-mechanically. Such lower-state atomic polarization is however in most cases of little practical importance. Other examples of non-classical phenomena are level-crossing interference (between levels of different total angular momentum J) and Raman scattering (initial and final levels of the scattering process are different). All these effects will be covered in Chapters 8 and 9 within the framework of quantum field theory.

CHAPTER 4

RADIATIVE TRANSFER WITHOUT SCATTERING

4.1. Introduction

The basic physics laid out in the previous two chapters will now be used to develop a formalism for the transfer of polarized radiation in a magnetized stellar atmosphere. The ordinary polarized transfer equation without scattering will be derived in the present chapter, while scattering and the Hanle effect will be taken care of in the next chapter.

The first successful formulation of a theory for the transfer of polarized radiation was developed in a phenomenological way by Unno (1956), while Stepanov (1958a) independently made a more stringent classical derivation that implicitly also included the magnetooptical effects (anomalous dispersion). Stepanov's approach was extended by Rachkovsky (1962a,b), who provided a firm classical foundation for the transfer problem in a magnetic field under the assumption of LTE (local thermodynamical equilibrium). Stepanov (1958b) and Rachkovsky (1963) also extended the LTE theory to include incoherent scattering. A new classical derivation of the LTE transfer equation without scattering has been done by Jefferies et al. (1989), who used an approach similar to the one adopted in the present chapter.

4.2. Coherent and Incoherent Superposition

In the preceding chapter the Jones matrices for absorption, dispersion, and scattering in a magnetic field were derived. The fundamental step when going from the Jones formalism to a full radiative-transfer treatment is the step from considering a single process to a statistical ensemble of processes.

Since in stellar atmospheres the radiative transitions in different atoms are to a very good approximation stochastically independent of each other, the phases of the photons from different atoms are uncorrelated with each other. This means that the contributions to the radiation field from the many individual interactions have to be superposed *incoherently*. Note that this incoherent superposition in no way affects coherency phenomena like the Hanle effect, since such coherency arises within each individual process, i.e., within each atom, due to the phase relations between the three coordinate components of the dipole oscillations.

Each individual photon that emerges from an interaction is always 100 % polarized. Due to the incoherent superposition of the photon states, however, the resulting radiation becomes *partially* polarized. To understand the difference between coherent and incoherent superposition, let us consider the superposition of two waves with electric fields $\boldsymbol{E}_1$ and $\boldsymbol{E}_2$, or two wave functions ψ_1 and ψ_2. What is measured is the intensity, proportional to $|\boldsymbol{E}|^2$, or the probability density $|\psi|^2$. Coherent superposition means that $|\boldsymbol{E}|^2 = |\boldsymbol{E}_1 + \boldsymbol{E}_2|^2$ or $|\psi|^2 = |\psi_1 + \psi_2|^2$, while for incoherent superposition we first square the individual contributions before summing them: $|\boldsymbol{E}|^2 = |\boldsymbol{E}_1|^2 + |\boldsymbol{E}_2|^2$ or $|\psi|^2 = |\psi_1|^2 + |\psi_2|^2$. The resulting difference between the two cases is of course that for coherent superposition one gets interference (the cross products), in contrast to the incoherent case.

The Jones formalism is unable to treat incoherent superposition and partial polarization, since the Jones vector, Eq. (2.28), consists of the unsquared electric field components. The coherency matrix and the Stokes vector on the other hand contain bilinear products of the electric field components, representing intensities (when observed with different polarizing filters, as described in Chapter 2). They are therefore suited for treating a statistical ensemble of incoherent photon states.

4.3. Radiative Transfer of the Coherency Matrix

The radiative transfer equation describing the differential change of the coherency matrix $\boldsymbol{D}$ can easily be obtained from the results of the preceding chapter for the Jones matrix. First we note that the defining Eq. (2.33) of the coherency matrix in terms of the Jones vector $\boldsymbol{J}$ implies that

$$\frac{\mathrm{d}\boldsymbol{D}}{\mathrm{d}s} = \frac{\mathrm{d}\boldsymbol{J}}{\mathrm{d}s}\boldsymbol{J}^\dagger + \boldsymbol{J}\frac{\mathrm{d}\boldsymbol{J}^\dagger}{\mathrm{d}s}\,. \tag{4.1}$$

The derivative of the Jones vector was given in Eq. (3.69), which with the defining Eq. (2.28) may be written as

$$\frac{\mathrm{d}\boldsymbol{J}}{\mathrm{d}s} = i\frac{\omega}{c}\boldsymbol{C}\boldsymbol{J}\,. \tag{4.2}$$

Inserting this in Eq. (4.1), we immediately obtain

$$\frac{\mathrm{d}\boldsymbol{D}}{\mathrm{d}s} = i\frac{\omega}{c}(\boldsymbol{C}\boldsymbol{D} - \boldsymbol{D}\boldsymbol{C}^\dagger) + \boldsymbol{F}_g\,, \tag{4.3}$$

where we have added an emission matrix $\boldsymbol{F}_g$ that will be explicitly given below. Index g indicates that $\boldsymbol{F}_g$ refers to a geometrical depth scale s. Later we will use $\boldsymbol{F}$ without an index, when it is defined in terms of an optical depth scale. The two versions of $\boldsymbol{F}$ differ by a factor that represents the continuous absorption coefficient.

The Jones matrix $\boldsymbol{C}$ was expanded in terms of the Pauli spin matrices in Eq. (3.77). Similarly we can expand its adjoint, making use of the property $\boldsymbol{\sigma}_k = \boldsymbol{\sigma}_k^\dagger$:

$$\boldsymbol{C}^\dagger = \sum_{k=0}^{3} n^{k*}\boldsymbol{\sigma}_k\,. \tag{4.4}$$

Eq. (4.3) thus becomes

$$\frac{\mathrm{d}\boldsymbol{D}}{\mathrm{d}s} = i\frac{\omega}{c}\sum_{k=0}^{3}(n^k\boldsymbol{\sigma}_k\boldsymbol{D} - n^{k*}\boldsymbol{D}\boldsymbol{\sigma}_k) + \boldsymbol{F}_g\,. \tag{4.5}$$

In the case of thermodynamic equilibrium there is detailed balance between absorption and emission (i.e., $\mathrm{d}\boldsymbol{D}/\mathrm{d}s = 0$), and the radiation field is unpolarized with intensity $B_\nu(T)$, the Planck function (where T is the temperature). In this case $\boldsymbol{D} = B_\nu(T)\boldsymbol{\sigma}_0/2$. Then Eq. (4.5) immediately gives us the explicit expression for the emission matrix $\boldsymbol{F}_g$.

This approach to deriving the expression for $\boldsymbol{F}_g$ can be generalized to the case when scattering also contributes to the emission, i.e., when the excitation is not only determined by the local temperature. When the incident radiation from non-local sources does not generate any atomic polarization (e.g. due to isotropy of the radiation field or depolarizing collisions), the polarization of the emitted radiation is the same as when the atom has been collisionally excited. The expression for the polarization then is the same as if it were due to negative absorption. Accordingly, detailed balance between absorption and emission can be achieved for an unpolarized radiation field of intensity S, which may differ from the local Planck function $B_\nu(T)$. Inserting

$$\boldsymbol{D} = \tfrac{1}{2}S\boldsymbol{\sigma}_0 \tag{4.6}$$

in Eq. (4.5), and letting the gradient on the left-hand side be zero (balance condition), we obtain

$$\boldsymbol{F}_g = -\tfrac{1}{2}i\,S\frac{\omega}{c}\sum_{k=0}^{3}(n^k - n^{k*})\boldsymbol{\sigma}_k\,, \tag{4.7}$$

or, with Eq. (3.74),

$$\boldsymbol{F}_g = S\frac{\omega}{c}(\boldsymbol{\sigma}_0\,\mathrm{Im}\,n_I + \boldsymbol{\sigma}_1\,\mathrm{Im}\,n_Q + \boldsymbol{\sigma}_2\,\mathrm{Im}\,n_U + \boldsymbol{\sigma}_3\,\mathrm{Im}\,n_V)\,. \tag{4.8}$$

So far only line processes (bound-bound transitions) have been considered. The continuum terms will be added later, after we have taken care of the line broadening.

4.4. Broadening of the Complex Refractive Index

Let us now consider a statistical ensemble of atoms or oscillators, each characterized by its own complex refractive index. If we want to obtain the total effect of the system, we have to add all the individual contributions to the transfer equation. Since the refractive index enters linearly in the transfer Eq. (4.5), and since the contributions to the coherency matrix can be linearly superposed (incoherent superposition of the photon states), the total effect is obtained by a simple integration over a distribution of properties.

Line broadening by collisions and Doppler motions is treated this way, by integrating over a distribution. This integration is actually a convolution of the intrinsic natural profile of the complex refractive index with a broadening distribution. The natural broadening is due to radiative damping, i.e., the radiative reaction force in the oscillator equation (3.33), and is a property of each individual process.

We recall that according to Eq. (3.54) the refractive index n_q that corresponds to the spherical vector component d_q of the dipole moment $\boldsymbol{d}$ can be written

$$\begin{aligned} \mathrm{Re}\, n_q &= 1 + \frac{k_N}{\pi} \frac{v_q}{v_q^2 + a_N^2} , \\ \mathrm{Im}\, n_q &= \frac{k_N}{\pi} \frac{a_N}{v_q^2 + a_N^2} , \end{aligned} \tag{4.9}$$

where a_N is the natural or radiative damping parameter. According to Eqs. (3.55) and (3.50)

$$k_N = \frac{e^2 N f}{16\pi\epsilon_0\, m\nu_0\, \Delta\nu_D} , \tag{4.10}$$

where we have included an oscillator strength f to phenomenologically account for quantum effects according to the prescription of Sect. 3.13. The angular frequencies ω have been replaced by frequencies ν $(= \omega/2\pi)$. According to Eq. (3.53)

$$v_q = v - q v_H , \tag{4.11}$$

with

$$\begin{aligned} v &= (\nu_0 - \nu)/\Delta\nu_D , \\ q v_H &= \Delta\nu_H/\Delta\nu_D , \end{aligned} \tag{4.12}$$

and the Zeeman splitting $\Delta\nu_H = \Delta\omega_H/(2\pi)$ is given by Eq. (3.89). The q value corresponds to a given value of M_u in Eq. (3.89) according to the quantum interpretation. For a $J_\ell = 0 \to J_u = 1$ transition, which corresponds to the classical oscillator case, $q = -M_u$ (cf. Sect. 6.5).

ν_0, and thus also v_q and v, refer to the rest frame of the undisturbed atoms. Line broadening can occur either because the rest frame of the observer does not coincide with the rest frame of the atom, or because the atom is disturbed by collisions. This can be treated (for the absorption-dispersion case — the general redistribution problem for radiative scattering will be dealt with in Chapters 5 and 10) in terms of a simple convolution of the profiles in Eq. (4.9) with a Lorentz profile (for collisional broadening) and a Gaussian profile (for the distribution of Doppler shifts). As the convolution operation is associative, we can consider each process separately.

Convolution corresponds to multiplication in the Fourier domain. It is readily seen that the convolution of the complex refractive index with a collisional Lorentz profile described by damping parameter a_c will preserve the form of Eq. (4.9), and the only thing needed is to replace a_N by $a = a_N + a_c$. That this is so can be seen by noting that the Fourier transform of a Lorentz profile (i.e., of $\mathrm{Im}\,(n_q - 1)$ in Eq. (4.9)) is proportional to $\exp(a_N t)$ for $t < 0$, $\exp(-a_N t)$ for $t \geq 0$, while the Fourier transform of $\mathrm{Re}\,(n_q - 1)$ is proportional to $i \exp(a_N t)$

for $t < 0$, $-i\exp(-a_N t)$ for $t \geq 0$. Multiplication of the Fourier transform of a collisional Lorentz profile with the transforms of either the real or imaginary part of the refractive index in Eq. (4.9) will thus merely have the effect of adding the exponents.

Next we convolve the collisionally broadened refractive index with a Gaussian $\exp(-v^2)/\sqrt{\pi}$ due to the distribution of line-of-sight velocities. The result, which we denote $\bar{n}_q$, can be written as

$$\bar{n}_q - 1 = i\,\frac{k_N}{\sqrt{\pi}}\mathcal{H}(a, v_q)\,, \tag{4.13}$$

where

$$\mathcal{H}(a, v) = H(a, v) - 2iF(a, v)\,. \tag{4.14}$$

The real part is the Voigt function

$$H(a, v) = \frac{a}{\pi}\int_{-\infty}^{+\infty}\frac{e^{-y^2}\,\mathrm{d}y}{(v-y)^2 + a^2}\,, \tag{4.15}$$

which when integrated over v has an area of $\sqrt{\pi}$ (so defined to make $H(0,0) = 1$), while the imaginary part contains the so-called line dispersion function

$$F(a, v) = \frac{1}{2\pi}\int_{-\infty}^{+\infty}\frac{(v-y)\,e^{-y^2}\,\mathrm{d}y}{(v-y)^2 + a^2}\,. \tag{4.16}$$

The Voigt function H describes the absorption effects, while the line dispersion function F represents the so-called magnetooptical effects (also called anomalous dispersion).

As unpolarized radiation of intensity S is represented by the coherency matrix $S\boldsymbol{\sigma}_0/2$, we see by comparison with the expression (4.8) for $\boldsymbol{F}_g$ or with the absorption term in Eq. (4.5) that an absorption coefficient describing the contribution from the spherical vector component q can be defined by

$$\kappa_q = \frac{2\omega}{c}\,\mathrm{Im}\,\bar{n}_q\,. \tag{4.17}$$

This will become clearer when the explicit versions of the transfer equation for the coherency matrix and the Stokes vector have been derived.

Eq. (4.17) implies that

$$\kappa_q = \kappa_0 H(a, v_q)\,, \tag{4.18}$$

where

$$\kappa_0 = \frac{2\omega k_N}{\sqrt{\pi}c}\,. \tag{4.19}$$

$\kappa_0 H(a, 0)$ is the absorption coefficient at the center of the line.

Next we define $\mathcal{H}_{I,Q,U,V} \equiv \mathcal{H}^k$, $k = 0, 1, 2, 3$, and $H_{I,Q,U,V} \equiv H^k$ with expressions of exactly the same form as those defining $n_{I,Q,U,V} \equiv n^k$ in Eq. (3.74), e.g. $\mathcal{H}_V = (\mathcal{H}_+ - \mathcal{H}_-)\cos\gamma\,/2$. From Eq. (4.13) we then obtain

$$\bar{n}^k = \delta_{0k} + i\,\frac{k_N}{\sqrt{\pi}}\,\mathcal{H}^k\,, \tag{4.20}$$

where $\delta_{0k} = 1$ if $k = 0$, zero otherwise. If we now replace n^k in Eq. (3.77) by $\bar{n}^k$ and use Eq. (4.19), we get the line-broadened absorption matrix

$$i\,\frac{\omega}{c}\boldsymbol{C} = i\,\frac{\omega}{c}\boldsymbol{\sigma}_0 - \tfrac{1}{2}\kappa_0 \sum_{k=0}^{3} \mathcal{H}^k \boldsymbol{\sigma}_k\,. \tag{4.21}$$

Only the second term on the right hand side is relevant for the radiative transfer problem, since the first one subtracts out when forming $\boldsymbol{CD} - \boldsymbol{DC}^\dagger$ in Eq. (4.3).

Accordingly we can write the absorption term in the transfer equation (4.3) as

$$i\,\frac{\omega}{c}(\boldsymbol{CD} - \boldsymbol{DC}^\dagger) = -(\boldsymbol{A}_g\boldsymbol{D} + \boldsymbol{DA}_g^\dagger)\,, \tag{4.22}$$

where

$$\boldsymbol{A}_g = \frac{\kappa_0}{2}\sum_{k=0}^{3}\mathcal{H}^k\boldsymbol{\sigma}_k = \frac{\kappa_0}{2}\begin{pmatrix} \mathcal{H}_I + \mathcal{H}_Q & \mathcal{H}_U + i\,\mathcal{H}_V \\ \mathcal{H}_U - i\,\mathcal{H}_V & \mathcal{H}_I - \mathcal{H}_Q \end{pmatrix}, \tag{4.23}$$

where index g refers to the geometrical depth scale.

For the emission term we obtain according to Eqs. (4.8) and (4.14)

$$\boldsymbol{F}_g = \tfrac{1}{2}\kappa_0 S(H_I\boldsymbol{\sigma}_0 + H_Q\boldsymbol{\sigma}_1 + H_U\boldsymbol{\sigma}_2 + H_V\boldsymbol{\sigma}_3)\,, \tag{4.24}$$

or

$$\boldsymbol{F}_g = S\,\frac{\kappa_0}{2}\begin{pmatrix} H_I + H_Q & H_U + i\,H_V \\ H_U - i\,H_V & H_I - H_Q \end{pmatrix}. \tag{4.25}$$

4.5. Stokes Formulation of the Transfer Equation

The relation between the Stokes parameters S_k and the coherency matrix was given by Eq. (2.43). Since the derivative of this expression commutes with the trace operation, we get

$$\frac{\mathrm{d}S_k}{\mathrm{d}s} = \mathrm{Tr}\,(\boldsymbol{\sigma}_k\frac{\mathrm{d}\boldsymbol{D}}{\mathrm{d}s})\,, \tag{4.26}$$

Using $\mathrm{d}\boldsymbol{D}/\mathrm{d}s$ from the transfer Eq. (4.3), $\boldsymbol{D}$ from the expansion (2.46), and $\boldsymbol{A}_g$ and $\boldsymbol{F}_g$ from Eqs. (4.23) and (4.24), we obtain

$$\begin{aligned}\frac{\mathrm{d}S_k}{\mathrm{d}s} &= -\tfrac{1}{4}\kappa_0\sum_{i,j}[\mathcal{H}^i\,\mathrm{Tr}\,(\boldsymbol{\sigma}_k\boldsymbol{\sigma}_i\boldsymbol{\sigma}_j) + \mathcal{H}^{i*}\,\mathrm{Tr}\,(\boldsymbol{\sigma}_k\boldsymbol{\sigma}_j\boldsymbol{\sigma}_i)]S_j \\ &+ \tfrac{1}{2}\kappa_0 S\sum_i H^i\,\mathrm{Tr}\,(\boldsymbol{\sigma}_k\boldsymbol{\sigma}_i)\,,\end{aligned} \tag{4.27}$$

where we have made use of the relation $\boldsymbol{\sigma}_i^\dagger = \boldsymbol{\sigma}_i$.

The transfer equation (4.27) may be recast in the form

$$\frac{\mathrm{d}S_k}{\mathrm{d}s} = -\sum_j M_{kj} S_j + j_k \,, \tag{4.28}$$

or in matrix notation as

$$\frac{\mathrm{d}\boldsymbol{S}}{\mathrm{d}s} = -\boldsymbol{M}\boldsymbol{S} + \boldsymbol{j} \,, \tag{4.29}$$

where $\boldsymbol{M}$ is the Mueller absorption matrix, and $\boldsymbol{j}$ is the emission vector.

According to Eq. (4.27) the Mueller matrix is

$$M_{kj} = \tfrac{1}{4}\kappa_0 \sum_i [\mathcal{H}^i \,\mathrm{Tr}\,(\boldsymbol{\sigma}_k \boldsymbol{\sigma}_i \boldsymbol{\sigma}_j) + \mathcal{H}^{i*} \,\mathrm{Tr}\,(\boldsymbol{\sigma}_k \boldsymbol{\sigma}_j \boldsymbol{\sigma}_i)] \,. \tag{4.30}$$

It can readily be evaluated by making use of the following relations for the Pauli spin matrices:

$$\begin{aligned} &\boldsymbol{\sigma}_j^2 = \boldsymbol{\sigma}_0 \,, \quad \boldsymbol{\sigma}_0 \boldsymbol{\sigma}_j = \boldsymbol{\sigma}_j \,, \\ &\boldsymbol{\sigma}_i \boldsymbol{\sigma}_j = -i\,\epsilon_{ijk} \boldsymbol{\sigma}_k \ \text{ for } i,j = 1,2,3, \ i \neq j \,, \\ &\mathrm{Tr}\,\boldsymbol{\sigma}_0 = 2 \,, \quad \mathrm{Tr}\,\boldsymbol{\sigma}_k = 0 \,, \quad k = 1,2,3 \,. \end{aligned} \tag{4.31}$$

ϵ_{ijk} is the Levi-Civita 3-index symbol (the factor i before ϵ_{ijk} in Eq. (4.31) is the imaginary unit, not index i).

Applying these relations we obtain for the diagonal elements

$$M_{kk} = \tfrac{1}{2}\kappa_0 (\mathcal{H}^0 + \mathcal{H}^{0*}) = \kappa_0 H_I \,, \tag{4.32}$$

for the first row and first column

$$M_{0k} = M_{k0} = \tfrac{1}{2}\kappa_0 (\mathcal{H}^k + \mathcal{H}^{k*}) = \kappa_0 H^k \,, \tag{4.33}$$

and for the off-diagonal terms $(k, j \neq 0 \,, \ k \neq j)$

$$M_{kj} = -\tfrac{1}{2} i\,\epsilon_{ijk} \kappa_0 (\mathcal{H}^i - \mathcal{H}^{i*}) = -2\epsilon_{ijk} \kappa_0 F^i \,. \tag{4.34}$$

From these expressions we see that $\boldsymbol{M}$ can be described in terms of two components: (1) A symmetric matrix containing the H terms, which represent the effects of *absorption.* (2) An anti-symmetric matrix containing the F terms, which represent the effects of *anomalous dispersion*, also called the *magnetooptical effects.* Explicitly

$$\begin{aligned} \boldsymbol{M} = \kappa_0 &\begin{pmatrix} H_I & H_Q & H_U & H_V \\ H_Q & H_I & 0 & 0 \\ H_U & 0 & H_I & 0 \\ H_V & 0 & 0 & H_I \end{pmatrix} \\ +2\kappa_0 &\begin{pmatrix} 0 & 0 & 0 & 0 \\ 0 & 0 & F_V & -F_U \\ 0 & -F_V & 0 & F_Q \\ 0 & F_U & -F_Q & 0 \end{pmatrix} \end{aligned} \tag{4.35}$$

Similarly, the emission vector is obtained from Eqs. (4.29) and (4.27):

$$
\boldsymbol{j} = \kappa_0 S \begin{pmatrix} H_I \\ H_Q \\ H_U \\ H_V \end{pmatrix} . \tag{4.36}
$$

This relation is valid quite generally for non-LTE problems without coherent scattering or atomic polarization. The non-LTE physics enters in the calculation of the value of the scalar source function S. The special case of local thermodynamic equilibrium (LTE) is obtained by setting $S = B_\nu(T)$, the Planck function, which is determined by the local temperature of the stellar atmosphere.

Coherent scattering will be treated in the next chapter. It contains the Hanle effect and gives an emission vector that depends on the polarization and angular distribution of the incident radiation field.

4.6. Addition of a Continuum and an Optical Depth Scale

The magnetic field strengths present in ordinary stellar atmospheres are far too small to generate any significant continuum polarization. Rayleigh and Thomson scattering may produce small amounts of linear polarization in the continuum, depending on the anisotropy of the incident radiation field (cf. Sect. 5.16). As this small continuous background polarization is generally unimportant for magnetic-field diagnostics, we will disregard it here, and instead assume that the continuum can be treated in terms of LTE (which is a good approximation, since the continuum is formed in the lower, denser layers of the atmosphere).

Let us denote the coefficient of continuos absorption by κ_c. In the transfer equation for the coherency matrix, the absorption term given by Eq. (4.22) (which is the sum of two parts) should equal $-\kappa_c \boldsymbol{D}$ for the continuum. This is achieved if we let the absorption matrix $\boldsymbol{A}_g$ for the continuum be given by

$$
\boldsymbol{A}_{g,c} = \tfrac{1}{2}\kappa_c \boldsymbol{\sigma}_0 . \tag{4.37}
$$

Similarly, since $S\boldsymbol{\sigma}_0/2$ represents unpolarized radiation of intensity S, the emission term $\boldsymbol{F}_g$ in the coherency matrix transfer equation (4.3), which for line radiation was given by Eq. (4.24), becomes for the continuum

$$
\boldsymbol{F}_{g,c} = \tfrac{1}{2}\kappa_c B_\nu(T) \boldsymbol{\sigma}_0 , \tag{4.38}
$$

if we assume that the continuum source function S_c is given by the Planck function $B_\nu(T)$.

With the same type of considerations for the Stokes radiative transfer equation (4.29), we readily obtain the Mueller matrix and emission vector for the continuum:

$$
\begin{aligned}
\boldsymbol{M}_c &= \kappa_c \boldsymbol{E} , \\
\boldsymbol{j}_c &= \kappa_c B_\nu(T)\, \mathbf{1} .
\end{aligned} \tag{4.39}
$$

$\boldsymbol{E}$ is the identity matrix with the diagonal elements equal to unity while the off-diagonal elements are zero. $\mathbf{1}$ is a four-vector representing unpolarized light with an intensity of unity, i.e., the first position is unity while the remaining three positions are zero.

It is convenient to introduce an optical depth scale τ_c based on the continuous absorption. The differential optical depth can be defined by

$$\mathrm{d}\tau_c = -\kappa_c\,\mathrm{d}s\,, \tag{4.40}$$

where $\mathrm{d}s$ is the differential geometrical depth.

4.7. Coherency Matrix Transfer Equation without Coherent Scattering

Including both line and continuum terms, and introducing the continuum optical depth scale of Eq. (4.40), the original line transfer equation (4.3) with its line-broadened components given by Eqs. (4.22)–(4.25) becomes

$$\frac{\mathrm{d}\boldsymbol{D}}{\mathrm{d}\tau_c} = \boldsymbol{A}\boldsymbol{D} + \boldsymbol{D}\boldsymbol{A}^\dagger - \boldsymbol{F}\,, \tag{4.41}$$

where

$$\begin{aligned} \boldsymbol{A} &= \boldsymbol{A}_L + \boldsymbol{A}_c\,, \\ \boldsymbol{F} &= \boldsymbol{F}_L + \boldsymbol{F}_c \end{aligned} \tag{4.42}$$

represent the sums of the line (index L) and continuum (index c) contributions. They relate to the previously used primed quantities through

$$\begin{aligned} \kappa_c\boldsymbol{A} &= \boldsymbol{A}_g\,, \\ \kappa_c\boldsymbol{F} &= \boldsymbol{F}_g\,. \end{aligned} \tag{4.43}$$

According to Eqs. (4.37) and (4.38) we thus have

$$\begin{aligned} \boldsymbol{A}_c &= \tfrac{1}{2}\boldsymbol{\sigma}_0\,, \\ \boldsymbol{F}_c &= B_\nu \boldsymbol{A}_c\,. \end{aligned} \tag{4.44}$$

Using Eqs. (4.23), (4.14), and (4.43), the line absorption matrix becomes

$$\begin{aligned} \boldsymbol{A}_L/\eta_0 = \frac{1}{2}&\begin{pmatrix} H_I + H_Q & H_U + i\,H_V \\ H_U - i\,H_V & H_I - H_Q \end{pmatrix} \\ -i&\begin{pmatrix} F_I + F_Q & F_U + i\,F_V \\ F_U - i\,F_V & F_I - F_Q \end{pmatrix}, \end{aligned} \tag{4.45}$$

where

$$\eta_0 = \kappa_0/\kappa_c\,. \tag{4.46}$$

Similarly, using Eqs. (4.25) and (4.43), the line emission matrix becomes

$$\boldsymbol{F}_L/\eta_0 = S_L\,\frac{1}{2}\begin{pmatrix} H_I + H_Q & H_U + i\,H_V \\ H_U - i\,H_V & H_I - H_Q \end{pmatrix}, \tag{4.47}$$

where S_L is a scalar line source function, which may quite generally be determined by non-LTE processes, excluding coherent scattering (the Hanle effect).

The explicit expressions for $H_{I,Q,U,V}$ are, in analogy with Eq. (3.74),

$$\begin{aligned} H_I &= H_\Delta \sin^2\gamma + \tfrac{1}{2}(H_+ + H_-)\,, \\ H_Q &= H_\Delta \sin^2\gamma \cos 2\chi\,, \\ H_U &= H_\Delta \sin^2\gamma \sin 2\chi\,, \\ H_V &= \tfrac{1}{2}(H_+ - H_-)\cos\gamma\,, \end{aligned} \tag{4.48}$$

with

$$\begin{aligned} H_q &= H(a, v - qv_H)\,, \quad q = 0, \pm 1\,, \\ H_\Delta &= \tfrac{1}{2}[H_0 - \tfrac{1}{2}(H_+ + H_-)]\,. \end{aligned} \tag{4.49}$$

The corresponding expressions for $F_{I,Q,U,V}$ are obtained if we simply replace H in Eqs. (4.48) and (4.49) by F.

4.8. Stokes Transfer Equation without Coherent Scattering

Denoting the Stokes vector by $\boldsymbol{I}_\nu$ instead of by $\boldsymbol{S}$, and using Eqs. (4.35), (4.36), (4.39), and (4.40), the Stokes transfer equation (4.29) may be written

$$\frac{\mathrm{d}\boldsymbol{I}_\nu}{\mathrm{d}\tau_c} = (\boldsymbol{\eta} + \boldsymbol{E})\boldsymbol{I}_\nu - (S_L\boldsymbol{\eta}\mathbf{1} + B_\nu\mathbf{1})\,. \tag{4.50}$$

While the continuum is here assumed to be formed in LTE, the line radiation may be formed by non-LTE processes (cf. Chapter 6).

$$\boldsymbol{\eta}/\eta_0 = \begin{pmatrix} H_I & H_Q & H_U & H_V \\ H_Q & H_I & 2F_V & -2F_U \\ H_U & -2F_V & H_I & 2F_Q \\ H_V & 2F_U & -2F_Q & H_I \end{pmatrix}, \tag{4.51}$$

as follows from Eqs. (4.35) and (4.46).

To compare theory with observations we have to solve the transfer equation to obtain the Stokes parameters emerging from the atmosphere at optical depth $\tau_c = 0$. Analytical solutions are available for idealized model atmospheres (cf. Chapter 11), but for more detailed, quantitative analysis, numerical solutions are needed with as realistic model atmospheres as possible. An example of a numerical solution of Eq. (4.50) is given in Fig. 4.1, while Fig. 4.2 shows an example of Stokes profiles observed on the real sun.

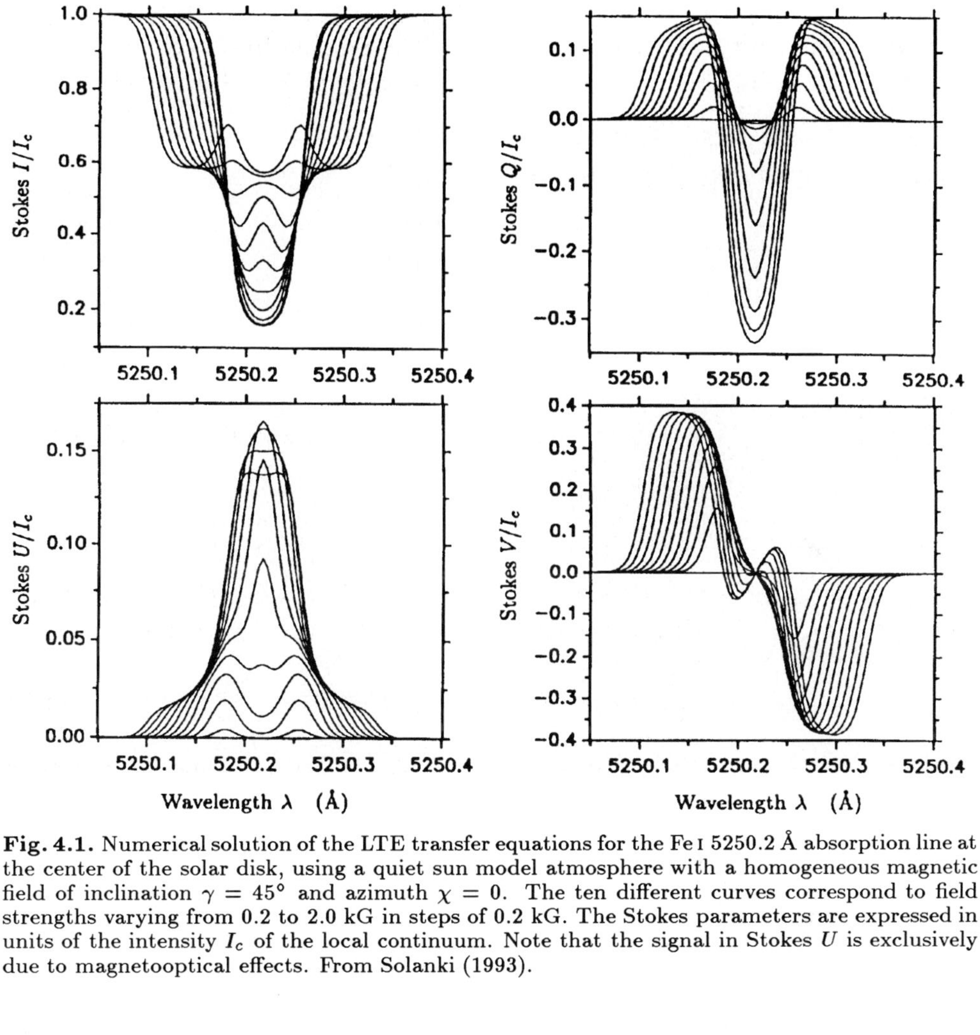

Fig. 4.1. Numerical solution of the LTE transfer equations for the Fe I 5250.2 Å absorption line at the center of the solar disk, using a quiet sun model atmosphere with a homogeneous magnetic field of inclination $\gamma = 45°$ and azimuth $\chi = 0$. The ten different curves correspond to field strengths varying from 0.2 to 2.0 kG in steps of 0.2 kG. The Stokes parameters are expressed in units of the intensity I_c of the local continuum. Note that the signal in Stokes U is exclusively due to magnetooptical effects. From Solanki (1993).

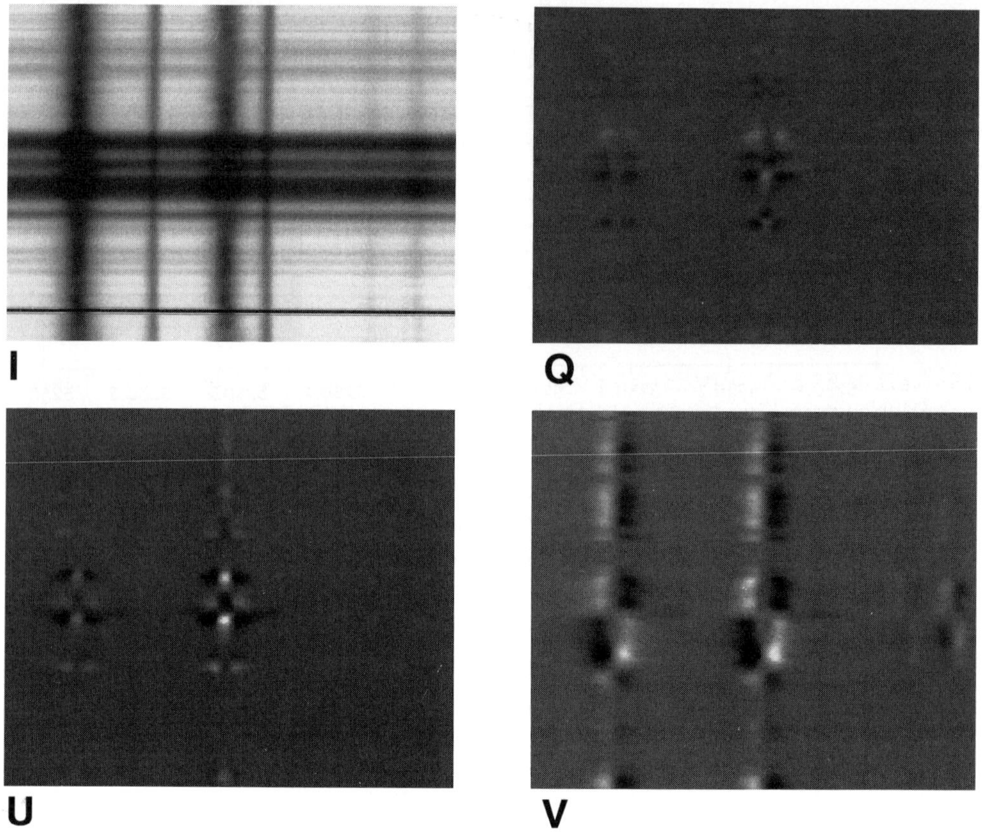

Fig. 4.2. Images of the four Stokes parameters, recorded simultaneously in the focal plane of a spectrograph with the Advanced Stokes Polarimeter (Lites et al., 1991) at the National Solar Observatory/Sacramento Peak. The field of view and spectral range are the same as in Fig. 1.14. The spectrograph slit cuts across a sunspot and surrounding plages. Positive and negative values are represented by brighter and darker grey-scale levels. Courtesy B.W. Lites, High Altitude Observatory, Boulder, Colorado.

CHAPTER 5

CLASSICAL SCATTERING AND THE HANLE EFFECT

5.1. Coherent Scattering in the Rest Frame

Derivation of the scattering emission vector to be used in the Stokes transfer equation requires integration over all the angles of incidence, which implies that we make a superposition of many scattering events. As the different scattering events are stochastically independent of each other, without any preferred phase relations between them, the superposition is incoherent. Still this incoherent superposition applies to the treatment of coherent scattering, since the coherence is between the three oscillating dipole components (in the classical description) *within each single scattering event.* This coherence is fully described by the Jones matrix for a single scattering process, which was derived in Chapter 3 and given by Eqs. (3.84), (3.85), and (3.88).

The incoherent superposition of many scattering events cannot be done within the Jones formalism, but requires use of the coherency matrix or the Stokes formalisms. For our discussion of scattering, we will stay within the Stokes formalism. The Mueller matrix $\boldsymbol{M}_{\rm sc}$ for a single scattering process can be directly derived from the corresponding Jones matrix, which according to Eq. (3.84) is

$$\frac{\omega^2}{2\pi c^2}\frac{e^{ikr}}{r}\,\boldsymbol{w}\,. \tag{5.1}$$

Using Eqs. (2.39) and (2.48) we then obtain

$$\boldsymbol{M}_{\rm sc}=\frac{\omega^4}{4\pi^2 c^4}\frac{1}{r^2}\,\boldsymbol{T}\,(\boldsymbol{w}\otimes\boldsymbol{w}^*)\,\boldsymbol{T}^{-1}\,, \tag{5.2}$$

where the transformation matrices $\boldsymbol{T}$ and $\boldsymbol{T}^{-1}$ are given by (2.49).

If the incident Stokes vector (per unit solid angle) is $\boldsymbol{I}'$, propagating in a direction characterized by colatitude θ' and azimuth ϕ', then the Stokes vector per unit solid angle, scattered in the direction defined by θ and ϕ and denoted by the vector $\boldsymbol{j}_{\rm sc}$, is obtained by integrating over all the incident directions, thus adding up the stochastically independent contributions from all the incident solid angles $\mathrm{d}\Omega'$:

$$\boldsymbol{j}_{\rm sc}=\int r^2\boldsymbol{M}_{\rm sc}\boldsymbol{I}'\,\mathrm{d}\Omega'\,. \tag{5.3}$$

The r^2 factor compensates for the $1/r^2$ factor in $\boldsymbol{M}_{\rm sc}$, such that $r^2\boldsymbol{M}_{\rm sc}$ represents scattering from one unit solid angle into another unit solid angle.

Inserting Eq. (5.2) into (5.3), we thus obtain

$$\boldsymbol{j}_{\mathrm{sc}} = \frac{\omega^4}{\pi c^4} \int \boldsymbol{T}(\boldsymbol{w} \otimes \boldsymbol{w}^*)\boldsymbol{T}^{-1}\, \boldsymbol{I}'\, \frac{\mathrm{d}\Omega'}{4\pi}\,. \tag{5.4}$$

$\boldsymbol{w}$, given by Eq. (3.85), is a sum over q (index for the three spherical vector components) containing the magnetic-field dependent dispersion factor $n_q - 1$ and the two q-dependent ε factors describing the geometry of the incident and scattered rays. The tensor product $\boldsymbol{w} \otimes \boldsymbol{w}^*$ in Eq. (5.4), explicitly given by Eq. (2.39), therefore contains not only the squared terms, but also the cross products connecting different q values with each other. It is these terms that represent the interference or coherence effects, and the magnitude of these effects depends on the strength and direction of the magnetic field. The magnetic-field dependent polarization phenomena resulting from this interference are commonly called the *Hanle effect.*

5.2. Frequency Redistribution of Polarized Radiation

Eq. (5.4) is valid in the rest frame of the scattering atom, in the absence of collisions or Doppler shifts. In scattering problems (as opposed to pure absorption problems) the effect of Doppler motions is not merely a matter of line broadening but of *redistribution*, since scattering is a two-photon process with the frequency shift being coupled to the angular redistribution. Polarization adds a new dimension of complexity, since the polarization properties are directly linked to the angular redistribution and are thereby, via the Doppler motions, coupled to the frequency redistribution.

In contrast to the scalar redistribution function used in current radiative transfer literature, the treatment of polarization problems requires the introduction of a 4×4 redistribution matrix $\boldsymbol{R}(\nu, \boldsymbol{n}; \nu', \boldsymbol{n}')$, defined as follows: If the incident Stokes vector is $\boldsymbol{I}_{\nu'}(\nu', \boldsymbol{n}')$, propagating in direction $\boldsymbol{n}'$ at frequency ν', then the probable contribution to the Stokes vector scattered in direction $\boldsymbol{n}$ at frequency ν from differential solid angle $\mathrm{d}\Omega'$ and frequency interval $\mathrm{d}\nu'$ is

$$\sigma \boldsymbol{R}(\nu, \boldsymbol{n}; \nu', \boldsymbol{n}')\, \boldsymbol{I}_{\nu'}(\nu', \boldsymbol{n}')\, \frac{\mathrm{d}\Omega'}{4\pi}\, \mathrm{d}\nu'\,. \tag{5.5}$$

σ is the scattering coefficient, the magnitude of which follows from the above definition and the normalization condition for $\boldsymbol{R}$:

$$\int \frac{\mathrm{d}\Omega'}{4\pi} \int \frac{\mathrm{d}\Omega}{4\pi} \int \mathrm{d}\nu' \int \mathrm{d}\nu\, R_{11} = 1\,, \tag{5.6}$$

where R_{11} is the matrix component in the first row and column of $\boldsymbol{R}$. It can be expressed as the vector product

$$R_{11} = \boldsymbol{1}^T \boldsymbol{R} \boldsymbol{1}\,, \tag{5.7}$$

where $\boldsymbol{1}$ is the unit four-vector (introduced in Eq. (4.39)), and $\boldsymbol{1}^T$ is its transpose. This type of product will be of later use below.

Note that in our redistribution matrix $\boldsymbol{R}$ we place the arguments representing the incident radiation to the right of those representing the scattered radiation. We have chosen this convention since it corresponds to the standard ordering of indices in matrix calculus. The convention used in previous literature for scalar redistribution problems (e.g. Mihalas, 1978) is however the opposite.

If ν' and ν are the frequencies in the observer's frame, the matrix $\boldsymbol{R}$ contains the combined effect of polarization, collisional, and Doppler redistribution. The scattering emission vector is obtained by summing the contributions (5.5) over all incident directions and frequencies:

$$\boldsymbol{j}_{\mathrm{sc}} = \sigma \int \frac{\mathrm{d}\Omega'}{4\pi} \int \mathrm{d}\nu' \, \boldsymbol{R} \boldsymbol{I}_{\nu'} \,. \tag{5.8}$$

Let us by $\boldsymbol{R}_0$ denote the redistribution matrix $\boldsymbol{R}$ that applies to the rest frame of the atom. While the frequencies in the observer's frame are ν and ν', we denote the rest frame frequencies by ξ and ξ'. When going from the rest frame to the observer's frame we first have to replace ξ' and ξ in $\boldsymbol{R}_0$ by their Doppler-shifted values (cf. Mihalas, 1978):

$$\begin{aligned} \xi &= \nu - \nu_0(\boldsymbol{v} \cdot \boldsymbol{n})/c \,, \\ \xi' &= \nu' - \nu_0(\boldsymbol{v} \cdot \boldsymbol{n}')/c \,. \end{aligned} \tag{5.9}$$

ν_0 is the central line frequency, $\boldsymbol{v}$ the velocity vector, and c the speed of light. The thermal and turbulent velocities may be assumed to have a Maxwellian velocity distribution given by

$$P_{\mathrm{Mx}}(v_x, v_y, v_z) \, \mathrm{d}v_x \, \mathrm{d}v_y \, \mathrm{d}v_z = \exp[-(v_x^2 + v_y^2 + v_z^2)/v_t^2\,] \, \mathrm{d}v_x \, \mathrm{d}v_y \, \mathrm{d}v_z \,, \tag{5.10}$$

where v_t is a Doppler width due to the combined effect of thermal and turbulent motions. We have to multiply $\boldsymbol{R}_0$ by P_{Mx} and integrate over the 3-dimensional velocity space. This is not just a line broadening integral, since the frequencies ν' and ν are coupled to the directions $\boldsymbol{n}$ and $\boldsymbol{n}'$.

For the rest frame redistribution matrix $\boldsymbol{R}_0(\xi, \boldsymbol{n}; \xi', \boldsymbol{n}')$ we may distinguish between two different cases: (1) The scattering process is undisturbed by collisions. In this case $\xi' = \xi$, i.e., we have frequency coherence in the rest frame (but not in the observer's frame). The redistribution matrix $\boldsymbol{R}_0$ contains the interference terms that characterize the Hanle effect. (2) The scattering process is disturbed by collisions. The effect of the collisions is two-fold: (a) We get complete frequency redistribution in the rest frame, since the absorption and emission processes become uncorrelated. (b) The interference terms and thus the Hanle effect are destroyed, due to the loss of phase coherence between the incident and scattered radiation. A more complete discussion of collisional effects and redistribution will be postponed until Chapter 10.

We will use the terminology *coherent scattering* for the collisionally undisturbed case regardless of the Doppler redistribution, since the phase coherence is preserved. This case generally does not correspond to frequency-coherent scattering, since in the observer's frame ν is generally $\neq \nu'$. The term *incoherent scattering* is reserved for the collisionally disturbed case.

According to Eqs. (5.4) and (5.8) the rest frame redistribution matrix in the coherent scattering case is given by

$$\sigma \boldsymbol{R}_0 = \frac{\omega^4}{\pi c^4} \boldsymbol{T}(\boldsymbol{w} \otimes \boldsymbol{w}^*) \boldsymbol{T}^{-1} \, \delta(\xi - \xi') \,. \tag{5.11}$$

$\delta(\xi - \xi')$ is the Dirac delta function. The redistribution matrix $\boldsymbol{R}_{0,\mathrm{ic}}$ for incoherent scattering can be derived from Eq. (5.11) through integration, as we will see below.

In scalar redistribution problems, when the radiation field is almost isotropic, it is a good approximation to average out the dependence of the redistribution function on $\boldsymbol{n}$ and $\boldsymbol{n}'$ by integrating over all solid angles of the incident radiation. Such straightforward averaging cannot be used in the matrix case, since the different matrix elements of $\boldsymbol{R}$ have different angular dependence due to the polarizing properties of $\boldsymbol{R}$. Direct angular averaging would wipe out the polarization properties and thus the physics that we are interested in!

We have to remember that the actual aim of the angular averaging is to average out the effects of $\boldsymbol{n}'$ and $\boldsymbol{n}$ due to the terms $\boldsymbol{v} \cdot \boldsymbol{n}'$ and $\boldsymbol{v} \cdot \boldsymbol{n}$ in (5.9), which are the source of the *Doppler coupling* between ν' and ν, while leaving the polarization properties intact. Only when this Doppler coupling can be factorized out from the polarization problem can a separate angular averaging over it be carried out, which then leads to a significant simplification of the overall redistribution problem. This will be done in Sects. 5.6 and 5.9.

5.3. The Absorption and Emission Matrices

We may introduce an absorption matrix $\boldsymbol{R}_{\mathrm{abs}}$ obtained from the redistribution matrix $\boldsymbol{R}$ through integration over all scattered directions and frequencies:

$$\boldsymbol{R}_{\mathrm{abs}}(\nu', \boldsymbol{n}') = \int \frac{\mathrm{d}\Omega}{4\pi} \int \mathrm{d}\nu \, \boldsymbol{R}(\nu, \boldsymbol{n}; \nu', \boldsymbol{n}') \,. \tag{5.12}$$

Similarly an emission matrix $\boldsymbol{R}_{\mathrm{em}}$ is defined by integration of $\boldsymbol{R}$ over all incident directions and frequencies:

$$\boldsymbol{R}_{\mathrm{em}}(\nu, \boldsymbol{n}) = \int \frac{\mathrm{d}\Omega'}{4\pi} \int \mathrm{d}\nu' \, \boldsymbol{R}(\nu, \boldsymbol{n}; \nu', \boldsymbol{n}') \,. \tag{5.13}$$

From the normalization condition (5.6) for $\boldsymbol{R}$ the normalization conditions for $\boldsymbol{R}_{\mathrm{abs}}$ and $\boldsymbol{R}_{\mathrm{em}}$ follow:

$$\begin{aligned} \int \frac{\mathrm{d}\Omega'}{4\pi} \int \mathrm{d}\nu' \, \boldsymbol{1}^T \boldsymbol{R}_{\mathrm{abs}} \boldsymbol{1} &= 1 \,, \\ \int \frac{\mathrm{d}\Omega}{4\pi} \int \mathrm{d}\nu \, \boldsymbol{1}^T \boldsymbol{R}_{\mathrm{em}} \boldsymbol{1} &= 1 \,. \end{aligned} \tag{5.14}$$

Applying Eq. (5.12) to the coherent scattering matrix in Eq. (5.11), we find for the rest frame absorption matrix

$$\sigma \boldsymbol{R}_{0,\mathrm{abs}} = \frac{\omega^4}{\pi c^4} \boldsymbol{T} \left[\int \frac{\mathrm{d}\Omega}{4\pi} (\boldsymbol{w} \otimes \boldsymbol{w}^*) \right] \boldsymbol{T}^{-1} . \tag{5.15}$$

The corresponding expression for the emission matrix is obtained if Ω is replaced by Ω' in the integral.

Using expressions (3.87) and (3.88) for the Jones scattering matrix $\boldsymbol{w}$, the above integrations may readily be carried out. When averaging over the azimuth angles the interference terms $b_q b^*_{q'}$, $q \neq q'$, vanish, and only the squared terms $|b_q|^2$ survive. The integrations show that

$$\sigma(\boldsymbol{R}_{0,\mathrm{abs}}\mathbf{1}) = \sigma(\boldsymbol{R}_{0,\mathrm{em}}\mathbf{1}) = \boldsymbol{M}_0 \mathbf{1} , \tag{5.16}$$

where $\boldsymbol{M}_0$ represents the absorption Mueller matrix $\boldsymbol{M}$ in Eq. (4.35) before Doppler and collisional broadening have been applied while letting the absorption coefficient κ_0 in (4.35) represent the effect of scattering. κ_0 is clearly proportional to the scattering coefficient σ, which will be evaluated explicitly in the next section. The proportionality constant unity for $\boldsymbol{M}_0$ in Eq. (5.16) follows from the expression (5.8) for the scattering emission vector and its comparison with Eq. (4.36) when the incident radiation is unpolarized and isotropic with intensity S. From the derivation of σ in the next section follows without any reference to Eq. (4.36) or assumption about the incident radiation field that Eq. (5.16) is of general validity.

5.4. The Scattering Coefficient

If Doppler and collisional broadening are applied separately to $\boldsymbol{R}_{0,\mathrm{abs}}$ and $\boldsymbol{R}_{0,\mathrm{em}}$ instead of performing the full redistribution treatment of $\boldsymbol{R}$ (see the following sections concerning the circumstances under which this is a good approximation), then index zero can be removed from the terms in Eq. (5.16), and we obtain

$$\sigma = \kappa_0 \sqrt{\pi} \Delta\nu_D \tag{5.17}$$

as a consequence of the normalization conditions (5.14).

Let us now evaluate the scattering coefficient σ explicitly in terms of fundamental constants, and check if the result is consistent with Eq. (5.17) and expression (4.19) for κ_0. This consistency is not obvious, since the physical considerations used to derive σ are entirely different from those that were used to derive κ_0.

The scattering coefficient σ is obtained by evaluating the right hand side of Eq. (5.15) and applying the normalization condition (5.14). The main part of the calculation is to perform the angular averaging of $\boldsymbol{w} \otimes \boldsymbol{w}^*$, using (3.87) and (3.88) for $\boldsymbol{w}$. As noted in the previous sections, only the squared terms $|b_q|^2$ survive this averaging. Using Eq. (3.54) for n_q, these terms are, after Doppler and collisional broadening have been applied,

$$|b_q|^2 = \frac{k_N^2}{\pi a} H(a, v_q) / \sqrt{\pi} . \tag{5.18}$$

The broadened version of $|b_q|^2$ is here only used for convenience of notation (with the Voigt function), and is not needed for physical or mathematical reasons. We can let Eq. (5.18) represent the unbroadened case by letting the Doppler broadening $\Delta\nu_D$ and collisional damping parameter in the Voigt function go to zero. Thus no approximation in treating the redistribution is involved. With some straightforward algebra (cf. also Eq. (8.38)) we finally find

$$\sigma = \frac{2}{3}\frac{\omega^4}{\pi^2 c^4}\frac{k_N^2}{a}\Delta\nu_D\,. \tag{5.19}$$

Before Eq. (5.19) is evaluated further, we need to make a fundamental reinterpretation of k_N, since insertion of the expression (4.10) valid for absorption processes would lead to entirely incorrect results. The reason for this consideration is that the Mueller matrix (5.2) formed from the Jones scattering matrix $\boldsymbol{w}$ is only valid for a *single*, classical scattering process. This implies that the expression (4.10) for k_N should be used with Nf (the number of classical oscillators involved) set equal to unity. However, *after* having formed the Mueller matrix, we are allowed to sum over all the scattering events in the volume element considered, which implies multiplication with Nf (where the oscillator strength f phenomenologically extends the classical expression to the quantum domain, cf. Sect. 9.15). The fundamental conclusion of this discussion is that the factor Nf appears only once in the expression for σ, not twice as it would at first appear due to the presence of k_N^2.

We thus insert in Eq. (5.19) expression (4.10) for k_N without the Nf factor, and subsequently multiply once by Nf. If we further insert the expression (3.53) for the damping parameter a and expression (3.22) for the damping constant γ, we obtain after some algebra

$$\sigma = \frac{e^2 Nf}{4\epsilon_0\, m} = \tfrac{1}{4}\,\omega_A^2 f\,, \tag{5.20}$$

where we have used the frequency ω_A defined by Eq. (3.50). This is consistent with Eqs. (5.17) and (4.19). The general validity of Eq. (5.16) is confirmed as a byproduct of this detailed derivation of σ.

5.5. The Incoherent Scattering Matrix

Incoherent scattering in the rest frame can be regarded as a two-step process: (1) Absorption, leading to an excited atomic state, the polarization of which is subsequently destroyed by a collision. (2) Emission from an unpolarized atomic state.

Due to the collisional destruction of the atomic polarization, only the absorbed energy $E_{\rm abs}$ per unit time needs to be considered. It is obtained by integrating the emitted intensity $j_{\rm sc}$ over all outgoing directions and frequencies. $j_{\rm sc}$ is obtained from the emission vector $\boldsymbol{j}_{\rm sc}$ by

$$j_{\rm sc} = \mathbf{1}^T \boldsymbol{j}_{\rm sc}\,. \tag{5.21}$$

Thus

$$E_{\mathrm{abs}} = \int j_{\mathrm{sc}}\, \mathrm{d}\Omega\, \mathrm{d}\nu\,, \tag{5.22}$$

or, with Eqs. (5.8) and (5.12),

$$E_{\mathrm{abs}} = 4\pi\sigma \int \frac{\mathrm{d}\Omega'}{4\pi} \int \mathrm{d}\nu'\, (\mathbf{1}^T \boldsymbol{R}_{\mathrm{abs}}) \boldsymbol{I}_{\nu'}\,. \tag{5.23}$$

The scattered radiation behaves as if the excitation had been done by isotropic, unpolarized radiation of intensity $E_{\mathrm{abs}}/(4\pi)$, i.e., we obtain the emission vector if we replace $\boldsymbol{I}_{\nu'}$ in Eq. (5.8) by $\mathbf{1}\, E_{\mathrm{abs}}/(4\pi)$. Hence

$$\boldsymbol{j}_{\mathrm{sc}} = \int \frac{\mathrm{d}\Omega''}{4\pi} \int \mathrm{d}\nu''\, (\boldsymbol{R}\mathbf{1})\, E_{\mathrm{abs}}/(4\pi)\,. \tag{5.24}$$

The scattering coefficient σ is already contained in E_{abs}. Double prime is used to distinguish the integration variables from the single prime used in E_{abs}.

With the emission matrix defined in Eq. (5.13) we then obtain

$$\boldsymbol{j}_{\mathrm{sc}} = (\boldsymbol{R}_{\mathrm{em}}\mathbf{1})\, E_{\mathrm{abs}}/(4\pi)\,. \tag{5.25}$$

Inserting expression (5.23) for E_{abs}, we finally obtain

$$\boldsymbol{j}_{\mathrm{sc}} = \sigma \int \frac{\mathrm{d}\Omega'}{4\pi} \int \mathrm{d}\nu'\, \boldsymbol{R}_{\mathrm{ic}} \boldsymbol{I}_{\nu'}\,, \tag{5.26}$$

where $\boldsymbol{R}_{\mathrm{ic}}$ is the *incoherent redistribution matrix* given by

$$\boldsymbol{R}_{\mathrm{ic}} = (\boldsymbol{R}_{\mathrm{em}}\mathbf{1})(\mathbf{1}^T \boldsymbol{R}_{\mathrm{abs}})\,. \tag{5.27}$$

As is evident from Eq. (5.16),

$$\mathbf{1}^T \boldsymbol{R}_{\mathrm{abs}} = (\boldsymbol{R}_{\mathrm{em}}\mathbf{1})^T\,. \tag{5.28}$$

5.6. Doppler Redistribution by Incoherent Scattering

Let us replace the absorption Mueller matrix $\boldsymbol{M}$ of Eq. (4.35) by a normalized version $\boldsymbol{\Phi}$ obeying the same normalization condition (5.14) as $\boldsymbol{R}_{\mathrm{abs}}$. According to Eqs. (5.16) and (5.17)

$$\boldsymbol{\Phi} = \boldsymbol{M}/(\sqrt{\pi}\, \Delta\nu_D\, \kappa_0)\,. \tag{5.29}$$

From Eqs. (5.27) and (5.16) we then obtain the incoherent redistribution matrix in the rest frame (index 0):

$$\boldsymbol{R}_{0,\mathrm{ic}}(\xi, \boldsymbol{n}; \xi', \boldsymbol{n}') = [\boldsymbol{\Phi}_0(\xi, \boldsymbol{n})\mathbf{1}][\mathbf{1}^T \boldsymbol{\Phi}_0(\xi', \boldsymbol{n}')]\,. \tag{5.30}$$

$\boldsymbol{\Phi}_0$ is the Doppler unbroadened version of $\boldsymbol{\Phi}$, but it includes collisional broadening, since this type of broadening applies to the rest frame.

To obtain the Doppler-broadened version $\boldsymbol{R}_{\rm ic}$ of $\boldsymbol{R}_{0,\rm ic}$ we need to insert expressions (5.9) for ξ and ξ', which include the $\boldsymbol{n}$, $\boldsymbol{n}'$ dependent Doppler shifts, multiply by the Maxwellian velocity distribution $P_{\rm Mx}$ of Eq. (5.10), and integrate over the 3-dimensional velocity space. Since the velocity $\boldsymbol{v}$ appears in both the $\boldsymbol{\Phi}$ factors of Eq. (5.30), both factors end up under the same broadening integral and cannot be factorized as a product of two integrals, one for $\boldsymbol{\Phi}_0(\xi, \boldsymbol{n})$, the other for $\boldsymbol{\Phi}_0(\xi', \boldsymbol{n}')$.

In scalar (unpolarized) problems a great simplification is achieved by using angular averages of the redistribution functions:

$$\bar{R}(\nu, \nu') = \int \frac{\mathrm{d}\Omega'}{4\pi} R(\nu, \boldsymbol{n}; \nu', \boldsymbol{n}') . \tag{5.31}$$

They turn out to give very good approximations to the full redistribution problem when the radiation field is nearly independent of angle (over the half sphere in the direction of the stellar exterior), which is most often the case in stellar atmospheres (as demonstrated by the small limb darkening on the sun).

In matrix (polarized) redistribution problems, however, a straightforward angular averaging of the redistribution matrix is not allowed, since it would wipe out the polarization information of the scattering process. Angular averaging may only be considered when it can be applied to the angular dependence of the Doppler redistribution separately without affecting the angular dependence of the polarization properties.

In the incoherent scattering case the excitation and emission processes are decoupled from each other, since the collisions wipe out the "atomic memory" of the way in which the excitation occurred. If we now treat the Doppler broadening as being direction independent, which has proven to be a good approximation in scalar problems, then the decoupling between the absorption and emission processes implies that the absorption matrix $\boldsymbol{R}_{\rm abs}$ can be broadened independently of the emission matrix $\boldsymbol{R}_{\rm em}$. The Doppler redistribution then simply amounts to a Gaussian convolution of each matrix element in each of the $\boldsymbol{\Phi}_0$ factors in Eq. (5.30), without disturbing the angular properties that describe the polarization effects.

In the scalar redistribution case these arguments would give an incoherent redistribution matrix that is $R_{\rm ic} = \varphi_\nu \varphi_{\nu'}$, i.e., the product of the area normalized absorption and emission profiles. In the matrix case we instead obtain

$$\boldsymbol{R}_{\rm ic}(\nu, \boldsymbol{n}; \nu', \boldsymbol{n}') = [\boldsymbol{\Phi}(\nu, \boldsymbol{n})\mathbf{1}][\mathbf{1}^T \boldsymbol{\Phi}(\nu', \boldsymbol{n}')] , \tag{5.32}$$

which is similar to Eq. (5.30), with the difference that the normalized absorption matrix $\boldsymbol{\Phi}$ has been obtained from $\boldsymbol{\Phi}_0$ by convolution with a Gaussian.

The emission vector for incoherent scattering can now be explicitly written as

$$\boldsymbol{j}_{\rm ic} = \sigma e \boldsymbol{\Phi} \mathbf{1} , \tag{5.33}$$

where

$$e = \int \frac{\mathrm{d}\Omega'}{4\pi} \int \mathrm{d}\nu' \, (\Phi'_I I_{\nu'} + \Phi'_Q Q_{\nu'} + \Phi'_U U_{\nu'} + \Phi'_V V_{\nu'}) . \tag{5.34}$$

$\Phi_{I,Q,U,V} = H_{I,Q,U,V}/(\sqrt{\pi}\Delta\nu_D)$ are the components of the vector $\boldsymbol{\Phi}\mathbf{1}$. $H_{I,Q,U,V}$ have been given by Eq. (4.48).

In any other case when the emission process is uncorrelated with the way in which the upper level of the line transition has been populated (e.g. for collisional excitation), the emission vector has the same form as in Eq. (5.33), i.e., it is proportional to $\boldsymbol{\Phi}\mathbf{1}$.

5.7. Doppler Redistribution by Coherent Scattering

According to Eq. (5.11) the redistribution matrix for coherent scattering in the atomic rest frame can be written as

$$\boldsymbol{R}_0(\xi, \boldsymbol{n}; \xi', \boldsymbol{n}') = \boldsymbol{R}_{0,\mathrm{coh}}(\boldsymbol{n}, \boldsymbol{n}'; \xi')\, \delta(\xi - \xi'), \tag{5.35}$$

where

$$\boldsymbol{R}_{0,\mathrm{coh}} = s\, \boldsymbol{T}(\boldsymbol{w} \otimes \boldsymbol{w}^*)\boldsymbol{T}^{-1} \tag{5.36}$$

and

$$s = \frac{\omega^4}{\pi c^4 \sigma}. \tag{5.37}$$

The damping constant to be used in the Jones matrix $\boldsymbol{w}$ is $\gamma_N + \gamma_c$, since this sum represents the inverse life time of the excited state for a coherent scattering process.

The directional $(\boldsymbol{n}, \boldsymbol{n}')$ dependence of $\boldsymbol{R}_{0,\mathrm{coh}}$ refers exclusively to its polarization properties. When we however insert the expressions (5.9) for ξ and ξ' in $\boldsymbol{R}_0$ of Eq. (5.35), a Doppler related $\boldsymbol{n}'$ dependence appears in $\boldsymbol{R}_{0,\mathrm{coh}}$, as well as a dependence on both $\boldsymbol{n}$ and $\boldsymbol{n}'$ of the Dirac δ function. This means that the Doppler redistribution of $\boldsymbol{R}_0$ cannot in general be factorized so that the incident and scattered frequencies become independently broadened. It is however possible to obtain good and simple approximations of the redistribution integral, i.e., the integral of the product $P_{\mathrm{Mx}}\boldsymbol{R}_{0,\mathrm{coh}}\,\delta$ over 3-dimensional velocity space, by considering the situation in two different frequency regimes, the Doppler core and the dispersion wings.

5.7.1. Line wings

Let us first consider the wings, and let ξ_0 and ν_0 denote the central line frequency. If the incident radiation in the observer's frame is in the wings, $|\nu' - \nu_0|$ is large. Then, for rest frame frequencies for which $\boldsymbol{R}_{0,\mathrm{coh}}$ is large, i.e., for $\xi' \approx \xi_0$, the integrand $P_{\mathrm{Mx}}\boldsymbol{R}_{0,\mathrm{coh}}\,\delta$ is small, since large velocities are needed to shift the frequency to the line center, and the Maxwell distribution function P_{Mx} falls off exponentially to become very small outside the Gaussian bell. Due to this exponential behavior of P_{Mx}, the product $P_{\mathrm{Mx}}\boldsymbol{R}_{0,\mathrm{coh}}\,\delta$ will have a maximum near the maximum of P_{Mx}, i.e., for small velocities. This means that as long as $|\nu' - \nu_0| \gg \Delta\nu_D$, it is a good approximation to assume frequency coherence, i.e., $\nu \approx \nu'$, and to replace $\delta(\xi - \xi')$ by $\delta(\nu - \nu')$. The δ function thus decouples from the velocity field, which allows the $\boldsymbol{R}_{0,\mathrm{coh}}$ factor to be broadened separately. As only the incident frequencies are involved in the broadening of $\boldsymbol{R}_{0,\mathrm{coh}}$, the broadening simply becomes a convolution of the matrix elements with a Gaussian, resulting in the convolved matrix $\boldsymbol{R}_{\rm coh}$. The wing redistribution matrix thus becomes

$$\boldsymbol{R}_{\rm wings}(\nu,\boldsymbol{n};\nu',\boldsymbol{n}') = \boldsymbol{R}_{\rm coh}(\boldsymbol{n},\boldsymbol{n}';\nu')\,\delta(\nu-\nu')\,. \tag{5.38}$$

This equation explicitly expresses that in the far line wings the frequency coherence is preserved even after Doppler redistribution has been applied. The directional $(\boldsymbol{n},\boldsymbol{n}')$ dependence of the polarization properties is conserved in this redistribution treatment.

5.7.2. Line core

Let us now turn to the line core. There, inside the Gaussian bell of $P_{\rm Mx}$, the velocities needed to make $\xi' \approx \xi_0$ (core absorption) are small or only moderately large. Then, if the matrix elements of $\boldsymbol{R}_{0,\rm coh}$ have terms that are sharply (in comparison with the width of the Gaussian) peaked around some frequency $\xi_{0,q}$, only the incident frequencies ξ' near $\xi_{0,q}$ will contribute significantly to the redistribution integral. For these terms ξ' may be replaced by $\xi_{0,q}$ in the δ function, so that their contributions $\boldsymbol{R}_{0,q}$ to the redistribution matrix $\boldsymbol{R}_0$ become

$$\boldsymbol{R}_{0,q} \approx \boldsymbol{R}_{0,{\rm coh},q}(\boldsymbol{n},\boldsymbol{n}';\xi')\,\delta(\xi-\xi_{0,q})\,. \tag{5.39}$$

With this approximation we have achieved a factorization of the functional dependence on ξ and ξ', which allows us to apply angular averaging of the Doppler redistribution separately for the $\boldsymbol{R}_{0,{\rm coh},q}$ and δ factors (while leaving the angular dependence of the polarization properties untouched). This then implies that we can carry out the broadening of the incident and scattered frequencies independently of each other, which amounts to a straightforward Gaussian convolution of the individual matrix elements of $\boldsymbol{R}_{0,{\rm coh},q}$ as well as of the δ function. The full redistribution function is the sum of the various contributing terms:

$$\boldsymbol{R}_{\rm core}(\nu,\boldsymbol{n};\nu',\boldsymbol{n}') \approx \sum_q \boldsymbol{R}_{{\rm coh},q}(\boldsymbol{n},\boldsymbol{n}';\nu')\,\varphi(\nu-\nu_{0,q})\,. \tag{5.40}$$

As φ_ν in Eq. (5.40) is the result of a Gaussian convolution of the δ function, it should be a Gaussian profile (normalized to unit area), but as the above approximation is only valid for the Doppler core we may as well let φ_ν represent the normalized Voigt profile (including radiative and collisional damping, which do not significantly affect the Doppler core).

In the weak-field limit when the separation between the different q components is much smaller than the Doppler width ($\Delta\nu_H \ll \Delta\nu_D$), we can replace $\xi_{0,q}$ in Eq. (5.39) by ξ_0 (the resonant frequency for zero magnetic field), and obtain

$$\boldsymbol{R}_{\rm core,\,weak\;field}(\nu,\boldsymbol{n};\nu',\boldsymbol{n}') \approx \boldsymbol{R}_{\rm coh}(\boldsymbol{n},\boldsymbol{n}';\nu')\,\varphi(\nu-\nu_0)\,. \tag{5.41}$$

$\boldsymbol{R}_{\rm coh}$ represents a straightforward Gaussian convolution of $\boldsymbol{R}_{0,\rm coh}$ and is the same as the $\boldsymbol{R}_{\rm coh}$ in Eq. (5.38).

For intermediate frequencies, in the transition region between core and wing, we generally need a full treatment of the partial redistribution problem, taking into account the angular dependence of the Doppler redistribution. This will be dealt with in Chapter 10.

5.8. Polarization Effects in the Dispersion Wings

In addition to the frequency coherence in the line dispersion wings as expressed by Eq. (5.38) another simplification occurs at frequencies far from the line core: The redistribution matrix becomes increasingly independent of the magnetic field. When both $|\Delta\nu| \gg \Delta\nu_D$ and $|\Delta\nu| \gg \Delta\nu_H$, then the frequency variation due to the factors $(n_q - 1)(n^*_{q'} - 1)$ that appear in $\boldsymbol{w} \otimes \boldsymbol{w}^*$ in Eq. (5.36) (cf. Eqs. (3.87) and (3.88)) approaches that of $|n_0 - 1|^2$, which is independent of the magnetic field vector $\boldsymbol{B}$. The Hanle effect thus vanishes in the wings. For the redistribution matrix $\boldsymbol{R}_{0,\mathrm{coh}}(\boldsymbol{n}, \boldsymbol{n}'; \nu')$ in Eq. (5.38) one may then use the matrix that is representative of the non-magnetic coherent scattering case.

5.9. Introduction of a Polarization Phase Matrix

In unpolarized redistribution problems an angular phase function $g(\boldsymbol{n}, \boldsymbol{n}')$ describing the probability that a photon incident in direction $\boldsymbol{n}'$ will be scattered into direction $\boldsymbol{n}$ can be factorized out from the full redistribution function R, since this probability is frequency independent (cf. Eq. (10.14)). In *non-magnetic* polarization problems a 4×4 phase matrix $\boldsymbol{P}(\boldsymbol{n}, \boldsymbol{n}')$ can for the same reason be factorized out from the redistribution matrix $\boldsymbol{R}$, which means that we may write

$$\boldsymbol{R}(\nu, \boldsymbol{n}; \nu', \boldsymbol{n}') = R(\nu, \boldsymbol{n}; \nu', \boldsymbol{n}')\boldsymbol{P}(\boldsymbol{n}, \boldsymbol{n}') . \tag{5.42}$$

The frequency redistribution is thus reduced to a scalar problem. Examples of non-magnetic scattering polarization observed in the solar spectrum are shown in Fig. 5.1.

Eq. (5.42) implies that all the matrix elements of $\boldsymbol{R}$ have the same frequency profile, which is the case in the absence of Zeeman splitting (and in the absence of quantum-mechanical interference between fine-structure components of different total angular momentum). The polarization properties are then exclusively contained in the phase matrix $\boldsymbol{P}$, which is decoupled from the frequency redistribution problem. The normalization that will be used for $\boldsymbol{P}$ is

$$\int \frac{\mathrm{d}\Omega'}{4\pi} \, \mathbf{1}^T \boldsymbol{P} \mathbf{1} = 1 . \tag{5.43}$$

It follows from Eq. (5.42) that the Doppler broadened (through straightforward Gaussian convolution) redistribution matrix $\boldsymbol{R}_{\mathrm{coh}}(\boldsymbol{n}, \boldsymbol{n}'; \nu')$ that appears both in the wing and core redistribution expressions, Eqs. (5.38) and (5.41), may in the non-magnetic case be factorized as

$$\boldsymbol{R}_{\mathrm{coh}}(\boldsymbol{n}, \boldsymbol{n}'; \nu') \approx \varphi_{\nu'} \, \boldsymbol{P}(\boldsymbol{n}, \boldsymbol{n}') . \tag{5.44}$$

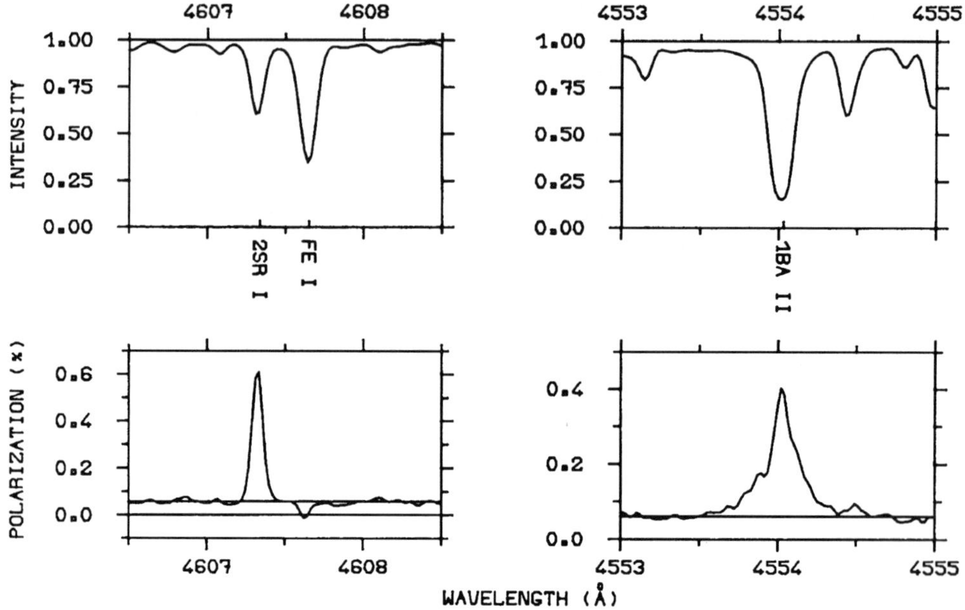

Fig. 5.1. Non-magnetic linear polarization Q/I recorded 10 arcsec inside the solar limb. Positive polarization means that the electric vector of the scattered radiation is parallel to the limb. The polarization of the local continuum (due to Rayleigh scattering at neutral hydrogen and Thomson scattering at free electrons) is indicated by a horizontal line. The polarization of the Sr I and Ba I lines is due exclusively to coherent scattering. From Stenflo et al. (1983b).

Detailed arguments will be given in Sect. 5.11. The factor $\varphi_{\nu'}$ appears because $\boldsymbol{w} \otimes \boldsymbol{w}^*$ of Eq. (5.36) contains $|n_0 - 1|^2$, which is $\sim \varphi_{\nu'}$.

When the magnetic field is not zero, this type of factorization is generally not possible, since the different matrix elements of $\boldsymbol{R}$ then have a different frequency dependence due to the Zeeman splitting. In the wings, however, we have in the previous section seen that the magnetic field effects disappear. If we by $\boldsymbol{P}_0$ mean $\boldsymbol{P}$ for $\boldsymbol{B} = 0$, then according to Eqs. (5.38) and (5.44)

$$\boldsymbol{R}_{\rm wings}(\nu, \boldsymbol{n}; \nu', \boldsymbol{n}') = \varphi_{\nu'}\, \delta(\nu - \nu') \boldsymbol{P}_0(\boldsymbol{n}, \boldsymbol{n}')\,, \tag{5.45}$$

independent of the strength of the magnetic field (as long as wing frequencies for which $|\Delta\nu| \gg \Delta\nu_H$ are considered).

5.10. Weak Field Redistribution in the Line Core

In the *weak-field limit*, when $\Delta\nu_H \ll \Delta\nu_D$, the Doppler broadening effectively performs a frequency averaging of the redistribution matrix (see next section), which means that the polarizing properties become frequency independent. This implies that a frequency-independent phase matrix $\boldsymbol{P}$ may be factorized out from the redistribution matrix $\boldsymbol{R}$ as we did in Eq. (5.44). Inserting Eq. (5.44) into

Eq. (5.41), we get

$$\boldsymbol{R}_{\text{core, weak field}}(\nu, \boldsymbol{n}; \nu', \boldsymbol{n}') \approx \varphi_\nu \, \varphi_{\nu'} \, \boldsymbol{P}(\boldsymbol{n}, \boldsymbol{n}') \,. \tag{5.46}$$

As the magnetic field is not zero, the phase matrix $\boldsymbol{P}$ now contains the magnetic-field effects, in particular the effects of the interference terms due to the products $(n_q - 1)(n^*_{q'} - 1)$, $q' \neq q$, which constitute the Hanle effect. Our next task will be to derive the structure of $\boldsymbol{P}(\boldsymbol{n}, \boldsymbol{n}')$, through frequency averaging of the terms in the scattering matrix.

The weak-field approximation is of special interest, because it is in this field-strength regime that the Hanle effect normally has its greatest magnetic-field sensitivity and diagnostic potential. This is also the case in which most effects of practical interest due to coherent scattering occur.

5.11. Frequency Averaging of the Scattering Matrix

Physically the frequency averaging is done by the Doppler broadening. The frequency-dependent terms to be smoothed in the tensor product $\boldsymbol{w} \otimes \boldsymbol{w}^*$ of the redistribution matrix $\boldsymbol{R}_{0,\text{coh}}(\boldsymbol{n}, \boldsymbol{n}'; \xi')$ in Eq. (5.36) are of the type $(n_q - 1)(n^*_{q'} - 1)$. These terms need to be convolved with a Gaussian $\exp(-v'^2_q)/\sqrt{\pi}$ due to the distribution of Doppler velocities (the factor $1/\sqrt{\pi}$ is for the area normalization of the Gaussian). If as is generally the case the damping constant $\gamma \ll \Delta\omega_D$, then averaging $(n_q - 1)(n^*_{q'} - 1)$ over the Gaussian distribution in the case that $q = q'$ can in the Doppler core be calculated with the approximation

$$\int_{-\infty}^{+\infty} e^{-(v'-y)^2} (n_q - 1)(n^*_{q'} - 1)\, \mathrm{d}y \approx e^{-v'^2_q} \int_{-\infty}^{+\infty} (n_q - 1)(n^*_{q'} - 1)\, \mathrm{d}y \,, \tag{5.47}$$

where v'_q is defined by Eq. (4.11). According to Eqs. (3.54) and (4.11) the integrand in Eq. (5.47) can be written

$$(n_q - 1)(n^*_{q'} - 1) = \left(\frac{k_N}{\pi}\right)^2 [(y - q v_H - i\,a)(y - q' v_H + i\,a)]^{-1} \,. \tag{5.48}$$

This expression is valid for the frequencies in the Doppler core of each q component, since the natural frequency spread of $(n_q - 1)(n^*_{q'} - 1)$ can be disregarded in comparison with the broadening function when $q' = q$.

For the interference terms with $q \neq q'$ the frequency spread of $(n_q - 1)(n^*_{q'} - 1)$ may however not be small in comparison with the Doppler width (since components with different Zeeman shifts are connected), unless we require that also the Zeeman splitting $\Delta\nu_H \ll \Delta\nu_D$. In this *weak-field case* the approximation given by Eq. (5.47) is of general validity for the entire scattering matrix in the Doppler core, and $v'_q \approx v'$.

In the far dispersion wings of the line the opposite approximation of the left hand side of Eq. (5.47) applies. As the wings decline slowly, with v^{-2}, as compared with the exponentially limited Gaussian profile, the width of the Gaussian may be disregarded in the convolution integral. Then the $(n_q - 1)(n^*_{q'} - 1)$ factor may be moved outside the integral sign, while the Gaussian is integrated to give $\sqrt{\pi}$. Further, as pointed out in Sect. 5.8, $(n_q - 1)(n^*_{q'} - 1)$ may be approximated by $|n_0 - 1|^2$ in the wings, for all values of q' and q, which means that the magnetic-field effects (the Hanle effect) vanish, and we get frequency-coherent, non-magnetic scattering.

In the dispersion wings the frequency dependence of $\boldsymbol{R}_{\rm coh}$, the convolved $\boldsymbol{R}_{0,\rm coh}$, can be described by the area-normalized Voigt profile $\varphi_{\nu'}$. The reason is that $|n_0(v') - 1|^2 \sim \varphi_{\nu'}$ in the dispersion wings, since the effect of Doppler smoothing can be disregarded there. The shape of the $\varphi_{\nu'}$ profile in the far wings is practically independent of the velocity distribution. In the Doppler core the frequency dependence of $\boldsymbol{R}_{\rm coh}$ can also be described by $\varphi_{\nu'}$, but for another reason: In the core, in the weak-field case, the Gaussian factor on the right hand side of Eq. (5.47) represents $\sqrt{\pi}\varphi_{\nu'}$. We have thus shown that Eq. (5.44) is valid, in the dispersion wings quite generally, in the line core in the case of the weak-field approximation.

Let us now carry out the frequency averaging, which will lead us to the expression for the frequency-independent phase matrix $\boldsymbol{P}(\boldsymbol{n}, \boldsymbol{n}')$. To solve the integral on the right hand side of Eq. (5.47) we will first introduce the following notations:

$$\begin{aligned} z &= y - q v_H \,, \\ t &= i\, a \,, \\ f(z) &= [z - (q' - q) v_H + i\, a]^{-1} \,. \end{aligned} \tag{5.49}$$

With Eqs. (5.48) and (5.49) we can then rewrite the integral as follows:

$$\int_{-\infty}^{+\infty} [(y - q v_H - i\, a)(y - q' v_H + i\, a)]^{-1}\, \mathrm{d}y = \oint \frac{f(z)}{z - t}\, \mathrm{d}z \,. \tag{5.50}$$

The contour integral should be made along a closed loop in the complex plane, along the real axis from $-\infty$ to $+\infty$, and back along a half circle with a radius approaching infinity, in the counter-clockwise direction. Due to the $|z|^{-2}$ dependence of the integrand, the contribution to the integral from the half circle vanishes.

The reason for closing the integration loop in the complex plane is to allow us to make use of the powerful Cauchy's residue theorem. As the loop encloses the singular point $z = t$, this theorem states that

$$f(t) = \frac{1}{2\pi\, i} \oint \frac{f(z)}{z - t}\, \mathrm{d}z \,. \tag{5.51}$$

This immediately gives us the solution to the integral of Eq. (5.47) as

$$\begin{aligned} \int_{-\infty}^{+\infty} (n_q - 1)(n^*_{q'} - 1)\, \mathrm{d}y &= 2\pi\, i \left(\frac{k_N}{\pi}\right)^2 [2i\, a + (q - q') v_H]^{-1} \\ &= \frac{2k_N^2}{\pi} \frac{2a + i\,(q - q') v_H}{(q - q')^2 v_H^2 + 4a^2} \,. \end{aligned} \tag{5.52}$$

This solution can be written in a more elegant way if we define an angle $\alpha_{q-q'}$ by

$$\tan \alpha_{q-q'} = \frac{(q-q')v_H}{2a} = \frac{(q-q')g_u\omega_L}{\gamma}\,, \tag{5.53}$$

where Eqs. (3.53) and (3.89) have been used in the last part. Only the Landé factor g_u of the excited level appears in this expression, not the one of the lower level, as one might have expected from Eqs. (4.12) and (3.89). The reason is that in the quantum-mechanical interpretation q and q' correspond to different magnetic quantum numbers $-M_u$ of the excited state. According to Eq. (4.12) qv_H and $q'v_H$ equal $\Delta\nu_H/\Delta\nu_D$ and $\Delta\nu'_H/\Delta\nu_D$, which however only differ from each other in their M_u values, not in their M_ℓ values. Therefore the contribution from g_ℓ vanishes when the difference $(q-q')v_H$ is formed.

Our frequency average then becomes

$$\int_{-\infty}^{+\infty} (n_q - 1)(n^*_{q'} - 1)\,dy = \frac{k_N^2}{\pi a}\cos\alpha_{q-q'}\,e^{i\,\alpha_{q-q'}}\,. \tag{5.54}$$

In the next section we will discuss some of the physical insights that one can gain from this expression.

5.12. Origin and Nature of the Hanle Effect

The essence of the Hanle effect is contained in the angle $\alpha_{q-q'}$. Only when interference effects are present, i.e., when $q-q' \neq 0$, does this angle differ from zero, provided that the field strength, which enters via v_H or ω_L, is also non-zero. Since $q,\ q' = 0, \pm 1$, $q-q'$ may assume the values ± 2, ± 1, or 0.

The Hanle effect (magnetic-field dependence of the interference terms) manifests itself via the two factors $\cos\alpha$ and $e^{i\alpha}$. The $\cos\alpha$ factor is a scaling factor that decreases from unity towards smaller values when the field strength increases from zero to larger values. This factor corresponds to *depolarization* caused by the magnetic field. The phase factor $e^{i\alpha}$ on the other hand gives rise to a *rotation of the plane of linear polarization.* The rotation angle increases with the strength of the magnetic field.

These effects have their optimum range of sensitivity to the magnetic field when $\tan\alpha \approx 1$, i.e., when the Larmor frequency ω_L is comparable to the damping constant γ, or, in other words, when the precession period in the magnetic field is comparable to the lifetime of the excited state. If B_0 denotes the field strength for which $\tan\alpha_2 = 1$, then the field-strength range over which the Hanle effect has adequate sensitivity for diagnostic purposes extends over typically a factor of five on either side of B_0, i.e., over $0.2B_0 \lesssim B \lesssim 5B_0$.

To see how these effects explicitly appear in the scattering matrix, let us calculate the component W_{44} of the scattering matrix $\boldsymbol{W} = \boldsymbol{w} \otimes \boldsymbol{w}^*$. For the geometry defined by Fig. 3.3, we have according to Eq. (3.88)

$$W_{44} = w_{22}w^*_{22} = \frac{1}{4}(|b_+|^2 + |b_-|^2 + b_+b^*_- + b^*_+b_-)\,, \tag{5.55}$$

where b_q is given by Eq. (3.87).

In the case of the non-interference terms $|b_\pm|^2$, not only the angle $\alpha_{q-q'}$, but also the phase term in b_q involving the azimuth angles, vanishes. The frequency averaged value of W_{44} is obtained by inserting the result (5.54) for the different terms in Eq. (5.55), which gives

$$W_{44} = \frac{k_N^2}{2\pi a}\{1 + \cos\alpha_2 \cos[2(\phi - \phi') - \alpha_2]\}\,. \tag{5.56}$$

The first, constant term within the outer brackets represents the incoherent contribution from the squared terms, while the second term in Eq. (5.56) represents the coherent contribution from the interference terms. Here we see explicitly how α_2 is coupled to the azimuth difference $\phi - \phi'$, leading to a rotation of the plane of linear polarization, and how the depolarization factor $\cos\alpha_2$ appears in front. The factor of proportionality containing k_N^2 can be expressed in terms of a scattering cross section σ via Eq. (5.19).

The other elements of the $\boldsymbol{W}$ matrix contain these two effects (rotation and depolarization) coupled in similar but different ways to the geometrical factors. In the limit of zero magnetic field the non-magnetic phase matrix for Rayleigh scattering is retrieved from the general Hanle-effect scattering matrix.

5.13. The Hanle Effect Phase Matrix

From the previous sections we have seen how the frequency-averaged elements of the $\boldsymbol{W}$ matrix can be derived for the geometry described by Fig. 3.3. Through multiplication with the $\boldsymbol{T}$ matrices of Eq. (2.49y) as in Eq. (5.4) and application of the normalization condition (5.43) we obtain after some tedious but straightforward calculations the full Hanle-effect phase matrix $\boldsymbol{P}(\boldsymbol{n}, \boldsymbol{n}')$.

As all the frequency-averaged terms in $\boldsymbol{P}$ contain the factors $\cos\alpha_{q-q'}\, e^{i\alpha_{q-q'}}$, they may be grouped according to the value of $|q - q'|$. The factor $e^{-i(q-q')(\phi-\phi')}$ that has its origin in Eq. (3.87) for b_q combines with the factor $e^{i\alpha_{q-q'}}$ to form terms with the factors $\cos[(q-q')(\phi-\phi') - \alpha_{q-q'}]$ and $\sin[(q-q')(\phi-\phi') - \alpha_{q-q'}]$.

Let us by $\boldsymbol{E}_{jj}$ denote a matrix that has the single element $E_{jj} = 1$, with all the remaining elements being zero. With the colatitudes θ and θ' and azimuths ϕ and ϕ' defined by Fig. 3.3, the phase matrix can be written

$$\boldsymbol{P}(\theta, \phi; \theta', \phi') = \boldsymbol{E}_{11} + \tfrac{3}{4}(\boldsymbol{P}^1 + \boldsymbol{P}^2)\,. \tag{5.57}$$

The matrix $\boldsymbol{P}^1$ is proportional to $\boldsymbol{E}_{44}$ and thus scatters the circular polarization without any coupling to the linear polarization, while the matrix $\boldsymbol{P}^2$ accounts for the scattering of the linear polarization, without any coupling to the circular polarization. The first matrix, $\boldsymbol{E}_{11}$, represents unpolarized, isotropic scattering.

$$\boldsymbol{P}^1 = [\mu\mu' + 2\cos\alpha_1\, \sin\theta\, \sin\theta'\, \cos(\phi - \phi' - \alpha_1)]\boldsymbol{E}_{44}\,, \tag{5.58}$$

and

$$\boldsymbol{P}^2 = \boldsymbol{P}_0^2 + 2\cos\alpha_1\,\sin\theta\,\sin\theta'\,[\boldsymbol{P}_1^2\cos(\phi-\phi'-\alpha_1) + \boldsymbol{P}_{-1}^2\sin(\phi-\phi'-\alpha_1)] + \cos\alpha_2\,\{\boldsymbol{P}_2^2\cos[2(\phi-\phi')-\alpha_2] + \boldsymbol{P}_{-2}^2\sin[2(\phi-\phi')-\alpha_2]\}\,, \tag{5.59}$$

where

$$\boldsymbol{P}_0^2 = \frac{1}{2}\begin{pmatrix} \frac{1}{3}(1-3\mu^2)(1-3\mu'^2) & (1-3\mu^2)(1-\mu'^2) & 0 & 0 \\ (1-\mu^2)(1-3\mu'^2) & 3(1-\mu^2)(1-\mu'^2) & 0 & 0 \\ 0 & 0 & 0 & 0 \\ 0 & 0 & 0 & 0 \end{pmatrix}, \tag{5.60}$$

$$\boldsymbol{P}_1^2 = \mu\mu'\begin{pmatrix} 1 & 1 & 0 & 0 \\ 1 & 1 & 0 & 0 \\ 0 & 0 & 0 & 0 \\ 0 & 0 & 0 & 0 \end{pmatrix} + \boldsymbol{E}_{33}\,, \tag{5.61}$$

$$\boldsymbol{P}_{-1}^2 = \begin{pmatrix} 0 & 0 & \mu & 0 \\ 0 & 0 & \mu & 0 \\ -\mu' & -\mu' & 0 & 0 \\ 0 & 0 & 0 & 0 \end{pmatrix}, \tag{5.62}$$

$$\boldsymbol{P}_2^2 = \frac{1}{2}\begin{pmatrix} (1-\mu^2)(1-\mu'^2) & -(1-\mu^2)(1+\mu'^2) & 0 & 0 \\ -(1+\mu^2)(1-\mu'^2) & (1+\mu^2)(1+\mu'^2) & 0 & 0 \\ 0 & 0 & 4\mu\mu' & 0 \\ 0 & 0 & 0 & 0 \end{pmatrix}, \tag{5.63}$$

$$\boldsymbol{P}_{-2}^2 = \begin{pmatrix} 0 & 0 & -(1-\mu^2)\mu' & 0 \\ 0 & 0 & (1+\mu^2)\mu' & 0 \\ \mu(1-\mu'^2) & -\mu(1+\mu'^2) & 0 & 0 \\ 0 & 0 & 0 & 0 \end{pmatrix}. \tag{5.64}$$

Although the phase matrix appears quite complicated with so many terms, all terms are not equally important. If the incident radiation field were unpolarized, only the first column of $\boldsymbol{P}$ would actually be used. In most practical situations the polarization of the incident radiation field is indeed quite small. In such cases the other columns of $\boldsymbol{P}$ provide only second-order contributions as compared with the first-order contributions of the first column. The often reasonable approximation of calculating the polarization assuming that the incident radiation is unpolarized is equivalent to assuming that the source of the polarization of the radiation emerging from a stellar atmosphere is the *last scattering process.*

The phase matrix that we have given here refers to the special choice of coordinate system when the polar (z) axis is parallel to the magnetic-field vector $\boldsymbol{B}$, as described by Fig. 3.3. Let us now consider the case of an arbitrary coordinate system, i.e., when $\boldsymbol{B}$ has an arbitrary direction with respect to the polar axis. The scattering geometry with the angles involved is defined in Fig. 5.2. While the colatitudes and azimuth angles in the previous, special coordinate system are θ, θ' and ϕ, ϕ', they are ϑ, ϑ' and φ, φ' in the new, arbitrary system.

The transformation of the phase matrix from the old to the new system is accomplished through the application of Mueller rotation matrices $\boldsymbol{L}(i_{1,2})$, where

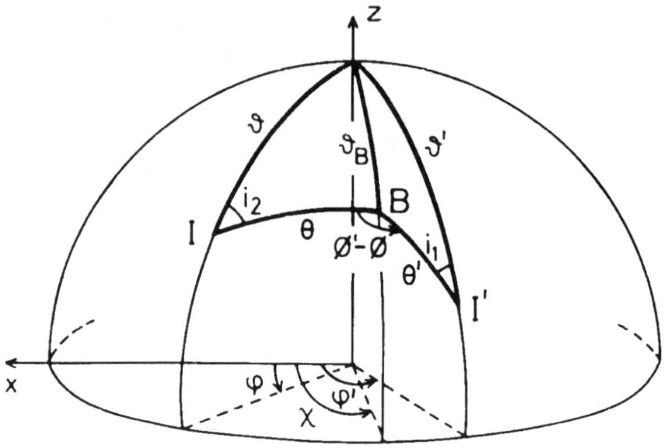

Fig. 5.2. Definition of the angles used to describe a scattering process when the magnetic field vector $\boldsymbol{B}$ has an arbitrary direction. From Stenflo (1978b).

$$\boldsymbol{L}(i) = \begin{pmatrix} 1 & 0 & 0 & 0 \\ 0 & \cos 2i & \sin 2i & 0 \\ 0 & -\sin 2i & \cos 2i & 0 \\ 0 & 0 & 0 & 1 \end{pmatrix}. \tag{5.65}$$

The angles $i_{1,2}$ are defined in Fig. 5.2. We first have to rotate the incident Stokes vector by angle i_1 to the special coordinate system, in which the phase matrix $\boldsymbol{P}$ given by Eqs. (5.57)–(5.64) is valid. The so obtained scattered Stokes vector has to be rotated by angle i_2 to refer to the general coordinate system. The new phase matrix is therefore

$$\boldsymbol{P}(\vartheta, \varphi; \vartheta', \varphi') = \boldsymbol{L}(i_2)\boldsymbol{P}(\theta, \phi; \theta', \phi')\boldsymbol{L}(i_1). \tag{5.66}$$

5.14. Inclusion of Atomic and Collisional Depolarization

The effect of collisions is to introduce additional broadening (via the collisional damping constant γ_c), and to depolarize the scattering process by scrambling the relative phases of the spherical vector components of the radiating oscillator. This led us in the previous sections to consider two different cases for the scattering process: coherent and incoherent scattering. Let us by k_c denote the fraction of all the scattering processes that occur as coherent scattering, which means that the fraction $1 - k_c$ occurs as incoherent scattering. Then, in the dispersion wings,

$$k_c = \frac{\gamma_N}{\gamma_N + \gamma_c}, \tag{5.67}$$

where γ_N and γ_c are the radiative and collisional damping constants, respectively.

If the total scattering coefficient is σ, given by Eq. (5.20), then the coefficient to be used to describe the coherent emission vector $\boldsymbol{j}_{\mathrm{coh}}$ is $k_c\sigma$, while the incoherent scattering coefficient is $(1 - k_c)\sigma$.

In the more complete account of collisional effects and partial redistribution in Chapter 10 we will see that although Eq. (5.67) remains valid in the dispersion wings, the expressions are different in the line core or at intermediate frequencies. Thus in the line core γ_c has to be multiplied by a numerical factor of 0.4–0.5. The collisional damping also combines with the natural damping γ_N in a similar way in the γ that is used in Eq. (5.53) for the "Hanle mixing angle" $\alpha_{q-q'}$. This aspect will also be treated in Chapter 10.

The collisionally undisturbed fraction k_c of the scattering processes does not only represent polarizing scattering when the quantum-mechanical properties of the atom are taken into account, in contrast to the classical oscillator case. The type of phase matrix that we have depends on the type of atomic transition considered, or, more precisely, on the J (total angular momentum) quantum numbers of the upper and lower levels. It is however possible to describe the phase matrix $\boldsymbol{P}$ for any atomic transition as consisting of one fraction W_2 that scatters the linear polarization in accord with the classical oscillator theory (which implies a dipole-scattering or Rayleigh phase matrix in the non-magnetic case), while the remaining fraction $1 - W_2$ is being scattered unpolarized with an isotropic angular distribution. The part $\boldsymbol{P}^2$ of the phase matrix that describes the linearly polarized scattering should therefore be multiplied by W_2. The part $\boldsymbol{P}^1$ of the phase matrix that describes scattering of the circular polarization is reduced by a factor W_1 with respect to the classical oscillator case. The factors W_1 and W_2 are functions of the J quantum numbers of the upper and lower levels, and have been explicitly given by Chandrasekhar (1950), who calls them E_3 and E_1. They will be derived in Chapter 9.

The phase matrix for coherent scattering (for the coordinate system of Fig. 3.3) can thus be written in the form

$$\boldsymbol{P}(\theta, \phi; \theta', \phi') = \boldsymbol{E}_{11} + \tfrac{3}{4}(W_1\boldsymbol{P}^1 + W_2\boldsymbol{P}^2)\,, \tag{5.68}$$

where $\boldsymbol{P}^1$ and $\boldsymbol{P}^2$ are given by Eqs. (5.58)–(5.64).

5.15. Explicit Examples of Hanle-Effect Scattering

One can get a better insight into the behavior of the phase matrix by considering some explicit examples of the polarization produced by coherent scattering. For simplicity we will assume that the incident radiation is unpolarized (in the solar photosphere the scattering polarization rarely exceeds 1 %). This means that only the first column of the phase matrix $\boldsymbol{P}$ contributes, and that the scattered radiation does not contain any circular polarization (in the weak-field limit described by Eqs. (5.68) and (5.58)–(5.64)). Let us also limit ourselves to the special case that the incident radiation is propagating in the vertical direction with an intensity that is unity, and that the magnetic-field vector $\boldsymbol{B}$ lies in the horizontal plane (if $\boldsymbol{B}$ would be parallel to the incident radiation or symmetrically oriented with respect to the angular distribution of the incident radiation, then the Hanle effect would be zero, i.e., the scattering would be non-magnetic). $\boldsymbol{B}$ makes an angle χ with respect to the direction of the scattered radiation. For the so defined geometry we can from our previous equations easily calculate the Stokes parameters I, Q, and U (V is equal to zero) of the scattered radiation, and obtain

$$\begin{aligned} I &= 1 + \tfrac{3}{8}\, W_2 \,[\, \tfrac{1}{3}(1 - 3\cos^2\chi) - \sin^2\chi \cos^2\alpha_2 \,] \,, \\ Q &= -\tfrac{3}{8}\, W_2 \,[\sin^2\chi + (1+\cos^2\chi)\cos^2\alpha_2 \,] \,, \\ U &= -\tfrac{3}{8}\, W_2 \cos\chi \, \sin 2\alpha_2 \,. \end{aligned} \tag{5.69}$$

Let us define the fractional polarization as

$$\begin{aligned} p_Q &= Q/I \,, \\ p_U &= U/I \,, \end{aligned} \tag{5.70}$$

and a field-strength parameter Ω as

$$\Omega = \tan\alpha_2 = 2 g_u \omega_L/\gamma \tag{5.71}$$

(cf. Eq. (5.53)). Figure 5.3 gives the fractional polarization as a function of Ω, for different values of the field angle χ. The two diagrams to the left in Fig. 5.3 represent the scattering case described by Eq. (5.69), while the two diagrams to the right represent the case when the magnetic field vector makes an angle of 45° with the vertical direction (instead of 90°).

Another way of graphical representation of the scattered radiation is in terms of the two "Hanle parameters" $p/p_{\max}$, describing the amount of Hanle-effect depolarization, and β, the angle of rotation of the plane of linear polarization. They are defined by

$$p = \sqrt{Q^2 + U^2}\,/I \tag{5.72}$$

and

$$\tan 2\beta = U/Q \,. \tag{5.73}$$

p is the total fractional polarization, and $p_{\max}$ is the value of p for zero magnetic field. The rotation law (5.73) follows from the property of the Mueller rotation matrix (5.65).

Explicit expressions for the "Hanle parameters" are obtained for the special scattering geometry that was considered above (and in the two diagrams to the left in Fig. 5.3) by inserting the expressions of Eq. (5.69) in Eqs. (5.72) and (5.73), which gives

$$p/p_{\max} = \frac{[\,1 + \Omega^2 + (\,\tfrac{1}{2}\Omega^2 \sin^2\chi\,)^2\,]^{1/2}}{1 + \Omega^2 + \tfrac{1}{2}\, p_{\max}\, \Omega^2 \sin^2\chi} \tag{5.74}$$

and

$$\tan 2\beta = \frac{\Omega\cos\chi}{1 + \tfrac{1}{2}\,\Omega^2 \sin^2\chi} \,. \tag{5.75}$$

These expressions are illustrated graphically in Fig. 5.4. In the upper diagram the case when $p_{\max} = 1$ is shown, which corresponds to the case given by the diagram to the upper left in Fig. 5.3, for $W_2 = 1$. The lower diagram represents the limiting case when $p_{\max} \to 0$ (the two cases differ via the last term in the

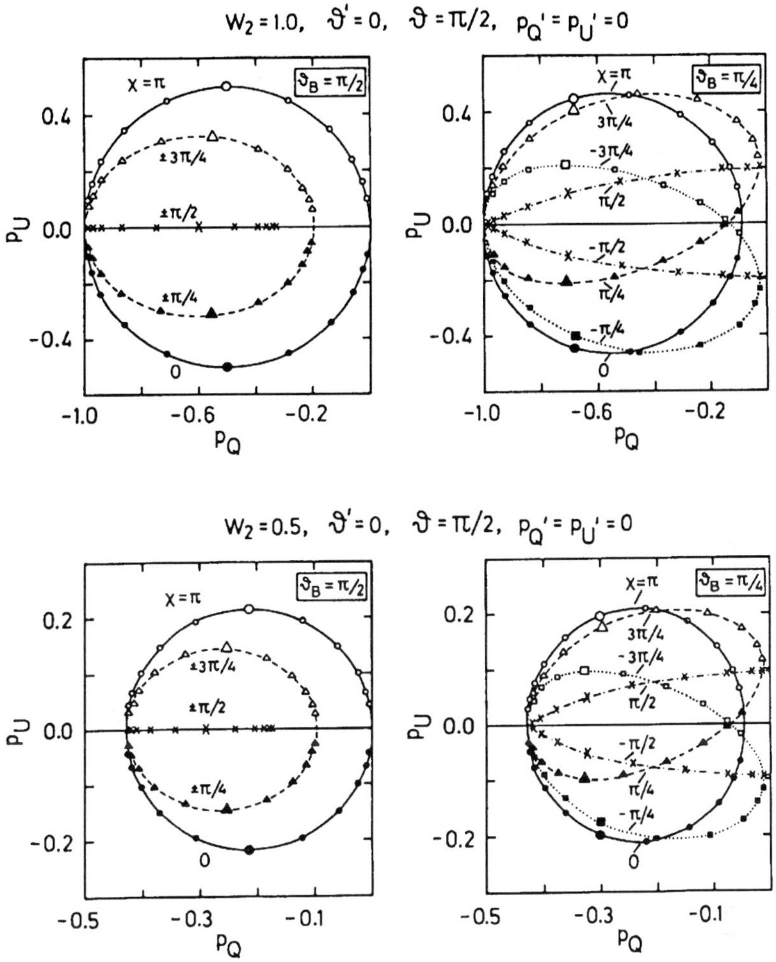

Fig. 5.3. Fractional polarization $p_Q = Q/I$ and $p_U = U/I$ as a function of field-strength parameter Ω, for various values of the magnetic field azimuth angle χ, inclination angle ϑ_B with the vertical direction, and atomic depolarization parameter W_2. The incident radiation is unpolarized and parallel to the vertical axis ($\vartheta' = 0$), while the scattered radiation is in the horizontal plane ($\vartheta = 90°$). The symbols along the various curves correspond to various values of the field-strength parameter Ω, increasing in steps of 0.2 in $\log \Omega$, from left to right. The larger symbols correspond to $\Omega = 1$. From Stenflo (1978b).

denominator of Eq. (5.74)). This second case is more representative of the real sun, since due to geometrical depolarization (from the wide angular distribution of the incident radiation field in a stellar atmosphere) $p_{\rm max}$ is generally $\ll 1$. As the polarization p has been normalized by $p_{\rm max}$ in this type of representation, the geometrical depolarization that occurs for both p and $p_{\rm max}$ divides out. This means that in the general case when integrating over a wide range of angles for the incident radiation we obtain diagrams that do not differ much from the special

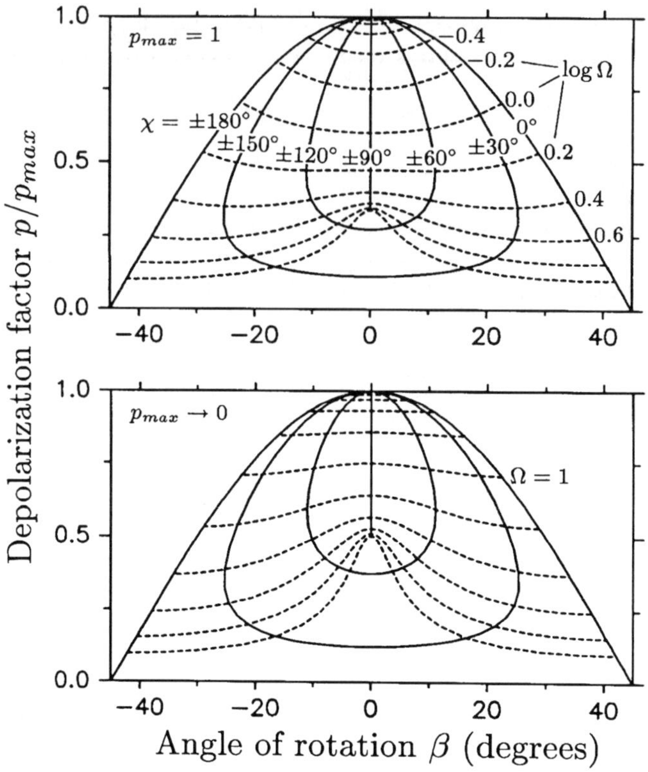

Fig. 5.4. Hanle-effect depolarization factor $p/p_{\rm max}$ and angle of rotation of the plane of polarization β, as functions of the field-strength parameter Ω and the field azimuth angle χ, for the same scattering geometry as the two diagrams to the left in Fig. 5.2. While the upper diagram represents the same case as the diagram to the upper left of Fig. 5.2, the lower diagram represents the case when the non-magnetic polarization $p_{\rm max}$ goes to zero (as is the case when the anisotropy of the incident radiation field is small). From Stenflo (1991a).

90° scattering case described by the lower diagram of Fig. 5.4 (provided that the angular distribution averaged over is symmetric around the vertical direction, and that the magnetic field is in the horizontal plane).

Although it may be appealing to represent the observed polarization in diagrams directly illustrating the two "Hanle parameters", the representation in terms of p_Q and p_U as in Fig. 5.3 is more adequate for quantitative analysis of observational data. The reason for this preference lies in the way in which the observational errors behave in the two representations. When the polarization p goes to zero, the error in the determined value of the polarization angle goes to infinity. In the "Hanle parameter" $\beta - p$ diagram the error boxes therefore become increasingly elongated as we move down in the diagram. In contrast, the error boxes in the $p_Q - p_U$ diagram are rectangles with a size that is practically independent of their location in the diagram (a small variation arises due to the Ω and χ dependence of Stokes I). This is so because Q and U are the directly observed quantities, and the errors ΔQ and ΔU are independent of Q and U. If Q and U are treated on an equal basis in the observations, the error boxes are simply squares.

5.16. Polarization of the Continuous Spectrum

There are two main physical processes contributing to the polarization of the continuous spectrum: Thomson scattering at free electrons and Rayleigh scattering at neutral hydrogen. Both processes can be treated as classical dipole-type scattering in terms of our present formalism.

As the free electrons are not subject to any external potential (like the central potential from the nucleus in atomic systems), we see from Eq. (3.31) that the resonant frequency ω_0 in the classical oscillator equation is zero. We can therefore use the same expressions as before and just insert $\omega_0 = 0$ to describe Thomson scattering.

In the case of neutral hydrogen the dominating resonance is that of the Lyman α transition with a resonant wavelength $\lambda_0 = 2\pi/\omega_0 = 1216\,\text{Å}$, but the other lines of the Lyman series, up to the Lyman limit at 912 Å, also contribute. We thus have the case of scattering in the very distant dispersion wings of the lines of the Lyman series (when wavelengths $\lambda \gtrsim 1230\,\text{Å}$ are considered). As the Lyman lines represent $S \rightarrow P$ transitions, and $|\lambda - \lambda_0| \gg$ the separation between the fine-structure components, the scattering in the distant wings approximates that of classical dipole-type scattering (as will be proven in Chapter 9). To treat scattering at neutral hydrogen we can thus simply use the expression for classical resonant scattering while omitting the unimportant (in the dispersion wings) damping constant, and for the resonant frequency λ_0 use the center of gravity of the Lyman series, which is about 1026 Å (cf. Bethe, 1933). The contribution from each Lyman line has to be weighted by its oscillator strength. Since the sum of the absolute oscillator strengths is unity for atomic hydrogen, the Lyman series may be treated as a single line with $\lambda_0 = 1026\,\text{Å}$.

For continuous scattering the frequency redistribution problem disappears, which means that we can replace $\boldsymbol{R}\,\mathrm{d}\nu'$ in expression (5.5) with the classical phase matrix $\boldsymbol{P}_{\rm cl}$ (which is $\boldsymbol{P}$ without magnetic-field effects and with the W_K coeffients of Eq. (5.68) equal to unity), and we do not need any frequency integration. Therefore the scattering vector becomes

$$\boldsymbol{j} = \sigma_c \int \frac{\mathrm{d}\Omega'}{4\pi}\, \boldsymbol{P}_{\rm cl} \boldsymbol{I}_{\nu'} \,. \tag{5.76}$$

As seen from Eq. (5.11) σ_c is obtained by performing angular averaging of $\omega^4/(\pi c^4)(\boldsymbol{w} \otimes \boldsymbol{w}^*)$, as was done to obtain σ in Eqs. (5.19) and (5.20). The components $w_{\alpha\beta}$ are obtained from Eq. (3.85), in which we may replace n_q with n_0 (since the Zeeman splitting effects can be neglected). $n_0 - 1$ has to be replaced by $\frac{1}{2}(n_0^2 - 1)$, since we have to use Eq. (3.49) to get the refractive index and not Eq. (3.51), which was based on the assumption $|\omega_0 - \omega| \ll \omega_0$ that is not valid for the continuum problem, as we are far from the resonance frequency. Thus

$$n_0^2 - 1 \approx \frac{\omega_A^2}{\omega_0^2 - \omega^2} , \tag{5.77}$$

where we have omitted the insignificant damping constant. The frequency ω_A is given by Eq. (3.50).

Like in the derivation of Eq. (5.19) the angular averaging of the product of the four geometric ϵ factors occurring in the bilinear products $w_{\alpha\beta} w^*_{\alpha'\beta'}$ and the summation of the two polarization states give a factor 2/3 (cf. the explicit expression for this procedure in Eq. (8.38)). We therefore get

$$\sigma_c = \frac{\omega^4}{\pi c^4} \, \tfrac{2}{3} [\, \tfrac{1}{2} (n_0^2 - 1)]^2 , \tag{5.78}$$

or

$$\sigma_c = \frac{8\pi}{3} r_e^2 \frac{\omega^4}{(\omega_0^2 - \omega^2)^2} N , \tag{5.79}$$

where

$$r_e = \frac{e^2}{4\pi \, \epsilon_0 \, m \, c^2} \tag{5.80}$$

is the classical electron radius. Note here that N is not squared, for the same reasons as were given in Sect. 5.4, since we first have to compute σ_c for a *single* scattering process, using $N = 1$ in the expression for $n_0^2 - 1$, and then multiply by the number of scatterers N.

In the special case of Thomson scattering $\omega_0 = 0$. We then get for the scattering coefficient

$$\sigma_e = \frac{8\pi}{3} r_e^2 \, N_e , \tag{5.81}$$

where N_e is the number of electrons per unit volume element.

In the case of Rayleigh scattering at neutral hydrogen, $\omega_0 = 2\pi/1026$Å. With Eqs. (5.79) and (5.81) we obtain the Rayleigh scattering cross section

$$\sigma_H = (\sigma_e/N_e) \frac{\omega^4}{(\omega_0^2 - \omega^2)^2} N_H , \tag{5.82}$$

where N_H is the neutral hydrogen number density. At wavelengths where $\omega^2 \ll \omega_0^2$, e.g. for $\lambda \gtrsim 3000$ Å,

$$\sigma_H \approx (\sigma_e/N_e) \left(\frac{\lambda_0}{\lambda} \right)^4 N_H , \tag{5.83}$$

where $\lambda_0 = 1026$ Å. We notice that σ_H increases steeply towards shorter wavelengths. This general wavelength dependence is in the case of molecular Rayleigh scattering responsible for the blue color of the sky.

The total continuous scattering coefficient is

$$\sigma_c = \sigma_e + \sigma_H . \tag{5.84}$$

It is convenient to introduce a "mean Stokes vector"

$$\boldsymbol{J}_\nu = \int \frac{\mathrm{d}\Omega'}{4\pi} \, \boldsymbol{P} \boldsymbol{I}_{\nu'} \,. \tag{5.85}$$

The total continuous emission vector can then be written as

$$\boldsymbol{j}_c = \sigma_c \boldsymbol{J}_{\nu,\mathrm{cl}} + \kappa_c B_\nu \, \boldsymbol{1} \,, \tag{5.86}$$

where the classical phase matrix $\boldsymbol{P}_{\mathrm{cl}}$ is used to represent $\boldsymbol{P}$ in $\boldsymbol{J}_\nu$. Coefficient κ_c accounts for all the non-scattering continuous emission processes, which are assumed to be unpolarized and in LTE (local thermodynamic equilibrium).

5.17. Stokes Transfer Equation, Including the Zeeman and Hanle Effects

Let us now wrap up the results of the previous sections and assemble them into a transfer equation. The line emission vector can be described as the sum of three contributions:

$$\boldsymbol{j}_L = \boldsymbol{j}_{\mathrm{coh}} + \boldsymbol{j}_{\mathrm{ic}} + \boldsymbol{j}_{\mathrm{ns}} \,. \tag{5.87}$$

$\boldsymbol{j}_{\mathrm{coh}}$ represents coherent scattering (undisturbed by collisions), $\boldsymbol{j}_{\mathrm{ic}}$ incoherent (collisionally disturbed) scattering, and $\boldsymbol{j}_{\mathrm{ns}}$ non-scattering emission.

The coherent scattering vector can quite generally be written as

$$\boldsymbol{j}_{\mathrm{coh}} = k_c \sigma \int \frac{\mathrm{d}\Omega'}{4\pi} \int \mathrm{d}\nu' \, \boldsymbol{R} \boldsymbol{I}_{\nu'} \tag{5.88}$$

(cf. Eq. (5.8)). The scattering coefficient σ is given by Eq. (5.20). The coherent fraction k_c of the scattering processes is given by Eq. (5.67). Most of the physics is contained in the redistribution matrix $\boldsymbol{R}$ for coherent scattering, which was discussed in Eqs. (5.35)–(5.41) above.

In general the redistribution matrix $\boldsymbol{R}$ has to account for partial frequency redistribution in addition to the polarization effects, which makes the problem technically very complicated if treated in its full generality. It is however possible to identify regimes in which greatly simplified expressions can be derived. In particular we have seen how a frequency-independent phase matrix $\boldsymbol{P}$, introduced in Eq. (5.42) with the normalization of Eq. (5.43), may be factorized out from the redistribution matrix $\boldsymbol{R}$ in the dispersion wings as well as in the line core in the case of weak magnetic fields.

Let $\boldsymbol{J}_\nu^0$ denote the value of the "mean Stokes vector" $\boldsymbol{J}_\nu$ of Eq. (5.85) when the magnetic field is zero. Then, for frequencies so far in the line wings that $|\Delta\nu| \gg \Delta\nu_D$ *and* $|\Delta\nu| \gg \Delta\nu_H$, the magnetic-field effects (Hanle effect) vanish, and the scattering is frequency coherent. The wing emission vector thus becomes (when Eq. (5.45) is inserted into Eq. (5.88))

$$\boldsymbol{j}_{\mathrm{coh,\,wings}} \approx k_c \sigma \, \varphi_\nu \, \boldsymbol{J}_\nu^0 \,. \tag{5.89}$$

In the line core, on the other hand, when the magnetic fields are weak ($\Delta\nu_H \ll \Delta\nu_D$), we get (when Eq. (5.46) is inserted into Eq. (5.88))

$$\boldsymbol{j}_{\text{coh,core,weak field}} \approx k_c \sigma\, \varphi_\nu \int \varphi_{\nu'} \boldsymbol{J}_{\nu'}\, \mathrm{d}\nu' \,. \tag{5.90}$$

The weak-field phase matrix $\boldsymbol{P}$ to be used in $\boldsymbol{J}_{\nu'}$ has been given by Eqs. (5.68) and (5.58)–(5.64).

From Eq. (5.33) we see that the incoherent emission vector may be written as

$$\boldsymbol{j}_{\text{ic}} = (1 - k_c)\sigma e\, \boldsymbol{\Phi} \mathbf{1} \,, \tag{5.91}$$

where e is given by Eq. (5.34). For the non-scattering emission vector $\boldsymbol{j}_{\text{ns}}$ it is natural to assume, as an approximation of the general situation, that the atom has no "memory" of how it had been excited when it emits the photon. This means that atomic polarization of the upper level is neglected for these non-scattering transitions. The polarization of $\boldsymbol{j}_{\text{ns}}$ is then the same as that of $\boldsymbol{j}_{\text{ic}}$, and $\boldsymbol{j}_{\text{ns}}$ has the form

$$\boldsymbol{j}_{\text{ns}} = j_{\text{ns}} \sigma\, \boldsymbol{\Phi} \mathbf{1} \,. \tag{5.92}$$

A full treatment of the general case when atomic polarization is taken into account for any emission-absorption process will be given in Chapter 8.

$\boldsymbol{\Phi}$, which was introduced in Eq. (5.29), is the normalized version of the Mueller absorption matrix $\boldsymbol{M}$ of Eq. (4.35) or $\boldsymbol{\eta}$ of Eq. (4.51). According to Eqs. (5.29), (5.17), (4.51), (4.35), and (4.46),

$$\sigma \boldsymbol{\Phi} = \kappa_c\, \boldsymbol{\eta} \,. \tag{5.93}$$

The incoherent scattering and non-scattering terms may be grouped together, since they have the same polarization:

$$(\boldsymbol{j}_{\text{ic}} + \boldsymbol{j}_{\text{ns}})/\kappa_c \approx [\,(1 - k_c)e + j_{\text{ns}}\,]\, \boldsymbol{\eta} \mathbf{1} \,. \tag{5.94}$$

We can now summarize our results by incorporating them into a final radiative transfer equation, obtained if we replace the emission vector $S_L \boldsymbol{\eta} \mathbf{1}$ in Eq. (4.50) with $(\boldsymbol{j}_L + \boldsymbol{j}_c)/\kappa_c$ and use Eqs. (5.86), (5.87), and (5.94). The transfer equation then becomes

$$\begin{aligned} \frac{\mathrm{d}\boldsymbol{I}_\nu}{\mathrm{d}\tau_c} &= (\boldsymbol{\eta} + \boldsymbol{E})\boldsymbol{I}_\nu - B_\nu \mathbf{1} - \sigma_c \boldsymbol{J}_{\nu,\text{cl}}/\kappa_c \\ &\quad - [\,(1 - k_c)e + j_{\text{ns}}\,]\, \boldsymbol{\eta} \mathbf{1} - \boldsymbol{j}_{\text{coh}}/\kappa_c \,. \end{aligned} \tag{5.95}$$

The coherent scattering vector $\boldsymbol{j}_{\text{coh}}$, which generally has a very complicated dependence on the scattering geometry, the magnetic field vector, and the partial redistribution processes, can in the weak-field limit be approximated by

$$\boldsymbol{j}_{\text{coh}} \approx k_c \sigma \varphi_\nu \left[(1 - a_\nu) \int \varphi_{\nu'}\, \boldsymbol{J}_{\nu'}\, \mathrm{d}\nu' + a_\nu \boldsymbol{J}^0_\nu \right], \tag{5.96}$$

where $a_\nu \approx 0$ in the line core ($|\Delta\nu| \lesssim 3\Delta\nu_D$), and $a_\nu \approx 1$ in the wings. $\boldsymbol{J}^0_\nu$ is the non-magnetic counterpart of the magnetic-field dependent $\boldsymbol{J}_\nu$. While the core expression (the first term) is only valid in the weak-field regime ($\Delta\nu_H \ll \Delta\nu_D$), the wing expression (the second term) is decoupled from the field strength, as long as frequencies with $|\Delta\nu| \gg \Delta\nu_H$ are considered. At intermediate frequencies, or

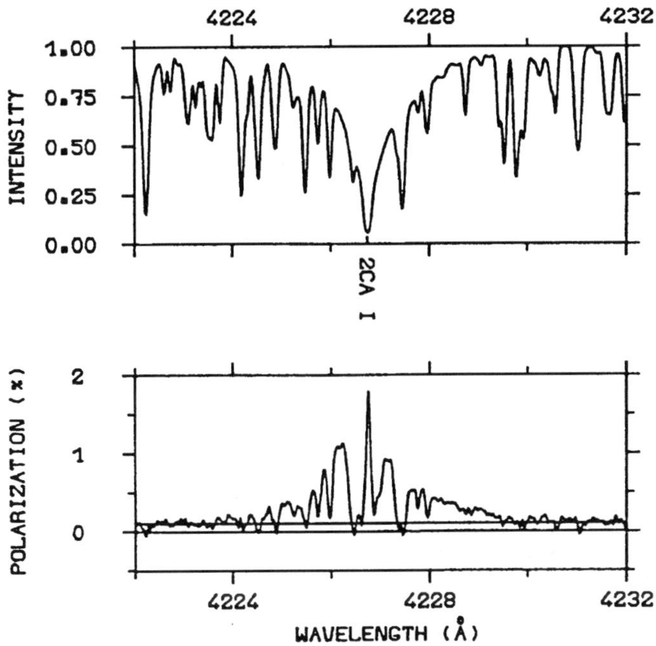

Fig. 5.5. Non-magnetic linear polarization Q/I recorded as in Fig. 5.1 10 arcsec inside the solar limb. Note the narrow polarization peak in the Doppler core of the Ca I 4227 Å line, which is surrounded first by minima and then by polarization maxima in the dispersion wings of the line. The many blend lines in this crowded portion of the spectrum depolarize by absorbing the polarized Ca I photons. From Stenflo et al. (1983b).

for arbitrary field strengths, a more complete partial redistribution treatment is needed, which will be given in Chapter 10.

The qualitatively different behavior of the core and wings is illustrated for the real sun in Figs. 5.5 and 5.6 for the Ca I 4227 Å line, which is the most strongly polarized line (due to coherent scattering) in the whole visible solar spectrum. The non-magnetic Q/I is shown in Fig. 5.5 so that it can be compared with the shapes of the Stokes profiles of the same line in a magnetic region in Fig. 5.6. As the limb distance of the recording of Fig. 5.6 is larger than that of Fig. 5.5, the polarization amplitudes are smaller (since the average scattering angle deviates more from 90°), but the shape of the Q/I profile should not change much, except for the influence of the magnetic field. While the linear polarization in the dispersion wings is unaffected by the Hanle effect, the plane of polarization in the Doppler core is rotated such that most of the core peak now appears in U/I (which would be zero in the absence of magnetic fields), whereas the Q/I core peak is suppressed. There is no significant U/I signal in the wings. The anti-symmetric profile signatures in V/I have nothing to do with the Hanle effect. They are due to the longitudinal Zeeman effect, which affects both the Ca I line core and the blend lines. The transverse Zeeman effect is too weak in this particular case to compete with the linear polarization caused by scattering.

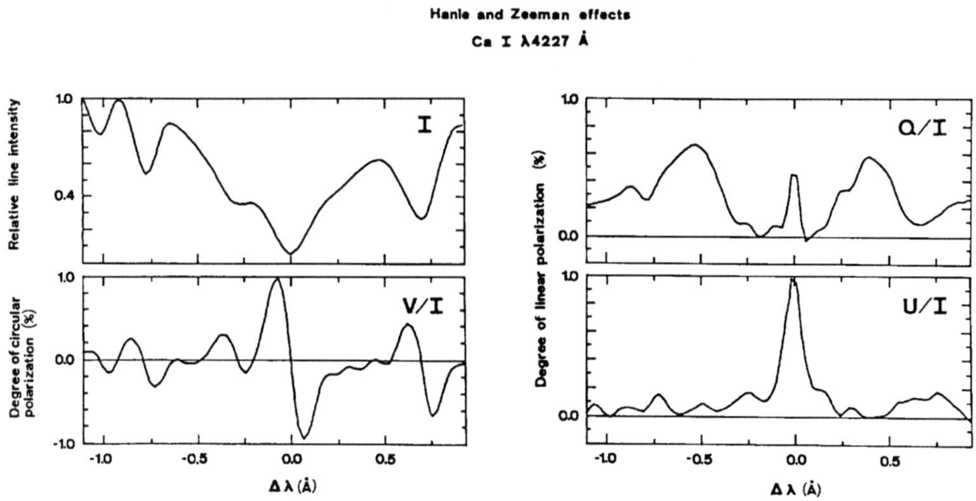

Fig. 5.6. The central part of the Ca I 4227 Å line recorded in a magnetic field region 70 arcsec inside the solar limb. This is an example of the Hanle effect in action in the line core, having rotated the plane of linear polarization such that the core signal is larger in U/I than in Q/I. The strong wing polarization in Q/I is however unaffected by the Hanle effect, in agreement with the theoretical expectation. From Stenflo (1982).

CHAPTER 6

NON-LTE RADIATIVE TRANSFER: PHENOMENOLOGICAL TREATMENT

6.1. Introduction

In the previous chapters we have seen how the classical theory, when phenomenologically extended to include a Landé factor g and an oscillator strength f, can account for the most important cases that occur in the transport of polarized radiation. Even a typical coherence phenomenon like the Hanle effect can be handled by the classical theory for the case of a $J_\ell = 0 \rightarrow J_u = 1$ transition (J_ℓ and J_u are the total angular momentum quantum numbers of the lower and upper levels, respectively). The Hanle effect for arbitrary atomic transitions can be calculated by phenomenologically introducing the W_1 and W_2 factors into the classical theory, as done in the previous chapter.

These successes of a phenomenologically generalized classical theory have been made possible by borrowing a number of concepts from the quantum mechanical theory, including parameters like g, f, and $W_{1,2}$, the values of which can only be derived within the framework of quantum mechanics. The semi-classical theory can easily be further generalized to allow arbitrary anomalous Zeeman splitting patterns to be treated. Similarly, a non-LTE formalism for the transfer of polarized radiation can be developed, using Einstein's phenomenological radiative rate coefficients A_{ij} and B_{ij}, extended with the collisional rate coefficient C_{ij}.

To put all these phenomenological features on a solid and selfconsistent theoretical foundation, it is not sufficient to quantize only the atomic system, the radiation field has to be quantized as well. We also need the framework of quantum field theory to derive non-LTE populations of the magnetic substates, level crossings, lower-level coherences, fluorescence effects, etc. We will however postpone the field theory treatment until the next two chapters, where the theory will be developed from first principles. In the present chapter we introduce the basic theoretical concepts of the Zeeman effect, including a derivation of the Landé factor and a treatment of anomalous splitting patterns, and then develop a general non-LTE formalism for the transfer of polarized radiation for the case that atomic coherences are neglected. The missing aspects and logical gaps in this largely phenomenological treatment will afterwards be filled in by the quantum field theory of the next two chapters. The present phenomenological treatment will help us to intuitively grasp the often abstruse quantum field formalism, and it is needed for a large variety of practical applications.

The phenomenological non-LTE theory has been developed by Domke and Staude (1973), Dolginov and Pavlov (1973), Šidlikovský (1974), and House and Steinitz (1975). Non-LTE transfer with scattering of polarized radiation in multi-dimensional media was first treated by Stenholm and Stenflo (1978). The Hanle effect was incorporated in the transfer equation by Stenflo (1978b).

6.2. The Hamiltonian of the System

The classical Hamiltonian Eq. (3.23) for an electron in an external electromagnetic field that is characterized by the vector potential $\boldsymbol{A}$ and a central electrostatic potential $V(r)$ is

$$H = \frac{1}{2m}(\boldsymbol{p} + e\boldsymbol{A})^2 + V(r)\,. \tag{6.1}$$

$\boldsymbol{A}$ may include both the contribution $\boldsymbol{A}_B$ from the external magnetic field and the contribution $\boldsymbol{A}_R$ from the time-varying radiation field:

$$\boldsymbol{A} = \boldsymbol{A}_B + \boldsymbol{A}_R\,. \tag{6.2}$$

When we later make the transition to quantum mechanics we have to replace the classical vectors $\boldsymbol{p}$ and $\boldsymbol{A}$ with the vector operators $\hat{\boldsymbol{p}}$ and $\hat{\boldsymbol{A}}$ and make sure that these operators commute, before we can unambiguously expand the squared bracket in Eq. (6.1). Let us therefore prove the commutation of $\hat{\boldsymbol{p}}$ and $\hat{\boldsymbol{A}}$ before we go on rewriting the classical Hamiltonian further. We will consistently use the convention that a "hat" ^ above a symbol denotes an operator.

From the quantum-mechanical definition of the x component of the momentum operator,

$$\hat{p}_x = -i\hbar\frac{\partial}{\partial x}\,, \tag{6.3}$$

follows that

$$[\hat{\boldsymbol{A}}, \hat{\boldsymbol{p}}] = i\hbar\nabla\cdot\hat{\boldsymbol{A}}\,. \tag{6.4}$$

For an external magnetic field that can be considered to be homogeneous over atomic length scales, $\nabla\cdot\boldsymbol{A}_B = 0$. With the previously introduced radiation gauge of Eq. (2.12), $\nabla\cdot\boldsymbol{A}_R = 0$. Thus

$$[\hat{\boldsymbol{A}}, \hat{\boldsymbol{p}}] = 0\,, \tag{6.5}$$

i.e., the vector potential commutes with the momentum operator.

As we now expand the classical Hamiltonian Eq. (6.1),

$$H = \frac{1}{2m}\boldsymbol{p}^2 + V(r) + \frac{e}{m}\boldsymbol{A}_B\cdot\boldsymbol{p} + \frac{e}{m}\boldsymbol{A}_R\cdot\boldsymbol{p} + \frac{e^2}{2m}\boldsymbol{A}^2\,, \tag{6.6}$$

Eq. (6.5) allows us to later reinterpret the various quantities as quantum-mechanical operators. The last term may be neglected for field strengths $B \ll 10^4$ T, which is valid for all stars except white dwarfs and neutron stars, and when photon self-interaction is unimportant, which is also the case for stellar atmospheres except in extreme situations.

The magnetic-field contribution to the vector potential can be expressed as

$$\boldsymbol{A}_B = -\tfrac{1}{2}(\boldsymbol{r} \times \boldsymbol{B}), \tag{6.7}$$

which can be verified by forming $\nabla \times \boldsymbol{A}_B$, which gives $\boldsymbol{B}$ as required by Eq. (2.7) when the spatial derivatives of $\boldsymbol{B}$ can be disregarded (the external field is homogeneous over atomic dimensions). We then obtain (after a standard vector transformation)

$$\boldsymbol{A}_B \cdot \boldsymbol{p} = \tfrac{1}{2}\boldsymbol{B} \cdot (\boldsymbol{r} \times \boldsymbol{p}). \tag{6.8}$$

Introducing the orbital angular momentum

$$\boldsymbol{L} = \boldsymbol{r} \times \boldsymbol{p}, \tag{6.9}$$

the magnetic-field or "Zeeman" term H_Z of the Hamiltonian becomes

$$H_Z = \frac{e}{m}\boldsymbol{A}_B \cdot \boldsymbol{p} = \frac{e}{2m}\boldsymbol{L} \cdot \boldsymbol{B}, \tag{6.10}$$

where $-(e/2m)\boldsymbol{L}$ is the magnetic moment of the orbital motion.

The classical treatment needs to be generalized to include the spin angular momentum $\boldsymbol{S}$ of the electron, to make H_Z represent the total (negative) magnetic moment. The relativistic Dirac theory for the electron shows however that the gyromagnetic ratio, which is the ratio between the magnetic moment and the angular momentum, is twice as large for the electron spin as compared with the orbital angular momentum. Therefore we get for the "Zeeman term"

$$H_Z = \frac{e}{2m}(\boldsymbol{L} + 2\boldsymbol{S}) \cdot \boldsymbol{B}. \tag{6.11}$$

Another effect of the spin is its interaction with the orbital motion, leading to an additional term (LS-coupling term) H_{LS} in the Hamiltonian, which is responsible for the fine-structure splitting of the atomic energy levels. The reason for this coupling is that the moving electron in its own reference frame sees a magnetic field $\boldsymbol{B}'$ that is due to the relativistic transformation of the electrostatic field $\boldsymbol{E}$:

$$\boldsymbol{B}' = -\boldsymbol{v} \times \boldsymbol{E}. \tag{6.12}$$

Since

$$\boldsymbol{E} = -\frac{\boldsymbol{r}}{r}\frac{\partial \Phi}{\partial r} \tag{6.13}$$

for a static, central potential Φ according to Eq. (2.7), we obtain for non-relativistic velocities $\boldsymbol{v}$ (relativistic effects become important only for elements with high atomic number Z)

$$\boldsymbol{B}' = -\frac{1}{mr}\frac{\partial \Phi}{\partial r}\boldsymbol{L}. \tag{6.14}$$

To get the Hamiltonian LS-coupling term H_{LS} we have to multiply $\boldsymbol{B}'$ with the negative spin magnetic moment $-(e/m)\boldsymbol{S}$ and in addition attach a factor 1/2 due to the relativistic Thomas precession:

$$H_{\mathrm{LS}} = \frac{1}{2m^2 r}\frac{\partial V}{\partial r}\boldsymbol{L} \cdot \boldsymbol{S}, \tag{6.15}$$

where we have used the relation $V = -e\Phi$ for the electrostatic potential energy V, given by Eq. (3.25).

We can now express the total Hamiltonian as

$$H_{\text{tot}} = H_A + H_R + H' \,, \tag{6.16}$$

where

$$H' = \frac{e}{m}\boldsymbol{A}_R \cdot \boldsymbol{p} \tag{6.17}$$

describes the interaction between the radiation field and the atomic system. In Eq. (6.17) we have added H_R, the Hamiltonian for the independent radiation system (which was not part of Eq. (6.1)). In the present chapter we will only discuss H_A and postpone the treatment of H_R and H' until the next two chapters. For a consistent treatment *both* the radiation field *and* the atomic system need to be quantized, which requires the formalism of quantum field theory.

The atomic Hamiltonian H_A in the presence of an external magnetic field can be represented as a sum of the Hamiltonian $H_{A,0}$ in the absence of the magnetic field and the interaction Hamiltonian H_Z describing the Zeeman effect:

$$H_A = H_{A,0} + H_Z \,, \tag{6.18}$$

where

$$H_{A,0} = \frac{1}{2m}\boldsymbol{p}^2 + V(r) + H_{\text{LS}} \,. \tag{6.19}$$

6.3. Multi-electron Atomic Systems

In the case of a one-electron system the orbital and spin angular momenta $\boldsymbol{L}$ and $\boldsymbol{S}$ combine vectorially to form a total angular momentum vector $\boldsymbol{J}$. For multi-electron systems, in the case when relativistic effects can be disregarded, the orbital angular momentum vectors $\boldsymbol{\ell}_i$ of the individual electrons superpose to form a combined orbital angular momentum vector $\boldsymbol{L}$, and so do the individual spin angular momentum vectors $\boldsymbol{s}_i$ to form $\boldsymbol{S}$. It is in this case possible to define vectors $\boldsymbol{L}$, $\boldsymbol{S}$, and their combination $\boldsymbol{J}$ as being representative of the whole electron configuration. Such a coupling scheme for the angular momenta in a multi-electron system is called LS or Russel-Saunders coupling.

When relativistic effects become important, as they do with increasing atomic number or nuclear charge Z, and when one is dealing with highly excited levels with little electrostatic interaction between the individual electrons, $\boldsymbol{\ell}_i$ and $\boldsymbol{s}_i$ first couple to form a total angular momentum vector $\boldsymbol{j}_i$ for each individual electron, after which the various $\boldsymbol{j}_i$ vectors interact with each other. Such a coupling scheme is called jj coupling. It represents an idealization that can never be fully realized but only approached.

In most cases LS coupling is a good approximation that will be used in the following. It means that we can take our Hamiltonian that was developed for a one-electron system and apply it to a multi-electron system as well, since the vectors $\boldsymbol{L}$, $\boldsymbol{S}$, and $\boldsymbol{J}$ retain their physical meanings as orbital, spin, and total angular momentum vectors.

6.4. Zeeman Effect

Introducing the total angular momentum vector $\boldsymbol{J}$, we can write the interaction Hamiltonian of Eq. (6.11) that contains the Zeeman effect as

$$H_Z = \frac{e}{2m}(\boldsymbol{J} + \boldsymbol{S}) \cdot \boldsymbol{B}. \tag{6.20}$$

It is convenient to choose the coordinate system such that the z axis is along the magnetic field vector $\boldsymbol{B}$, which allows us to write

$$H_Z = \frac{eB}{2m}(J_z + S_z). \tag{6.21}$$

Both the vectors $\boldsymbol{J}$ and $\boldsymbol{S}$ will precess around the magnetic field $\boldsymbol{B}$, but in addition the spin vector $\boldsymbol{S}$ will precess around $\boldsymbol{J}$. Due to axial symmetry, J_z remains constant in time. S_z on the other hand is time varying, since its precession around $\boldsymbol{J}$ breaks the symmetry with respect to $\boldsymbol{B}$. The time averaged Zeeman Hamiltonian can therefore be written

$$H_Z = \omega_L(J_z + \langle S_z \rangle), \tag{6.22}$$

where the Larmor frequency ω_L that was defined in Eq. (3.44) has been introduced, and $\langle\ \rangle$ means time averaging. The problem has now been reduced to the calculation of $\langle S_z \rangle$.

We can distinguish between three regimes depending on the magnitude of the Zeeman splitting H_Z of the energy levels with respect to the fine-structure splitting H_{LS}:

A. $H_Z \ll H_{\mathrm{LS}}$. In this case the precession rate of $\boldsymbol{S}$ is much faster around $\boldsymbol{J}$ than around $\boldsymbol{B}$. We have a clean separation of the two precession time scales, and may therefore make the time averaging in two successive steps, first averaging out the faster precession around $\boldsymbol{J}$.

B. H_Z is comparable to H_{LS}. This is the so-called Paschen-Back regime.

C. $H_Z \gg H_{\mathrm{LS}}$, which is called the strong-field regime. $\boldsymbol{S}$ precesses around $\boldsymbol{B}$ almost independently of the precession of $\boldsymbol{J}$ around $\boldsymbol{B}$.

For normal stars (except for the collapsed late evolutionary stages) it is with few exceptions only case A that is relevant. This case generally results in anomalous splitting patterns, which we will treat below. In the strong-field regime (case C), on the other hand, the precessing system greatly simplifies, such that the splitting pattern always is that of a so-called "normal Zeeman triplet".

Let us now turn to the relevant case A and the calculation of $\langle S_z \rangle$. As the two time scales are well separated, we may average over a time window long in comparison with the $\boldsymbol{S}$-$\boldsymbol{J}$ precession period, but short as compared with the Larmor precession period. In this case $\langle \boldsymbol{S} \rangle$ becomes proportional to $\boldsymbol{J}$:

$$\langle \boldsymbol{S} \rangle = k\boldsymbol{J}\,, \tag{6.23}$$

where k is a constant of proportionality. As $\boldsymbol{J}$ can be regarded as constant over the averaging time window and therefore is unaffected by the averaging bracket, scalar multiplication of Eq. (6.23) with $\boldsymbol{J}$ gives

$$\langle \boldsymbol{J} \cdot \boldsymbol{S} \rangle = k\boldsymbol{J}^2\,, \tag{6.24}$$

which will be used below to eliminate k.

As Eq. (6.23) implies that

$$\langle S_z \rangle = kJ_z\,, \tag{6.25}$$

Eq. (6.22) may be written

$$H_Z = gJ_z\omega_L\,, \tag{6.26}$$

where the *Landé factor*

$$g = 1 + \frac{\langle \boldsymbol{J} \cdot \boldsymbol{S} \rangle}{\boldsymbol{J}^2}\,. \tag{6.27}$$

As

$$\boldsymbol{L}^2 = (\boldsymbol{J} - \boldsymbol{S})^2 = \boldsymbol{J}^2 + \boldsymbol{S}^2 - 2\boldsymbol{J} \cdot \boldsymbol{S}\,, \tag{6.28}$$

$$\boldsymbol{J} \cdot \boldsymbol{S} = \tfrac{1}{2}(\boldsymbol{J}^2 - \boldsymbol{L}^2 + \boldsymbol{S}^2)\,. \tag{6.29}$$

Since J_z, $\boldsymbol{J}^2$, $\boldsymbol{L}^2$, and $\boldsymbol{S}^2$ have common eigenfunctions, we may in the transition from vectors $\boldsymbol{J}$ to vector operators $\hat{\boldsymbol{J}}$ replace them with their eigenvalues, as is done when solving the time-independent Schrödinger equation. Thus we make the replacements $\hat{\boldsymbol{J}}^2 \to \hbar^2 J(J+1)$, $\hat{\boldsymbol{J}}_z \to \hbar M$, etc., where M is the magnetic quantum number, J the total angular momentum quantum number, etc. The Landé factor then becomes

$$g = 1 + \frac{J(J+1) - L(L+1) + S(S+1)}{2J(J+1)}\,, \tag{6.30}$$

while the Zeeman splitting H_Z of the energy levels is

$$H_Z = gM\hbar\omega_L\,. \tag{6.31}$$

With these results the transition frequency between two atomic energy levels can be written

$$\nu_{M_u M_\ell} = \nu_0 + \nu_L(g_u M_u - g_\ell M_\ell)\,, \tag{6.32}$$

where ν_0 is the non-magnetic resonant frequency of the transition, $\nu_L = \omega_L/2\pi$, g_u, g_ℓ the Landé factors, M_u, M_ℓ the magnetic quantum numbers of the upper and lower levels, respectively.

The selection rules for allowed dipole transitions (which as will be seen in the next chapter follow from the Wigner-Eckart theorem and the mathematical properties of 3-j symbols) are:

$\Delta J = 0, \pm 1$, but $J = 0 \to 0$ is forbidden,
$\Delta L = 0, \pm 1$, $\Delta S = 0$,
$\Delta M = 0, \pm 1$.

The spectral line components due to transitions with $\Delta M = 0$ are called π components, those due to transitions with $\Delta M = \pm 1$ are called σ components.

6.5. Anomalous Splitting

The strength of a transition is proportional to the relative population of the initial sublevel, which is labeled by the quantum number M_ℓ in the case of an absorption transition, by M_u in the case of emission. It is also proportional to the squared matrix element $|\langle M_\ell | \hat{r}_q | M_u \rangle|^2$ in the case of a dipole transition, where $\hat{r}_q$ is the spherical vector component (cf. Sect. 3.5) of the position vector operator of the electron. This will become clear in the next two chapters, where we will treat the interaction matter-radiation in quantum field theory. The phenomenological expressions for anomalous Zeeman splitting that we will give in the present section represent only a special case of the more general treatment of the next two chapters, because the Zeeman substates are generally characterized not only by their populations, but also by the coherences or phase relations between their wave functions. The phases are lost when squaring the matrix elements like in $|\langle M_\ell | \hat{r}_q | M_u \rangle|^2$.

For the special case of no atomic coherences and complete redistribution of the populations among the Zeeman sublevels, the transition strength is obtained from the Wigner-Eckart theorem, which will be described in the next chapter, Eq. (7.95). The strength is thus found to be

$$S_q(M_\ell, M_u) = 3 \begin{pmatrix} J_\ell & J_u & 1 \\ -M_\ell & M_u & q \end{pmatrix}^2 . \tag{6.33}$$

The selection rules for the quantum numbers follow from the symmetry properties of the 3-j symbol in Eq. (6.33), since it is non-zero only when $q = M_\ell - M_u$ and if $J_u - J_\ell = 0, \pm 1$. The q defined by Eq. (6.33) corresponds to the q defined in the classical theory, used for instance in Eqs. (3.39), (3.51), or (4.11). It is directly related to the definitions (3.34) and (3.35) of the spherical vector components. The factor of 3 in Eq. (6.33) assures normalization,

$$\sum_{M_\ell, M_u} S_q(M_\ell, M_u) = 1 , \tag{6.34}$$

which follows from the orthogonality properties of 3-j symbols (Brink and Satchler, 1968).

Our definition of the transition strengths $S_q(M_\ell, M_u)$ is angular invariant (in contrast to other definitions in the literature, e.g. by Beckers, 1969), and does not depend on the direction of the field vector $\boldsymbol{B}$ or on the line of sight. The angular dependence of the intensities of the π and σ components will be discussed below.

Explicit expressions for the transition strengths $S_q(M_\ell, M_u)$ are given in Table 6.1, but the normalization used is different from that of Eqs. (6.33) and (6.34) to make the expressions more compact. Correct normalization is obtained if we simply divide each entry by the sum formed as done on the left hand side of Eq. (6.34), keeping the values of q, J_u, and J_ℓ fixed.

Table 6.1. Unnormalized expressions for the transition strengths $S_q(M_\ell, M_u)$

$J_u - J_\ell$	$q{=}0$	$q{=}{-1}$	$q{=}{+1}$
0	$2M_u^2$	$(J_u+M_u)(J_u-M_u+1)$	$(J_u-M_u)(J_u+M_u+1)$
+1	$2(J_u^2-M_u^2)$	$(J_u+M_u)(J_u+M_u-1)$	$(J_u-M_u)(J_u-M_u-1)$
−1	$2[(J_u+1)^2-M_u^2]$	$(J_u-M_u+1)(J_u-M_u+2)$	$(J_u+M_u+1)(J_u+M_u+2)$

The polarization properties of the transitions will be derived with the quantum field theory presented in the next two chapters, but we will summarize the results here. The angular dependence and polarization properties of the π and σ components turn out to be the same as those obtained before with the classical theory. The concept of anomalous splitting, where several separate transitions may contribute several π or σ components of the spectral line, is however outside the classical framework, in which each of these three components is treated as due to one transition only, i.e., due to one of the three independent spherical vector components of the classical oscillator. The classical case corresponds quantum-mechanically to a $J_\ell = 0 \rightarrow J_u = 1$ transition and produces what is called a "normal Zeeman triplet". Nevertheless, the group of transitions that generates the π components in an anomalous splitting pattern has the same angular and polarization properties as the classical π component, and the same applies to the σ components.

When the field $\boldsymbol{B}$ is perpendicular to the line of sight (*transverse Zeeman effect*), both the π ($q = 0$) and σ ($q = \pm 1$) components are linearly polarized. In emission, the plane of polarization is parallel to $\boldsymbol{B}$ for the π components, perpendicular to $\boldsymbol{B}$ for the σ components.

When the magnetic field is parallel to the line of sight (*longitudinal Zeeman effect*), the σ components are circularly polarized, while the π components vanish. For an emission line the σ component with $q = +1$ is redshifted and right-handed circularly polarized (rotation of the electric vector clockwise in a fixed plane, as seen from the observer), when $\boldsymbol{B}$ is directed towards the observer. The $q = -1$ component is then blueshifted and has left-handed circular polarization.

The angular dependence *and* polarization of the emitted components can be obtained directly from the classically derived Eqs. (4.48)–(4.51), as will be shown with the general theory of the next two chapters. According to the classical expressions, the emitted Stokes vector is proportional to

$$\begin{pmatrix} H_I \\ H_Q \\ H_U \\ H_V \end{pmatrix}, \tag{6.35}$$

with $H_{I,Q,U,V}$ defined by Eqs. (4.48), (4.49), and (4.15).

Let us consider an optically thin emission line. Without polarization analysers we measure the intensity, which is proportional to H_I in the classical case (i.e., for a normal Zeeman triplet). According to Eqs. (4.48) and (4.49)

$$H_I = \tfrac{1}{2}[\, H_0 \sin^2 \gamma + \tfrac{1}{2}(H_+ + H_-)(1 + \cos^2 \gamma)\,] \,, \tag{6.36}$$

where γ is the angle between the directions of the magnetic field and the line of sight. Indices $0, +, -$ correspond to $q = 0$ (the π component) and $q = +1$ or -1 (the σ components). H_q are Voigt functions having the same shape but different wavelength displacements for different q values. In the anomalous Zeeman splitting case, however, we have to weight the H_q contributions from the various individual transitions with the corresponding transition strengths S_q, and sum over all the contributions. The measured intensity is then obtained by replacing H_q with a weighted average profile, which we denote ϕ_q:

$$\phi_q = \frac{1}{\sqrt{\pi}\Delta\nu_D} \sum_{M_u} S_q(M_\ell, M_u)\, H_q(a, v_q) \,, \tag{6.37}$$

subject to the constraint $M_\ell = M_u + q$. The factor $1/\sqrt{\pi}$ has been introduced to make the area of ϕ_q as a function of frequency ν normalized to unity.

$$v_q = v - v_H(M_\ell, M_u) \,, \tag{6.38}$$

where

$$\begin{aligned} v &= (\nu_0 - \nu)/\Delta\nu_D = \Delta\lambda/\Delta\lambda_D \,, \\ \Delta\lambda &= \lambda - \lambda_0 \,. \end{aligned} \tag{6.39}$$

It follows from Eq. (6.32) that

$$v_H(M_\ell, M_u) = (g_\ell M_\ell - g_u M_u)\,\omega_L/\Delta\omega_D \,. \tag{6.40}$$

Some different types of anomalous splitting patterns are illustrated in Fig. 6.1.

For a $J_\ell = 0 \to J_u = 1$ transition (normal Zeeman triplet), $M_\ell = 0$ and thus $q = -M_u$. Therefore $q = 1$ means a redshifted profile, which also follows directly from Eq. (6.32), since $q = 1$ means $M_u = -1$ and thus $\nu < \nu_0$. This redshifted profile has right-handed circular polarization when the magnetic field is parallel to the line of sight and points towards the observer, which follows from Eq. (6.42) below. The classical expression for v_+ in (4.11) also implies a redshift, so our generalized definition of ϕ_q is consistent with the classical theory.

Disregarding a constant of proportionality, we obtain for the emitted intensity, if we make the required replacements in Eq. (6.36),

$$I = \phi_0 \sin^2 \gamma + \tfrac{1}{2}(\phi_+ + \phi_-)(1 + \cos^2 \gamma) \,. \tag{6.41}$$

Similarly, if we use a polarizing filter transmitting right-handed $(I + V)$ or left-handed $(I - V)$ circular polarization, we measure in the classical case (for an optically thin emission line) a signal that is proportional to $(H_I \pm H_V)$. We therefore get in the anomalous splitting case (disregarding a constant of proportionality)

$$I \pm V = \phi_0 \sin^2 \gamma + \tfrac{1}{2}\phi_+(1 \pm \cos\gamma)^2 + \tfrac{1}{2}\phi_-(1 \mp \cos\gamma)^2 \,. \tag{6.42}$$

Eqs. (6.41) and (6.42) describe the polarization and angular dependence of the π $(q = 0)$ and σ $(q = \pm 1)$ components. In the special case when $\gamma = 0$, Eq. (6.42)

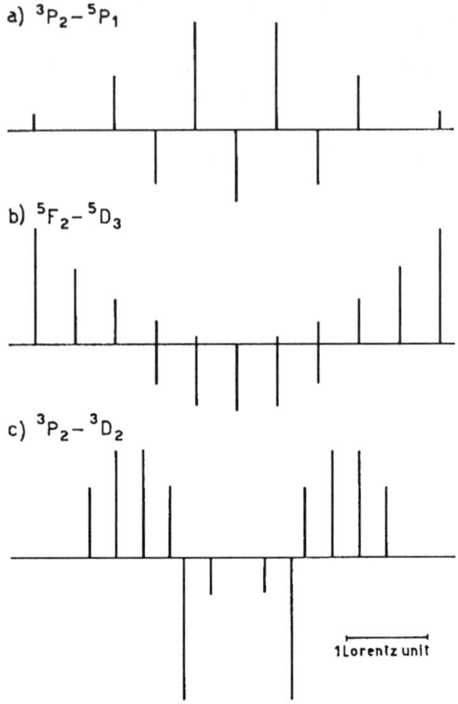

Fig. 6.1. Illustration of different anomalous Zeeman splitting patterns. The lengths of the vertical bars are proportional to the strengths S_q of the transitions, their positions indicate their wavelength shifts. The π components point downwards, the σ components upwards. From Landi Degl'Innocenti (1985a).

expresses the polarization rule that we have already mentioned for the σ components.

The detailed calculation of the π and σ component profiles ϕ_q should use Eq. (6.37) to superpose the weighted contributions from all the allowed transitions between the Zeeman sublevels. This is no problem with modern computers. In many cases it is however useful to approximate the anomalous splitting pattern with that of a normal Zeeman triplet, for which the wavelength positions of the σ components are the same as the positions of the centers of gravity of the $q = \pm 1$ components in the anomalous splitting pattern. The Landé factor of the normal triplet that would have this splitting is $g_{\rm eff}$, representing the *effective Landé factor* of the anomalous splitting pattern. From this definition it follows that

$$g_{\rm eff} = \sum_{M_u} S_1(M_\ell, M_u)\,(g_\ell M_\ell - g_u M_u)\,. \tag{6.43}$$

With some algebra one finds (using $M_\ell = M_u + 1$ for $q = 1$)

$$g_{\rm eff} = \tfrac{1}{2}(g_u + g_\ell) + \tfrac{1}{4}(g_u - g_\ell)[\,J_u(J_u + 1) - J_\ell(J_\ell + 1)\,]\,. \tag{6.44}$$

For very asymmetric anomalous splitting patterns it may be useful to describe the pattern by more parameters than just $g_{\rm eff}$. The most useful parameters are the various moments of the pattern. The kth order moment of the π and σ components is defined by

$$\mu_q^{(k)} = \sum_{M_u} S_q(M_\ell, M_u)\,[\,(g_\ell M_\ell - g_u M_u - q g_{\rm eff})\Delta\lambda_H\,]^k\,, \qquad (6.45)$$

where as before $M_\ell = M_u + q$. $\mu_q^{(0)} = 1$ according to Eq. (6.34) and $\mu_q^{(1)} = 0$ according to Eq. (6.43).

The moments $\mu_q^{(k)}$ have been tabulated for the various anomalous splitting patterns that can occur in LS coupling by Mathys and Stenflo (1987a, b), who also list the Zeeman null transitions (with zero Landé factor) and the normal triplet transitions.

The Zeeman splitting factor $\Delta\lambda_H$ used in Eq. (6.45) is defined by

$$\Delta\lambda_H/\lambda = \omega_L/\omega\,, \qquad (6.46)$$

which according to Eq. (3.44) means that

$$\Delta\lambda_H = \frac{e}{4\pi mc}\,\lambda^2\,B\,. \qquad (6.47)$$

Note that this definition does not contain the Landé factor g, in contrast to definitions often used by others. Our definition allows us to use it in general expressions of the type of Eq. (6.45), which do not contain a single Landé factor but a combination of the factors for the lower and upper levels.

We should caution that the LS-coupling scheme used here does not always give reliable Landé factors. Figure 6.2 shows a case where it works, Fig. 6.3 a case where it does not. The recordings shown were made in a magnetic region at the center of the solar disk.

6.6. Non-LTE Populations of the Sublevels and the Equation of Radiative Transfer

The relative sublevel populations are given by the diagonal elements of the atomic density matrix operator $\hat\rho_A$, which is normalized such that the sum of the diagonal elements is unity:

$$\mathrm{Tr}\,\hat\rho_A = \sum_M \rho_{MM} = 1\,. \qquad (6.48)$$

Here M can represent the substate of any atomic level (upper or lower). The theory underlying the concept of a density matrix will be described in the next chapter, including the off-diagonal elements.

If N is the number density of the considered ion, the above definitions imply that $N\rho_{MM}$ is the population of the substate with label M. The population of a given J state, e.g. that of a lower level with total angular momentum quantum

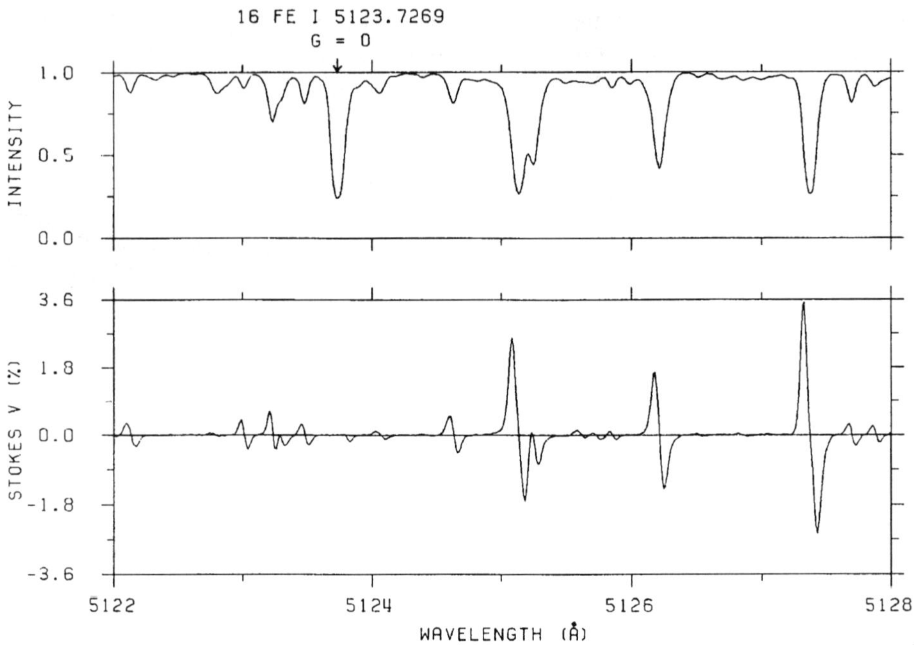

Fig. 6.2. Example of a $g = 0$ line (Fe I 5123.7 Å) that does not show any circular polarization, as expected from LS-coupling theory. Stokes V is given in units of the intensity of the local continuum. From Stenflo et al. (1984).

number J_ℓ, is obtained by summing over the populations of all the sublevels belonging to that state. In the case of complete redistribution, i.e., equal population of the various sublevels, we have

$$N_{J_\ell} = N \sum_{M_\ell} \rho_{M_\ell M_\ell} = N(2J_\ell + 1)\rho_{M_\ell M_\ell} \,. \tag{6.49}$$

The absorption and stimulated emission transition rates between two substates M_ℓ and M_u cannot be given directly in terms of the Einstein coefficients $B_{J_\ell J_u}$ and $B_{J_u J_\ell}$, since the sublevel transition rates only couple to certain polarization states of the radiation field. The Einstein coefficients are defined in terms of the unpolarized intensity I_ν, with which they have to be multiplied to give the transition rate between two J states. A treatment of the transition-rate problem, including the quantum-mechanical interferences between the levels (off-diagonal terms of the density matrix) will be postponed till the next two chapters. In the present chapter we will limit ourselves to a phenomenological treatment, the correctness of which will be verified in the following two chapters. The phenomenological treatment has the virtue of allowing us to expose in a more transparent way the direct link between the formalisms of polarized and standard unpolarized radiative transfer, and to establish how the usual Einstein rate coefficients can be used in the more general expressions.

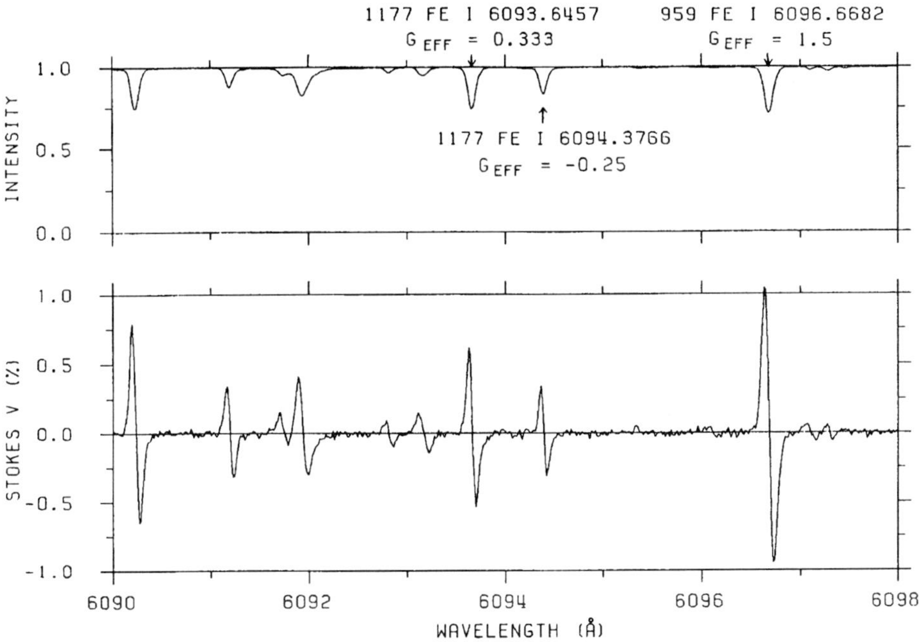

Fig. 6.3. Example when LS coupling fails. The observed Stokes V profile of the Fe I 6094.4 Å line has the same antisymmetric signature as the surrounding lines with positive Landé factors, although its LS-coupling effective Landé factor is negative, -0.25. A comparison of its polarization amplitude with the amplitudes of the other lines suggests that its true Landé factor should instead be about $+0.3$. From Stenflo et al. (1984).

In our previous classical expressions for the absorption matrix, either in the form of the radiative coherency matrix $\boldsymbol{C}$ of Eq. (4.21) or the Mueller matrix $\boldsymbol{\eta}$ of Eq. (4.51), the matrix elements consist of various combinations of $\mathcal{H}_q = H_q - 2iF_q$, where

$$\begin{aligned} H_q &= H(a, v - qv_H)\,, \\ F_q &= F(a, v - qv_H) \end{aligned} \tag{6.50}$$

for $q = 0, \pm 1$ as in Eqs. (4.49) and (4.14). The generalization of these expressions to the case of anomalous splitting patterns when there is complete redistribution between the Zeeman sublevels can be achieved directly if we replace H_q by ϕ_q of Eq. (6.37) (while accounting for the $\sqrt{\pi}$ difference in normalization) and do a corresponding replacement for F_q. As we did when introducing $\mathcal{H}_q$ in Eq. (4.14), we can generalize ϕ_q to a complex function that includes anomalous dispersion in its imaginary part. We call this generalized, complex profile function Φ_q. Its normalization to unit area makes it in the non-magnetic case become the usual, unpolarized line profile function φ_ν (cf. Sect. 5.17), which allows the conventional non-LTE equation for unpolarized radiation,

$$\frac{\mathrm{d}I_\nu}{\mathrm{d}s} = -(N_i B_{ij} - N_j B_{ji})\,\frac{h\nu}{4\pi}\,\varphi_\nu I_\nu + \frac{h\nu}{4\pi}\,\varphi_\nu N_j A_{ji}\,, \tag{6.51}$$

to be retrieved when the magnetic field vanishes, as we will see below. The three terms on the right hand side of Eq. (6.51) represent absorption, stimulated emission, and spontaneous emission, respectively. This standard formulation of the transfer equation (cf. Mihalas, 1978) follows directly from the definitions of the Einstein coefficients B_{ij}, B_{ji}, and A_{ji}.

When we have deviations from complete redistribution among the Zeeman sublevels, we have to introduce the diagonal density matrix elements ρ_{MM} as weights in Eq. (6.37). In this case the complex profile function becomes

$$\varPhi_{q,J_\ell} = \frac{N(2J_\ell+1)}{N_{J_\ell}\sqrt{\pi}\Delta\nu_D} \sum_{M_\ell,M_u} \rho_{M_\ell M_\ell}\, S_q(M_\ell,M_u)\,\mathcal{H}_q \tag{6.52}$$

for the J_ℓ level, and similarly for the function $\varPhi_{q,J_u}$ that represents the J_u level. The normalization factor in Eq. (6.52) assures that

$$\int_0^\infty \varPhi_{q,J_{\ell,u}}\,\mathrm{d}\nu = 1\,. \tag{6.53}$$

With this definition we see from Eq. (6.49) that in the case of complete redistribution among the sublevels, ϕ_q of Eq. (6.37) is retrieved from $\varPhi_{q,J_{\ell,u}}$ as its real part.

These definitions allow us to write the non-LTE equation for the transfer of polarized radiation in the matrix form

$$\frac{\mathrm{d}\boldsymbol{I}_\nu}{\mathrm{d}s} = -(N_{J_\ell}B_{J_\ell J_u}\boldsymbol{\varPhi}_{J_\ell} - N_{J_u}B_{J_u J_\ell}\,\boldsymbol{\varPhi}_{J_u})\,\frac{h\nu}{4\pi}\,\boldsymbol{I}_\nu\; + \boldsymbol{j}\,, \tag{6.54}$$

where in the case of non-scattering processes the emission vector can be approximated by (cf. Eq. (5.92))

$$\boldsymbol{j} = \frac{h\nu}{4\pi}\,\boldsymbol{\varPhi}_{J_u}\mathbf{1}\,N_{J_u}A_{J_u J_\ell}\,. \tag{6.55}$$

The 4×4 matrix profile functions $\boldsymbol{\varPhi}_{J_{\ell,u}}$ introduced in Eqs. (6.54) and (6.55) are obtained from the complex scalar profile functions $\varPhi_{q,J_{\ell,u}}$ defined in Eq. (6.52) as follows: Similar to the definition

$$\mathcal{H}_q = H_q - 2iF_q \tag{6.56}$$

obtained from Eq. (4.14), we define an anomalous dispersion profile $\psi_{q,J}$ via

$$\varPhi_{q,J} = \phi_{q,J} - i\psi_{q,J}\,. \tag{6.57}$$

Note that we have omitted a factor of 2 in the imaginary part in this definition, to obtain a simpler expression for the profile matrix $\boldsymbol{\varPhi}$. Analogous to the definitions of $n_{I,Q,U,V}$ in Eq. (3.74) and $H_{I,Q,U,V}$ in Eq. (4.48), we define

$$
\begin{aligned}
\phi_I &= \phi_\Delta \sin^2\gamma + \tfrac{1}{2}(\phi_+ + \phi_-)\,,\\
\phi_Q &= \phi_\Delta \sin^2\gamma \cos 2\chi\,,\\
\phi_U &= \phi_\Delta \sin^2\gamma \sin 2\chi\,,\\
\phi_V &= \tfrac{1}{2}(\phi_+ - \phi_-)\cos\gamma\,,\\
\phi_\Delta &= \tfrac{1}{2}[\,\phi_0 - \tfrac{1}{2}(\phi_+ + \phi_-)]\,,
\end{aligned} \tag{6.58}
$$

and correspondingly for $\psi_{I,Q,U,V}$. We then see from Eqs. (4.34) and (4.50) that we can write

$$
\boldsymbol{\Phi} = \begin{pmatrix} \phi_I & \phi_Q & \phi_U & \phi_V \\ \phi_Q & \phi_I & \psi_V & -\psi_U \\ \phi_U & -\psi_V & \phi_I & \psi_Q \\ \phi_V & \psi_U & -\psi_Q & \phi_I \end{pmatrix}, \tag{6.59}
$$

where we have left out index $J_{\ell,u}$, which should be applied to every component of Eq. (6.59) to obtain $\boldsymbol{\Phi}_{J_{\ell,u}}$. In the special case of a normal Zeeman triplet with complete redistribution among the sublevels the $\boldsymbol{\Phi}$ of Eq. (6.59) is the same as the $\boldsymbol{\Phi}$ introduced in Eq. (5.29) and used throughout Chapter 5.

In the complete redistribution case we can omit the $J_{\ell,u}$ index, since we then have $\boldsymbol{\Phi}_{J_\ell} = \boldsymbol{\Phi}_{J_u} = \boldsymbol{\Phi}$. This gives us

$$
\frac{\mathrm{d}\boldsymbol{I}_\nu}{\mathrm{d}s} = -(N_{J_\ell}B_{J_\ell J_u} - N_{J_u}B_{J_u J_\ell})\,\frac{h\nu}{4\pi}\,\boldsymbol{\Phi}\boldsymbol{I}_\nu + \frac{h\nu}{4\pi}\,\boldsymbol{\Phi}\mathbf{1}\,N_{J_u}A_{J_u J_\ell}\,. \tag{6.60}
$$

From this expression we immediately see how the standard, unpolarized scalar transfer equation (6.51) is retrieved as a special case of our more general matrix equation when the magnetic field vanishes. We also see how the forms of the absorption and emission vectors in Eq. (6.60) agree with the corresponding vectors in the LTE transfer equation (4.50).

6.7. Relations between the Rate Coefficients

Let us introduce the statistical weight g_J of a level:

$$
g_J = 2J + 1\,. \tag{6.61}
$$

We then obtain for the case of complete redistribution, as follows from Eq. (6.49),

$$
N_{J_\ell}B_{J_\ell J_u} = N\rho_{M_\ell M_\ell}\, g_{J_\ell}\, B_{J_\ell J_u}\,. \tag{6.62}
$$

Since $N\rho_{M_\ell M_\ell}$ is the population of a given sublevel, the rate per unit population density is proportional to $g_{J_\ell}\, B_{J_\ell J_u}$. Due to the symmetry properties of the atomic system (see explicit expressions in the next chapter), this rate is the same in the upward and downward directions, from which follows that

$$
g_{J_\ell}\, B_{J_\ell J_u} = g_{J_u}\, B_{J_u J_\ell} \tag{6.63}
$$

(or, in the more common notation, $g_iB_{ij} = g_jB_{ji}$). Eq. (6.63) provides the relation between the coefficients for absorption and stimulated emission.

In the case of thermodynamic equilibrium, there are no intensity gradients ($\mathrm{d}I_\nu/\mathrm{d}s = 0$), and the intensity I_ν is given by the Planck function B_ν. It then follows from the radiative transfer equation (6.51) that

$$\frac{N_j A_{ji}}{N_i B_{ij} - N_j B_{ji}} = B_\nu = \frac{2h\nu^3}{c^2} \frac{1}{e^{h\nu/kT} - 1} \,. \tag{6.64}$$

As Boltzmann statistics applies in thermodynamic equilibrium,

$$\frac{N_j}{N_i} = \frac{g_j}{g_i} \, e^{-h\nu/kT} \,, \tag{6.65}$$

which when used in Eq. (6.64) gives the relation between the coefficients for spontaneous and stimulated emission:

$$A_{ji} = \frac{2h\nu^3}{c^2} \, B_{ji} \,. \tag{6.66}$$

This is basically the way in which Einstein derived this relation.

It is in fact possible to make a "classical" derivation of the factor $2h\nu^3/c^2$ that represents the ratio between A_{ji} and B_{ji}, without making use of the concept of thermodynamic equilibrium, and in spite of the circumstance that this factor contains Planck's constant. This derivation will be done in the next section.

Let us however first conclude this section by deriving the relation between the collisional rate coefficients C_{ij} (for collisional excitation) and C_{ji} (for collisional deexcitation), which will be needed later for the formulation of the statistical equilibrium equations. In thermodynamic equilibrium the balance between the upward and downward transition rates is satisfied independently for the radiative and collisional processes. The detailed balance condition for the collisional processes gives

$$N_i C_{ij} = N_j C_{ji} \,. \tag{6.67}$$

Using the Boltzmann relation (6.65), we get the desired relation

$$g_i C_{ij} = g_j C_{ji} \, e^{-h\nu/kT} \,. \tag{6.68}$$

6.8. Semi-Classical Derivation of the Factor $2h\nu^3/c^2$

Classically, spontaneous emission corresponds to the dipole radiation from an oscillating charge. Radiative absorption on the other hand is due to the work done by the radiation field on the oscillating charge, a process that can be expressed in terms of the imaginary part of a complex refractive index. The classical theory for these two processes was developed in Chapter 3, from which "classical" expressions for the rate coefficients A_{ji} and B_{ji} can be derived, as will be shown below. The ratio between these coefficients then gives us the factor $2h\nu^3/c^2$. As will be seen, the classically derived ratio is actually $2\Delta E/\lambda^2$, where ΔE is the energy difference between the two atomic levels considered. Planck's constant enters by making the identification $\Delta E = h\nu$.

Let $\mathcal{E}_A$ be the energy density of the oscillating dipoles. The classically derived rate of energy loss $\mathrm{d}\mathcal{E}_A/\mathrm{d}t$ per unit volume due to dipole radiation has been given by Eq. (3.14). The relation between this quantity and the rate coefficient A_{21} (if we consider the two levels $j = 2$ and $i = 1$) is *by definition*

$$\frac{\mathrm{d}\mathcal{E}_A}{\mathrm{d}t} = -N_2 A_{21}\,\Delta E\,. \tag{6.69}$$

N_2 is the number density of atoms excited to state no. 2, $N_2 A_{21}$ represents the number of transitions per unit time and volume, and ΔE is the energy per transition (energy difference between levels 1 and 2).

From the definition of ΔE it follows that the excitation energy per unit volume is

$$E_{\text{exc}} = N_2\,\Delta E\,. \tag{6.70}$$

Thus

$$\frac{\mathrm{d}\mathcal{E}_A}{\mathrm{d}t} = -A_{21} E_{\text{exc}}\,. \tag{6.71}$$

A classical harmonic oscillator with oscillation amplitude $x_0 = d_0/e$, where d_0 is the amplitude of the electric dipole moment, has a total energy (which we can regard as its excitation energy) of

$$E_{\text{exc}} = \tfrac{1}{2} m\,\omega^2\, d_0^2/e^2\,. \tag{6.72}$$

The time average (over an oscillation period) of the energy loss due to dipole radiation is according to Eq. (3.14)

$$\frac{\mathrm{d}\mathcal{E}_A}{\mathrm{d}t} = -\frac{\mu_0}{12\pi} k\,\omega^3 d_0^2\,, \tag{6.73}$$

where $k = \omega/c$ is the wave number. From Eqs. (6.71)–(6.73) we can solve for A_{21}, and thus obtain the classical value for the spontaneous transition rate:

$$A_{21,\text{classical}} = \frac{e^2\omega^2}{6\pi\epsilon_0 m c^3}\,. \tag{6.74}$$

Note that μ_0 has been replaced by $1/(\epsilon_0 c^2)$ to make the expression conform with standard choices of variables in the literature. Also note that the above expression does not contain Planck's constant.

Let us now turn to radiative absorption. Electric currents (due to matter) with density $\boldsymbol{j}$ interacting with an electric field $\boldsymbol{E}$ (due to radiation) cause Joule energy losses at the rate $\boldsymbol{j}\cdot\boldsymbol{E}$ per volume element according to Eq. (2.5). If $\boldsymbol{E}$ is a complex, plane electromagnetic wave, the time-averaged Joule loss rate is $\mathrm{Re}\,\boldsymbol{j}\cdot\boldsymbol{E}^*/2$ according to Eq. (2.6). The current density $\boldsymbol{j}$ is given by $\dot{\boldsymbol{d}}$ according to Eqs. (2.20) and (2.23). $\boldsymbol{E}$ consists of a real amplitude $\boldsymbol{E}_0$ that is not affected by the complex conjugation, and a phase factor. As according to Eqs. (2.25) and (2.18) $\dot{\boldsymbol{d}}$ and $\boldsymbol{E}$ are proportional to each other, the phase factors in the product $\boldsymbol{j}\cdot\boldsymbol{E}^*$ cancel each other, leaving us with a scalar product between the two real, linear vector amplitudes $\boldsymbol{j}_0$ and $\boldsymbol{E}_0$ (we use the notation $\boldsymbol{j}_0$ instead of $\dot{\boldsymbol{d}}_0$ to represent the amplitude of $\dot{\boldsymbol{d}}$, since the latter notation has a dot that may give the wrong impression that we are making a time derivative of a constant amplitude). Decomposing these linear vectors in their spherical vector components (cf. Sect. 3.6), we obtain for the loss rate of the energy density $\mathcal{E}_R$ in the radiation field due to absorption:

$$\frac{\mathrm{d}\mathcal{E}_R}{\mathrm{d}t} = -\tfrac{1}{2}\,\mathrm{Re}\left(\sum_q j_{0q} E_{0q}^*\right). \tag{6.75}$$

From Eqs. (2.25) or (3.48) and (3.41) we get

$$j_{0q} = -2i\epsilon_0\,\omega\,(n_q - 1)E_{0q}\,, \tag{6.76}$$

which when inserted in Eq. (6.75) gives

$$\frac{\mathrm{d}\mathcal{E}_R}{\mathrm{d}t} = -\epsilon_0\omega \sum_q |E_{0q}|^2\,\mathrm{Im}\,(n_q - 1)\,. \tag{6.77}$$

Next we need to express the square of the electric vector, $\sum_q |E_{0q}|^2$, in terms of the mean intensity J_ν, and insert the previously derived solution for n_q in Eq. (6.77). First we note that the energy density u_ν of the radiation field per unit frequency interval is given by

$$u_\nu\,\mathrm{d}\nu = \tfrac{1}{2}\epsilon_0 \sum_q |E_{0q}|^2 \tag{6.78}$$

for a monochromatic wave with a frequency within the infinitesimal interval $\mathrm{d}\nu$, according to Eq. (2.6). (Note that the amplitudes of the electric and magnetic energy terms in Eq. (2.6) are equally large for an electromagnetic wave, which means that the energy averaged over an oscillation period equals the energy amplitude for one of the two terms, e.g. the electric term.) The energy density u_ν is related to the mean intensity J_ν by

$$u_\nu = \frac{4\pi}{c} J_\nu\,, \tag{6.79}$$

where J_ν is the mean intensity defined by

$$J_\nu = \int I_{\nu'} \frac{\mathrm{d}\Omega'}{4\pi}\,, \tag{6.80}$$

which may be considered as a special case of the general matrix definition given by Eq. (5.85).

To compute the energy absorption rate $\mathrm{d}\mathcal{E}_R/\mathrm{d}t$ in Eq. (6.77) we need to insert the frequency-dependent expression for n_q that was derived in Sect. 3.8, and then sum (integrate) over the contributions from all the monochromatic wave trains (i.e., over the independent oscillation modes or Fourier components of the electromagnetic field). The imaginary part of the classical solution for n_q given by Eq. (3.51) or (3.54) can be written in the form

$$\operatorname{Im}(n_q - 1) = \frac{e^2 N_1}{8\epsilon_0 m\omega} \varphi_\nu , \tag{6.81}$$

where N_1 is the number density of absorbing atoms (in the lower state), and φ_ν is the frequency profile that satisfies the usual normalization condition

$$\int \varphi_\nu \, \mathrm{d}\nu = 1 . \tag{6.82}$$

Inserting Eqs. (6.78)–(6.81) in Eq. (6.77) and integrating over all frequencies, we obtain

$$\frac{\mathrm{d}\mathcal{E}_R}{\mathrm{d}t} = -\frac{\pi e^2 N_1}{\epsilon_0 c m} \int \varphi_\nu J_\nu \, \mathrm{d}\nu . \tag{6.83}$$

An alternative expression for $\mathrm{d}\mathcal{E}_R/\mathrm{d}t$ in terms of the Einstein rate coefficient B_{12} for radiative absorption can be given, since the rate per atom for radiative excitation is $B_{12} \int \varphi_\nu J_\nu \, \mathrm{d}\nu$ *by definition.* Multiplication with N_1 gives the excitation rate per volume element, and further multiplication with the energy ΔE per excitation gives the energy loss rate $\mathrm{d}\mathcal{E}_R/\mathrm{d}t$:

$$\frac{\mathrm{d}\mathcal{E}_R}{\mathrm{d}t} = -N_1 \left(B_{12} \int \varphi_\nu J_\nu \mathrm{d}\nu \right) \Delta E . \tag{6.84}$$

Identification of Eq. (6.84) with Eq. (6.83) gives the "classical" expression for the Einstein absorption-rate coefficient:

$$B_{12,\mathrm{classical}} = \frac{\pi e^2}{\epsilon_0 c m \Delta E} . \tag{6.85}$$

As the excited atomic oscillator has three superposed substates, corresponding to the three degrees of freedom or the three dipole moment vector components labeled by the three q values, while the ground state has only a single substate or one degree of freedom (the classical case, which corresponds to a $J_\ell = 0 \rightarrow J_u = 1$ transition), it follows from Eq. (6.63) that the rate coefficient for transitions in the opposite direction (stimulated emission) is

$$B_{21,\mathrm{classical}} = B_{12,\mathrm{classical}}/3 . \tag{6.86}$$

We can now combine Eq. (6.74) with Eqs. (6.86) and (6.85) to obtain

$$A_{21,\mathrm{classical}}/B_{21,\mathrm{classical}} = 2\Delta E/\lambda^2 , \tag{6.87}$$

where we have used the relation $\omega = 2\pi c/\lambda$.

If we make the identification

$$\Delta E = h\nu \tag{6.88}$$

for the energy of a single transition, then

$$\frac{A_{21,\mathrm{classical}}}{B_{21,\mathrm{classical}}} = \frac{2h\nu^3}{c^2} , \tag{6.89}$$

as previously obtained in Eq. (6.66) via the use of Planck's law. We note that the present derivation of Eq. (6.89) does not make use of Boltzmann statistics, Planck's law, or detailed radiative balance between the atomic levels. Instead, it simply uses the classical expressions for dipole radiation and complex refractive index, but adds the identification (6.88), via which Planck's constant enters.

6.9. Phenomenological Treatment of Emission

In the previous chapter, Eq. (5.87), the emission vector $\boldsymbol{j}_L$ was described as the sum of three separate vectors: $\boldsymbol{j}_{\rm coh}$, representing coherent scattering processes, $\boldsymbol{j}_{\rm ic}$, representing incoherent scattering (when the phases of the Zeeman sublevels are scrambled due to a collision during the excited state), and $\boldsymbol{j}_{\rm ns}$, representing non-scattering emission processes. When scattering processes ($\boldsymbol{j}_{\rm coh}$ and $\boldsymbol{j}_{\rm ic}$) can be neglected, the emission vector may be approximated by Eq. (6.55) (cf. also Eq. (5.92)). It would thus have the form $\boldsymbol{j} \sim \boldsymbol{\Phi}\mathbf{1}$, which is the form it also has in the case of incoherent scattering, as shown by Eq. (5.91).

To phenomenologically separate non-scattering and scattering processes from each other, we need to estimate the fraction α of all the spontaneous emission processes that are part of scattering transitions between levels J_ℓ and J_u. To make this estimate we disregard atomic polarization (off-diagonal terms in the atomic density matrix and deviations from complete redistribution between the Zeeman sublevels), which allows us to express α as

$$\alpha = N_{J_\ell}\, B_{J_\ell J_u} \int \varphi_\nu J_\nu \,\mathrm{d}\nu \Big/ \sum_k N_k\, R_{k J_u} \,. \tag{6.90}$$

It represents the ratio between the radiative absorption rate from the lower level J_ℓ to the upper level J_u (the nominator) and the total transition rate to level J_u from all other levels k by any process (the denominator). The rate coefficient $R_{k J_u}$ implicitly includes all collisional and radiative transitions both from levels lower and higher than J_u.

As the fraction of non-scattering spontaneous emission processes assumed to come from an unpolarized J_u state is $1-\alpha$, we see from Eq. (6.55) that

$$\boldsymbol{j}_{\rm ns} = (1-\alpha)\,\frac{h\nu}{4\pi}\,\boldsymbol{\Phi}_{J_u}\mathbf{1}\,N_{J_u}\,A_{J_u J_\ell} \,. \tag{6.91}$$

The scattering emission vector, regardless of whether it is coherent or incoherent, can formally be written as in Eq. (5.8):

$$\boldsymbol{j}_{\rm sc} = \sigma \int \frac{\mathrm{d}\Omega'}{4\pi} \int \mathrm{d}\nu' \,\boldsymbol{R}\boldsymbol{I}_{\nu'} \,. \tag{5.8}$$

$\boldsymbol{R}$ is a normalized redistribution matrix. If all the radiative absorption processes were followed by a spontaneous transition back to the initial level (i.e., were part of Rayleigh scattering processes), then the scattering cross section σ would be given by $N_{J_\ell} B_{J_\ell J_u}\,(h\nu/4\pi)$. However, of all the transitions away from level J_u, only the spontaneous transitions to level J_ℓ are part of scattering events. If we disregard atomic polarization when calculating this fraction (as we did for α), then we obtain for the scattering cross section

$$\sigma = \frac{h\nu}{4\pi} N_{J_\ell} B_{J_\ell J_u} N_{J_u} A_{J_u J_\ell} \Big/ \sum_k N_{J_u} R_{J_u k} \,. \tag{6.92}$$

Statistical equilibrium in the population of level J_u requires that

$$\sum_k N_k R_{k J_u} = \sum_k N_{J_u} R_{J_u k} \,. \tag{6.93}$$

The left-hand side represents the transition rate to level J_u (the gains), the right-hand side the rate away from level J_u (the losses). Eq. (6.93) shows that the denominators in expressions (6.90) and (6.92) for α and σ are equal, which allows us to express the cross section in terms of α:

$$\sigma = \alpha \frac{h\nu}{4\pi} N_{J_u} A_{J_u J_\ell} \Big/ \int \varphi_\nu J_\nu \, \mathrm{d}\nu \,. \tag{6.94}$$

As concerns the division of the scattering processes in coherent and incoherent parts, we recall that according to Eq. (5.67) the fraction k_c of the scattering processes that are undisturbed by collisions and therefore are coherent is given by

$$k_c = \frac{\gamma_N}{\gamma_N + \gamma_c} \tag{5.67}$$

in the dispersion wings, where γ_N and γ_c represent the radiative and collisional damping constants, respectively. $\boldsymbol{j}_{\mathrm{coh}}$ is therefore obtained from Eq. (5.8) through multiplication of the right-hand side with k_c, while we for $\boldsymbol{R}$ use the redistribution matrix $\boldsymbol{R}_{\mathrm{coh}}$ for coherent scattering. $\boldsymbol{j}_{\mathrm{ic}}$ is obtained through multiplication with $1 - k_c$ and use of the redistribution matrix $\boldsymbol{R}_{\mathrm{ic}}$ for incoherent scattering.

The rest-frame redistribution matrix $\boldsymbol{R}_{0,\mathrm{coh}}$ for coherent scattering is according to Eq. (5.11)

$$\boldsymbol{R}_{0,\mathrm{coh}} \sim \boldsymbol{T}\boldsymbol{W}\boldsymbol{T}^{-1}\, \delta(\xi - \xi') \,. \tag{6.95}$$

In the classical theory $\boldsymbol{W} = \boldsymbol{w} \otimes \boldsymbol{w}^*$, where $\boldsymbol{w}$ is the Jones matrix for radiative scattering that was given by Eq. (3.85). This expression can be generalized to any quantum-mechanical configuration if we disregard off-diagonal terms in the density matrix of the lower, initial state, and let the scattering amplitudes $\boldsymbol{w}$ be given by the Kramers-Heisenberg dispersion formula. If we let index β symbolize the polarization state of the incident photons, α that of the scattered photons, as we did when calculating the scattering Jones matrix $\boldsymbol{w}$ in Sect. 3.12, then the matrix elements of $\boldsymbol{w}$ that represent the scattering amplitudes from a given initial substate characterized by quantum numbers J_ℓ and M' into a final substate characterized by J_ℓ and M are given by

$$w_{\alpha\beta} \sim \sum_{J_u} \sum_{M_u} \frac{\langle J_\ell M \, | \hat{\boldsymbol{r}} \cdot \boldsymbol{e}_\alpha | \, J_u M_u \rangle \langle J_u M_u \, | \hat{\boldsymbol{r}} \cdot \boldsymbol{e}_\beta | \, J_\ell M' \rangle}{\omega_{M_u M} - \omega - i\gamma/2} \,. \tag{6.96}$$

To be as general as possible, we have included not only the coherent superposition of the different Zeeman sublevels M_u (the second sum), but also the coherent superposition of states of different total angular momentum J_u (the first sum). We will encounter examples of such interference later.

$\boldsymbol{e}_{\alpha,\beta}$ are the unit polarization vectors, while the position operator $\hat{\boldsymbol{r}}$ is proportional to the dipole moment vector operator $-e\hat{\boldsymbol{r}}$. Derivations of Eq. (6.96) will be given in Chapters 8 and 9.

$$\omega_{M_u M} = \omega_0 + \omega_L(g_u M_u - g_\ell M) \tag{6.97}$$

(cf. Eq. (6.32)).

To obtain the elements of the $\boldsymbol{W}$ matrix in Eq. (6.95) we have to sum $\boldsymbol{w} \otimes \boldsymbol{w}^*$ over all the possible final M substates that are allowed in the emission process, as well as over all the initial M' substates, weighted by the population of the respective initial substate (these populations may deviate from complete redistribution). The matrix elements of $\boldsymbol{W}$ are thus proportional to

$$\sum_{M'} N_{J_\ell M'} \sum_{M} w_{\alpha\beta} w^*_{\alpha'\beta'} \tag{6.98}$$

(for the organization of these elements in the $\boldsymbol{W}$ 4×4 matrix, see Eq. (2.39)).

When going from the rest frame to the observer's frame to obtain the coherent redistribution matrix $\boldsymbol{R}_{\rm coh}$, we have to introduce Doppler shifts and integrate over a Maxwellian velocity distribution as indicated in Eqs. (5.9) and (5.10). For coherent scattering it is in general not practical to write down an explicit, analytical expression for the general case, because of the complicated coupling between polarization properties, angular redistribution, and Zeeman splitting. In two special cases it is however possible to factorize the redistribution matrix into two parts, one matrix factor $\boldsymbol{P}$ that is frequency independent and only deals with the transformation of the state of polarization, and one scalar factor that exclusively deals with the frequency redistribution, independent of the polarization state. This factorization was introduced in Eq. (5.42). The polarization matrix is the phase matrix $\boldsymbol{P}$, normalized according to Eq. (5.43) and explicitly given by Eqs. (5.68) and (5.58)–(5.64) for the case of weak magnetic fields. The two special wavelength regimes for which this factorization is possible are: (1) Line core, as long as the fields are weak ($\Delta\nu_H \ll \Delta\nu_D$). $\boldsymbol{P}$ is then given by Eq. (5.68). (2) Dispersion wings of the line, for any field strength as long as $|\Delta\nu| \gg \Delta\nu_H$. $\boldsymbol{P}$ for the zero-field case should then be used.

The redistribution treatment in Chapter 5 which was based on the classical theory, and which for the special cases mentioned above led to the compact analytical expression (5.96) for the coherent emission vector $\boldsymbol{j}_{\rm coh}$, is also valid in the general quantum-mechanical case (if we disregard the off-diagonal density matrix elements of the lower atomic level). It will be shown in Chapter 8 that the phase matrix $\boldsymbol{P}$ acquires the form given by Eqs. (5.68) and (5.58)–(5.64) in the weak-field limit. The purely classical treatment had given the form of Eq. (5.57), but we generalized this expression to Eq. (5.68) by introducing the parameters W_1 and W_2, thereby covering all the cases that can occur quantum-mechanically. $W_{1,2}$ can only be derived quantum-mechanically, which will be done in Chapter 8. They can be given as algebraic expressions in terms of the J_u and J_ℓ quantum numbers of the upper and lower levels of the transition, as done by Chandrasekhar (1950) for Rayleigh scattering transitions.

Turning now to the incoherent scattering case, the expressions drastically simplify, since collisions decouple the absorption and emission processes. As was shown in Eqs. (5.27) and (5.32), we then get

$$\boldsymbol{R}_{\rm ic} = (\boldsymbol{\Phi}\,\mathbf{1})_{\rm em}\,(\mathbf{1}^T\,\boldsymbol{\Phi})_{\rm abs}\,. \tag{6.99}$$

Integrating over all the incident frequencies and directions, we obtain as in Eqs. (5.33) and (5.91)

$$\boldsymbol{j}_{\rm ic} = (1-k_c)\,\sigma e\,\boldsymbol{\Phi}\mathbf{1}\,, \tag{5.91}$$

where e is given by Eq. (5.34):

$$e = \int \frac{{\rm d}\Omega'}{4\pi} \int {\rm d}\nu' \,(\Phi'_I I_{\nu'} + \Phi'_Q Q_{\nu'} + \Phi'_U U_{\nu'} + \Phi'_V V_{\nu'})\,. \tag{5.34}$$

$\Phi_{I,Q,U,V} = \phi_{I,Q,U,V}$ are the components of the vector $\boldsymbol{\Phi}\mathbf{1}$ and have been defined by Eqs. (6.58) and (6.59).

6.10. Polarized Transfer Equation with Scattering

Non-LTE effects become prominent for strong lines, which are formed high in the solar atmosphere, while the LTE assumption works well for lines formed in the lower and middle photosphere. The non-LTE transfer theory that we are developing is therefore needed when we want to diagnose the magnetic-field structure in the higher atmospheric layers. An example of a strong and diagnostically useful line is shown in Fig. 6.4.

We summarize our present results by formulating a polarized transfer equation that includes coherent and incoherent scattering, non-scattering emission, continuous absorption, and an optical depth scale. The simplest form that also conforms with the standard formalism of unpolarized (scalar) non-LTE radiative transfer is obtained if we make a generalization of the previously derived Eq. (6.60) and include our results for the scattering $\boldsymbol{j}$ vector. The assumption behind Eq. (6.60) that we have complete redistribution of the populations of the Zeeman sublevels in the absorption-stimulated emission term allows us to introduce a line absorption coefficient κ_L of the form

$$\kappa_L = (N_\ell B_{\ell u} - N_u B_{u\ell})\,\frac{h\nu}{4\pi} \tag{6.100}$$

and a line source function $S_{u\ell}$ as

$$S_{u\ell} = \frac{N_u A_{u\ell}}{N_\ell B_{\ell u} - N_u B_{u\ell}}\,. \tag{6.101}$$

We further assume that the continuum can be described by a scalar absorption coefficient κ_c, and that the continuum emission occurs in LTE with emission vector $B_\nu\mathbf{1}$, where B_ν is the Planck function. Let us define the differential optical depth

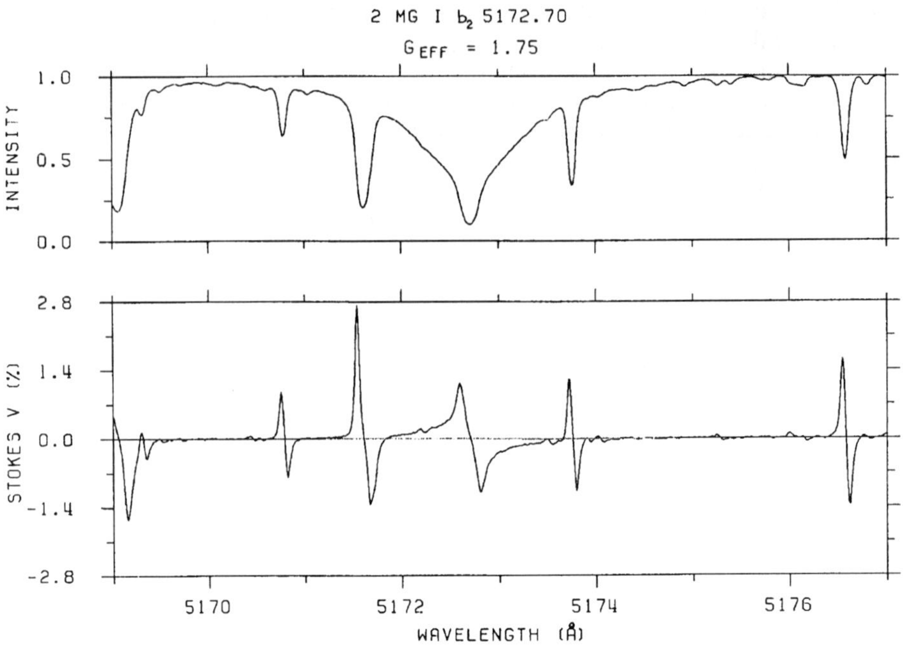

Fig. 6.4. Stokes I and V profiles of the Mg I b_2 5172.7 Å line, observed at the center of the solar disk. Deviations from LTE become large for strong lines like this one, needed for the diagnostics of the higher layers of the solar atmosphere. From Stenflo et al. (1984).

$$\mathrm{d}\tau_\nu = -\kappa_L \,\mathrm{d}s \tag{6.102}$$

and the absorption coefficient ratio

$$r = \kappa_c/\kappa_L\,. \tag{6.103}$$

The generalized Eq. (6.60) then becomes, if we include the continuous emission vector of Eq. (5.86) as in Eq. (5.95) and use Eqs. (5.88) and (6.94),

$$\begin{aligned}\frac{\mathrm{d}\boldsymbol{I}_\nu}{\mathrm{d}\tau_\nu} =&(\boldsymbol{\Phi} + r\boldsymbol{E})\boldsymbol{I}_\nu - (S_L\boldsymbol{\Phi} + rB_\nu)\mathbf{1} - \sigma_c \boldsymbol{J}_{\nu,\mathrm{cl}}/\kappa_L \\ &- \alpha k_c S_{u\ell} \int \frac{\mathrm{d}\Omega'}{4\pi} \int \mathrm{d}\nu' \,\boldsymbol{R}_{\mathrm{coh}} \boldsymbol{I}_{\nu'} \Big/ \int \varphi_\nu J_\nu \,\mathrm{d}\nu\,,\end{aligned} \tag{6.104}$$

where

$$S_L = S_{u\ell}\left[1 - \alpha + \alpha e\,(1 - k_c)\Big/ \int \varphi_\nu J_\nu \,\mathrm{d}\nu\right] \tag{6.105}$$

as follows from Eqs. (5.91) and (6.94).

If more than one spectral line contributes at frequency ν, the contributions from the various lines to absorption and emission should first be added in the transfer equation (6.60) that is based on a geometrical depth scale, before an optical depth scale and line source functions are defined, but the general procedure is the same as used to obtain Eq. (6.104).

We note that Eq. (6.104) has the same form as the classically derived transfer equation (5.95), except that the optical depth scale used in Eq. (6.104) is based on the line absorption coefficient instead of the continuous absorption coefficient. In contrast to Eq. (5.95) we have now been able to obtain explicit expressions for the various parameters in terms of the atomic transition rates. Let us also note that although the assumption of complete redistribution between the Zeeman sublevels has been used to define an optical depth scale and a source function $S_{u\ell}$, this assumption is *not* used for the calculation of the polarization effects due to coherent scattering. When calculating the matrix elements (6.98) to obtain the coherent redistribution matrix $\boldsymbol{R}_{\rm coh}$, any deviations from complete redistribution of the sublevel populations are allowed. The only restrictive assumption is that there are no lower-level coherences, i.e., the off-diagonal terms of the density matrix for the lower atomic level are zero. The effect of lower-level coherences are small in most practical situations, but Eq. (6.98) can be generalized to include them, as will be shown in Chapter 8.

If we disregard atomic polarization when solving the statistical equilibrium equations for the level populations, we obtain in the case of a 2-level model atom

$$S_{u\ell} = \frac{\int \varphi_\nu J_\nu \,\mathrm{d}\nu + \epsilon B_\nu}{1+\epsilon} \tag{6.106}$$

with

$$\epsilon = \frac{C_{u\ell}}{A_{u\ell}} \left(1 - e^{-h\nu/kT}\right) , \tag{6.107}$$

where $C_{u\ell}$ is the collisional deexcitation rate coefficient. This is a well-known result from unpolarized non-LTE transfer theory (cf. Mihalas, 1978).

From the 2-level model further follows that

$$\alpha = \int \varphi_\nu J_\nu \,\mathrm{d}\nu \Big/ \left(\int \varphi_\nu J_\nu \,\mathrm{d}\nu + \epsilon B_\nu\right) . \tag{6.108}$$

Inserting Eqs. (6.106) and (6.108) in Eq. (6.105), we get

$$S_L = \frac{e\,(1-k_c) + \epsilon B_\nu}{1+\epsilon} . \tag{6.109}$$

The approximation of an unpolarized ground state and unpolarized rate equations is useful also when coherent scattering is included in the transfer problem, as we have seen above, and it greatly simplifies the transfer problem. The above version of the transfer equation, for a 2-level atom in the case when $k_c = 0$ (no coherent scattering), has been used by Stenholm and Stenflo (1978) to treat the multi-dimensional transfer of polarized radiation in small-scale solar magnetic flux tubes.

6.11. Non-LTE Transfer of the Coherency Matrix

In the previous sections the non-LTE transfer problem has been formulated in terms of the Stokes parameters and Mueller matrices. A formulation in terms of the 2×2 complex coherency matrix $\boldsymbol{D}$, defined by Eq. (2.33), is however also useful, since in quantum field theory (next two chapters) the polarization of the radiation field is best described in terms of a coherency matrix operator.

In Chapter 4 we developed the classical theory of polarized radiative transfer, both in terms of the coherency matrix and in terms of the Stokes parameters. Comparison between these two formulations, Eqs. (4.41) and (4.50), shows how one can obtain the matrix components in one formulation directly from the components in the other formulation. From such a comparison we can from the Stokes non-LTE transfer equation (6.104) immediately write down the non-LTE coherency matrix transfer equation as

$$\frac{\mathrm{d}\boldsymbol{D}}{\mathrm{d}\tau_\nu} = \tfrac{1}{2}(\boldsymbol{\Phi}_D + r\boldsymbol{\sigma}_0)\boldsymbol{D} + \tfrac{1}{2}\boldsymbol{D}\,(\boldsymbol{\Phi}_D^\dagger + r\boldsymbol{\sigma}_0) - \tfrac{1}{2}(S_L\boldsymbol{\phi}_D + rB_\nu\boldsymbol{\sigma}_0) \tag{6.110}$$

if we disregard coherent and continuous scattering.

$$\boldsymbol{\Phi}_D = \boldsymbol{\phi}_D - i\boldsymbol{\psi}_D\,, \tag{6.111}$$

where

$$\boldsymbol{\phi}_D = \begin{pmatrix} \phi_I + \phi_Q & \phi_U + i\phi_V \\ \phi_U - i\phi_V & \phi_I - \phi_Q \end{pmatrix}. \tag{6.112}$$

We obtain the corresponding expression for $\boldsymbol{\psi}_D$, if we replace ϕ in Eq. (6.112) with ψ (cf. Eqs. (6.57)–(6.59)).

CHAPTER 7

INTRODUCTION TO QUANTUM FIELD THEORY OF POLARIZED RADIATIVE TRANSFER

7.1. Introduction

The quantum theory of the preceding chapter has been introduced in a largely phenomenological way, to provide an overview and intuitive grasp of the subject. The framework of quantum field theory with a quantization of the electromagnetic field should however be used for a consistent treatment and a logical development of the theory of polarized radiative transfer in a magnetic field from first principles.

Quantum field theory is not only needed for a deepened physical understanding, but also for the practical purpose of being able to formulate the statistical equilibrium with atomic polarization and coherences between any combinations of Zeeman sublevels. It allows us to treat coherent scattering when it is Rayleigh or Raman scattered, when it involves arbitrary level crossings (including coherences between states of different total angular momentum), etc. The algebraic expressions of the W_1 and W_2 coefficients that we introduced in the scattering phase matrix, Eq. (5.68), can be derived in terms of the quantum numbers of the transition.

For a general treatment in quantum field theory of a statistical ensemble of matter and radiation, the density matrix formalism is needed. The density matrix is governed by the Liouville equation, which is incorporated in the equation for the evolution of the expectation values, out of which the equations of statistical equilibrium and radiative transfer emerge as special cases.

An overview of the relevant physical concepts of the quantum theory of light can be found in the monograph by Loudon (1983). The first general derivation of the equation of polarized radiative transfer in a magnetic field from first principles in quantum field theory was made by Landi Degl'Innocenti (1983). Our approach is similar to his, although the formalisms used are different.

7.2. The Density Matrix and the Liouville Equation

Starting from the time-dependent Schrödinger equation

$$i\hbar\,\frac{\partial\psi(t)}{\partial t}=\hat{H}\,\psi(t)\,,\tag{7.1}$$

where $\hat{H}$ is the Hamiltonian operator, it is straightforward to make a formal derivation of the Liouville equation. First we note that the formal solution for the wave function in Eq. (7.1) is

$$\psi(t) = e^{-i\hat{H}t/\hbar}\, \psi(0)\,. \tag{7.2}$$

Next we introduce the definition of the density matrix operator $\hat{\rho}$ for a statistical ensemble in formal Dirac notation as

$$\hat{\rho} = |\,\psi(t)\,\rangle\langle\,\psi(t)\,|\,. \tag{7.3}$$

The physical contents of this definition will become clearer later.

Inserting the formal Schrödinger solution (7.2) in Eq. (7.3), we get

$$\hat{\rho} = e^{-i\hat{H}t/\hbar}\, |\,\psi(0)\,\rangle\langle\,\psi(0)\,|\, e^{i\hat{H}t/\hbar}\,. \tag{7.4}$$

Partial derivation with respect to time then gives

$$\frac{\partial\hat{\rho}}{\partial t} = \frac{1}{i\hbar}[\hat{H},\hat{\rho}\,]\,, \tag{7.5}$$

where the square brackets on the right hand side denote a commutator. This is the Liouville equation, which serves as the foundation of quantum statistics, and from which the general statistical equilibrium and radiative transfer equations will be derived.

The word density *matrix* refers to a matrix representation based on an expansion of the wavefunction ψ in terms of the eigenstates of the system. The expansion

$$\psi = \sum_n c_n\,|\,n\rangle \tag{7.6}$$

is normally made in terms of time-independent eigenfunctions $|n\rangle$, which are the solutions of the time-independent Schrödinger equation. In the general case when the Hamiltonian is time dependent (which happens when the atomic system interacts with an electromagnetic field), expansion (7.6) implies that the coefficients c_n are time dependent. As the time dependent (interaction) part of the Hamiltonian is generally small in comparison with the time-independent (undisturbed) part, the evolution of the system can be treated with perturbation theory. As we will see below, it turns out that the 1st order perturbations average out to zero, which makes it necessary to go to 2nd (or higher) order to treat the interaction between matter and radiation.

A formalism in terms of the complex c_n coefficients alone is unable to describe the superposition of states in a statistical ensemble in which the phases of separate events are uncorrelated. We need to use bilinear products between all the various combinations of coefficients, since these products fix the relative phases within each individual quantum system of the ensemble, *before* the different uncorrelated individual quantum systems are statistically superposed.

The reason why we need bilinear products is the same as the reason, stated in Sect. 2.6.1 and elucidated in Sect. 4.1, why Jones vectors are unable to describe a statistical ensemble of uncorrelated photons. In the classical treatment of a statistical ensemble of uncorrelated wave trains, which we are dealing with in radiative-transfer theory, we had to introduce bilinear products of the complex electric vectors, which led us to the definition of the coherency matrix and the

Stokes parameters. In this analogy, the counterpart of the complex state vector formed by the coefficients c_n is the Jones vector, while the counterpart of the density matrix is the classical coherency matrix.

The elements ρ_{mn} of the density matrix operator $\hat{\rho}$ are naturally defined by

$$\rho_{mn} = \langle m \,|\, \hat{\rho} \,|\, n \rangle \,. \tag{7.7}$$

From the definition (7.3) and the expansion (7.6) it then follows that

$$\rho_{mn} = c_m c_n^* \,, \tag{7.8}$$

i.e., the matrix elements are bilinear products of two complex, time-dependent coefficients.

One fundamental use of the density matrix is for the derivation of the expectation value of an operator for a statistical ensemble. The expectation value of the operator $\hat{X}$ is

$$\begin{aligned} \langle \hat{X} \rangle = \langle \psi | \hat{X} \,|\, \psi \rangle &= \sum_{m,n} c_m c_n^* \langle n \,| \hat{X} \,|\, m \rangle \\ &= \sum_{m,n} \rho_{mn} X_{nm} = \mathrm{Tr}\,(\hat{\rho}\hat{X}) = \mathrm{Tr}\,(\hat{X}\hat{\rho}) \,. \end{aligned} \tag{7.9}$$

The trace Tr is the sum of the diagonal elements (of the product matrix $\hat{\rho}\hat{X}$) over the complete set of states. Since the sum of the probabilities is unity, we have the normalization condition

$$\mathrm{Tr}\,\hat{\rho} = 1 \,. \tag{7.10}$$

Note that as the density matrix represents an average over the states of the entire statistical ensemble, which includes a spatial average, we only need to consider its time dependence. We may therefore replace $\partial/\partial t$ by $\mathrm{d}/\mathrm{d}t$. In particular, we will write the Liouville equation as

$$\frac{\mathrm{d}\hat{\rho}}{\mathrm{d}t} = \frac{1}{i\hbar}[\hat{H}, \hat{\rho}] \,. \tag{7.11}$$

7.3. The Schrödinger and Interaction Pictures

A quantum-mechanical system can be described in different *representations*. The expressions for the operators and wave functions in the different representations are connected by unitary transformations. Although all representations give (as they must) identical predictions for all measurable quantities, the choice of a certain representation may make the calculations simpler and more tranparent than they would be in other representations.

The representation used in the preceding section was the so-called Schrödinger picture, in which the operator $\hat{H}$ in the Liouville equation is the total Hamiltonian operator, which is dominated by the time-independent part $\hat{H}_0$. We can write

$$\hat{H} = \hat{H}_0 + \hat{H}' \,, \tag{7.12}$$

where $\hat{H}'$ is the perturbation or interaction Hamiltonian.

A most useful representation for treating perturbation problems is the *interaction representation*, in which the wave functions ψ_I are related to the Schrödinger wave functions ψ by the unitary transformation

$$\psi_I(t) = e^{i\hat{H}_0 t/\hbar}\, \psi(t). \tag{7.13}$$

Note that this is not merely the inverse of Eq. (7.2), since only the unperturbed Hamiltonian $\hat{H}_0$ is involved in the transformation, not the total Hamiltonian.

Differentiation of Eq. (7.13) gives the equation of motion

$$i\hbar \frac{\partial \psi_I}{\partial t} = \hat{H}_I\, \psi_I\,, \tag{7.14}$$

where

$$\hat{H}_I = e^{i\hat{H}_0 t/\hbar}\, \hat{H}'\, e^{-i\hat{H}_0 t/\hbar}. \tag{7.15}$$

Eq. (7.15) indicates the general rule for the transformation of operators from the Schrödinger to the interaction picture. It also follows from Eq. (7.13) that

$$\hat{\rho}_I = e^{i\hat{H}_0 t/\hbar}\, \hat{\rho}\, e^{-i\hat{H}_0 t/\hbar}, \tag{7.16}$$

where $\hat{\rho}_I$ is defined as $\hat{\rho}$ if we in Eq. (7.3) add index I on both sides of the equation.

A general operator $\hat{X}$ that represents an observable is time independent in the Schrödinger picture. If we use the transformation relation (7.15) for $\hat{X}_I$ and differentiate, we obtain the equation governing the time evolution of the operator in the interaction picture:

$$\frac{\mathrm{d}\hat{X}_I}{\mathrm{d}t} = \frac{1}{i\hbar}\,[\hat{X}_I, \hat{H}_0\,]. \tag{7.17}$$

The density matrix operator $\hat{\rho}$ on the other hand is *not* time independent in the Schrödinger picture (it does not represent an observable), but is governed by the Liouville equation (7.11). If this is taken into account when differentiating Eq. (7.16), we obtain the Liouville equation in the interaction picture:

$$\frac{\mathrm{d}\hat{\rho}_I}{\mathrm{d}t} = \frac{1}{i\hbar}[\hat{H}_I, \hat{\rho}_I\,]. \tag{7.18}$$

We see that the dynamics is now entirely determined by the perturbation, while the unperturbed Hamiltonian is not involved at all.

Finally we note that the expectation values are independent of the representation used, as seen explicitly from

$$\langle \hat{X}_I \rangle = \mathrm{Tr}\,(\hat{\rho}_I\, \hat{X}_I) = \mathrm{Tr}\,(\hat{\rho}\, \hat{X}) = \langle \hat{X} \rangle\,. \tag{7.19}$$

7.4. Quantization of the Radiation Field

For a self-consistent treatment not only the matter but also the radiation field has to be quantized. In particular the resulting zero-point energy of the vacuum is essential, without which spontaneous emission processes would not be possible.

Quantization of the electromagnetic field is done via a description of the field fluctuations in terms of the expressions used for a harmonic oscillator. The fundamental difference between classical and quantum physics becomes manifest for a harmonic oscillator, for which the total energy oscillates between the forms of potential energy (described by the position coordinate) and kinetic energy (described by the momentum coordinate). Since in quantum mechanics position and momentum are represented by operators that do not commute (which leads to the Heisenberg uncertainty relation), the lowest energy eigenvalue (ground state) is not zero.

The steps that need to be taken to quantize a classical vector field are: (i) Fourier decomposition of the classical field into discrete wave modes. (ii) Coordinate transformation such that the classical wave equation for the wave modes assumes the same form as that of a harmonic oscillator. This allows us to define a "mode position" and a "mode momentum". (iii) Transition to quantum mechanics by letting the mode position and momentum be represented by non-commuting operators.

A consequence of this procedure is that each quantized vector field has a ground state, the "vacuum state", of non-zero energy. The vacuum fluctuations induce spontaneous radiative deexcitation of the excited atomic states at predictable rates, which are found to agree with the observed rates and with the rates calculated with semi-classical theory, as will be seen below.

According to the quantization procedure, we start by expanding the vector potential $\boldsymbol{A}$ of the classical electromagnetic field in a Fourier series:

$$\boldsymbol{A} = \sum_{\boldsymbol{k}} \left[\boldsymbol{A}_{\boldsymbol{k}}(t)\, e^{i\boldsymbol{k}\cdot\boldsymbol{r}} + \boldsymbol{A}_{\boldsymbol{k}}^{*}(t)\, e^{-i\boldsymbol{k}\cdot\boldsymbol{r}} \right] . \tag{7.20}$$

The sum is taken over all the wave modes in a cubic cavity of space with side L, volume $V = L^3$. The size of the cavity is arbitrary, but to have well-defined discrete modes, we impose periodic boundary conditions such that the wave number vector $\boldsymbol{k}$ has the component values

$$\begin{aligned} k_{x,y,z} &= 2\pi\, n_{x,y,z}/L\,, \\ n_x,\, n_y,\, n_z &= 0, \pm 1, \pm 2, \ldots, \end{aligned} \tag{7.21}$$

assuming that the axes of the Cartesian coordinate system are along the sides of the cube. In the limit when L is much larger than the mode wavelength, the mode sum may be replaced by a mode integral, as will be done later.

The classical wave equation in vacuum is

$$\nabla^2 \boldsymbol{A} - \frac{1}{c^2}\frac{\partial^2 \boldsymbol{A}}{\partial t^2} = 0 \tag{7.22}$$

according to Eq. (2.17). Inserting the mode expansion (7.20) we get for each separate mode

$$\frac{\partial^2 \boldsymbol{A_k}}{\partial t^2} + \omega_{\boldsymbol{k}}^2 \boldsymbol{A_k} = 0 \tag{7.23}$$

with

$$\omega_{\boldsymbol{k}} = ck\,, \tag{7.24}$$

since the superposed Fourier components are independent of each other. $\boldsymbol{A}_{\boldsymbol{k}}^*$ is given by the same equation, which describes the harmonic oscillation of each wave mode.

To make the transition to quantum mechanics we need to define from the mode vector $\boldsymbol{A_k}$ a mode position $q_{\boldsymbol{k}}$ and momentum $p_{\boldsymbol{k}}$, such that the mode energy has the standard form (a bar above a symbol denotes average value)

$$\bar{\mathcal{E}}_{\boldsymbol{k}} = \frac{1}{2}(p_{\boldsymbol{k}}^2 + \omega_{\boldsymbol{k}}^2\, q_{\boldsymbol{k}}^2) \tag{7.25}$$

for a harmonic oscillator. For the classical electromagnetic field

$$\bar{\mathcal{E}} = \sum_{\boldsymbol{k}} \bar{\mathcal{E}}_{\boldsymbol{k}} = \int \left(\frac{\epsilon_0}{2} \overline{\boldsymbol{E}^2} + \frac{1}{2\mu_0} \overline{\boldsymbol{B}^2} \right) \mathrm{d}V \tag{7.26}$$

according to Eq. (2.5). The wave equation (7.23) has the solution

$$\boldsymbol{A_k}(t) = \boldsymbol{A_k}\, e^{\mp i\omega t}\,, \tag{7.27}$$

where the plus sign in $\mp$ can be discarded, since it represents advanced potentials, which are not allowed by the "arrow of time". Inserting this solution in Eq. (7.20) and using Eq. (2.18), we obtain

$$\boldsymbol{E} = \sum_{\boldsymbol{k}} i\,\omega_{\boldsymbol{k}} \left(\boldsymbol{A_k}\, e^{-i\omega_{\boldsymbol{k}} t + i\boldsymbol{k}\cdot\boldsymbol{r}} - \boldsymbol{A}_{\boldsymbol{k}}^*\, e^{i\omega_{\boldsymbol{k}} t - i\boldsymbol{k}\cdot\boldsymbol{r}} \right), \tag{7.28}$$

which when inserted in Eq. (7.26) allows the identification

$$\bar{\mathcal{E}}_{\boldsymbol{k}} = 2\epsilon_0 V \omega_{\boldsymbol{k}}^2\, \boldsymbol{A_k} \cdot \boldsymbol{A}_{\boldsymbol{k}}^*\,, \tag{7.29}$$

since the second, magnetic energy term in Eq. (7.26) has the same magnitude as the electric energy term for an electromagnetic wave. This expression can be brought to the standard form for a harmonic oscillator if we let

$$\boldsymbol{A_k} = (\omega_{\boldsymbol{k}}\, q_{\boldsymbol{k}} + ip_{\boldsymbol{k}})\, \boldsymbol{e_k} \,/\, \sqrt{4\epsilon_0 V \omega_{\boldsymbol{k}}^2}\,, \tag{7.30}$$

where $\boldsymbol{e_k}$ is the unit polarization vector.

The classical mode vector $\boldsymbol{A_k}$ becomes a quantum-mechanical operator and thus the field becomes quantized if we now replace $q_{\boldsymbol{k}}$ and $p_{\boldsymbol{k}}$ by operators $\hat{q}_{\boldsymbol{k}}$ and $\hat{p}_{\boldsymbol{k}}$ obeying the fundamental commutation relation between position and momentum:

$$[\hat{q}_{\boldsymbol{k}}, \hat{p}_{\boldsymbol{k}}] = i\hbar\,. \tag{7.31}$$

Instead of working with the operators $\hat{\boldsymbol{A}}_{\boldsymbol{k}}$ and $\hat{\boldsymbol{A}}^*_{\boldsymbol{k}}$, it has been found more convenient to use the so-called annihilation and creation operators $\hat{a}_{\boldsymbol{k}}$ and $\hat{a}^\dagger_{\boldsymbol{k}}$, which are defined by

$$\begin{aligned}\hat{\boldsymbol{A}}_{\boldsymbol{k}} &= v\,\hat{a}_{\boldsymbol{k}}\,\boldsymbol{e}_{\boldsymbol{k}}\,,\\ \hat{\boldsymbol{A}}^*_{\boldsymbol{k}} &= v\,\hat{a}^\dagger_{\boldsymbol{k}}\,\boldsymbol{e}^*_{\boldsymbol{k}}\,,\end{aligned} \tag{7.32}$$

where

$$v = \sqrt{\frac{\hbar}{2\epsilon_0 V \omega_{\boldsymbol{k}}}}\,. \tag{7.33}$$

It follows from Eqs. (7.30), (7.32), and (7.33) that

$$\begin{aligned}\hat{a}_{\boldsymbol{k}} &= (\omega_{\boldsymbol{k}}\,\hat{q}_{\boldsymbol{k}} + i\hat{p}_{\boldsymbol{k}})/\sqrt{2\hbar\omega_{\boldsymbol{k}}}\,,\\ \hat{a}^\dagger_{\boldsymbol{k}} &= (\omega_{\boldsymbol{k}}\,\hat{q}_{\boldsymbol{k}} - i\hat{p}_{\boldsymbol{k}})/\sqrt{2\hbar\omega_{\boldsymbol{k}}}\,.\end{aligned} \tag{7.34}$$

As a consequence of the commutation relation (7.31) between position and momentum we then obtain the fundamental commutation relation between annihilation and creation operators

$$[\,\hat{a}_{\boldsymbol{k}}, \hat{a}^\dagger_{\boldsymbol{k}}] = 1\,. \tag{7.35}$$

This simple form has been made possible by a proper choice of the normalization factor v, Eq. (7.33), in the definition (7.32) of the relation between the operators $\hat{\boldsymbol{A}}_{\boldsymbol{k}}$ and $\hat{a}_{\boldsymbol{k}}$.

The Hamiltonian operator $\hat{H}_{\boldsymbol{k}}$ for a single mode of the electromagnetic field is obtained as the operator version of Eq. (7.25) for $\bar{\mathcal{E}}_{\boldsymbol{k}}$:

$$\hat{H}_{\boldsymbol{k}} = \frac{1}{2}(\hat{p}^2_{\boldsymbol{k}} + \omega^2_{\boldsymbol{k}}\,\hat{q}^2_{\boldsymbol{k}})\,. \tag{7.36}$$

The total Hamiltonian for the free radiation field is then obtained as the sum over all the modes:

$$\hat{H}_R = \sum_{\boldsymbol{k}} \hat{H}_{\boldsymbol{k}}\,. \tag{7.37}$$

If we express $\hat{p}_{\boldsymbol{k}}$ and $\hat{q}_{\boldsymbol{k}}$ in terms of the creation and annihilation operators via Eq. (7.34) and make use of the commutation relation (7.35), we get

$$\hat{H}_R = \sum_{\boldsymbol{k}} \hbar\omega_{\boldsymbol{k}}\,(\hat{a}^\dagger_{\boldsymbol{k}}\hat{a}_{\boldsymbol{k}} + \tfrac{1}{2})\,. \tag{7.38}$$

It can be shown that the energy eigenvalues of $\hat{H}_{\boldsymbol{k}}$ are

$$\bar{\mathcal{E}}_{\boldsymbol{k}} = (n_{\boldsymbol{k}} + \tfrac{1}{2})\,\hbar\omega_{\boldsymbol{k}}\,, \tag{7.39}$$

where $n_{\boldsymbol{k}} = 0, 1, 2, \ldots$, i.e., an integer ≥ 0. The bilinear product $\hat{a}^\dagger_{\boldsymbol{k}}\hat{a}_{\boldsymbol{k}}$ is therefore called the number operator. $\hbar\omega_{\boldsymbol{k}}$ is the photon energy for the mode with wave number vector $\boldsymbol{k}$, the number of photons is given by $n_{\boldsymbol{k}}$, while $\frac{1}{2}$ represents the zero-point energy of the vacuum state (for which $n_{\boldsymbol{k}} = 0$). The presence of this zero-point energy is a consequence of the circumstance that $\hat{a}_{\boldsymbol{k}}$ and $\hat{a}^{\dagger}_{\boldsymbol{k}}$ do not commute.

Finally we give the expression for the electric-field operator, obtained from Eqs. (7.28) and (7.32):

$$\hat{\boldsymbol{E}}_{\boldsymbol{k}} = i v\, \omega_{\boldsymbol{k}}\, \boldsymbol{e}_{\boldsymbol{k}} \left(\hat{a}_{\boldsymbol{k}} e^{-i\omega_k t + i\boldsymbol{k}\cdot\hat{\boldsymbol{r}}} - \hat{a}^{\dagger}_{\boldsymbol{k}} e^{i\omega_k t - i\boldsymbol{k}\cdot\hat{\boldsymbol{r}}} \right). \tag{7.40}$$

7.5. Mode Counting

It is conceptually convenient for the quantization procedure to represent the electromagnetic field as a sum over discrete modes as we did in Eq. (7.20). For practical calculations of physical quantities we however need to use continuous functions and replace the mode sums by integrals over frequency and solid angle (in addition to a sum over the two orthogonal polarization states). To make the necessary conversion we need to count the modes, to determine the mode density in $\boldsymbol{k}$ space.

From the periodic boundary conditions (7.21) that we introduced when making the modal decomposition (7.20) of the electromagnetic field, it immediately follows that there is exactly one mode per volume $(2\pi/L)^3$ in 3-dimensional $\boldsymbol{k}$ space. When increasing the size L of the cavity used for the modal decomposition, the mode number density in $\boldsymbol{k}$ space thus increases to infinity, and we approach the case of a continuous instead of a discrete mode distribution.

Since the momentum $p = \hbar k$, the $\boldsymbol{k}$ space volume $(2\pi/L)^3$ corresponds to a volume of h^3/L^3 in $\boldsymbol{p}$ space, i.e., a phase space volume of h^3. There is thus always one mode per volume element h^3 in phase space.

When $L \to \infty$ we can make the transition to differentials, and count the number of modes in a spherical shell of radius k and infinitesimal thickness $\mathrm{d}k$ in $\boldsymbol{k}$ space. If $\mathcal{N}_k$ is the mode number per unit interval of k, we obtain

$$\mathcal{N}_k \,\mathrm{d}k = 4\pi k^2 \,\mathrm{d}k \,(2\pi/L)^{-3} . \tag{7.41}$$

Since

$$k = \frac{2\pi}{\lambda} = \frac{2\pi\nu}{c} , \tag{7.42}$$

we obtain the corresponding mode number in frequency interval $\mathrm{d}\nu$ as

$$\mathcal{N}_\nu \,\mathrm{d}\nu = \mathcal{N}_k \,\mathrm{d}k = \frac{4\pi\nu^2 \,\mathrm{d}\nu}{c^3} V , \tag{7.43}$$

where $V = L^3$.

The above expression represents (by considering a spherical shell) an average over all directions in $\boldsymbol{k}$ space, which has to be accounted for when making mode sums over anisotropic quantities. To sum over the modes over a volume in $\boldsymbol{k}$ space, we further have to integrate over k (or ν). Finally we have to take into account that a vector field can be decomposed into two orthogonal polarization components in a plane perpendicular to $\boldsymbol{k}$ (the propagation direction). We will indicate the two separate polarization components by index α. In the limit of a continuous mode distribution we should therefore make the substitution

$$\sum_{\boldsymbol{k}} \longrightarrow \frac{4\pi V}{c^3} \int \frac{\mathrm{d}\Omega}{4\pi} \int \nu^2 \,\mathrm{d}\nu \sum_{\alpha} . \tag{7.44}$$

Since for resonance transitions and most other cases of interest the relevant frequency interval over which the integration is carried out in practice is $\ll \nu$, we can move ν^2 outside the integral sign, which will always be done in the following.

Using the substitution (7.44) when summing over the energy eigenvalues in Eq. (7.39) and the obvious relation

$$\sum_{\alpha} \int \frac{\mathrm{d}\Omega}{4\pi} = 2 , \tag{7.45}$$

we find the energy density per unit volume and frequency interval to be

$$u_{\nu,\,\mathrm{total}} = \frac{4\pi h\nu^3}{c^3} \left(1 + \sum_{\alpha} \int \frac{\mathrm{d}\Omega}{4\pi}\, n_\alpha \right) . \tag{7.46}$$

Index 'total' is used to point out that not only the usual energy density u_ν of a real radiation field but also the zero-point vacuum energy density (represented by the 1 inside the brackets) contributes. n_α is the photon number density per unit frequency interval and polarization state.

Since the energy density u_ν is related to the mean intensity J_ν and the specific intensity I_ν by

$$u_\nu = \frac{4\pi}{c} J_\nu = \frac{4\pi}{c} \int \frac{\mathrm{d}\Omega}{4\pi} I_\nu , \tag{7.47}$$

we have in intensity units

$$I_\nu = \frac{h\nu^3}{c^2} \sum_{\alpha} n_\alpha \tag{7.48}$$

for the "real" photons, while the "intensity of the vacuum" (which is isotropic and unpolarized) is

$$I_{\nu,\,\mathrm{vacuum}} = \frac{h\nu^3}{c^2} . \tag{7.49}$$

7.6. Radiation Coherency Matrix

In the classical theory an arbitrary polarization state of the radiation field could be fully characterized by the coherency matrix $\boldsymbol{D}$, which according to Eq. (2.33) was defined as

$$D_{\alpha\alpha'} = E_\alpha E^*_{\alpha'} , \tag{7.50}$$

where

$$E_\alpha = E_{0\alpha}\, e^{-i\omega t} \tag{7.51}$$

represents one Fourier and polarization component of the electric field vector (cf. Eqs. (2.27) and (2.28)). In this definition we did not concern ourselves with the normalization of $\boldsymbol{D}$, since it was not needed for the polarization theory that we developed.

The coherences refer to phase relations between the two different polarization components of an electromagnetic wave mode travelling in a given direction and having a given frequency. Contributions from coherences between different frequencies or directions are disregarded (cf. (7.80) below). We will therefore in the following only use the polarization state indices α, α' for the creation and annihilation operators, and not the wave number vector $\boldsymbol{k}$. According to Eq. (7.28) $\hat{E}_\alpha \sim \hat{A}_\alpha$, while $\hat{E}^*_{\alpha'} \sim \hat{A}^*_{\alpha'}$. It then follows from Eq. (7.32) that the natural operator version of the coherency matrix is

$$\hat{D}_{\alpha\alpha'} \sim \hat{E}^*_{\alpha'} \hat{E}_\alpha \sim \hat{a}^\dagger_{\alpha'} \hat{a}_\alpha \,. \tag{7.52}$$

This particular ordering of the operators, and not the opposite, has to be used, since $\hat{a}^\dagger_\alpha \hat{a}_\alpha$ represents the number operator with eigenvalue n_α, and for correspondence with the classical theory we require according to Eq. (2.35) that $\mathrm{Tr}\,\boldsymbol{D} \sim \sum_\alpha n_\alpha$.

According to Eq. (7.48) the coherency matrix operator can be normalized to intensity units such that its trace represents the intensity of the radiation field if we apply the factor $h\nu^3/c^2$. The intensity-normalized coherency matrix operator thus becomes

$$\hat{D}_{\alpha\alpha'} = \frac{h\nu^3}{c^2} \, \hat{a}^\dagger_{\alpha'} \hat{a}_\alpha \,. \tag{7.53}$$

We will later derive the transfer equation for the expectation value of $\hat{D}_{\alpha\alpha'}$

7.7. The Interaction Hamiltonian

The part of the classical Hamiltonian that describes the interaction between an atom and the radiation field (described by the vector potential $\boldsymbol{A}_R$) was given by Eq. (6.17):

$$H' = \frac{e}{m} \boldsymbol{A}_R \cdot \boldsymbol{p} \,. \tag{6.17}$$

Using our mode decomposition (7.20) for $\boldsymbol{A}_R$ and annihilation and creation operators introduced via Eq. (7.32), we get for the operator version of the interaction Hamiltonian

$$\hat{H}' = \frac{e}{m} \sum_{\boldsymbol{k}} v(\hat{\boldsymbol{p}} \cdot \boldsymbol{e}_{\boldsymbol{k}} \, \hat{a}_{\boldsymbol{k}} \, e^{i\boldsymbol{k}\cdot\hat{\boldsymbol{r}}} + \hat{\boldsymbol{p}}^\dagger \cdot \boldsymbol{e}^*_{\boldsymbol{k}} \, \hat{a}^\dagger_{\boldsymbol{k}} \, e^{-i\boldsymbol{k}\cdot\hat{\boldsymbol{r}}}) \,. \tag{7.54}$$

Note that $\hat{a}_{\boldsymbol{k}}$ and $\hat{\boldsymbol{p}}$ commute, and since $\boldsymbol{p}$ is a real vector, we have been able to replace $\hat{\boldsymbol{p}}$ in the second term by $\hat{\boldsymbol{p}}^\dagger$ to make the expression manifestly symmetric. The symbol † denotes the Hermitian adjoint of an operator.

Let $|n\rangle$ represent an eigenstate of the atomic Hamiltonian, which we write as

$$\hat{H}_A = \frac{1}{2m}\hat{\boldsymbol{p}}^2 + \hat{V}. \tag{7.55}$$

The potential $\hat{V}$ may formally include the energy perturbations due to spin-orbit coupling and an external magnetic field (Zeeman effect).

It is convenient to introduce the *atomic projection operator* $|j\rangle\langle\ell|$. Then if $\boldsymbol{p}_{j\ell}$ are the matrix elements of the momentum operator $\hat{\boldsymbol{p}}$, i.e.,

$$\boldsymbol{p}_{j\ell} = \langle j\,|\hat{\boldsymbol{p}}|\,\ell\rangle, \tag{7.56}$$

the operator $\hat{\boldsymbol{p}}$ can be expanded in terms of the projection operators as

$$\hat{\boldsymbol{p}} = \sum_{j,\ell} \boldsymbol{p}_{j\ell}|\,j\rangle\langle\ell\,|. \tag{7.57}$$

To evaluate the matrix elements $\boldsymbol{p}_{j\ell}$ of the momentum operator, we need to transform them to the matrix elements $\boldsymbol{r}_{j\ell}$ of the position operator $\hat{\boldsymbol{r}}$, for which powerful mathematical tools have been developed (like the Wigner-Eckart theorem, see below). In the case when the potential $\hat{V}$ can be assumed to be a function of position only, it commutes with $\hat{\boldsymbol{r}}$, which has the consequence that

$$[\hat{\boldsymbol{r}}, \hat{H}_A] = i\hbar\,\hat{\boldsymbol{p}}/m, \tag{7.58}$$

which is seen from Eq. (7.55) if we use the commutation relation

$$[\hat{\boldsymbol{r}}, \hat{\boldsymbol{p}}] = i\hbar. \tag{7.59}$$

Using Eqs. (7.58) and (7.59) to calculate the matrix element $\boldsymbol{p}_{j\ell}$ via Eq. (7.56), we obtain

$$\boldsymbol{p}_{j\ell} = \frac{m}{i\hbar}\langle j\,|\hat{\boldsymbol{r}}|\,\ell\rangle\,(E_\ell - E_j), \tag{7.60}$$

where $E_{\ell,j}$ are the eigenvalues of $\hat{H}_A$. Introducing for convenience the frequency

$$\omega_{\ell j} = (E_\ell - E_j)/\hbar, \tag{7.61}$$

we get

$$\boldsymbol{p}_{j\ell} = im\omega_{j\ell}\boldsymbol{r}_{j\ell}. \tag{7.62}$$

This expression implies that it is the electric dipole moment operator, proportional to $\hat{\boldsymbol{r}}$, that governs the transition rate. The dominating contribution to $\hat{H}'$ comes in fact from the potential energy due to the electric dipole moment of the electron (our assumption above that $\hat{V}$ is a function of position only), while the contributions from the electric quadrupole and magnetic dipole interactions (obtained from a multipole expansion of the Hamiltonian) are smaller by a factor on the order of the fine structure constant ($\approx 1/137$).

Let us recall that E_ℓ and E_j, and therefore also $\omega_{\ell j}$, vary linearly with field strength (or Larmor frequency), as shown for instance by Eqs. (6.31) and (6.32). This dependence can lead to level crossings for certain field strengths, when levels

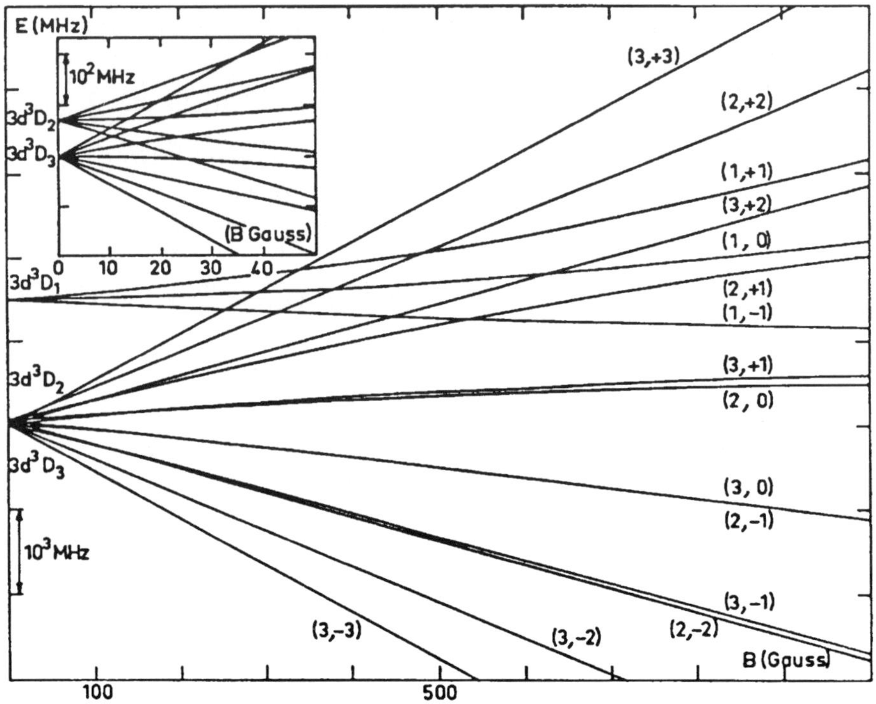

Fig. 7.1. Energies E_j of the excited sublevels (with label j) in He I, involved in the formation of the He I D3 5876 Å line ($3d\ ^3D_{3,2,1} \rightarrow 2p\ ^3P_{2,1,0}$). Each line in the diagram is labeled by its (J_u, M_u) quantum numbers. The blown-up portion for the $J_u = 2$ and 3 fine-structure components shows that level-crossing occurs already for field strengths of about 10 G, while crossings with the $J_u = 1$ level requires fields stronger than 200 G. From Bommier (1980).

having different total angular momentum quantum numbers cross. An example of this is given in Fig. 7.1.

When computing the matrix elements of the interaction Hamiltonian $\hat{H}'$, we would get complications from the $\exp(\pm i\hat{\boldsymbol{k}}\cdot\hat{\boldsymbol{r}})$ factors, unless we introduce the normally valid dipole approximation (cf. Eq. (3.7)), for which these factors are unity. This assumption implies that the wavelength of the radiation is considered to be long in comparison with the atomic dimensions, which is the case for wavelengths longer than those in the X-ray range. With this approximation we can write Eq. (7.54) as

$$\hat{H}' = \sum_{j,\ell}\sum_{\boldsymbol{k}}(\, d_{j\ell\boldsymbol{k}}\,|\,j\rangle\langle\ell\,|\hat{a}_{\boldsymbol{k}} + d^{\dagger}_{j\ell\boldsymbol{k}}\,|\,j\rangle\langle\ell\,|\hat{a}^{\dagger}_{\boldsymbol{k}}\,)\,, \tag{7.63}$$

where

$$\begin{aligned} d_{j\ell\boldsymbol{k}} &= ie\,v\,\omega_{j\ell}\,(\boldsymbol{r}\cdot\boldsymbol{e}_{\boldsymbol{k}})_{j\ell}\,,\\ d^{\dagger}_{j\ell\boldsymbol{k}} &= (d_{\ell j\boldsymbol{k}})^* = -ie\,v\,\omega_{\ell j}\,(\boldsymbol{r}\cdot\boldsymbol{e}_{\boldsymbol{k}})^*_{\ell j} = d_{j\ell\boldsymbol{k}}\,. \end{aligned} \tag{7.64}$$

The last equality holds, since the scalar product $\boldsymbol{r}\cdot\boldsymbol{e}_{\boldsymbol{k}}$ is real, and $\omega_{\ell j} = -\omega_{j\ell}$.

Let us now transform the interaction Hamiltonian from the Schrödinger to the interaction picture. We will use index I to mark operators referring to the interaction picture. From the transformation rule Eq. (7.15) follows that

$$\hat{H}_I = \sum_{j,\ell}\sum_{\boldsymbol{k}}(\, d_{j\ell\boldsymbol{k}}\, |\, j\rangle\langle \ell\, |_I\, \hat{a}_{I,\boldsymbol{k}} + d^{\dagger}_{j\ell\boldsymbol{k}}\, |\, j\rangle\langle \ell\, |_I\, \hat{a}^{\dagger}_{I,\boldsymbol{k}}\,)\,, \tag{7.65}$$

We now need to find expressions for the time dependence of the creation and annihilation operators and the projection operator in the interaction picture. For this we will make use of Eq. (7.17) describing the evolution of the operators.

First we note that the total, non-interacting Hamiltonian can be written as a sum of the atomic and radiative Hamiltonians:

$$\hat{H}_0 = \hat{H}_A + \hat{H}_R\,. \tag{7.66}$$

While the radiation Hamiltonian $\hat{H}_R$ commutes with the atomic projection operator, the atomic Hamiltonian $\hat{H}_A$ commutes with the annihilation and creation operators. When inserting the annihilation operator in Eq. (7.17) we therefore obtain

$$\frac{\mathrm{d}\hat{a}_I}{\mathrm{d}t} = \frac{1}{i\hbar}\,[\,\hat{a}_I, \hat{H}_R]\,, \tag{7.67}$$

since the atomic Hamiltonian commutes with $\hat{a}_I$. In Eq. (7.67) we have omitted index $\boldsymbol{k}$ for simplicity of notation.

Creation and annihilation operators have a non-zero commutator given by Eq. (7.35) only if their wave number vectors $\boldsymbol{k}$ agree. Therefore only one wave mode in the expression for the radiation Hamiltonian of Eq. (7.38) contributes in Eq. (7.67), and we get

$$\frac{\mathrm{d}\hat{a}_I}{\mathrm{d}t} = -i\omega\,[\,\hat{a}_I, (\hat{a}^{\dagger}_I\hat{a}_I + \tfrac{1}{2})\,]\,. \tag{7.68}$$

Using the commutation relation (7.35), the expression simplifies to

$$\frac{\mathrm{d}\hat{a}_I}{\mathrm{d}t} = -i\omega\,\hat{a}_I\,, \tag{7.69}$$

which leads to the solutions

$$\begin{aligned} \hat{a}_{I,\boldsymbol{k}} &= \hat{a}_{\boldsymbol{k}}\, c^{-i\omega_k t}\,, \\ \hat{a}^{\dagger}_{I,\boldsymbol{k}} &= \hat{a}^{\dagger}_{\boldsymbol{k}}\, e^{i\omega_k t}\,. \end{aligned} \tag{7.70}$$

Similarly, by inserting the atomic projection operator in the evolution equation (7.17) it is straightforward to obtain

$$|\, j\rangle\langle \ell\, |_I = e^{i\omega_{j\ell}t}\; |\, j\rangle\langle \ell\, |\,. \tag{7.71}$$

The solutions (7.70) and (7.71) can now be inserted in Eq. (7.65) to give the interaction Hamiltonian $\hat{H}_I$ expressed in terms of time-independent radiative and atomic operators:

$$\hat{H}_I = \sum_{j,\ell}\sum_{\boldsymbol{k}}(\,\hat{h}_{j\ell\boldsymbol{k}} + \hat{h}^\dagger_{j\ell\boldsymbol{k}}\,)\,, \tag{7.72}$$

where

$$\begin{aligned}\hat{h}_{j\ell\boldsymbol{k}} &= d_{j\ell\boldsymbol{k}}\,\hat{a}_{\boldsymbol{k}}\;e^{i(\omega_{j\ell}-\omega_k)t}\;|\,j\rangle\langle\ell\,| \\ \hat{h}^\dagger_{j\ell\boldsymbol{k}} &= d^\dagger_{j\ell\boldsymbol{k}}\,\hat{a}^\dagger_{\boldsymbol{k}}\;e^{-i(\omega_{\ell j}-\omega_k)t}\;|\,j\rangle\langle\ell\,|\,.\end{aligned} \tag{7.73}$$

7.8. Evolution of the Expectation Values

The expectation value of an operator $\hat{X}$ is given by

$$\langle\hat{X}\rangle = \mathrm{Tr}\,(\hat{X}_I\,\hat{\rho}_I\,) \tag{7.74}$$

according to Eqs. (7.9) and (7.19). Its evolution is obtained by derivation while making use of the Liouville equation (7.18):

$$\frac{\mathrm{d}\langle\hat{X}\rangle}{\mathrm{d}t} = \mathrm{Tr}\left(\frac{\mathrm{d}\hat{X}_I}{\mathrm{d}t}\,\hat{\rho}_I\right) + \frac{1}{i\hbar}\,\mathrm{Tr}\,(\,\hat{X}_I\,[\,\hat{H}_I,\hat{\rho}_I\,]\,)\,. \tag{7.75}$$

If $\hat{X} = \hat{D}_{\alpha\alpha'}$ given by Eq. (7.53), we obtain the equation of radiative transfer for the radiation coherency matrix, which contains all the polarization information. If $\hat{X} = |\,j\rangle\langle\ell\,|$, we obtain the statistical equilibrium equations for the atomic density matrix.

The problem is however not so straightforward as it may look, since $\hat{\rho}_I$ is a function of time and of $\hat{H}_I$, as governed by the Liouville equation. The problem is therefore nonlinear, but as the radiative interaction is weak in ordinary stellar atmospheres, a perturbation approach is feasible. Thus the last term in Eq. (7.75) can be expanded in terms of the perturbation $\hat{H}_I$ to increasing orders.

Since in stellar atmospheres (in contrast to lasers) the radiation field and the atomic system can be regarded as uncorrelated, i.e., the phases of $\hat{a}_{\boldsymbol{k}}$ and $|\,j\rangle\langle\ell\,|$ are uncorrelated, the average of $\hat{H}_I$ for a statistical ensemble would be zero. Therefore there can be no first-order contributions from the interaction between matter and radiation. We have to go to higher even orders, but since the interaction is weak in stellar atmospheres, it is sufficient here to go to second order.

The absence of correlations between the atomic system and the radiation field can be stated explicitly in the form

$$\hat{\rho}_I = \hat{\rho}_A \otimes \hat{\rho}_R\,, \tag{7.76}$$

where $\hat{\rho}_A$ and $\hat{\rho}_R$ are the density matrices (in the interaction representation) of the atomic and radiative system, respectively. The second-order perturbation approach implies that $\hat{\rho}_A$ and $\hat{\rho}_R$ remain uncorrelated not only before, but also during the time that the interaction takes place.

If $\hat{A}$ is an operator that only acts on the atomic system, while $\hat{R}$ only acts on the radiation field, Eq. (7.76) implies that

$$\mathrm{Tr}\,(\hat{A}\hat{R}\,\hat{\rho}_I) = \mathrm{Tr}\,(\hat{A}\,\hat{\rho}_A)\;\mathrm{Tr}\,(\hat{R}\,\hat{\rho}_R)\,. \tag{7.77}$$

This property will be used extensively in the following.

For the radiation field

$$\mathrm{Tr}\,(\hat{a}_{\boldsymbol{k}}\,\hat{\rho}_R) = \mathrm{Tr}\,(\hat{a}^\dagger_{\boldsymbol{k}}\,\hat{\rho}_R) = 0\,, \tag{7.78}$$

since the phases of the different photons in a statistical ensemble are uncorrelated. This is also the case for the classical Jones matrices. The average of the Jones matrices of an ensemble of (uncorrelated) photons is zero. For this reason we have to make use of bilinear products, which is done in the coherency matrix and Stokes vector formulations.

Bilinear products of either the annihilation or the creation operators however give zero contributions, i.e.,

$$\mathrm{Tr}\,(\,\hat{a}_{\boldsymbol{k}}\hat{a}_{\boldsymbol{k}'}\,\hat{\rho}_R) = \mathrm{Tr}\,(\,\hat{a}^\dagger_{\boldsymbol{k}}\hat{a}^\dagger_{\boldsymbol{k}'}\,\hat{\rho}_R) = 0\,, \tag{7.79}$$

since the phase factors of such bilinear products are random and uncorrelated (as are the phase factors of each individual operator). The only bilinear products that may survive ensemble averaging are of the kind $\hat{a}^\dagger_{\boldsymbol{k}}\,\hat{a}_{\boldsymbol{k}'}$, where the two operators refer to the same photon. Then the phase factors cancel, except for a systematic phase difference (determining the state of polarization) that does not average out to zero. This corresponds to the classical case, where $E_\alpha E_{\alpha'}$ averages out to zero while $E_\alpha E^*_{\alpha'}$ does not.

The trace

$$\mathrm{Tr}\,(\hat{a}^\dagger_{\boldsymbol{k}}\,\hat{a}_{\boldsymbol{k}'}\,\hat{\rho}_R) \tag{7.80}$$

is thus different from zero only when the wave number vectors $\boldsymbol{k} = \boldsymbol{k}'$, or more precisely, when the frequency and direction of propagation are identical, while the state of polarization may be different (this distinction is essential here, since for simplicity of notation we have let the *indices* $\boldsymbol{k}$ and $\boldsymbol{k}'$ symbolize not only the wave number vector, but *also* the state of polarization). There can only be correlations between the different polarization states of the same photon.

To obtain the second-order equation for the evolution of the expectation values we first solve the Liouville equation (7.18) for $\hat{\rho}_I$ formally to first order, and then insert this first-order solution in the last term of Eq. (7.75). The first-order solution would vanish if we were to make an ensemble average *before* it has been inserted in Eq. (7.75). We should consider the first-order solution as applying to each individual atom, and make the ensemble average *after* it has been inserted in Eq. (7.75). Bilinear products of the type (7.80) will then appear in the second-order solution, and they are the ones that can survive the averaging.

7.9. First-order Solution of the Liouville Equation

Our task is to solve the Liouville equation

$$\frac{\mathrm{d}\hat{\rho}_I}{\mathrm{d}t} = \frac{1}{i\hbar}[\hat{H}_I, \hat{\rho}_I] \tag{7.18}$$

to first order so that it can later be inserted in Eq. (7.75) for the evolution of the expectation values. The formalism can be greatly simplified if we instead of performing an integration and carrying along integrals in the different expressions can obtain a steady-state solution of Eq. (7.18) in closed form. Steady-state solutions are however only possible when a decay term balances the growth term, and in a rigorous theory damping only enters when one goes to higher orders. For this reason we will here use heuristic arguments to introduce a damping term, to allow us to obtain a steady-state "first-order" solution (containing the heuristical damping, "borrowed" from higher orders). The procedure may retroactively be justified in the case of radiative damping when we later calculate the higher orders.

The damping can be due to two types of effects: radiative and collisional processes. A quantum-mechanical treatment of collisions is outside the scope of the present treatment, so collisional damping can only be introduced heuristically anyways. Radiative damping is primarily due to spontaneous emission from the considered atomic level, which has the effect of limiting the lifetime of the state.

For density matrix elements $\rho_{mm'}$, where m and m' refer to the magnetic sublevels of a given excited level, it is natural that the damping term should be $-\gamma\rho_{mm'}$, where γ is the inverse lifetime of the excited state. In the first-order problem, however, the Liouville equation does not describe the evolution of $\rho_{mm'}$ but instead deals with the matrix elements $\rho_{u\ell}$ and $\rho_{\ell u}$, where indices ℓ and u refer to the lower and upper levels, respectively, which are connected by a radiative transition. The reason for this is that the operator $\hat{H}_I$ that acts on the zero-order density matrix $\hat{\rho}_I$ is the sum of two terms, each of which contains either one annihilation or one creation operator, as shown by Eqs. (7.72) and (7.73), but not a product of both operators. Therefore the operator $\hat{H}_I$ can only induce single-photon transitions and thus density matrix elements between levels connected by a single-photon transition. In the second-order treatment on the other hand we get bilinear products of the annihilation and creation operators, which allows us to obtain the evolution of $\rho_{mm'}$ and $\rho_{\ell\ell'}$, for which the two levels involved are not connected by a radiative transition.

The decay rate of $-\gamma\rho_{mm'}$ for $\rho_{mm'} = c_m c^*_{m'}$ would be reproduced by a decay rate of $-(\gamma/2)c_m$ for the coefficients c_m. Since the decay of the lower level can usually be neglected in comparison with that of the upper level, $\rho_{u\ell} = c_u c^*_\ell$ decays at the rate $-(\gamma/2)\rho_{u\ell}$. Adding this decay term to the Liouville equation (7.18) and inserting expression (7.72) for $\hat{H}_I$, we obtain the first-order equation

$$\frac{\mathrm{d}\hat{\rho}_I}{\mathrm{d}t} = \frac{1}{i\hbar}\sum_{j,\ell}\sum_{\boldsymbol{k}}([\hat{h}_{j\ell\boldsymbol{k}}, \hat{\rho}_I] + [\hat{h}^\dagger_{j\ell\boldsymbol{k}}, \hat{\rho}_I]) - \gamma\hat{\rho}_I/2\,. \tag{7.81}$$

For steady-state solutions we make the Ansatz

$$\hat{\rho}_I = \sum_{j,\ell} \sum_{\boldsymbol{k}} \left[\hat{\rho}_{j\ell\boldsymbol{k}}\, e^{i(\omega_{j\ell}-\omega_k)t} + \hat{\rho}^\dagger_{j\ell\boldsymbol{k}}\, e^{-i(\omega_{\ell j}-\omega_k)t} \right] \tag{7.82}$$

and require that

$$\frac{\mathrm{d}\hat{\rho}_{j\ell\boldsymbol{k}}}{\mathrm{d}t} = \frac{\mathrm{d}\hat{\rho}^\dagger_{j\ell\boldsymbol{k}}}{\mathrm{d}t} = 0\,. \tag{7.83}$$

Insertion in Eq. (7.81) and use of Eq. (7.73) gives

$$i\sum_{j,\ell} \sum_{\boldsymbol{k}} \left[(\omega_{j\ell} - \omega_{\boldsymbol{k}} - i\gamma/2)\hat{\rho}_{j\ell\boldsymbol{k}}\, e^{i(\omega_{j\ell}-\omega_k)t} - (\omega_{\ell j} - \omega_{\boldsymbol{k}} + i\gamma/2)\hat{\rho}^\dagger_{j\ell\boldsymbol{k}}\, e^{-i(\omega_{\ell j}-\omega_k)t} \right]$$
$$= \frac{1}{i\hbar} \sum_{j,\ell} \sum_{\boldsymbol{k}} \left\{ [\hat{d}_{j\ell\boldsymbol{k}} |\, j\rangle\langle \ell\,| \hat{a}_{\boldsymbol{k}},\, \hat{\rho}_I] e^{i(\omega_{j\ell}-\omega_k)t} + [\hat{d}^\dagger_{j\ell\boldsymbol{k}} |\, j\rangle\langle \ell\,| \hat{a}^\dagger_{\boldsymbol{k}},\, \hat{\rho}_I]\, e^{-i(\omega_{\ell j}-\omega_k)t} \right\}. \tag{7.84}$$

This equation is satisfied only when the coefficients in front of each of the two types of oscillating factors are zero, which implies that

$$\begin{aligned} \hat{\rho}_{j\ell\boldsymbol{k}} &= -\frac{1}{\hbar} \frac{[\hat{d}_{j\ell\boldsymbol{k}} \,|\, j\rangle\langle \ell \,|\, \hat{a}_{\boldsymbol{k}}\,,\, \hat{\rho}_I]}{\omega_{j\ell} - \omega_{\boldsymbol{k}} - i\gamma/2}\,, \\ \hat{\rho}^\dagger_{j\ell\boldsymbol{k}} &= \frac{1}{\hbar} \frac{[\hat{d}^\dagger_{j\ell\boldsymbol{k}} \,|\, j\rangle\langle \ell \,|\, \hat{a}^\dagger_{\boldsymbol{k}}\,,\, \hat{\rho}_I]}{\omega_{\ell j} - \omega_{\boldsymbol{k}} + i\gamma/2}\,. \end{aligned} \tag{7.85}$$

Insertion in Eq. (7.82) and use of expressions (7.73) then gives

$$\hat{\rho}_I = -\frac{1}{\hbar} \sum_{j,\ell} \sum_{\boldsymbol{k}} \left(\frac{[\hat{h}_{j\ell\boldsymbol{k}}, \hat{\rho}_I]}{\omega_{j\ell} - \omega_{\boldsymbol{k}} - i\gamma/2} - \frac{[\hat{h}^\dagger_{j\ell\boldsymbol{k}}, \hat{\rho}_I]}{\omega_{\ell j} - \omega_{\boldsymbol{k}} + i\gamma/2} \right). \tag{7.86}$$

The frequency-dependent factors in Eq. (7.86) can be written in a more useful and compact form by introducing the normalized complex profile function Φ, which is basically the same as the Φ previously introduced in Eq. (6.57), except that we use different indices in the quantum field theory treatment and have not yet introduced Doppler broadening. Thus we define

$$\Phi(\nu)_{j\ell\boldsymbol{k}} = \frac{1}{\Delta\nu_D} \frac{1}{\pi} \frac{a - iv_{j\ell\boldsymbol{k}}}{v^2_{j\ell\boldsymbol{k}} + a^2}\,, \tag{7.87}$$

where

$$\begin{aligned} v_{j\ell\boldsymbol{k}} &= (\omega_{j\ell} - \omega_{\boldsymbol{k}})/\Delta\omega_D\,, \\ a &= \gamma/(2\Delta\omega_D) \end{aligned} \tag{7.88}$$

(cf. Eqs. (3.53) and (3.54)). This definition satisfies the normalization

$$\int_0^\infty \mathrm{Re}\ \Phi(\nu)_{j\ell\boldsymbol{k}}\, \mathrm{d}\nu = 1\,. \tag{7.89}$$

The frequency-dependent factors now become

$$
\begin{aligned}
\frac{1}{\omega_{j\ell} - \omega_{\boldsymbol{k}} - i\gamma/2} &= \frac{i}{2}\Phi(\nu)_{j\ell\boldsymbol{k}}\,, \\
\frac{1}{\omega_{\ell j} - \omega_{\boldsymbol{k}} + i\gamma/2} &= -\frac{i}{2}\Phi^\dagger(\nu)_{j\ell\boldsymbol{k}}\,,
\end{aligned}
\tag{7.90}
$$

where

$$
\Phi^\dagger(\nu)_{j\ell\boldsymbol{k}} = \Phi^*(\nu)_{\ell j\boldsymbol{k}}\,. \tag{7.91}
$$

When Doppler broadening is added,

$$
\Phi(\nu) = \frac{1}{\sqrt{\pi}\Delta\nu_D}\mathcal{H}(a,v)\,, \tag{7.92}
$$

where $\mathcal{H}(a,v)$ is the complex combination of the Voigt and line dispersion functions, given by Eqs. (4.14)–(4.16).

Using Eq. (7.90) in Eq. (7.86), we obtain a compact expression for the first-order solution of the Liouville equation:

$$
\hat\rho_I = -\frac{i}{2\hbar}\sum_{j,\ell}\sum_{\boldsymbol{k}}\Big([\hat h_{j\ell\boldsymbol{k}}, \hat\rho_I]\,\Phi_{j\ell\boldsymbol{k}} + [\hat h^\dagger_{j\ell\boldsymbol{k}}, \hat\rho_I]\,\Phi^\dagger_{j\ell\boldsymbol{k}}\Big)\,. \tag{7.93}
$$

7.10. Second-order Equation for the Expectation Values

We only need to focus on the second trace in Eq. (7.75) for the evolution of the expectation values, since this is where the non-linearities are contained and where the perturbation expansion has to be made. Inserting the first-order solution (7.93) for $\hat\rho_I$ and expression (7.72) for $\hat H_I$, this trace becomes

$$
\begin{aligned}
\mathrm{Tr}(\hat X_I[\hat H_I, \hat\rho_I]) = -\frac{i}{2\hbar}\sum_{j,\ell,r,s}\sum_{\boldsymbol{k},\boldsymbol{k}'}\{&\mathrm{Tr}(\hat X_I[\hat h^\dagger_{rs\boldsymbol{k}'}, [\hat h_{j\ell\boldsymbol{k}}, \hat\rho_I]]\,\Phi_{j\ell\boldsymbol{k}}) \\
&+\mathrm{Tr}(\hat X_I[\hat h_{rs\boldsymbol{k}'}, [\hat h^\dagger_{j\ell\boldsymbol{k}}, \hat\rho_I]]\,\Phi^\dagger_{j\ell\boldsymbol{k}})\}\,,
\end{aligned}
\tag{7.94}
$$

where we have made use of the property expressed by Eq. (7.79) that only bilinear products of annihilation and creation operators contribute, not bilinear products of the same kind of operator.

Next we will derive explicit forms of these traces. When Eqs. (7.73) and (7.64) are inserted, we get complicated expressions that however can be greatly simplified due to various circumstances, which we will now discuss one by one.

First we note that since there are no correlations between photons with different wave number vectors $\boldsymbol{k}$ and $\boldsymbol{k}'$, we have $\omega_{\boldsymbol{k}} = \omega_{\boldsymbol{k}'} = \omega$, which means that the exponential terms $\exp(\pm i\omega_{\boldsymbol{k}}t)$ that enter via expressions (7.73) mutually compensate each other and vanish. We then do not need to carry along index $\boldsymbol{k}$ in Φ, but can write $\Phi_{j\ell\boldsymbol{k}} = \Phi_{j\ell}$.

Although the wave number vectors are the same, the polarization states may be different. The double sum over $\boldsymbol{k}, \boldsymbol{k}'$ in Eq. (7.94) can therefore be replaced by a sum over $\boldsymbol{k}$ (i.e., over frequency and angle) and a sum over the two possible polarization states of the two photons. The polarization states will be marked by a greek letter. We may for instance write $\hat{a}_\alpha = \hat{a}_{\boldsymbol{k}}$.

The traces contain products of atomic and radiative operators, but they can be converted into products of atomic and radiative traces by making use of the important property expressed by Eq. (7.77). The atomic trace is best calculated in the Schrödinger representation (according to Eq. (7.19) we are allowed to switch between representations if all the operators within the trace are changed to the new representation). The advantage of doing this is that the oscillating factors $\exp(i\omega_{j\ell}t)$ and $\exp(i\omega_{rs}t)$ disappear from the expressions, so that there is no explicit time dependence remaining.

Since the frequency range of interest is never far from the relevant atomic resonances ($|\omega_0 - \omega| \ll \omega$), it is a good approximation to replace $\omega_{j\ell}$ in expressions (7.64) by ω, which simplifies the notation.

Let us next consider the factor $(\boldsymbol{r} \cdot \boldsymbol{e_k})_{j\ell}$ that occurs in expressions (7.64). In the classical case we saw in Eq. (3.39) that the vector equation of motion of the electron became diagonalized, i.e., the oscillations in the three coordinate directions became decoupled from each other, when the linear unit vectors were replaced by spherical unit vectors. Similarly it can be shown that a decoupling is achieved if spherical unit vectors are introduced as a basis for the position operator $\hat{\boldsymbol{r}}$ and the linear polarization vector $\boldsymbol{e_k}$. Let us denote the corresponding spherical vector components by $\hat{r}_q$ and ε_q^α, respectively. In quantum mechanics $\hat{r}_q$ has the function of being a *ladder operator*, having the same eigenfunctions as the angular momentum operators $\hat{J}_z$ and $\hat{\boldsymbol{J}}^2$, but with the property of increasing the M quantum number ($M\hbar$ being the eigenvalue of $\hat{J}_z$) by q, where $q = 0, \pm 1$.

These relations become manifest in the fundamental *Wigner-Eckart theorem*, which allows the geometric (M dependent) properties of a matrix element to become separated from the spherically symmetric (M independent) properties. According to this theorem,

$$
\begin{aligned}
(r_q)_{j\ell} &= \langle J_j M_j \,|\, \hat{r}_q \,|\, J_\ell M_\ell \rangle \\
&= (-1)^{J_\ell + M_j + 1} \sqrt{2J_j + 1}\, \langle J_j \,||\, \hat{\boldsymbol{r}} \,||\, J_\ell \rangle \begin{pmatrix} J_j & J_\ell & 1 \\ -M_j & M_\ell & q \end{pmatrix},
\end{aligned}
\tag{7.95}
$$

where the spherically symmetric, reduced matrix element $\langle J_j \,||\, \hat{\boldsymbol{r}} \,||\, J_\ell \rangle$ can be directly related to the oscillator strength of the transition, as will be seen in the next chapter. It is defined somewhat differently by different authors. In our definition the factor $\sqrt{2J_j + 1}$ has been broken out, which has some advantages by making the expressions more symmetric (see next chapter). A concise overview of the complex algebra and the various symmetry properties involved when evaluating the matrix elements can be found in Brink and Satchler (1968).

The 3-j symbol is non-zero only when $q = M_j - M_\ell$, which thus serves as a *selection rule* for the change ΔM of the magnetic quantum number for an electric dipole transition (which is determined by the matrix elements of $\hat{r}_q$). Another property of the 3-j symbols is that they vanish unless $J_j - J_\ell = 0$ or ± 1 ($J_j = J_\ell = 0$ is excluded). This represents another selection rule.

Because of these selection rules only one of the three vector components may contribute to a matrix element of the vector operator $\hat{\boldsymbol{r}}$. Thus the spherical vector components are decoupled from each other as they were in the classical case.

With these tools we are now in a position to give explicit and compact expressions for the matrix elements $(\boldsymbol{r}\cdot\boldsymbol{e_k})_{j\ell}$ that occur in Eq. (7.64). With Eq. (3.37) for the scalar product of two spherical vectors and the selection rule for ΔM, we get

$$\begin{aligned}(\boldsymbol{r}\cdot\boldsymbol{e_k})_{j\ell} &= \langle J_j M_j \mid \sum_q r_q \varepsilon_q^{\alpha *} \mid J_\ell M_\ell \rangle \\ &= (r_{M_j - M_\ell})_{j\ell}\, \varepsilon^{\alpha *}_{M_j - M_\ell} = (r^*_{M_j - M_\ell})_{j\ell}\, \varepsilon^{\alpha}_{M_j - M_\ell}\,,\end{aligned} \tag{7.96}$$

since $q = M_j - M_\ell$ and thus only one component of the sum contributes, and since both $\boldsymbol{r}$ and $\boldsymbol{e_k}$ are real vectors. Note that we are using polarization index α instead of wave number vector $\boldsymbol{k}$ for ε according to our previous discussion. ε_q^α represents the spherical vector component q of the linear unit vector $\boldsymbol{e_k}$ as defined by Eq. (3.61).

For simplicity of index notation we introduce the definition

$$\begin{aligned} r_{j\ell} &\equiv (r_q)_{j\ell} = (r_{M_j - M_\ell})_{j\ell}\,, \\ \varepsilon^\alpha_{j\ell} &\equiv (\varepsilon^\alpha_q)_{j\ell} = \varepsilon^\alpha_{M_j - M_\ell}\,. \end{aligned} \tag{7.97}$$

It is important in later use to remember the exact meaning of these symbols by recalling the above definition.

Eq. (7.64) can now be written

$$\begin{aligned} d_{j\ell\boldsymbol{k}} &= ie\,v\,\omega_{j\ell}\, r_{j\ell}\, \varepsilon^{\alpha *}_{j\ell}\,, \\ d^\dagger_{j\ell\boldsymbol{k}} &= (d_{\ell j\boldsymbol{k}})^* = -ie\,v\,\omega_{\ell j}\,(r_{\ell j})^*\,\varepsilon^\alpha_{\ell j} = d_{j\ell\boldsymbol{k}}\,, \end{aligned} \tag{7.98}$$

where the last equality holds since $r_q^*\,\varepsilon_q^\alpha = r_{-q}\,\varepsilon^{\alpha *}_{-q}$ according to Eq. (3.35). While $r_{j\ell}$ is evaluated via the Wigner-Eckart theorem, the geometrical factors $\varepsilon^\alpha_{j\ell}$ depend on the choice of coordinate system. With the choice defined by Fig. 3.2, $\varepsilon^\alpha_{j\ell}$ are given by Eq. (3.71).

Note that although $-\omega_{\ell j} = \omega_{j\ell}$, we have used $-\omega_{\ell j}$ in the second equation of (7.98) because of its association with $\Phi^\dagger_{j\ell} = \Phi^\dagger_{j\ell\boldsymbol{k}}$ in Eq. (7.93). According to Eq. (7.90), the $\omega_{\ell j}$ that appears in the expression for $\Phi^\dagger_{j\ell}$ has to be positive to represent a real transition. The ℓ level then has to be higher than the j level. For $\Phi_{j\ell}$ (and thus also for $d_{j\ell\boldsymbol{k}}$) the situation is the opposite. The j and ℓ indices in the two terms of Eq. (7.93) or in $d_{j\ell\boldsymbol{k}}$ and $d^\dagger_{j\ell\boldsymbol{k}}$ thus do not represent the same physical levels. In spite of this we keep the same labels in the two terms for the sake of economy of notation and to make the expressions formally symmetric.

Since the traces in Eq. (7.94) contain bilinear products of $d_{j\ell\boldsymbol{k}}$ and $d^\dagger_{j\ell\boldsymbol{k}}$, we get according to Eq. (7.98) factors $e^2v^2\omega^2$. With Eq. (7.33) we have

$$e^2 v^2 \omega^2 = \frac{e^2 \hbar\omega}{2\epsilon_0 V} = \frac{e^2 \hbar\omega N}{2\epsilon_0} \,. \tag{7.99}$$

Here we have introduced the number density of interacting atoms N. As there is only one interacting atom per volume V considered (as implied by the normalization condition $\mathrm{Tr}\,\rho_A = 1$ of Eq. (7.10)), $V = 1/N$, which explains the second part of Eq. (7.99).

The first trace on the right hand side of Eq. (7.94) can now be written, using Eqs. (7.73) and (7.98),

$$\begin{aligned}\mathrm{Tr}\,(\hat{X}_I\,[\,\hat{h}^{\dagger}_{rsk'}\,,[\,\hat{h}_{j\ell k}\,,\hat{\rho}_I]\,]\,\Phi_{j\ell k}\,) =& \frac{e^2\hbar\omega N}{2\epsilon_0}\, r_{j\ell}\, r^*_{sr}\, \varepsilon^{\beta *}_{j\ell}\, \varepsilon^{\beta'}_{sr}\, \Phi_{j\ell} \\ & \mathrm{Tr}\,(\hat{X}\,[\,|\,r\rangle\langle s\,|\,\hat{a}^{\dagger}_{\beta'}\,,\,[\,|\,j\rangle\langle \ell\,|\,\hat{a}_{\beta}\,,\hat{\rho}\,]\,]\,)\,,\end{aligned} \tag{7.100}$$

where we have changed the operators in the trace to the Schrödinger representation. The reason for choosing β instead of α as the polarization label will become clear in the next chapter, where α is needed as an index for the coherency matrix $\hat{D}_{\alpha\alpha'}$. Similarly we obtain for the second trace on the right hand side of Eq. (7.94)

$$\begin{aligned}\mathrm{Tr}\,(\hat{X}_I\,[\,\hat{h}_{rsk'}\,,[\,\hat{h}^{\dagger}_{j\ell k}\,,\hat{\rho}_I]\,]\,\Phi^{\dagger}_{j\ell k}\,) =& \frac{e^2\hbar\omega N}{2\epsilon_0}\, r^*_{\ell j}\, r_{rs}\, \varepsilon^{\beta}_{\ell j}\, \varepsilon^{\beta' *}_{rs}\, \Phi^{\dagger}_{j\ell} \\ & \mathrm{Tr}\,(\hat{X}\,[\,|\,r\rangle\langle s\,|\,\hat{a}_{\beta'}\,,\,[\,|\,j\rangle\langle \ell\,|\,\hat{a}^{\dagger}_{\beta}\,,\hat{\rho}\,]\,]\,)\,.\end{aligned} \tag{7.101}$$

The next step in the reduction is to make use of the assumption (7.76) that the radiation field and the atomic system are uncorrelated, which implies that the radiative and atomic operators commute with each other, and that the trace can be factorized in an atomic and a radiative trace as in Eq. (7.77). As however the operators generally do not commute with the density matrix operator $\hat{\rho}$, we first have to rearrange the traces by making use of their cyclic permutation property,

$$\mathrm{Tr}\,(\hat{A}\hat{B}\hat{C}) = \mathrm{Tr}\,(\hat{C}\hat{A}\hat{B})\,, \tag{7.102}$$

to bring them to a form with $\hat{\rho}$ to the far right, to allow Eq. (7.77) to be used. This form is also needed for the later evaluation of expectation values.

Expanding the nested commutators, using property (7.102) to move $\hat{\rho}$ to the right, and expressing the remaining operator (that does not contain $\hat{\rho}$) in terms of nested commutators, a trace with the same general structure as that of Eqs. (7.100) and (7.101) becomes

$$\mathrm{Tr}\,(\hat{X}\,[\,\hat{F}\,,\,[\,\hat{G}\,,\hat{\rho}\,]\,]\,) = \mathrm{Tr}\,(\,[\,[\hat{X}\,,\hat{F}\,]\,,\hat{G}\,]\,\hat{\rho}\,)\,. \tag{7.103}$$

Accordingly the trace in Eq. (7.100) can be expressed in the form

$$\begin{aligned}&\mathrm{Tr}\,(\hat{X}\,[\,|\,r\rangle\langle s\,|\,\hat{a}^{\dagger}_{\beta'}\,,\,[\,|\,j\rangle\langle \ell\,|\,\hat{a}_{\beta}\,,\hat{\rho}\,]\,]\,) \\ =&\mathrm{Tr}\,(\,[\,[\,\hat{X}_A\hat{X}_R\,,\,|\,r\rangle\langle s\,|\,\hat{a}^{\dagger}_{\beta'}\,]\,,\,|\,j\rangle\langle \ell\,|\,\hat{a}_{\beta}\,]\,\hat{\rho}_A\otimes\hat{\rho}_R\,)\,,\end{aligned} \tag{7.104}$$

where we have written the operator $\hat{X}$ as a product of atomic and radiative operators $\hat{X}_A$ and $\hat{X}_R$ and used Eq. (7.76) for the density matrix operator. The trace in Eq. (7.101) can be converted in the same way.

A further simplification can be achieved in two special cases: When the statistical equilibrium of the atomic system is considered, $\hat{X}_R = 1$, and the inner commutator on the right hand side of Eq. (7.104) becomes

$$[\hat{X}_A\,,\,|\,r\rangle\langle s\,|\,\hat{a}^\dagger_{\beta'}\,] = [\hat{X}_A\,,\,|\,r\rangle\langle s\,|\,]\,\hat{a}^\dagger_{\beta'}\,. \tag{7.105}$$

When the radiative transfer problem is considered, $\hat{X}_A = 1$, and we get

$$[\hat{X}_R\,,\,|\,r\rangle\langle s\,|\,\hat{a}^\dagger_{\beta'}\,] = |\,r\rangle\langle s\,|\,[\,\hat{X}_R\,,\,\hat{a}^\dagger_{\beta'}\,]\,. \tag{7.106}$$

As will be seen in the next chapter, the equation of radiative transfer is obtained if we in Eq. (7.75) for the evolution of the expectation values let $\hat{X} = \hat{X}_R = \hat{D}_{\alpha\alpha'}$, the radiative coherency matrix. Similarly, the statistical equilibrium equations are obtained if we let $\hat{X} = \hat{X}_A = |\,p\rangle\langle n\,|$, the atomic projection operator.

CHAPTER 8

MULTI-LEVEL RADIATIVE TRANSFER WITH COHERENCE EFFECTS

8.1. Introduction

In the preceding chapter we have developed from first principles a foundation and tools for treating the general problem of the interaction between matter and radiation with quantum field theory. In the present chapter we will build on this foundation and apply these tools to develop a general non-LTE theory of polarized radiative transfer. This theory should be able to account for any coherences that may occur between the atomic levels. The formulations developed in Chapters 3-6 may be retrieved as special cases, and parameters that were previously introduced heuristically can be derived.

The transfer equation for the radiation field and the statistical equilibrium equation for the atomic system are both derived from Eq. (7.75), which describes the evolution of the expectation values. The non-linear term is expanded with perturbation theory to second order, by inserting the first-order solution of the Liouville equation.

The polarization state of the radiation is completely described by the coherency matrix operator $\hat{D}_{\alpha\alpha'}$, which corresponds to the classical coherency matrix defined by Eq. (2.33). The equation describing the evolution of its expectation value is the transfer equation for polarized radiation.

The atomic coherences between levels j and ℓ are described by the density matrix elements $\rho_{j\ell}$ defined by Eqs. (7.7) and (7.8). The relative level populations are represented by the diagonal elements of the density matrix. The statistical equilibrium equation that at the same time describes the coherences (off-diagonal terms) and populations (diagonal terms) is obtained from the equation for the evolution of the expectation value of the atomic projection operator $|\,j\rangle\langle\ell\,|$.

The radiative transfer and statistical equilibrium equations form an equation system that jointly determines the transport of polarized radiation. This complex system does not admit closed solutions unless simplified assumptions are made. The most important simplification is the neglect of lower-level coherences. In this case an explicit expression for the scattering cross section that includes the Hanle effect and arbitrary upper level interferences (e.g. between levels of different total angular momentum J) can be derived. The detailed treatment of this case will however be postponed till the next chapter.

The first quantum-mechanical derivation of the LTE equations for polarized radiative transfer in a magnetic field was done by Landi Degl'Innocenti and Landi Degl'Innocenti (1972). The quantum theory of coherent scattering (Hanle effect) was developed in various forms by House (1970a,b, 1971), Lamb and ter Haar (1971), Omont et al. (1973), and Bommier and Sahal-Bréchot (1978). The first unified derivation of the transfer equations, including the Zeeman and Hanle effects, was done by Landi Degl'Innocenti (1983) within the framework of quantum field theory. Although the formalisms are different, our treatment here is similar to his approach. The theory has been further developed by Landi Degl'Innocenti et al. (1991a,b).

8.2. Radiative Transfer Equation

As mentioned in the introduction, the transfer equation is obtained if we insert the radiative coherency matrix operator $\hat{D}_{\alpha\alpha'}$ in Eq. (7.75) that describes the evolution of the expectation values. For simplicity of notation, we will write $D_{\alpha\alpha'}$ instead of $\langle\hat{D}_{\alpha\alpha'}\rangle$.

As the operator $\hat{D}_{\alpha\alpha'}$ has no time dependence, both in the Schrödinger and interaction representations, which is seen from Eqs. (7.53) and (7.70), its time derivative and thus also the first term on the right hand side of Eq. (7.75) vanishes. For steady-state situations the expectation value at a given spatial location does not fluctuate in time, i.e., $\partial D_{\alpha\alpha'}/\partial t = 0$, which means that we only have spatial derivatives. The evolution of $D_{\alpha\alpha'}$ is therefore only due to transfer effects, and we have

$$\frac{\mathrm{d}D_{\alpha\alpha'}}{\mathrm{d}t} = c\,\frac{\mathrm{d}D_{\alpha\alpha'}}{\mathrm{d}s}\,. \tag{8.1}$$

Eq. (7.75) then gives us the transfer equation

$$\frac{\mathrm{d}D_{\alpha\alpha'}}{\mathrm{d}s} = \frac{1}{i\hbar c}\,\mathrm{Tr}\,(\hat{D}_{\alpha\alpha'}\,[\hat{H}_I\,,\hat{\rho}_I]\,)\,. \tag{8.2}$$

As we will see in the following, the single term on the right hand side splits up into three terms, representing radiative absorption, stimulated emission, and spontaneous emission, respectively, when we go to second order in $\hat{H}_I$ by inserting the first-order solution for $\hat{\rho}_I$. Using the second-order results given by Eqs. (7.94), (7.100), and (7.101), and renaming the summation indices in Eq. (7.101) so that j,ℓ and r,s as well as β and β' are exchanged, we obtain

$$\begin{aligned}\frac{\mathrm{d}D_{\alpha\alpha'}}{\mathrm{d}s} = -\frac{e^2\,\omega N}{4\,c\,\epsilon_0\hbar}&\sum_{j,\ell,r,s}\;\sum_{\beta,\beta'} r_{j\ell}\,r^*_{sr}\,\varepsilon^{\beta *}_{j\ell}\,\varepsilon^{\beta'}_{sr}\\ &\{\,\mathrm{Tr}\,(\hat{D}_{\alpha\alpha'}\,[\,|\,r\rangle\langle s\,|\,\hat{a}^\dagger_{\beta'}\,,[\,|\,j\rangle\langle\ell\,|\,\hat{a}_\beta\,,\hat{\rho}\,]\,]\,)\,\Phi_{j\ell}\\ &+\mathrm{Tr}\,(\hat{D}_{\alpha\alpha'}\,[\,|\,j\rangle\langle\ell\,|\,\hat{a}_\beta\,,[\,|\,r\rangle\langle s\,|\,\hat{a}^\dagger_{\beta'}\,,\hat{\rho}\,]\,]\,)\,\Phi^\dagger_{rs}\,\}\,.\end{aligned} \tag{8.3}$$

Note that we sum over the polarization indices β, β' but not over wave number vector $\boldsymbol{k}$, since $\boldsymbol{k}$ is fixed by $D_{\alpha\alpha'}$, and other wave number vectors will not contribute for reasons given in connection with expression (7.80).

Our next step is to factorize out the atomic and radiative traces. According to Eqs. (7.104) and (7.106) the first trace in Eq. (8.3) becomes

$$\begin{aligned}\mathrm{Tr}_1 &= \mathrm{Tr}\,(\hat{D}_{\alpha\alpha'}\,[\,|\,r\rangle\langle s\,|\,\hat{a}^\dagger_{\beta'}\,,[\,|\,j\rangle\langle\ell\,|\,\hat{a}_\beta\,,\hat{\rho}\,]\,]\,)\,\Phi_{j\ell}\\ &= \mathrm{Tr}\,(\,[\,|\,r\rangle\langle s\,|\,[\,\hat{D}_{\alpha\alpha'}\,,\hat{a}^\dagger_{\beta'}\,],\;|\,j\rangle\langle\ell\,|\,\hat{a}_\beta\,]\,\hat{\rho}\,)\,\Phi_{j\ell}\,.\end{aligned} \tag{8.4}$$

Inserting expression (7.53) for $\hat{D}_{\alpha\alpha'}$ and applying the commutation rule (7.35) we get

$$[\,\hat{D}_{\alpha\alpha'}\,,\hat{a}^\dagger_{\beta'}\,] = \frac{h\nu^3}{c^2}\,\hat{a}^\dagger_{\alpha'}\,\delta_{\alpha\beta'}\,, \tag{8.5}$$

which when inserted in Eq. (8.4) gives

$$\mathrm{Tr}_1 = \frac{h\nu^3}{c^2}\,\delta_{\alpha\beta'}\,\mathrm{Tr}\,(\,[\,|\,r\rangle\langle s\,|\,\hat{a}^\dagger_{\alpha'}\,,\;|\,j\rangle\langle\ell\,|\,\hat{a}_\beta\,]\,\hat{\rho}_A\otimes\hat{\rho}_R\,)\,\Phi_{j\ell}\,, \tag{8.6}$$

where we have also used Eq. (7.76), since we are going to factorize the atomic and radiative traces.

As the atomic and radiative operators commute, the commutator in Eq. (8.6) can easily be expanded:

$$\begin{aligned}&[\,|\,r\rangle\langle s\,|\,\hat{a}^\dagger_{\alpha'}\,,\;|\,j\rangle\langle\ell\,|\,\hat{a}_\beta\,] = |\,r\rangle\langle s\,|\,j\rangle\langle\ell\,|\,\hat{a}^\dagger_{\alpha'}\hat{a}_\beta \;-\; |\,j\rangle\langle\,\ell|\,r\rangle\langle s\,|\,\hat{a}_\beta\hat{a}^\dagger_{\alpha'}\\ &= \delta_{js}\,|\,r\rangle\langle\ell\,|\,\left(\frac{h\nu^3}{c^2}\right)^{-1}\hat{D}_{\beta\alpha'} \;-\; \delta_{\ell r}\,|\,j\rangle\langle s\,|\left\{\,\delta_{\beta\alpha'} + \left(\frac{h\nu^3}{c^2}\right)^{-1}\hat{D}_{\beta\alpha'}\right\}.\end{aligned} \tag{8.7}$$

The three terms on the right hand side represent absorption, spontaneous emission, and stimulated emission, respectively.

Using the factorization (7.77) we now obtain

$$\begin{aligned}\mathrm{Tr}_1 = \Big\{\,&\delta_{js}\,\mathrm{Tr}\,(\,|\,r\rangle\langle\ell\,|\,\hat{\rho}_A\,)\,\mathrm{Tr}\,(\hat{D}_{\beta\alpha'}\,\hat{\rho}_R\,)\\ &-\,\delta_{\ell r}\,\mathrm{Tr}\,(\,|\,j\rangle\langle s\,|\,\hat{\rho}_A\,)\left(\delta_{\beta\alpha'}\frac{h\nu^3}{c^2} + \mathrm{Tr}\,(\hat{D}_{\beta\alpha'}\,\hat{\rho}_R\,)\right)\Big\}\,\delta_{\alpha\beta'}\,\Phi_{j\ell}\,.\end{aligned} \tag{8.8}$$

Since according to Eq. (7.9)

$$\mathrm{Tr}\,(\,|\,j\rangle\langle\ell\,|\,\hat{\rho}_A\,) = \sum_{m,n}\rho_{mn}\,\langle n\,|\,j\rangle\langle\ell\,|\,m\rangle = \delta_{nj}\,\delta_{\ell m}\,\rho_{mn} = \rho_{\ell j}\,, \tag{8.9}$$

and as the corresponding radiative traces give the expectation values of the coherency matrix operator, we obtain

$$\mathrm{Tr}_1 = \Big\{\,(\delta_{sj}\,\rho_{\ell r} - \delta_{\ell r}\,\rho_{sj})\,D_{\beta\alpha'} \;-\; \delta_{\ell r}\,\delta_{\beta\alpha'}\,\rho_{sj}\,\frac{h\nu^3}{c^2}\Big\}\,\delta_{\alpha\beta'}\,\Phi_{j\ell}\,, \tag{8.10}$$

where we have grouped absorption and stimulated emission together.

Similarly we obtain for the second trace in Eq. (8.3)

$$\mathrm{Tr}_2 = \Big\{\,(\delta_{sj}\,\rho_{\ell r} - \delta_{\ell r}\,\rho_{sj})\,D_{\alpha\beta'} \;-\; \delta_{\ell r}\,\delta_{\alpha\beta'}\,\rho_{sj}\,\frac{h\nu^3}{c^2}\Big\}\,\delta_{\beta\alpha'}\,\Phi^\dagger_{rs}\,. \tag{8.11}$$

Inserting the expressions (8.10) and (8.11) for the two traces in the radiative transfer equation (8.3), we obtain

$$\begin{aligned}\frac{\mathrm{d}D_{\alpha\alpha'}}{\mathrm{d}s} = -\frac{\pi^2 e^2 \nu N}{c\,\epsilon_0\, h} \sum_{j,\ell,r,s} \sum_{\beta,\beta'} r_{j\ell}\, r^*_{sr}\, \varepsilon^{\beta *}_{j\ell}\, \varepsilon^{\beta'}_{sr} \\ \Big\{ \Big[(\delta_{sj}\,\rho_{\ell r} - \delta_{\ell r}\,\rho_{sj}) D_{\beta\alpha'} - \delta_{\ell r}\,\delta_{\beta\alpha'}\,\rho_{sj}\,\frac{h\nu^3}{c^2} \Big]\, \delta_{\alpha\beta'}\, \Phi_{j\ell} \\ + \Big[(\delta_{sj}\,\rho_{\ell r} - \delta_{\ell r}\,\rho_{sj}) D_{\alpha\beta'} - \delta_{\ell r}\,\delta_{\alpha\beta'}\,\rho_{sj}\,\frac{h\nu^3}{c^2} \Big]\, \delta_{\beta\alpha'}\, \Phi^{\dagger}_{rs} \Big\}.\end{aligned} \tag{8.12}$$

This equation can be given a compact matrix formulation if we introduce the matrices $\boldsymbol{g}$ and $\boldsymbol{f}$ defined by

$$\begin{aligned} g_{\alpha\alpha'} &= \frac{\pi^2 e^2 \nu N}{c\,\epsilon_0\, h} \sum_{j,\ell,r} r_{j\ell}\, r^*_{jr}\, \varepsilon^{\alpha' *}_{j\ell}\, \varepsilon^{\alpha}_{jr}\, \rho_{\ell r}\, \Phi_{j\ell}\,, \\ f_{\alpha\alpha'} &= \frac{\pi^2 e^2 \nu N}{c\,\epsilon_0\, h} \sum_{j,\ell,s} r_{j\ell}\, r^*_{s\ell}\, \varepsilon^{\alpha' *}_{j\ell}\, \varepsilon^{\alpha}_{s\ell}\, \rho_{sj}\, \Phi_{j\ell}\,. \end{aligned} \tag{8.13}$$

Forming the Hermitian adjoint of Eq. (8.13) by complex conjugation and transposition while remembering that $\rho^*_{mn} = \rho_{nm}$ as seen from Eq. (7.8), and exchanging as index names ℓ with r and j with s, we obtain

$$\begin{aligned} g^{\dagger}_{\alpha\alpha'} &= \frac{\pi^2 e^2 \nu N}{c\,\epsilon_0\, h} \sum_{j,\ell,r} r_{j\ell}\, r^*_{jr}\, \varepsilon^{\alpha' *}_{j\ell}\, \varepsilon^{\alpha}_{jr}\, \rho_{\ell r}\, \Phi^{\dagger}_{rj}\,, \\ f^{\dagger}_{\alpha\alpha'} &= \frac{\pi^2 e^2 \nu N}{c\,\epsilon_0\, h} \sum_{j,\ell,s} r_{j\ell}\, r^*_{s\ell}\, \varepsilon^{\alpha' *}_{j\ell}\, \varepsilon^{\alpha}_{s\ell}\, \rho_{sj}\, \Phi^{\dagger}_{\ell s}\,. \end{aligned} \tag{8.14}$$

From these expressions we see that the radiative transfer equation (8.12) can be given the compact form

$$\frac{\mathrm{d}D_{\alpha\alpha'}}{\mathrm{d}s} = -\sum_{\beta} \Big\{ (g_{\alpha\beta} - f_{\alpha\beta}) D_{\beta\alpha'} + D_{\alpha\beta}(g^{\dagger}_{\beta\alpha'} - f^{\dagger}_{\beta\alpha'}) \Big\} + \frac{h\nu^3}{c^2}(f_{\alpha\alpha'} + f^{\dagger}_{\alpha\alpha'})\,. \tag{8.15}$$

In this form the direct correspondence between the classical forms (4.3) and (4.41) or the phenomenologically derived non-LTE form (6.110) becomes clear. While the $\boldsymbol{g}$ matrix represents radiative absorption, the $\boldsymbol{f}$ matrix represents stimulated emission in the first two terms on the right hand side of Eq. (8.15), spontaneous emission in the last term.

The index notation that we have used so far has been convenient for the derivation of formal expressions, but it is not so transparent if one wants to see what kind of physical level (e.g. upper or lower) a given index refers to. To distinguish between upper and lower levels in the notation, we will now introduce the convention that the upper, excited magnetic sublevels are denoted m and m', the lower sublevels μ and μ'. Later we will have to expand on this notation convention in connection with the treatment of the statistical equilibrium, but we will wait doing it until it is needed.

It is straightforward to identify what type of levels the indices in the previous equations refer to, due to the following physical constraints: From the definition of $\Phi_{j\ell}$ in Eq. (7.90) follows that the j level has to be above the ℓ level (since the frequency $\omega_{j\ell}$ has to be positive). Similarly, from the definition (7.90) of $\Phi^\dagger_{rs}$ follows that the s level is above the r level. A further index constraint comes from the circumstance that the $r_{j\ell}$ matrix elements are non-zero only when they connect levels of opposite parity (since the vector $\boldsymbol{r}$ is an odd function). Therefore, if j represents an upper level, ℓ has to represent a lower level, and vice versa.

From these considerations it follows that we can write

$$\begin{aligned} g_{\alpha\alpha'} &= \frac{\pi^2\, e^2\, \nu N}{c\,\epsilon_0\, h} \sum_{\mu,\mu',m} r_{m\mu}\, r^*_{m\mu'}\, \varepsilon^{\alpha'*}_{m\mu}\, \varepsilon^{\alpha}_{m\mu'}\, \rho_{\mu\mu'}\, \Phi_{m\mu}\,, \\ f_{\alpha\alpha'} &= \frac{\pi^2\, e^2\, \nu N}{c\,\epsilon_0\, h} \sum_{\mu,m,m'} r_{m'\mu}\, r^*_{m\mu}\, \varepsilon^{\alpha'*}_{m'\mu}\, \varepsilon^{\alpha}_{m\mu}\, \rho_{mm'}\, \Phi_{m'\mu}\,. \end{aligned} \tag{8.16}$$

A more compact but less explicit way of writing these expressions is in terms of the $d_{j\ell k}$ matrix elements given by Eqs. (7.98) and (7.33):

$$\begin{aligned} g_{\alpha\alpha'} &= \frac{1}{2c\,\hbar^2} \sum_{\mu,\mu',m} d_{\mu' m\alpha}\, d_{m\mu\alpha'}\, \rho_{\mu\mu'}\, \Phi_{m\mu}\,, \\ f_{\alpha\alpha'} &= \frac{1}{2c\,\hbar^2} \sum_{\mu,m,m'} d_{\mu m\alpha}\, d_{m'\mu\alpha'}\, \rho_{mm'}\, \Phi_{m'\mu}\,. \end{aligned} \tag{8.17}$$

For the Hermitian adjoints we get in the same way as in the transition from Eq. (8.13) to (8.14)

$$\begin{aligned} g^\dagger_{\alpha\alpha'} &= \frac{1}{2c\,\hbar^2} \sum_{\mu,\mu',m} d_{\mu' m\alpha}\, d_{m\mu\alpha'}\, \rho_{\mu\mu'}\, \Phi^\dagger_{\mu' m}\,, \\ f^\dagger_{\alpha\alpha'} &= \frac{1}{2c\,\hbar^2} \sum_{\mu,m,m'} d_{\mu m\alpha}\, d_{m'\mu\alpha'}\, \rho_{mm'}\, \Phi^\dagger_{\mu m}\,. \end{aligned} \tag{8.18}$$

8.3. Statistical Equilibrium Equation

The statistical equilibrium of the atomic system is described by the steady-state solution for the evolution of the elements ρ_{np} of the density matrix. As ρ_{np} is the expectation value of the projection operator $|\,p\rangle\langle n\,|$ according to Eqs. (8.9) and (7.9), the equations of statistical equilibrium are obtained by solving the evolution equation (7.75) for the atomic projection operator.

For an atomic system at rest, stationary matrix elements imply $\mathrm{d}\rho_{np}/\mathrm{d}t = 0$, i.e., the left hand side of Eq. (7.75) is zero. A stationary situation requires a balance between gain and loss terms, which however will become explicit only when we express the second-order solution in its various components, as we will see later.

Remembering that operators with index I refer to the interaction representation, those without index to the Schrödinger representation, the stationary version of Eq. (7.75) for the case that $\hat{X}$ is the atomic projection operator is

$$\mathrm{Tr}\left(\frac{\mathrm{d}\,|\,p\rangle\langle n\,|_I}{\mathrm{d}t}\;\hat{\rho}_I \right) = \frac{i}{\hbar}\,\mathrm{Tr}\left(|\,p\rangle\langle n\,|_I\,[\,\hat{H}_I\,,\hat{\rho}_I\,] \right). \tag{8.19}$$

As according to Eq. (7.71)

$$|\,p\rangle\langle n\,|_I = e^{i\omega_{pn}\,t}\,|\,p\rangle\langle n\,|\,, \tag{8.20}$$

where ω_{pn} is defined as in Eq. (7.61), and as the Schrödinger operator $|\,p\rangle\langle n\,|$ is time independent, Eq. (8.19) becomes

$$i\omega_{pn}\,\rho_{np} = \frac{i}{\hbar}\,\mathrm{Tr}\left(|\,p\rangle\langle n\,|_I\,[\,\hat{H}_I\,,\hat{\rho}_I\,] \right). \tag{8.21}$$

Expanding the right hand side to second order as we did when going from Eq. (8.2) to (8.3), but using here the more compact expressions allowed with the $d_{j\ell\boldsymbol{k}}$ coefficients of Eq. (7.98), we get

$$\begin{aligned} i\omega_{pn}\,\rho_{np} = \frac{1}{2\hbar^2} &\sum_{j,\ell,r,s}\,\sum_{\boldsymbol{k},\boldsymbol{k}'} d_{j\ell\boldsymbol{k}}\,d_{rs\boldsymbol{k}'} \\ &\{\,\mathrm{Tr}\,(\,|\,p\rangle\langle n\,|\,[\,|\,r\rangle\langle s\,|\,\hat{a}^\dagger_{\boldsymbol{k}'}\,,[\,|\,j\rangle\langle \ell\,|\,\hat{a}_{\boldsymbol{k}}\,,\hat{\rho}\,]\,]\,)\,\Phi_{j\ell\boldsymbol{k}} \\ &+\mathrm{Tr}\,(\,|\,p\rangle\langle n\,|\,[\,|\,j\rangle\langle \ell\,|\,\hat{a}_{\boldsymbol{k}}\,,[\,|\,r\rangle\langle s\,|\,\hat{a}^\dagger_{\boldsymbol{k}'}\,,\hat{\rho}\,]\,]\,)\,\Phi^\dagger_{rs\boldsymbol{k}}\,\}\,. \end{aligned} \tag{8.22}$$

Note that in this equation we have retained the wave vector index notation $\boldsymbol{k}$ (which implicitly includes the different polarization states as well) in contrast to Eq. (8.3), in which we only sum over the polarization indices β, β'. The reason for this difference in the radiative transfer and statistical equilibrium treatments is that in Eq. (8.3), $D_{\alpha\alpha'}$ defines a given wave vector (propagation direction and frequency). All radiative operators with a different propagation direction or frequency will give zero contributions to the sum, as explained in connection with expression (7.80). As the atomic projection operator in Eq. (8.22) does not define any $\boldsymbol{k}$, we have to retain a sum over all $\boldsymbol{k}$, but since the bilinear products between the creation and annihilation operators are only allowed to differ in polarization state but not in frequency, we can use greek polarization indices for these operators and replace the sum over $\boldsymbol{k}, \boldsymbol{k}'$ by a sum over $\boldsymbol{k}, \alpha, \alpha'$.

We proceed in the same fashion as when we derived Eq. (8.4) from (8.3), but this time we use Eq. (7.105) instead of (7.106). The first trace in Eq. (8.22) then becomes

$$\mathrm{Tr}_1 = \mathrm{Tr}\,(\,[\,[\,|\,p\rangle\langle n\,|\,,\,|\,r\rangle\langle s\,|\,]\hat{a}^\dagger_{\alpha'}\,,\,|\,j\rangle\langle \ell\,|\,\hat{a}_\alpha\,]\,\hat{\rho}\,)\,\Phi_{j\ell}\,, \tag{8.23}$$

while for the second trace we get

$$\mathrm{Tr}_2 = \mathrm{Tr}\big(\big[\big[\,|\,p\rangle\langle n\,|\,,\,|\,j\rangle\langle \ell\,|\,\big]\hat{a}_\alpha\,,\,|\,r\rangle\langle s\,|\,\hat{a}^\dagger_{\alpha'}\,\big]\,\hat{\rho}\,\big)\Phi^\dagger_{rs}\,. \tag{8.24}$$

If we exchange j,ℓ and r,s, Eq. (8.24) takes the form

$$\mathrm{Tr}_2 = \mathrm{Tr}\big(\big[\big[\,|\,p\rangle\langle n\,|\,,\,|\,r\rangle\langle s\,|\,\big]\hat{a}_{\alpha'}\,,\,|\,j\rangle\langle \ell\,|\,\hat{a}^\dagger_{\alpha}\,\big]\,\hat{\rho}\,\big)\Phi^\dagger_{j\ell}\,, \tag{8.25}$$

since the $d_{j\ell\boldsymbol{k}}\,d_{rs\boldsymbol{k}'}$ product in Eq. (8.22) requires that we then also exchange α and α'. Now it becomes manifest that Tr_1 and Tr_2 are the Hermitian adjoints of each other.

The inner commutator in Eqs. (8.23) and (8.25) is readily evaluated, using the Kronecker δ_{ij}:

$$\big[\,|\,p\rangle\langle n\,|\,,\,|\,r\rangle\langle s\,|\,\big] = \delta_{nr}\,|\,p\rangle\langle s\,| - \delta_{ps}\,|\,r\rangle\langle n\,|\,. \tag{8.26}$$

To obtain Tr_1 we first need to derive

$$\begin{aligned}\big[\,|\,p\rangle\langle s\,|\,\hat{a}^\dagger_{\alpha'}\,,\,|\,j\rangle\langle \ell\,|\,\hat{a}_\alpha\,\big] &= \delta_{js}\,|\,p\rangle\langle \ell\,|\left(\frac{h\nu^3}{c^2}\right)^{-1}\hat{D}_{\alpha\alpha'}\\ &\quad -\delta_{\ell p}\,|\,j\rangle\langle s\,|\left\{\delta_{\alpha\alpha'}+\left(\frac{h\nu^3}{c^2}\right)^{-1}\hat{D}_{\alpha\alpha'}\right\},\end{aligned} \tag{8.27}$$

where we have used Eqs. (7.35) and (7.53). This represents the contribution from the operator $|\,p\rangle\langle s\,|$ on the right hand side of Eq. (8.26). The contribution from the second operator $|\,r\rangle\langle n\,|$ is then obtained from Eq. (8.27) simply by replacing index p,s with r,n. Inserting these results into Eq. (8.23) we obtain

$$\begin{aligned}\mathrm{Tr}_1 = \left(\frac{h\nu^3}{c^2}\right)^{-1}\Big\{&(\delta_{nr}\,\delta_{js}\,\rho_{\ell p} - \delta_{ps}\,\delta_{nj}\,\rho_{\ell r})D_{\alpha\alpha'}\\ &-(\delta_{nr}\,\delta_{\ell p}\,\rho_{sj} - \delta_{ps}\,\delta_{\ell r}\,\rho_{nj})\Big(D_{\alpha\alpha'} + \frac{h\nu^3}{c^2}\,\delta_{\alpha\alpha'}\Big)\Big\}\,\Phi_{j\ell}\,.\end{aligned} \tag{8.28}$$

Of the two terms on the right hand side, the first represents radiative absorption processes, the second stimulated emission (via the $D_{\alpha\alpha'}$ term) and spontaneous emission (via the $\delta_{\alpha\alpha'}$ term) processes.

Similarly, we obtain from Eqs. (8.25) and (8.26), using Eqs. (7.35) and (7.53),

$$\begin{aligned}\mathrm{Tr}_2 = \left(\frac{h\nu^3}{c^2}\right)^{-1}\Big\{&(\delta_{nr}\,\delta_{js}\,\rho_{\ell p} - \delta_{ps}\,\delta_{nj}\,\rho_{\ell r})\Big(D_{\alpha'\alpha} + \frac{h\nu^3}{c^2}\,\delta_{\alpha'\alpha}\Big)\\ &-(\delta_{nr}\,\delta_{\ell p}\,\rho_{sj} - \delta_{ps}\,\delta_{\ell r}\,\rho_{nj})D_{\alpha'\alpha}\Big\}\,\Phi^\dagger_{j\ell}\,.\end{aligned} \tag{8.29}$$

Note that in comparison with Eq. (8.28), absorption and emission have traded places, and the order between α and α' has been reversed.

Inserting Eqs. (8.28) and (8.29) in Eq. (8.22) and making use of the Kronecker δ's to eliminate two of the summation indices, we obtain

$$
\begin{aligned}
i\omega_{pn}\,\rho_{np} =& \left(\frac{h\nu^3}{c^2}\right)^{-1} \frac{1}{2\hbar^2} \sum_{\alpha,\alpha',\boldsymbol{k}} \sum_{j,\ell} \Big\{ \\
& (d_{j\ell\alpha}\, d_{nj\alpha'}\, \rho_{\ell p}\, \Phi_{j\ell} - d_{n\ell\alpha}\, d_{jp\alpha'}\, \rho_{\ell j}\, \Phi_{n\ell} \\
& - d_{n\ell\alpha}\, d_{jp\alpha'}\, \rho_{\ell j}\, \Phi^{\dagger}_{jp} + d_{\ell p\alpha}\, d_{j\ell\alpha'}\, \rho_{nj}\, \Phi^{\dagger}_{j\ell}\,)\, D_{\alpha\alpha'} \\
& - (d_{jp\alpha}\, d_{n\ell\alpha'}\, \rho_{\ell j}\, \Phi_{jp} - d_{j\ell\alpha}\, d_{\ell p\alpha'}\, \rho_{nj}\, \Phi_{j\ell} \\
& - d_{nj\alpha}\, d_{j\ell\alpha'}\, \rho_{\ell p}\, \Phi^{\dagger}_{j\ell} + d_{jp\alpha}\, d_{n\ell\alpha'}\, \rho_{\ell j}\, \Phi^{\dagger}_{n\ell}\,) \\
& \left(D_{\alpha\alpha'} + \frac{h\nu^3}{c^2}\, \delta_{\alpha\alpha'} \right) \Big\} .
\end{aligned}
\tag{8.30}
$$

The four terms in the first bracket represent radiative absorption, those in the second bracket the combined effects of stimulated and spontaneous emission. The positive and negative signs correspond to losses and gains with respect to the density matrix element ρ_{np} on the left hand side. In absorption one has gains through transitions from lower levels, losses through transitions to upper levels. In emission one has gains through transitions from upper levels, losses through transitions to lower levels.

Summation over $\boldsymbol{k}$ means summation over all directions as well as over all frequencies. The profile function Φ and the coherency matrix $D_{\alpha\alpha'}$ vary with frequency, while $D_{\alpha\alpha'}$ in addition varies with direction. The $d_{j\ell}$ factors are also direction dependent via the ε factors they contain, as shown in Eq. (7.98).

Note that the terms having the same indices of for instance Φ and $\Phi^{\dagger}$ cannot be combined, since the indices of the different terms represent different physical levels although they have the same names. Furthermore, the first index of Φ represents a higher level than the second index, while for $\Phi^{\dagger}$ it is the opposite. To make the physical levels involved manifest in the expressions as we did in Sect. 8.2 for the radiative transfer equation, we change our index notation. Thus we let m, m', and m'' represent the magnetic sublevels of the level for which the equilibrium is considered, so we replace n, p in Eq. (8.30) by m, m'. Further we let μ and μ' represent magnetic sublevels of any lower level, while r and r' stand for sublevels of any higher level.

Next we let R_{abs} denote the absorption rate from the lower levels μ, μ', while R^{abs} is the absorption rate to the upper levels r, r'. Similarly R_{stim} and R^{stim} represent the transition rates of stimulated emission to lower and from upper levels, respectively, and correspondingly for the spontaneous transition rates R_{spon} and R^{spon}. For completeness we also introduce the collisional excitation and deexcitation rates $R_{\text{coll. exc}}$, $R^{\text{coll. exc}}$, $R_{\text{coll. deexc}}$, and $R^{\text{coll. deexc}}$, which will be treated phenomenologically later.

With these definitions, Eq. (8.30) becomes

$$
\begin{aligned}
i\omega_{mm'}\,\rho_{mm'} =& R_{\text{abs}} + R^{\text{stim}} + R^{\text{spon}} + R_{\text{coll. exc}} + R^{\text{coll. deexc}} \\
& - R^{\text{abs}} - R_{\text{stim}} - R_{\text{spon}} - R^{\text{coll. exc}} - R_{\text{coll. deexc}} .
\end{aligned}
\tag{8.31}
$$

$$
R_{\text{abs}} = \frac{2\pi^2 c^2}{h^3\nu^3} \sum_{\alpha,\alpha',\boldsymbol{k}} \sum_{\mu,\mu'} d_{m\mu\alpha}\, d_{\mu' m'\alpha'}\, \rho_{\mu\mu'}\, (\Phi_{m\mu} + \Phi^{\dagger}_{\mu' m'}\,)\, D_{\alpha\alpha'} , \tag{8.32}
$$

$$R^{\mathrm{abs}} = \frac{2\pi^2 c^2}{h^3\nu^3} \sum_{\alpha,\alpha',\boldsymbol{k}} \sum_{m'',r} \big(d_{rm''\alpha}\, d_{mr\alpha'}\, \rho_{m''m'}\, \Phi_{rm''} + d_{rm'\alpha}\, d_{m''r\alpha'}\, \rho_{mm''}\, \Phi^{\dagger}_{m''r}\big)\, D_{\alpha\alpha'} \,, \tag{8.33}$$

$$R_{\mathrm{stim}} + R_{\mathrm{spon}} = \frac{2\pi^2 c^2}{h^3\nu^3} \sum_{\alpha,\alpha',\boldsymbol{k}} \sum_{\mu,m''} \big(d_{m''\mu\alpha}\, d_{\mu m'\alpha'}\, \rho_{mm''}\, \Phi_{m''\mu} + d_{m\mu\alpha}\, d_{\mu m''\alpha'}\, \rho_{m''m'}\, \Phi^{\dagger}_{\mu m''}\big)\Big(D_{\alpha\alpha'} + \frac{h\nu^3}{c^2}\,\delta_{\alpha\alpha'}\Big) \,, \tag{8.34}$$

$$R^{\mathrm{stim}} + R^{\mathrm{spon}} = \frac{2\pi^2 c^2}{h^3\nu^3} \sum_{\alpha,\alpha',\boldsymbol{k}} \sum_{r,r'} d_{r'm'\alpha}\, d_{mr\alpha'}\, \rho_{rr'} \big(\Phi_{r'm'} + \Phi^{\dagger}_{mr}\big)\Big(D_{\alpha\alpha'} + \frac{h\nu^3}{c^2}\,\delta_{\alpha\alpha'}\Big) \,. \tag{8.35}$$

8.4. Derivation of the Einstein Transition Rates

Expressions (8.32)–(8.35) represent sums over the radiative transition rates that contribute to the density matrix element $\rho_{mm'}$. The Einstein coefficients A_{ji}, B_{ji}, and B_{ij} that we introduced phenomenologically in Chapter 6 can easily be extracted from these expressions, since they represent the special case when only the diagonal elements of the density matrix (the sublevel populations) are accounted for, while the off-diagonal elements (the coherences) are disregarded.

Let us demonstrate this extraction explicitly for the Einstein A_{ji} coefficient, representing spontaneous emission. We begin by rewriting the following portion of Eq. (8.34) for R_{spon}:

$$\frac{2\pi^2 c^2}{h^3\nu^3} \sum_{\alpha,\alpha',\boldsymbol{k}} d_{m''\mu\alpha}\, d_{\mu m'\alpha'} = \frac{4\pi^3 e^2}{\epsilon_0 c h^2} \int \frac{\mathrm{d}\Omega}{4\pi} \int \mathrm{d}\nu \sum_{\alpha,\alpha'} r_{m''\mu}\, r^*_{m'\mu}\, \varepsilon^{\alpha *}_{m''\mu}\, \varepsilon^{\alpha'}_{m'\mu} \tag{8.36}$$

according to Eqs. (7.44), (7.98), and (7.33), assuming that $|\nu_{m''\mu} - \nu|$ and $|\nu_{m'\mu} - \nu|$ are $\ll \nu$. All the corresponding expressions in Eqs. (8.32)–(8.35) can of course be rewritten in a similar fashion.

Next we make use of the Wigner-Eckart theorem (7.95), from which it follows that

$$r_{m''\mu}\, r^*_{m'\mu} = (-1)^{2J_\mu + m' + m''} (2J_m + 1)\, |\langle J_m \,||\, \hat{\boldsymbol{r}} \,||\, J_\mu\rangle|^2 \begin{pmatrix} J_m & J_\mu & 1 \\ -m' & \mu & m'-\mu \end{pmatrix} \begin{pmatrix} J_m & J_\mu & 1 \\ -m'' & \mu & m''-\mu \end{pmatrix} . \tag{8.37}$$

J_m is the total angular momentum quantum number of the state that has substates with magnetic quantum numbers m, m', and m''.

For spontaneous emission, $\alpha = \alpha'$ (no polarization), and the directional dependence is exclusively due to the ε factors. The frequency dependence only resides in the profile function Φ_q. The integrations and summations in Eq. (8.36) can therefore readily be carried out. Using the expressions (3.73) for ε_q^α, and remembering that Φ_q is area normalized, we get

$$\sum_\alpha \int \frac{d\Omega}{4\pi} \varepsilon_q^{\alpha *} \varepsilon_{q'}^\alpha \int d\nu\, \Phi_q = \tfrac{2}{3}\delta_{qq'} \,. \tag{8.38}$$

Due to the Kronecker $\delta_{qq'}$ in Eq. (8.38) we see that in the first term on the right hand side of Eq. (8.34) only the factors with $m'' = m'$ contribute to the sum, while in the second term only the factors with $m'' = m$ contribute. This allows us to make use of the orthogonality property of the 3-j symbols (Brink and Satchler, 1968),

$$\sum_{\mu,q} \begin{pmatrix} J_m & J_\mu & 1 \\ -m & \mu & q \end{pmatrix}^2 = \frac{1}{2J_m+1} \,. \tag{8.39}$$

Inserting the results of (8.36)–(8.39) in Eq. (8.34) we obtain

$$R_{\rm spon} = \sum_{J_\mu} \frac{16\pi^3 e^2}{3\epsilon_0 ch^2} \frac{h\nu^3}{c^2} \,|\langle J_m \,||\, \hat{\boldsymbol{r}} \,||\, J_\mu\rangle|^2 \rho_{mm'} \tag{8.40}$$

(the sign factor from Eq. (8.37) disappears, since $2(J_\mu + m)$ is an even number), or

$$R_{\rm spon} = \sum_{J_\mu} A_{J_m J_\mu} \rho_{mm'} \,, \tag{8.41}$$

where

$$A_{J_m J_\mu} = \frac{16\pi^3 e^2 \nu^3}{3\epsilon_0 c^3 h} \,|\langle J_m \,||\, \hat{\boldsymbol{r}} \,||\, J_\mu\rangle|^2 \tag{8.42}$$

is the Einstein coefficient for spontaneous emission. The sum over all lower levels with total angular momentum J_μ in Eqs. (8.40) and (8.41) appears because it is implicitly included in the $\sum_\mu$ that appears in Eq. (8.34), in contrast to the $\sum_\mu$ that appears in Eq. (8.39), which is done over the substates only, while J_μ is kept fixed. This difference in the understanding of $\sum_\mu$ in the two cases is due to our effort of keeping the notation as compact as possible, hoping that not fully spelled out notations become clear from the context in which they appear.

A similar derivation can be made for the Einstein B coefficients for stimulated emission and absorption, by considering the transition rates for isotropic, unpolarized radiation with $D_{\alpha\alpha'} = I_\nu/2$ (Tr $\boldsymbol{D} = I_\nu$ according to Eq. (2.35)). We then retrieve the relation (6.66) between the A and B coefficients that was obtained with the phenomenological theory, and obtain

$$\begin{aligned} B_{J_m J_\mu} &= \frac{8\pi^3 e^2}{3\epsilon_0 ch^2} \,|\langle J_m \,||\, \hat{\boldsymbol{r}} \,||\, J_\mu\rangle|^2 \,, \\ B_{J_\mu J_m} &= \frac{8\pi^3 e^2}{3\epsilon_0 ch^2} \,|\langle J_\mu \,||\, \hat{\boldsymbol{r}} \,||\, J_m\rangle|^2 \,, \end{aligned} \tag{8.43}$$

and

$$(2J_\mu + 1)B_{J_\mu J_m} = (2J_m + 1)B_{J_m J_\mu} \,. \tag{8.44}$$

The last relation follows from the symmetry

$$(2J_m + 1)\,|\,\langle J_m \,||\, \hat{\boldsymbol{r}} \,||\, J_\mu\rangle\,|^2 = (2J_\mu + 1)\,|\,\langle J_\mu \,||\, \hat{\boldsymbol{r}} \,||\, J_m\rangle\,|^2\,, \tag{8.45}$$

which is readily derived from the Wigner-Eckart theorem (7.95) and the symmetry properties of the 3-j symbols.

Expressions for other useful physical quantities like refractive index n_q and oscillator strength f_{ij} can be obtained by comparing our results with those of the classical treatment (f_{ij} should not be confused with the radiative $\boldsymbol{f}$ matrix in Eqs. (8.13)–(8.18)). Thus we see from a comparison between the radiative-transfer equations (4.3) and (8.15) that we can identify the absorption matrix $g_{\alpha\alpha'}$ given by Eq. (8.16) with $-i(\omega/c)C_{\alpha\alpha'}$. The matrix $\boldsymbol{C}$ is expressed in terms of the refractive indices n_q via Eqs. (3.70) or (3.76).

Our area normalized complex profile function $\Phi_{m\mu}$, which we for convenience will denote by Φ_q, corresponds to $\mathcal{H}(a, v_{-q})\Delta\nu_D/\sqrt{\pi}$. The reason for the minus sign in front of q becomes clear when we compare Eqs. (4.11) and (4.12) for the definition of v_q with the definition of $\Phi_{m\mu}$ in Eq. (7.90), recalling the definition of $\omega_{m\mu}$ in Eq. (7.61). According to Eqs. (4.13) and (4.10), the classical refractive index n_q for the spherical vector component q of the classical oscillator can be expressed as

$$n_q - 1 = \frac{ie^2 N_\ell f_{\ell u}}{16\pi\epsilon_0 m_e \nu}\,\Phi_{-q}\,. \tag{8.46}$$

Here we have written m_e instead of m for the electron mass to avoid confusion with the magnetic quantum number and have added indices to N and f to mark that N refers to the population of the initial, lower state (index ℓ), while the oscillator strength $f_{\ell u}$ connects the lower (index ℓ) and upper (index u) levels. The replacement of N in the classical theory by Nf can be regarded as a *definition* of the oscillator strength.

According to Eq. (8.16) when it is compared with Eqs. (3.70) and (4.3), the quantum-mechanical expression for n_q is given by

$$n_q - 1 = \frac{ic}{\omega}\,\frac{\pi^2 e^2\,\nu N}{c\,\epsilon_0\,h}\,\sum_{\mu,m}(2J_\mu + 1)\begin{pmatrix} J_\mu & J_m & 1 \\ -\mu & m & q \end{pmatrix}^2 |\,\langle J_\mu \,||\, \hat{\boldsymbol{r}} \,||\, J_m\rangle\,|^2\,\rho_{\mu\mu}\,\Phi_{-q}\,, \tag{8.47}$$

where $q = \mu - m$ is to be kept fixed in the summation. Note that the comparison with Eq. (3.70) appears to suggest that we should have n_q instead of $n_q - 1$ on the left hand side of Eq. (8.47). We are however free to add a constant imaginary number to $g_{\alpha\alpha'}$, since it subtracts out in the radiative transfer equation (8.15) when forming $g_{\alpha\beta}\,D_{\beta\alpha'} + D_{\alpha\beta}\,g^\dagger_{\beta\alpha'}$. Only the version (8.47) gives correspondence with the classical theory while being consistent with the general radiative-transfer formulation.

In the case of complete redistribution over the sublevels, we have according to Eq. (6.49)

$$N(2J_\mu + 1)\rho_{\mu\mu} = N_{J_\mu} \,. \tag{8.48}$$

According to the orthogonality property of the 3-j symbols

$$\sum_{\mu,m} \begin{pmatrix} J_\mu & J_m & 1 \\ -\mu & m & q \end{pmatrix}^2 = \frac{1}{3} \,. \tag{8.49}$$

This gives

$$n_q - 1 = \frac{i\pi e^2}{6\epsilon_0 h} \sum_{J_\mu} N_{J_\mu} \, | \langle J_\mu \, || \, \hat{\boldsymbol{r}} \, || \, J_m \rangle |^2 \, \Phi_{-q} \,. \tag{8.50}$$

Identification with Eq. (8.46) when there is only one lower level J_μ then gives us the expression for the oscillator strength (if we in Eq. (8.46) replace indices ℓ, u with J_μ, J_m):

$$f_{J_\mu J_m} = \frac{8\pi^2 \, m_e \nu}{3h} \, | \langle J_\mu \, || \, \hat{\boldsymbol{r}} \, || \, J_m \rangle |^2 \,. \tag{8.51}$$

Through exchange of indices μ and m on both sides of Eq. (8.51) we define $f_{J_m J_\mu}$. From the symmetry (8.45) it then follows that

$$(2J_\mu + 1) f_{J_\mu J_m} = (2J_m + 1) f_{J_m J_\mu} \,. \tag{8.52}$$

Previously, in Eq. (6.85) combined with the quantum condition (6.88), we had derived a classical expression for the Einstein coefficient B_{12} for absorption transitions. Comparing this with Eqs. (8.43) and (8.51) we see that

$$B_{J_\mu J_m} = B_{12,\text{classical}} \, f_{J_\mu J_m} \tag{8.53}$$

as expected.

8.5. Collisional Rates and the Damping Parameter

In the preceding section we obtained an explicit expression for one of the rate terms in Eq. (8.31), R_{spon}, in the form of Eqs. (8.41) and (8.42). In the present section we will give phenomenological expressions for the four collisional rate terms in Eq. (8.31) and relate them to the previously introduced (in Sect. 7.9) damping or broadening parameter γ. In addition we will give an approximate expression for the spontaneous emission rate R^{spon}. Expressions for the remaining four rate terms in Eq. (8.31), representing absorption and stimulated emission, will be given in the next section.

When treating the collisions, we have to distinguish between the collisional decay (or gain) of the diagonal and off-diagonal elements of $\rho_{mm'}$. The decay of the off-diagonal elements, at the collisional rate γ_c, is primarily due to *elastic* collisions, which scramble the relative phases of the magnetic sublevels, but do not alter the level populations. *Inelastic* collisions on the other hand induce transitions between the levels, thereby changing the populations. Introducing in a phenomenological way collisional rate coefficients C in analogy with Einstein's radiative rate coefficients A and B, we get for (inelastic) collisional transitions *away from* the J_m level

$$
\begin{aligned}
R_{\text{coll. deexc}} &= \sum_{J_\mu} C_{J_m J_\mu} \rho_{mm'} \,, \\
R^{\text{coll. exc}} &= \sum_{J_r} C_{J_m J_r} \rho_{mm'} \,.
\end{aligned}
\tag{8.54}
$$

In the absence of an external radiation field, collisions and spontaneous emission accordingly cause a decay of the density matrix elements $\rho_{mm'}$ at the rate

$$
\Gamma = \sum_{J_\mu} (A_{J_m J_\mu} + C_{J_m J_\mu}) + \sum_{J_r} C_{J_m J_r} + (1 - \delta_{mm'})\gamma_c \,. \tag{8.55}
$$

The Kronecker $\delta_{mm'}$ in the last term is subtracted, since the elastic collisions, represented by the rate γ_c, do not affect the diagonal elements. The remaining three terms follow from Eqs. (8.41) and (8.54).

Let us denote the rate due to natural radiative decay by γ_N (cf. Sect. 5.14), given by

$$
\gamma_N = \sum_{J_\mu} A_{J_m J_\mu} \,. \tag{8.56}
$$

Let us further denote

$$
\gamma = \sum_{J_\mu} (A_{J_m J_\mu} + C_{J_m J_\mu}) + \sum_{J_r} C_{J_m J_r} + \gamma_c \,. \tag{8.57}
$$

According to Eq. (8.55) we then have

$$
\Gamma = \gamma - \delta_{mm'} \gamma_c \,. \tag{8.58}
$$

γ of Eq. (8.57) clearly represents the decay rate of the off-diagonal terms of the density matrix $\rho_{mm'}$. Our previous heuristic arguments of Sect. 7.9 for describing the decay of the density matrix elements $\rho_{m\mu}$ in the first order treatment of the Liouville equation in terms of a decay rate $\gamma/2$ when the decay of the off-diagonal elements of $\rho_{mm'}$ are given by γ then imply that the γ used in Eq. (8.57) is the same as the γ used in the line profile function Φ of Eqs. (7.90)–(7.92) to describe the line broadening.

As the collisional rate coefficients C are generally small as compared with the A coefficients and the γ_c damping parameter, it is a good approximation to omit them in Eq. (8.57), which according to Eq. (8.56) means that

$$
\gamma \approx \gamma_N + \gamma_c \tag{8.59}
$$

(cf. Sect. 5.14). This is the expression for the damping parameter to be used in Φ.

So far we have discussed decay or loss terms, but we have collisional *gain* terms as well. Since collisions scramble the phases, the collisional gain terms due to transitions *to* the J_m level will only contribute to the diagonal elements ρ_{mm}.

As these contributions are small, it is normally not necessary for their calculation to account for deviations from complete redistribution in the population of the magnetic sublevels of the upper and lower levels J_r and J_μ. Then, with the complete redistribution assumption, the collisional gain terms become

$$\begin{aligned}(R^{\text{coll. deexc}})_0 &= \sum_{J_r} \frac{1}{2J_m+1} C_{J_r J_m} \rho_{J_r} \delta_{mm'} ,\\ (R_{\text{coll. exc}})_0 &= \sum_{J_\mu} \frac{1}{2J_m+1} C_{J_\mu J_m} \rho_{J_\mu} \delta_{mm'} ,\end{aligned} \tag{8.60}$$

where we by index 0 for the R rate mark that the assumption of complete redistribution has been used, and where

$$\begin{aligned}\rho_{J_r} &= \sum_r \rho_{rr} ,\\ \rho_{J_\mu} &= \sum_\mu \rho_{\mu\mu} .\end{aligned} \tag{8.61}$$

The $1/(2J_m+1)$ factor appears, because without it the above expressions would give the rates with respect to the (normalized) population of the whole level, which is $\sum_m \rho_{mm} = (2J_m+1)\rho_{mm}$ when we have complete redistribution. Instead, the rates are given with respect to ρ_{mm}.

Similar considerations apply to the gain term due to spontaneous radiative transitions from higher levels J_r, when the effect on the off-diagonal $\rho_{mm'}$ elements is small. If the J_m level is not the ground state, the R^{spon} contribution is often small in comparison with the other rates. The complete redistribution assumption of Eq. (8.60) may then be useful. It leads to the formally similar expression

$$(R^{\text{spon}})_0 = \sum_{J_r} \frac{1}{2J_m+1} A_{J_r J_m} \rho_{J_r} \delta_{mm'} . \tag{8.62}$$

More general expressions with the complete redistribution assumption removed will be given in the next section, both for the spontaneous and the collisional transitions.

8.6. Radiatively Induced Transitions

The only rate terms in the statistical equilibrium equation (8.31) that we have not yet evaluated are those representing radiative absorption and stimulated emission. Let us first consider the rate R_{abs}, describing absorption transitions from lower levels J_μ to the J_m level. Similar to the derivation of Eq. (8.36) we obtain from Eq. (8.32) according to Eqs. (7.44), (7.98), (7.33), and the Wigner-Eckart theorem (7.95) as in Eq. (8.37)

$$R_{\text{abs}} = \frac{4\pi^3 e^2}{\epsilon_0 c h^2} \sum_{J_\mu} (-1)^{2J_\mu+m+m'} (2J_m+1) \, | \langle J_m \, || \, \hat{\boldsymbol{r}} \, || \, J_\mu \rangle |^2$$

$$\sum_{\mu,\mu'} \begin{pmatrix} J_m & J_\mu & 1 \\ -m & \mu & m-\mu \end{pmatrix} \begin{pmatrix} J_m & J_\mu & 1 \\ -m' & \mu' & m'-\mu' \end{pmatrix} \rho_{\mu\mu'} \tag{8.63}$$

$$\sum_{\alpha,\alpha'} \int \mathrm{d}\nu \, (\Phi_{m\mu} + \Phi^{\dagger}_{\mu' m'}) \int \frac{\mathrm{d}\Omega}{4\pi} \, \varepsilon^{\alpha *}_{m\mu} \, \varepsilon^{\alpha'}_{m'\mu'} \, D_{\alpha\alpha'} \, .$$

The sign factor in Eq. (8.63) can be simplified, since for both integer and half-integer quantum number systems, $J_\mu + m'$ is an integer (*both* J_μ *and* m' are either integers or half integers), so that

$$(-1)^{2J_\mu+m+m'} = (-1)^{2(J_\mu+m')} (-1)^{m-m'} = (-1)^{m-m'} \, . \tag{8.64}$$

For Lorentz-type profile functions without Doppler broadening we obtain according to Eq. (7.90)

$$\Phi_{m\mu} + \Phi^{\dagger}_{\mu' m'} = \frac{2}{\pi \Delta\nu_D} \frac{a + i(v_{mm'} + v_{\mu\mu'})/2}{(v_{m\mu} - ia)(v_{m'\mu'} + ia)} \, , \tag{8.65}$$

where we as in Eq. (3.53) have introduced the dimensionless frequency variable $v_{m\mu}$ and damping parameter a, defined by

$$\begin{aligned} v_{m\mu} &= (\nu_{m\mu} - \nu)/\Delta\nu_D \, , \\ a &= \gamma/(4\pi\Delta\nu_D) \end{aligned} \tag{8.66}$$

and correspondingly for $v_{m'\mu'}$. In addition we have defined dimensionless frequency differences between magnetic sublevels not connected by a radiative transition as

$$\begin{aligned} v_{mm'} &= (\nu_m - \nu_{m'})/\Delta\nu_D \, , \\ v_{\mu\mu'} &= (\nu_\mu - \nu_{\mu'})/\Delta\nu_D \, . \end{aligned} \tag{8.67}$$

Using Eqs. (8.64) and (8.43), Eq. (8.63) now becomes

$$R_{\text{abs}} = (-1)^{m-m'} \, \tfrac{3}{2} \sum_{J_\mu} (2J_\mu + 1) B_{J_\mu J_m}$$

$$\sum_{\mu,\mu'} \begin{pmatrix} J_m & J_\mu & 1 \\ -m & \mu & m-\mu \end{pmatrix} \begin{pmatrix} J_m & J_\mu & 1 \\ -m' & \mu' & m'-\mu' \end{pmatrix} \rho_{\mu\mu'} \tag{8.68}$$

$$\sum_{\alpha,\alpha'} \int \mathrm{d}\nu \, (\Phi_{m\mu} + \Phi^{\dagger}_{\mu' m'}) \int \frac{\mathrm{d}\Omega}{4\pi} \, \varepsilon^{\alpha *}_{m\mu} \, \varepsilon^{\alpha'}_{m'\mu'} \, D_{\alpha\alpha'} \, .$$

For the special case of complete redistribution between the magnetic sublevels of the lower J_μ levels, and no off-diagonal elements of the density matrix, we can set $m' = m$ and $\mu' = \mu$ and replace $\rho_{\mu'\mu}$ by $\rho_{J_\mu}/(2J_\mu + 1)$ according to Eq. (8.61).

If we further assume that the radiation is unpolarized, so that $D_{\alpha\alpha'} = I\delta_{\alpha\alpha'}/2$, and that it is isotropic, Eq. (8.68) becomes $(R_{\text{abs}})_0 = \sum_{J_\mu} B_{J_\mu J_m} I\, \rho_{J_\mu}/(2J_m+1)$, where I is the intensity integrated over all frequencies, weighted by the real part of the profile function, which we in the absence of Zeeman shifts denote φ_ν. Using these assumptions, replacing I_ν by the mean intensity J_ν defined by Eq. (6.80) to allow for small anisotropies in the radiation field while leaving it unpolarized, we obtain

$$(R_{\text{abs}})_0 = \sum_{J_\mu} \left(B_{J_\mu J_m} \int \varphi_\nu J_\nu \, \mathrm{d}\nu \right) \rho_{J_\mu}/(2J_m+1) \,. \tag{8.69}$$

Let us next consider B_{stim}, the rate of stimulated emission transitions from level J_m to the lower levels J_μ. Since stimulated emission is small as compared with spontaneous emission and radiative absorption, it should normally be sufficient to limit oneself to consider its effect on the diagonal terms of the density matrix only, which means that we may set $m'' = m' = m$. The expression for B_{stim} then becomes identical to that of radiative absorption if we exchange m and μ in the various factors of Eq. (8.68), and we get according to Eqs. (8.34) and (8.68)

$$\begin{aligned} R_{\text{stim}} \approx 3 \sum_{J_\mu} (2J_m+1) B_{J_m J_\mu}\, \rho_{mm} \sum_{\mu} \begin{pmatrix} J_m & J_\mu & 1 \\ -m & \mu & m-\mu \end{pmatrix}^2 \\ \sum_{\alpha,\alpha'} \int \frac{\mathrm{d}\Omega}{4\pi}\, \varepsilon^{\alpha *}_{m\mu}\, \varepsilon^{\alpha'}_{m\mu} \int \mathrm{d}v \, \frac{(a/\pi)\, D_{\alpha\alpha'}}{v^2_{m\mu} + a^2} \,. \end{aligned} \tag{8.70}$$

For isotropic, unpolarized radiation the above expression becomes $B_{J_m J_\mu} I \rho_{mm}$, as expected. Replacing I by $\int \varphi_\nu \, J_\nu \, \mathrm{d}\nu$ as we did when deriving Eq. (8.69), we obtain

$$(R_{\text{stim}})_0 = \sum_{J_\mu} \left(B_{J_m J_\mu} \int \varphi_\nu \, J_\nu \, \mathrm{d}\nu \right) \rho_{mm} \,. \tag{8.71}$$

Similarly, for the usually small effects of radiative absorption from level J_m to the *upper* levels J_r as well as for stimulated emission in the opposite direction, it should often be sufficient to only consider the effects on the diagonal density matrix elements ρ_{mm} (in which case we may set $m'' = m' = m$ in Eqs. (8.33) and (8.35)), and disregard coherences in the J_r levels. Then, with the same procedure used to obtain Eq. (8.70), we get

$$\begin{aligned} R^{\text{abs}} - R^{\text{stim}} \approx 3 \sum_{J_r} (2J_m+1) B_{J_m J_r} \sum_{r} \begin{pmatrix} J_m & J_r & 1 \\ -m & r & m-r \end{pmatrix}^2 (\rho_{mm} - \rho_{rr}) \\ \sum_{\alpha,\alpha'} \int \frac{\mathrm{d}\Omega}{4\pi}\, \varepsilon^{\alpha *}_{rm}\, \varepsilon^{\alpha'}_{rm} \int \mathrm{d}v \, \frac{(a/\pi)\, D_{\alpha\alpha'}}{v^2_{rm} + a^2} \,. \end{aligned} \tag{8.72}$$

Under isotropic and unpolarized conditions, with the assumption of complete redistribution of the sublevel populations of the upper J_r levels, we obtain

$$
\begin{aligned}
(R^{\text{abs}} - R^{\text{stim}})_0 &= \sum_{J_r} \left(B_{J_m J_r} \int \varphi_\nu \, J_\nu \, \mathrm{d}\nu \right) (\rho_{mm} - \rho_{rr}) \\
&= \sum_{J_r} [B_{J_m J_r} \, \rho_{mm} - B_{J_r J_m} \, \rho_{J_r} / (2J_m + 1)] \int \varphi_\nu \, J_\nu \, \mathrm{d}\nu \, .
\end{aligned} \tag{8.73}
$$

If we replace $D_{\alpha\alpha'}$ in Eq. (8.72) with $(h\nu^3/c^2)\delta_{\alpha\alpha'}$, we obtain

$$
R^{\text{spon}} \approx \sum_{J_r} (2J_r + 1) A_{J_r J_m} \sum_r \begin{pmatrix} J_m & J_r & 1 \\ -m & r & m - r \end{pmatrix}^2 \rho_{rr} \, \delta_{mm'} \, , \tag{8.74}
$$

which allows for deviations from complete redistribution between the sublevels of J_r. In the special case when complete redistribution applies, Eq. (8.62) is obtained.

Guided by Eq. (8.74) we can now generalize Eq. (8.60) to the case when deviations from complete redistribution is taken into account in the J_r and J_μ levels for the collisional transitions. Thus we get

$$
\begin{aligned}
R^{\text{coll. deexc}} &= \sum_{J_r} (2J_r + 1) C_{J_r J_m} \sum_r \begin{pmatrix} J_m & J_r & 1 \\ -m & r & m - r \end{pmatrix}^2 \rho_{rr} \, \delta_{mm'} \, , \\
R_{\text{coll. exc}} &= \sum_{J_\mu} (2J_\mu + 1) C_{J_\mu J_m} \sum_\mu \begin{pmatrix} J_m & J_\mu & 1 \\ -m & \mu & m - \mu \end{pmatrix}^2 \rho_{\mu\mu} \, \delta_{mm'} \, .
\end{aligned} \tag{8.75}
$$

Eq. (8.60) is retrieved in the case of complete redistribution.

8.7. Solution for the Off-diagonal Terms

The off-diagonal density matrix elements $\rho_{mm'}$, $m' \neq m$, have their main source in radiative absorption from lower levels, and they decay primarily by spontaneous emission and collisional processes (elastic and inelastic). The equilibrium Eq. (8.31) can therefore be written as

$$
i\omega_{mm'} \, \rho_{mm'} \approx R_{\text{abs}} - \Gamma \rho_{mm'} - R^{\text{abs}} + R^{\text{stim}} - R_{\text{stim}} \, . \tag{8.76}
$$

Γ of Eq. (8.55) implicitly includes the remaining loss rate terms in Eq. (8.31), but according to Eqs. (8.58) and (8.59) it can be approximated as

$$
\Gamma = \gamma \approx \gamma_N + \gamma_c \tag{8.77}
$$

when $m' \neq m$. γ_N is given by Eq. (8.56). The remaining collisional gain terms in Eq. (8.31) only affect the diagonal elements and therefore do not appear in Eq. (8.76). Similarly we ignore the effect of the R^{spon} gain term on the off-diagonal elements (which is allowed in the approximation of Eq. (8.74)).

The last three terms on the right hand side of Eq. (8.76) are usually much smaller than the first two terms and may in most cases be neglected. Eqs. (8.76) and (8.77) then give us

$$(\gamma + i\omega_{mm'})\rho_{mm'} \approx R_{\rm abs}\,. \tag{8.78}$$

With our definitions of a and $v_{mm'}$ in Eqs. (8.66) and (8.67),

$$\gamma + i\omega_{mm'} = 4\pi\Delta\nu_D\,(a + iv_{mm'}/2)\,. \tag{8.79}$$

As in Eq. (5.53) in the classical treatment of the Hanle effect it is convenient to define a "mixing angle" $\alpha_{mm'}$ that describes the interference between the two substates m and m' by

$$\tan\alpha_{mm'} = \frac{(m-m')\,g_m\omega_L}{\gamma} = \frac{v_{mm'}}{2a}\,, \tag{8.80}$$

where g_m is the Landé factor of the J_m level. Eq. (8.79) can then be written

$$\gamma + i\omega_{mm'} = 4\pi a\Delta\nu_D\,(1 + i\tan\alpha_{mm'})\,. \tag{8.81}$$

Eq. (8.65) then becomes

$$\Phi_{m\mu} + \Phi^\dagger_{\mu'm'} = \frac{2}{\Delta\nu_D}\,\frac{a/\pi}{(v_{m\mu} - ia)(v_{m'\mu'} + ia)}\,(1 + i\tan\alpha_{mm'} + i\tan\alpha_{\mu\mu'})\,. \tag{8.82}$$

Let us next introduce the "magnetic depolarization factor" $k_{\mu\mu'\to mm'}$ that governs the radiative coherency transfer from levels J_μ to level J_m. It is defined by

$$k_{\mu\mu'\to mm'} = \frac{1 + i(\tan\alpha_{mm'} + \tan\alpha_{\mu\mu'})}{1 + i\tan\alpha_{mm'}}\,. \tag{8.83}$$

Using Eqs. (8.68), (8.81), and (8.82), we can now solve Eq. (8.78) for the off-diagonal density matrix element $\rho_{mm'}$ in terms of $\rho_{\mu\mu'}$:

$$\begin{aligned}\rho_{mm'} \approx\; & \frac{3}{4\pi}\,\frac{1}{a\Delta\nu_D}(-1)^{m-m'}\sum_{J_\mu}(2J_\mu+1)B_{J_\mu J_m}\\ & \sum_{\mu,\mu'} k_{\mu\mu'\to mm'}\begin{pmatrix} J_m & J_\mu & 1\\ -m & \mu & m-\mu\end{pmatrix}\begin{pmatrix} J_m & J_\mu & 1\\ -m' & \mu' & m'-\mu'\end{pmatrix}\rho_{\mu\mu'}\\ & \sum_{\alpha,\alpha'}\int\frac{\mathrm{d}\Omega}{4\pi}\,\varepsilon^{\alpha *}_{m\mu}\,\varepsilon^{\alpha'}_{m'\mu'}\int \mathrm{d}v\,\frac{(a/\pi)\,D_{\alpha\alpha'}}{(v_{m\mu} - ia)(v_{m'\mu'} + ia)}\,.\end{aligned} \tag{8.84}$$

Note that the effect of collisions enters in two different ways: in the damping parameter a and in the depolarization factor $k_{\mu\mu'\to mm'}$. When the collision rate increases, the mixing angles $\alpha_{mm'}$ and $\alpha_{\mu\mu'}$ go to zero.

To obtain $\rho_{mm'}$ from Eq. (8.84) we also need to know $\rho_{\mu\mu'}$, which in its full generality would mean that we have to solve the full system of statistical equilibrium equations with coherences. In the common case that the off-diagonal elements $\rho_{\mu\mu'}$, $\mu' \neq \mu$, are much smaller than the diagonal ones, we can simplify the problem greatly and obtain an explicit solution for the off-diagonal elements in terms of the diagonal ones, as will be shown below. The remaining task then becomes to solve the system of equations for the diagonal terms, the result of which can be plugged into our previous solutions to obtain both $\rho_{\mu\mu'}$ and $\rho_{mm'}$.

The solution for the off-diagonal $\rho_{\mu\mu'}$ could be written in the same way as that for $\rho_{mm'}$ in Eq. (8.84) after appropriate index changes, whereby $\rho_{\mu\mu'}$ is expressed in terms of the density matrix elements $\rho_{\ell\ell'}$ of all levels J_ℓ below J_μ, unless J_μ is the ground state. The most common case in practice is however that level J_μ is in fact the ground state. In this case there is no contribution from $R_{\rm abs}$, absorption from lower levels, in the rate equation that corresponds to Eq. (8.76), which means that we have to make use of the previously disregarded terms $R^{\rm abs}$, $R^{\rm stim}$, and $R^{\rm coll.\,exc}$ (from the expression for γ, cf. Eqs. (8.57) and (8.54)). If we still ignore the effect of $R^{\rm spon}$ on the off-diagonal terms as in Eqs. (8.74) and (8.76), we obtain from Eqs. (8.76) and (8.77)

$$i\omega_{\mu\mu'}\,\rho_{\mu\mu'} \approx -R^{\rm coll.\,exc} - R^{\rm abs} + R^{\rm stim} - \gamma_{c,\mu}\,\rho_{\mu\mu'}\,, \tag{8.85}$$

where index μ in $\gamma_{c,\mu}$ marks that this elastic collision rate refers to level J_μ, in contrast to our previous γ_c, which referred to level J_m.

Let us introduce

$$\gamma_\mu = \gamma_{c,\mu} + \sum_{J_m} C_{J_\mu J_m}\,, \tag{8.86}$$

which represents Eq. (8.57) for the ground state. Using Eq. (8.54), Eq. (8.85) can now be written

$$(\gamma_\mu + i\omega_{\mu\mu'})\rho_{\mu\mu'} \approx -(R^{\rm abs} - R^{\rm stim})\,. \tag{8.87}$$

Note that in Eq. (8.86) J_m represents not a single, selected level as in the case of the statistical equilibrium equations for $\rho_{mm'}$, but *all* the levels above J_μ.

With the definitions

$$a_\mu = \gamma_\mu/(4\pi\Delta\nu_D) \tag{8.88}$$

and

$$\tan\beta_{\mu\mu'} = \frac{(\mu-\mu')\,g_\mu\omega_L}{\gamma_\mu}\,, \tag{8.89}$$

Eq. (8.87) can be solved:

$$\rho_{\mu\mu'} \approx \frac{-1}{4\pi a_\mu\Delta\nu_D}\,\frac{1}{1+i\tan\beta_{\mu\mu'}}\,(R^{\rm abs} - R^{\rm stim})\,, \tag{8.90}$$

where

$$\begin{aligned} R^{\rm abs} - R^{\rm stim} \approx 3\sum_{J_m}(2J_\mu+1)B_{J_\mu J_m}\sum_m \begin{pmatrix} J_\mu & J_m & 1\\ -\mu & m & \mu-m\end{pmatrix}^2 \\ (\rho_{\mu\mu} - \rho_{mm})\sum_{\alpha,\alpha'}\int\frac{{\rm d}\Omega}{4\pi}\,\varepsilon^{\alpha *}_{m\mu}\,\varepsilon^{\alpha'}_{m\mu}\int{\rm d}v\,\frac{(a/\pi)\,D_{\alpha\alpha'}}{v^2_{m\mu}+a^2}\end{aligned} \tag{8.91}$$

is obtained by making the appropriate index changes in Eq. (8.72). (Note that Eq. (8.72) uses the approximation that the diagonal matrix elements dominate over the off-diagonal ones, an assumption that should be handled with caution.)

Eqs. (8.90) and (8.91) represent the solution for the off-diagonal $\rho_{\mu\mu'}$ in terms of the diagonal $\rho_{\mu\mu}$ and ρ_{mm}.

The isotropic part of $\rho_{\mu\mu'}$ is obtained if we replace Eq. (8.91) with its unpolarized, complete redistribution version

$$(R^{\rm abs} - R^{\rm stim})_0 = \sum_{J_m} [\, B_{J_\mu J_m}\, \rho_{\mu\mu} - B_{J_m J_\mu}\, \rho_{J_m} / (2J_\mu + 1)] \int \varphi_\nu\, J_\nu\, {\rm d}\nu\,, \quad (8.92)$$

which is obtained by index renaming in Eq. (8.73). Directional information is contained in the more complete expression (8.91). Let us in connection with these various approximations recall that the main purpose of deriving the small off-diagonal elements $\rho_{\mu\mu'}$ is limited to obtaining corrections to $\rho_{mm'}$ due to coherency transfer from the ground state to the J_m level.

We have shown how approximate expressions for the coherences (interferences between the magnetic substates) can be obtained directly, once the sublevel populations (the diagonal density matrix elements) are known. The remaining task is then to calculate these diagonal terms.

8.8. System of Equations for the Diagonal Terms

For the diagonal terms of the density matrix with $m' = m$, the left hand side of Eq. (8.31) vanishes, since $\omega_{mm'} = 0$. Contributions from the small off-diagonal terms to the diagonal ρ_{mm} may usually be neglected, so that we only have to deal with a system of equations relating the diagonal elements to each other.

For the special case of a nearly isotropic, unpolarized radiation field and complete redistribution among the substates of the levels *different* from the J_m level, we get from Eqs. (8.41), (8.54), (8.60), (8.62), (8.69), (8.71), and (8.73) the unpolarized statistical equilibrium equations

$$\begin{aligned}
&\Big[\sum_{J_\mu} \Big(A_{J_m J_\mu} + B_{J_m J_\mu} \int \varphi_\nu\, J_\nu\, {\rm d}\nu + C_{J_m J_\mu} \Big) \\
&+ \sum_{J_r} \Big(B_{J_m J_r} \int \varphi_\nu\, J_\nu\, {\rm d}\nu + C_{J_m J_r} \Big) \Big] (2J_m + 1) \rho_{mm} \\
&= \sum_{J_r} \Big(A_{J_r J_m} + B_{J_r J_m} \int \varphi_\nu\, J_\nu\, {\rm d}\nu + C_{J_r J_m} \Big) \rho_{J_r} \\
&+ \sum_{J_\mu} \Big(B_{J_\mu J_m} \int \varphi_\nu\, J_\nu\, {\rm d}\nu + C_{J_\mu J_m} \Big) \rho_{J_\mu} \,.
\end{aligned} \quad (8.93)$$

The left hand side contains the loss terms, the right hand side the gain terms with respect to the sublevel population ρ_{mm}. From Eq. (8.93) follows that the assumption of complete redistribution among the sublevels of the J_μ and J_r levels leads to complete redistribution for the J_m level as well. In Eq. (8.93) we have in fact retrieved the standard form of the statistical equilibrium equations in the theory of unpolarized non-LTE radiative transfer.

To obtain the deviations from complete redistribution among the ρ_{mm} elements we need to use the more general expressions that were given in Eqs. (8.68), (8.70), (8.72), (8.74), and (8.75). This gives us the general statistical equilibrium equations for the diagonal elements without the assumption of complete redistribution while accounting for the anisotropy and polarization of the radiation field. We however refrain from writing down this lengthy equation here, since it can readily be assembled by the reader.

8.9. The Radiative Transfer Matrices

Guided by our applications of the Wigner-Eckart theorem to obtain explicit expressions for the rate coefficients, e.g. as in Eq. (8.68), with the Einstein $B_{J_\mu J_m}$ and $B_{J_m J_\mu}$ coefficients given by Eqs. (8.43) and (8.44), we can now formulate explicit expressions for the radiative transfer matrices $g_{\alpha\alpha'}$ and $f_{\alpha\alpha'}$ of Eq. (8.16). They become

$$\begin{gathered} g_{\alpha\alpha'} = \frac{3}{2}\frac{h\nu}{4\pi} N \sum_{J_\mu, J_m} (2J_\mu + 1) B_{J_\mu J_m} \sum_{\mu,\mu',m} \\ \begin{pmatrix} J_m & J_\mu & 1 \\ -m & \mu & m-\mu \end{pmatrix} \begin{pmatrix} J_m & J_\mu & 1 \\ -m & \mu' & m-\mu' \end{pmatrix} \varepsilon^{\alpha'*}_{m\mu}\, \varepsilon^{\alpha}_{m\mu'}\, \rho_{\mu\mu'}\, \Phi_{m\mu} \end{gathered} \tag{8.94}$$

and

$$\begin{gathered} f_{\alpha\alpha'} = \frac{3}{2}\frac{h\nu}{4\pi} N \sum_{J_\mu, J_m} (2J_m + 1) B_{J_m J_\mu} \sum_{\mu,m,m'} (-1)^{m-m'} \\ \begin{pmatrix} J_m & J_\mu & 1 \\ -m & \mu & m-\mu \end{pmatrix} \begin{pmatrix} J_m & J_\mu & 1 \\ -m' & \mu & m'-\mu \end{pmatrix} \varepsilon^{\alpha'*}_{m'\mu}\, \varepsilon^{\alpha}_{m\mu}\, \rho_{mm'}\, \Phi_{m'\mu} \,. \end{gathered} \tag{8.95}$$

Since $f^\dagger_{\alpha\alpha'}$ is obtained from $f_{\alpha\alpha'}$ simply by replacing $\Phi_{m'\mu}$ with $\Phi^\dagger_{\mu m}$ as seen from a comparison of Eqs. (8.17) and (8.18) with each other (the relation between $g^\dagger_{\alpha\alpha'}$ and $g_{\alpha\alpha'}$ is similar), the emission term in the radiative transfer equation (8.15) becomes

$$\begin{gathered} \frac{h\nu^3}{c^2}(f_{\alpha\alpha'} + f^\dagger_{\alpha\alpha'}) = \frac{3}{4}\frac{h\nu}{4\pi} N \sum_{J_\mu, J_m} (2J_m + 1) A_{J_m J_\mu} \sum_{\mu,m,m'} (-1)^{m-m'} \\ \begin{pmatrix} J_m & J_\mu & 1 \\ -m & \mu & m-\mu \end{pmatrix} \begin{pmatrix} J_m & J_\mu & 1 \\ -m' & \mu & m'-\mu \end{pmatrix} \\ \varepsilon^{\alpha'*}_{m'\mu}\, \varepsilon^{\alpha}_{m\mu}\, \rho_{mm'}\, (\Phi_{m'\mu} + \Phi^\dagger_{\mu m}) \,, \end{gathered} \tag{8.96}$$

where we have used the relation (6.66) between the Einstein A and B coefficients.

In the case when there is only one J_μ and one J_m level to be considered (no overlapping of levels with different total angular momentum), and when the off-diagonal terms $\rho_{\mu\mu'}$ of the J_μ level can be neglected, we obtain from Eq. (8.94) and the orthogonality relation (8.49) for the 3-j symbols

$$g_{\alpha\alpha'} = \tfrac{1}{2} N_{J_\mu} \, B_{J_\mu J_m} \, \frac{h\nu}{4\pi} \sum_q \Phi_{-q} \, \varepsilon_q^{\alpha *} \, \varepsilon_q^{\alpha'} \,, \tag{8.97}$$

where N_{J_μ} is defined as in Eqs. (8.48) and (6.49), and $q = \mu - m$. Note that the ε factors in Eq. (8.97) are the complex conjugates of those in Eq. (8.94), a transformation that is allowed according to Eqs. (3.35) and (3.36) if we change the sign of the index and attach a $(-1)^q$ factor to each of the ε's. Since $(-1)^{2q} = 1$, this sign factor disappears, but the remaining result is that the index of the ε's will have a sign opposite to that of Φ. The reason for doing this transformation here is to make the equivalence between our $g_{\alpha\alpha'}$ and the corresponding classical absorption matrix $C_{\alpha\alpha'}$ of Eq. (3.70) manifest. We saw in Eq. (8.46) that the classically derived $n_q - 1 \sim \Phi_{-q}$, and that the addition of a constant to n_q (like -1) is of no consequence for the radiative transfer problem, since it subtracts out in the transfer equation (as seen when forming expression (4.22)). A comparison between Eqs. (8.97) and (3.70) thus shows that both the frequency and angular dependence of the absorption matrix are identical in the quantum and classical theories when coherences in the lower atomic state can be neglected.

Similarly, if the coherences in the upper state (off-diagonal elements $\rho_{mm'}$) can be neglected, we obtain from Eq. (8.95)

$$f_{\alpha\alpha'} = \tfrac{1}{2} N_{J_m} \, B_{J_m J_\mu} \, \frac{h\nu}{4\pi} \sum_q \Phi_{-q} \, \varepsilon_q^{\alpha *} \, \varepsilon_q^{\alpha'} \,. \tag{8.98}$$

When the off-diagonal elements of both the upper and lower levels are neglected, we may define an absorption coefficient κ_L as in our phenomenological treatment, Eq. (6.100):

$$\kappa_L = (N_{J_\mu} \, B_{J_\mu J_m} - N_{J_m} \, B_{J_m J_\mu}) \frac{h\nu}{4\pi} \,. \tag{8.99}$$

Then

$$g_{\alpha\alpha'} - f_{\alpha\alpha'} = \tfrac{1}{2} \kappa_L \sum_q \Phi_{-q} \, \varepsilon_q^{\alpha *} \, \varepsilon_q^{\alpha'} \,. \tag{8.100}$$

In the same way as when the classical absorption matrix $C_{\alpha\alpha'}$ was transformed from a dependence on n_q as in Eq. (3.70) to an expression in terms of $n_{I,Q,U,V}$ as in Eq. (3.75), we can transform $g_{\alpha\alpha'} - f_{\alpha\alpha'}$ from a dependence on Φ_{-q} as in Eq. (8.100) to an expression in terms of $\Phi_{I,Q,U,V}$, defined in Eqs. (6.57)–(6.59). This leads to a radiative transfer equation of the type (6.110). It is thus clear that the absorption-stimulated emission term of Eq. (8.100) is in full agreement with the corresponding term of the classical theory of Chapters 3-4 and with the phenomenological theory of Chapter 6.

The classical emission term $\boldsymbol{F}_g$ in Eq. (4.7) was derived by replacing the radiative coherency matrix $\boldsymbol{D}$ with a scalar, representing unpolarized radiation of intensity S as in Eq. (4.6), and then solving the radiative-transfer equation (4.3) assuming that there are no gradients in the radiation field, i.e., that there is detailed balance between emission and absorption. In our case this procedure is not needed, since an emission term, Eq. (8.96), is already completely specified by our quantum field theory. It is nevertheless of interest to turn the problem around and, with the use of our known emission term, see what the assumption of detailed balance between emission and absorption for an unpolarized radiation field of intensity S physically implies. Applying this assumption to the radiative-transfer equation (8.15) gives

$$\tfrac{1}{2}S\,(g_{\alpha\alpha'} - f_{\alpha\alpha'} + g^{\dagger}_{\alpha\alpha'} - f^{\dagger}_{\alpha\alpha'}) = \frac{h\nu^3}{c^2}(f_{\alpha\alpha'} + f^{\dagger}_{\alpha\alpha'})\,. \tag{8.101}$$

Denoting

$$\begin{aligned} f &= f_{\alpha\alpha'} + f^{\dagger}_{\alpha\alpha'}\,, \\ g &= g_{\alpha\alpha'} + g^{\dagger}_{\alpha\alpha'}\,, \end{aligned} \tag{8.102}$$

we can solve for S:

$$S = \frac{2h\nu^3}{c^2}\,\frac{1}{(g/f) - 1}\,. \tag{8.103}$$

g corresponds to radiative absorption, f to stimulated emission.

In the absence of off-diagonal elements in the density matrix of the upper and lower levels, we obtain from Eqs. (8.97), (8.98), and (8.44)

$$\frac{g}{f} = \frac{N_{J_\mu}\,B_{J_\mu J_m}}{N_{J_m}\,B_{J_m J_\mu}} = \frac{(2J_m+1)N_{J_\mu}}{(2J_\mu+1)N_{J_m}}\,. \tag{8.104}$$

In the case when the level populations obey Boltzmann statistics,

$$\frac{g}{f} = e^{h\nu/kT}\,, \tag{8.105}$$

which means that S equals the Planck function, as expected. This result is thus retrieved as a very special case of our general formulation of the radiative-transfer problem.

While the off-diagonal terms are often important in the emission term in the form of coherent scattering (see next section), they can normally be neglected in the absorption and stimulated emission terms, in which case the phenomenological theory of Chapter 6 applies. From this as well as from the classical theory we immediately see how the coherency matrix formulation of the radiative transfer problem can be transformed into a Stokes vector formulation. Using the basic relation (2.43) between the coherency matrix and the Stokes parameters, the transformation of the transfer equation was done in Eqs. (4.26)–(4.36). This transformation was taken over in the phenomenological non-LTE theory of Chapter 6. The relation between the two formulations is seen when comparing Eqs. (6.110) and (6.104) (without coherent scattering) with each other.

8.10. Scattering Amplitudes and the Phase Matrix

Radiative scattering can be regarded as a two-stage process: radiative absorption followed by spontaneous emission. Formally, the emission term Eq. (8.96) of the radiative transfer equation describes directly only the second stage of the process (the emission), while the first stage is implicitly contained in the density matrix elements $\rho_{mm'}$ of the excited state. Although $\rho_{mm'}$ is primarily determined by radiative absorption from lower levels, as shown by Eqs. (8.76) and (8.78), many other processes, including collisional transitions, spontaneous transitions from above, or stimulated transitions, may jointly determine the atomic polarization $\rho_{mm'}$ of the excited state.

Our formalism is capable of treating any such general situation, whereby the scattering process can be regarded as a special case. In general terms, the two groups of processes governing the polarization of the emitted radiation are: (i) the preparation of the atomic polarization of the excited level, and (ii) the spontaneous emission. Only in the case when (i) is due to radiative absorption alone are we dealing with scattering transitions.

If the initial and final levels (characterized by total angular momentum quantum numbers J_{μ_i} and J_{μ_f}, respectively) of a scattering process are the same, we speak of Rayleigh scattering, while if they are different we are dealing with Raman scattering.

According to Eqs. (8.17) and (8.18), the spontaneous emission term in the radiative transfer equation (8.15) is

$$\frac{h\nu^3}{c^2}(f_{\alpha\alpha'} + f^\dagger_{\alpha\alpha'}) = \frac{2\pi^2\,\nu^3}{hc^3} \sum_{\mu_f,m,m'} d_{\mu_f m\alpha}\, d_{m'\mu_f\alpha'}\, \rho_{mm'}\,(\Phi_{m'\mu_f} + \Phi^\dagger_{\mu_f m})\,, \tag{8.106}$$

where we by index f in μ_f mark that it refers to the magnetic substate of the final state, to distinguish it from the later used magnetic quantum number μ_i of the initial state.

From Eq. (8.82) and the area normalization of the profile functions it follows that

$$\Phi_{m'\mu_f} + \Phi^\dagger_{\mu_f m} = \frac{2\pi\Delta\nu_D}{a}\,(1 + i\tan\alpha_{mm'}\,)\,\Phi_{m'\mu_f}\,\Phi^\dagger_{\mu_f m}\,. \tag{8.107}$$

Next we need an expression for $\rho_{mm'}$ in Eq. (8.106). According to Eqs. (8.78) and (8.81), if we only consider the dominant process of radiative absorption,

$$\rho_{mm'} = \frac{R_{\rm abs}}{4\pi a\Delta\nu_D\,(1 + i\tan\alpha_{mm'}\,)}\,, \tag{8.108}$$

where $R_{\rm abs}$ is given by Eq. (8.32), with μ, μ' in the index notation replaced by μ_i, μ'_i. This expression was obtained without accounting for Doppler broadening. The main effect of the Doppler broadening is to make a frequency smoothing of $R_{\rm abs}$ such that spectral variations on a scale smaller than the Doppler width $\Delta\nu_D$ are filtered out. A more complete discussion of the general frequency-redistribution problem for polarized scattering in a magnetic field is given in Chapters 5 and 10.

Here we will limit the discussion to the *weak-field case*, when the Zeeman splitting is much smaller than the Doppler width ($\Delta\nu_H \ll \Delta\nu_D$). In this common case we can perform frequency averaging of Eq. (8.32) as in Sect. 5.11, which gives us (since $\int \mathrm{d}\nu \, (\Phi_{m\mu} + \Phi^\dagger_{\mu' m'}) = 2$)

$$\bar{R}_{\mathrm{abs}} = \frac{4\pi^2 c^2}{h^3 \nu^3} \sum_{\beta,\beta',\boldsymbol{k}} \sum_{\mu_i,\mu_i'} d_{m\mu_i\beta} \, d_{\mu_i' m' \beta'} \, \rho_{\mu_i\mu_i'} \, D_{\beta\beta'} \, . \tag{8.109}$$

We let indices β, β' refer to the incident, exciting radiation (while indices α, α' are reserved for the emitted radiation).

Let us now define *scattering amplitudes* $w_{\alpha\beta}$ and $w^*_{\alpha'\beta'}$ as

$$\begin{aligned} w_{\alpha\beta} &= \frac{2\pi^2}{ah^2} \sum_{J_m,m} d_{m\mu_i\beta} \, d_{\mu_f m\alpha} \, \Phi_{m\mu_f} \, , \\ w^*_{\alpha'\beta'} &= \frac{2\pi^2}{ah^2} \sum_{J_m,m} d_{m\mu_f\alpha'} \, d_{\mu_i m\beta'} \, \Phi^\dagger_{\mu_f m} \, , \end{aligned} \tag{8.110}$$

where we have written J_m under the summation symbol to emphasize that the sum can extend over the substates of several levels (to explicitly allow for the quantum-mechanical interference between states of different total angular momentum). Note that with this definition of the scattering amplitude,

$$w_{\alpha\beta} \sim \sum_{J_m,m} \frac{\langle J_{\mu_f}\mu_f \,|\, \hat{\boldsymbol{r}}\cdot\boldsymbol{e}_\alpha \,|\, J_m m\rangle \langle J_m m \,|\, \hat{\boldsymbol{r}}\cdot\boldsymbol{e}_\beta \,|\, J_{\mu_i}\mu_i\rangle}{\omega_{m\mu_f} - \omega - i\gamma/2} \tag{8.111}$$

as indicated in Eq. (6.96).

Using Eqs. (8.107)–(8.110) in Eq. (8.106) and disregarding the off-diagonal terms in the initial level (i.e., assuming $\mu_i' = \mu_i$), the emission term becomes

$$\frac{h\nu^3}{c^2}(f_{\alpha\alpha'} + f^\dagger_{\alpha\alpha'}) = \frac{1}{c} \sum_{\mu_i} \rho_{\mu_i\mu_i} \sum_{\mu_f} \sum_{\beta,\beta',\boldsymbol{k}} w_{\alpha\beta} \, w^*_{\alpha'\beta'} \, D_{\beta\beta'} \, . \tag{8.112}$$

Eq. (8.112) implies that the coherency matrix is transformed through

$$D_{\alpha\alpha'} = \sum_{\beta,\beta'} w_{\alpha\beta} \, D_{\beta\beta'} \, w^\dagger_{\beta'\alpha'} \tag{8.113}$$

by the $\boldsymbol{w}$ matrix, in agreement with Eq. (2.36).

As in Eq. (2.39) we can now construct a 4×4 matrix $\boldsymbol{W}$ out of the 2×2 complex scattering amplitude matrices $\boldsymbol{w}$:

$$\boldsymbol{W} = \sum_{\mu_i} \rho_{\mu_i\mu_i} \sum_{\mu_f} \boldsymbol{w} \otimes \boldsymbol{w}^* \tag{8.114}$$

with elements $w_{\alpha\beta}\, w^*_{\alpha'\beta'}$. This matrix is useful, because it allows us to obtain the Mueller matrix $\boldsymbol{M}$ for scattering of the Stokes vector $\boldsymbol{I}$ according to Eqs. (2.48), (5.2), and (5.11) via

$$\boldsymbol{M} = \boldsymbol{TWT}^{-1}\,, \tag{2.48}$$

where $\boldsymbol{T}$ is given by Eq. (2.49). The scattered Stokes vector is then proportional to $\boldsymbol{MI}$.

For Rayleigh scattering ($J_{\mu_i} = J_{\mu_f} = J_\mu$) we obtain from Eqs. (8.110), (8.51), (7.98), and (7.95), ignoring unimportant constants of proportionality,

$$w_{\alpha\beta} \sim \sum_{J_m, m} f_{J_\mu J_m} \begin{pmatrix} J_m & J_\mu & 1 \\ -m & \mu_i & m-\mu_i \end{pmatrix} \begin{pmatrix} J_m & J_\mu & 1 \\ -m & \mu_f & m-\mu_f \end{pmatrix} (-1)^{m-\mu_i}\, \varepsilon^\beta_{\mu_i m}\, \varepsilon^\alpha_{m\mu_f}\, \Phi_{m\mu_f}\,, \tag{8.115}$$

where $f_{J_\mu J_m}$ is the oscillator strength between the lower and upper levels. Note that a factor $(2J_\mu+1)$ has been omitted, since it remains constant and therefore can be absorbed in the proportionality constant if only one lower level J_μ is considered. When more than one initial or final J-state is considered, however, this factor has to be explicitly used (see below).

For the case that $J_\mu = 0$, we see from Eq. (8.115) that

$$w_{\alpha\beta} \sim \sum_q \Phi_{-q}\, \varepsilon^{\alpha *}_q\, \varepsilon^\beta_q\,, \tag{8.116}$$

which is identical to (when ignoring the constant of proportionality) the classical Jones scattering matrix that was given by Eq. (3.85), as can be seen from Eq. (8.46).

In the Raman scattering case ($J_{\mu_i} \neq J_{\mu_f}$),

$$w_{\alpha\beta} \sim \sum_{J_m, m} \langle J_{\mu_i} \| \hat{\boldsymbol{r}} \| J_m\rangle^* \langle J_{\mu_f} \| \hat{\boldsymbol{r}} \| J_m\rangle \begin{pmatrix} J_m & J_{\mu_i} & 1 \\ -m & \mu_i & m-\mu_i \end{pmatrix} \begin{pmatrix} J_m & J_{\mu_f} & 1 \\ -m & \mu_f & m-\mu_f \end{pmatrix} (-1)^{m-\mu_i}\, \varepsilon^\beta_{\mu_i m}\, \varepsilon^\alpha_{m\mu_f}\, \Phi_{m\mu_f}\,. \tag{8.117}$$

The reduced matrix elements of the real operator $\hat{\boldsymbol{r}}$ are always real (since the radial wave equation has only real terms, and the imaginary factor in the non-radial spherical harmonics vanishes when integrating over all directions, which is done when forming the matrix elements). They can therefore be replaced by $\pm\sqrt{f_{J_{\mu_i} J_m}}$ and $\pm\sqrt{f_{J_{\mu_f} J_m}}$, respectively (since the factor of proportionality is unimportant for $w_{\alpha\beta}$). The sign of + or − however depends in a complicated way on the J, L, and S quantum numbers of the transition. This will be dealt with in the next chapter (Sect. 9.11).

If there are more than one initial or final level, we have to attach the factor $[(2J_{\mu_i}+1)(2J_{\mu_f}+1)]^{1/2}$ to the expression in Eq. (8.117), since the sum over μ_i and μ_f in Eq. (8.112) implicitly also includes a sum over J_{μ_i} and J_{μ_f}.

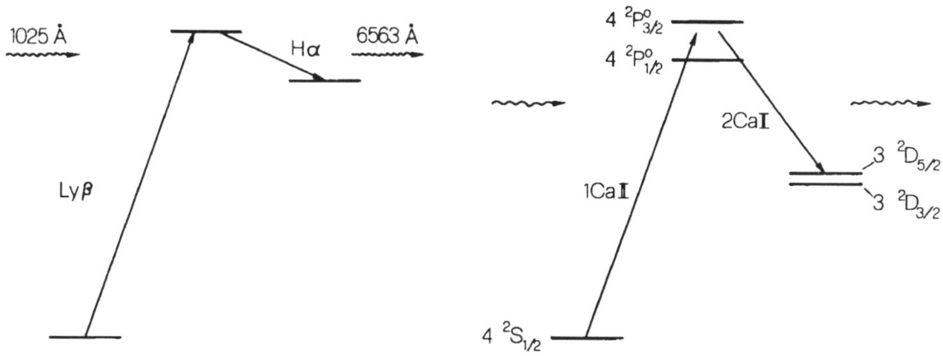

Fig. 8.1. Examples of Raman scattering in hydrogen (absorption in H Lyβ 1025 Å, emission in Hα 6563 Å) and Ca II (absorption in the ultraviolet H (3968 Å) or K (3933 Å) lines, emission in the infrared triplet at 8498, 8542, and 8662 Å). From Stenflo et al. (1983b).

Figure 8.1 illustrates two cases in which Raman scattering is likely to play a significant role, although the other processes populating the excited level must also be accounted for when computing the polarized line profiles.

When forming the tensor product $\boldsymbol{w} \otimes \boldsymbol{w}^*$, there will be cross products (interferences) between substates of different magnetic quantum number m, resulting in what we call the Hanle effect, as well as cross products between magnetic substates belonging to levels of different total angular momentum J_m (level-crossing interference). Our scattering matrix formulation is very convenient, because it avoids the statistical equilibrium problem, the solution of which is already implicitly included via the used expression for $\rho_{mm'}$. The formalism provides an intuitive understanding of the physics involved, and the calculation of the emitted radiation is straightforward. The symmetric form of Eq. (8.112) was however arrived at assuming that there are no off-diagonal elements $\rho_{\mu_i\mu_i'}$ for the initial level, which is generally a good aproximation.

Our scattering formalism can however easily be extended to the case when the symmetry is broken by the presence of off-diagonal elements $\rho_{\mu_i\mu_i'}$. Let us by $\boldsymbol{w}(\mu_i')$ define the scattering amplitude matrix that is obtained from Eq. (8.110) if we replace μ_i by μ_i'. Then, if we also transform the sum over $\boldsymbol{k}$ in Eq. (8.112) to integrals via Eq. (7.44), we get

$$\frac{h\nu^3}{c^2}(f_{\alpha\alpha'} + f^\dagger_{\alpha\alpha'}) \sim \int \frac{\mathrm{d}\Omega}{4\pi} \int \mathrm{d}\nu \sum_{\mu_i,\mu_i'} \rho_{\mu_i\mu_i'} \sum_{\mu_f} \sum_{\beta,\beta'} w_{\alpha\beta}(\mu_i)\, D_{\beta\beta'}\, w^\dagger_{\beta'\alpha'}(\mu_i'), \tag{8.118}$$

which accounts for all cases that can occur in Rayleigh and Raman scattering. For the off-diagonal components $\rho_{\mu_i\mu_i'}$ in Eq. (8.118) we have to insert the values obtained from the solution of the statistical equilibrium equations, e.g. from Eqs. (8.90) and (8.91). The $\boldsymbol{W}$ matrix can easily be generalized in the same way to account for coherences in the lower level.

For the He I D_3 5876 Å transition the polarization signature due to coherences among the Zeeman sublevels of the lower-level fine-structure components $^3P_{2,1,0}$

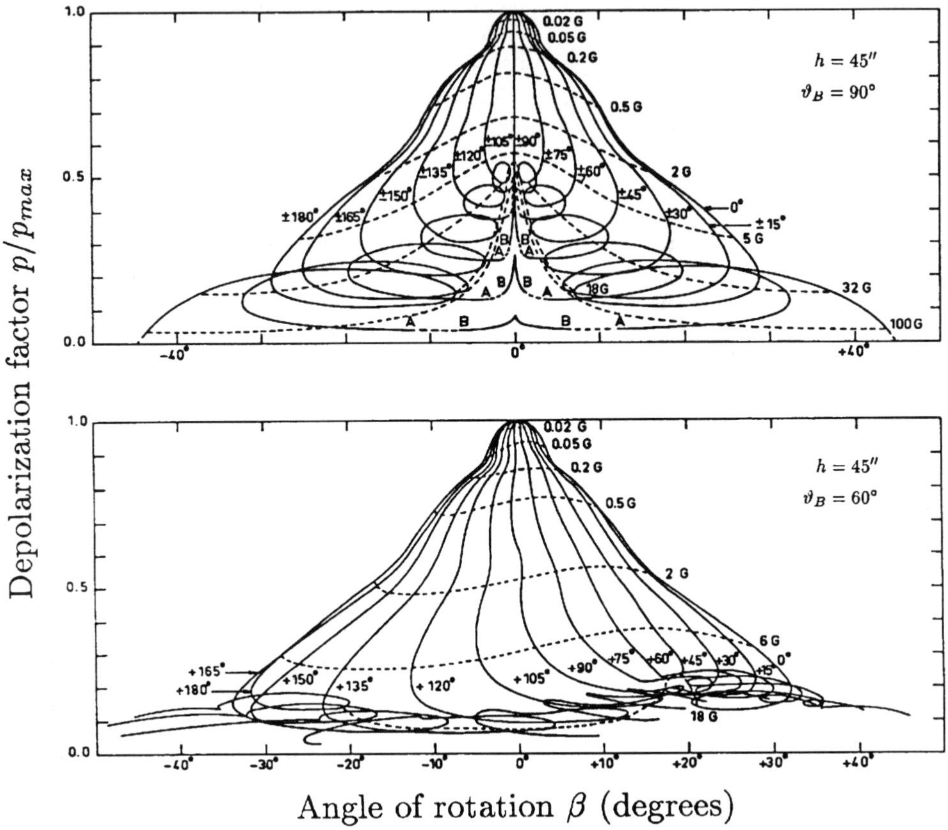

Fig. 8.2. Example of the effect of level crossings (which cause the loops in the diagrams) and coherences among the lower sublevels of the transition (which cause the undulations of the outer contours). The upper diagram, for which the magnetic field vector is assumed to lie in the horizontal plane, is to be compared with the corresponding diagrams in Fig. 5.4 for a classical oscillator. In the lower diagram the symmetry is broken, since the field vector is assumed to be at an angle of 60° with respect to the vertical direction. From Bommier (1980).

can be discerned and separated from the entirely different signature of the level crossings between the fine-structure components ${}^3D_{3,2,1}$ of the upper level. These signatures are illustrated in the Hanle-effect diagrams of Fig. 8.2. As was shown in the energy-level diagram of Fig. 7.1, level crossing starts already for fields of about 10 G, within the range where the Larmor precession frequency is comparable to the inverse life time of the excited states. This gives rise to prominent loop-like features in the diagrams. The signature of the lower-state coherences is more modest, in the form of mild undulations of the contours.

For the weak-field case that we are considering here, with the assumption $\Delta\nu_H \ll \Delta\nu_D$ (cf. the discussion of Eq. (8.109)), the angular-dependent part of the scattering Mueller matrix $\boldsymbol{TWT}^{-1}$ that describes the polarization properties factorizes out from the frequency redistribution problem that is described by a scalar function, as explained in Sects. 5.9 and 5.10. This allows us to form a scattering *phase matrix* $\boldsymbol{P}$, which is normalized as in Eq. (5.43). The results of the present section show that it is possible to form a phase matrix that contains all the information on the angular variation of the polarization properties of the scattered radiation, including any coherence effects in the upper and lower levels, level crossings, fluorescence effects, etc., while being decoupled from the frequency redistribution problem. Hence $\boldsymbol{P} \sim \boldsymbol{T}\boldsymbol{W}\boldsymbol{T}^{-1}$ without the frequency-dependent factor. All the physics governing $\boldsymbol{P}$ is contained in the matrix $\boldsymbol{W}$, which will be explored in the next chapter.

CHAPTER 9

RAYLEIGH AND RAMAN SCATTERING

9.1. Introduction

The statistical multi-level approach of the preceding chapter that is based on the density matrix formalism provides a foundation for physical understanding and for the treatment of quite general problems. It can account for the effects of coherences of any atomic level, although the effects of collisions are included in a phenomenological and incomplete way (this will be improved on in the next chapter). In practice, however, coherence phenomena play a significant role only in connection with scattering processes (Rayleigh, resonant, Raman, and fluorescent scattering). The dominant polarization effects come from the coherences in the intermediate, excited states of the scattering process, while lower-state coherences only result in minor perturbations of the scattering polarization, effects that would be very hard to measure in astrophysical situations. It is therefore a good approximation to neglect atomic polarization and coherences (phase relations between the magnetic substates) of the initial level of the scattering process. This greatly simplifies the problem, since the scattering polarization may be calculated without solving the polarized statistical equilibrium equations.

The reason why initial-level coherences are expected to be small is that the lower level is generally long-lived so that the magnetic substates become well separated even for very weak fields (which are present everywhere in the solar atmosphere), and that the rate of collisions, which scramble the relative phases, is generally much higher than the inverse life-time of the initial atomic state, at least in the part of the solar atmosphere where most of the spectral lines in the visible part of the solar spectrum are formed.

Instead of using the cumbersome general formalism it is for the above reasons preferable to develop a special scattering formalism, which is much simpler and more transparent, and with which closed analytical expressions can be developed. In most cases this is the only formalism that needs to be used for treating coherence effects in the solar atmosphere, without having to resort to the general theory. In the end of the previous chapter we indicated how the scattering processes could be formulated in terms of a scattering phase matrix $\boldsymbol{P}$, which constitutes the connection between the quantum field theory and the classical treatment of Chapter 5. As we will show below, $\boldsymbol{P}$ always has the form given by Eq. (5.68), regardless of whether we are dealing with Rayleigh or Raman scattering, as long as we neglect coherences and atomic polarization in the initial atomic state. The physics of coherence phenomena on the sun (in the rest frame, excluding collisional effects) then boils down to the calculation of the expressions for the polarizability factors $W_{1,2}$ in the phase matrix of Eq. (5.68).

The physics of scattering may be treated in a unified way, to cover the various cases that can occur. The two main categories of scattering processes are *Rayleigh* and *Raman* scattering. In our terminology Rayleigh scattering refers to the case when the n, L, and J quantum numbers are the same for the initial and final states of the scattering process, Raman scattering to the case when any one of these numbers is different. In the special case of Rayleigh scattering when the incident photon has a frequency that differs from the resonant frequency of an atomic transition by an amount that is not much larger than the natural damping width of the transition one usually speaks of *resonant* scattering. The corresponding special case of Raman scattering is called *fluorescent* scattering. For Rayleigh and Raman scattering in the distant dispersion wings of a transition the scattering is frequency coherent as a consequence of energy conservation, but for resonant and fluorescent scattering one has to deal with the more complex partial redistribution problem (see next chapter).

After we have explored the symmetry properties of the phase matrix we will derive algebraic expressions for the polarizabilities $W_{1,2}$ in terms of the L, S, and J quantum numbers, for both Rayleigh and Raman scattering, including the case when there are arbitrary quantum-mechanical interferences in the excited state between levels with different total angular momentum quantum numbers J_m. These expressions are then checked by invoking a generalized principle of spectroscopic stability (requiring that the expressions should become independent of the spin quantum number S when the fine-structure splitting goes to zero).

One aspect that has not been accounted for by the statistical formalism of the preceding chapter is the frequency correlation between the absorption and scattering events. As the frequency redistribution was given in terms of a profile function Φ and the individual transitions in the statistical equilibrium were treated as stochastically independent of each other, the special case of complete frequency redistribution (cf. Chapter 5) was implicitly assumed. The most direct avenue of making frequency coherence between absorption and scattering manifest and to obtain the rest frame scattering redistribution matrix is via time-dependent quantum-mechanical perturbation theory, which we will introduce in the next section (in the absence of collisions and atomic polarization of the initial atomic state).

9.2. Transition Rates in Time-dependent Perturbation Theory

If the initial state (at time $t = 0$) of the atom is the eigenstate $|\,a\,\rangle$, then the solution (7.2) of the Schrödinger equation (7.1) for the wave function at time t is given by

$$\psi(t) = e^{-i\hat{H}t/\hbar}\,|\,a\,\rangle\,. \tag{9.1}$$

The probability of finding the system at time t in the final state $|f\,\rangle$ is

$$|\,\langle\, f\,|\,\psi(t)\,\rangle\,|^2 = |\,\langle\, f\,|\,e^{-i\hat{H}t/\hbar}\,|\,a\,\rangle\,|^2\,. \tag{9.2}$$

The transition rate $1/\tau$ per atom for transitions from state $|\,a\rangle$ to all possible states $|\,f\,\rangle$ is

$$\frac{1}{\tau} = \frac{\mathrm{d}}{\mathrm{d}t} \sum_f |\langle f \,|\, \psi(t) \rangle|^2 \,, \tag{9.3}$$

where we have summed over all the final substates. As we will see, Eq. (9.3) gives us to first order of the perturbation (of the Hamiltonian $\hat{H}$) the Einstein A_{ji} or B_{ij} coefficients, to second order the Kramers-Heisenberg scattering cross section.

According to Eq. (7.12) we can write $\hat{H}$ as a sum of a time-independent part $\hat{H}_0$ and a time-dependent perturbation $\hat{H}'$. Only the latter is capable of changing the state of the system. Since $\hat{H}_0$ and $\hat{H}'$ do not commute, we cannot expand the exponential $\exp(-i\hat{H}t/\hbar)$ in the normal, straightforward way, but with some mathematical transformations a perturbation expansion is still possible. We will here not list these mathematical steps, since they are part of standard time-dependent perturbation theory (cf. Loudon, 1983), but instead give the results for τ to second order in the perturbation $\hat{H}'$:

$$\begin{aligned} \frac{1}{\tau} \approx \frac{2\pi}{\hbar^2} \sum_f & \Bigg| \langle f \,|\, \hat{H}' \,|\, a \rangle \\ & + \sum_b \frac{\langle f \,|\, \hat{H}' \,|\, b \rangle \langle b \,|\, \hat{H}' \,|\, a \rangle}{E_a - E_b} \Bigg|^2 \, \delta\Big(\frac{E_a - E_f}{\hbar}\Big) \,. \end{aligned} \tag{9.4}$$

It is important to remember that the energies $E_{a,b,f}$ involved here represent matter *and* radiation together, not just the atomic system. When the atomic parts of a and f are different, a photon exchange is therefore necessary to make the δ function, which represents energy conservation, different from zero. The first term on the right-hand side of Eq. (9.4) contains $\hat{H}'$ only once (first-order contribution) and consequently represents the transition rate with the exchange of a single photon. The second term, where $\hat{H}'$ occurs twice (second-order contribution), represents a two-photon process with an intermediate state b, i.e., a scattering event.

9.3. Explicit Expressions for the Transition Rates

To explicitly evaluate Eq. (9.4) we can use expressions (7.63) and (7.64) for $\hat{H}'$ and as in Sect. 8.4 calculate all the proportionality factors and integrate over all the photon states. For the first-order contribution to Eq. (9.4) we then obtain

$$\Big(\frac{1}{\tau}\Big)_{\text{first order}} = \frac{2\pi}{\hbar^2} \sum_f \Big| \langle f \,|\, \hat{H}' \,|\, a \rangle \Big|^2 \, \delta\Big(\frac{E_a - E_f}{\hbar}\Big) = \sum_f A_{af} \,, \tag{9.5}$$

where A_{af} is the Einstein coefficient of spontaneous emission if we assume that the atomic state $|\, a \rangle$ is above the $|f\rangle$ state.

The second-order contribution that represents scattering involves two photons, the incident one with frequency ω', and the scattered one with frequency ω. If we denote the atomic frequencies of the initial and final states by ω_a and ω_f, we have

$$\begin{aligned} E_a &= \hbar(\omega' + \omega_a)\,, \\ E_f &= \hbar(\omega + \omega_f)\,. \end{aligned} \tag{9.6}$$

The scattering transition rate then becomes

$$\left(\frac{1}{\tau}\right)_{\text{scattering}} = \frac{2\pi}{\hbar^2} \sum_{f_{\text{atom}}} \sum_{\boldsymbol{k}} |w|^2\, \delta(\omega' - \omega + \omega_a - \omega_f)\,, \tag{9.7}$$

where

$$w = \frac{1}{\hbar} \sum_b \frac{\langle f \,|\, \hat{H}' \,|\, b \rangle \langle b \,|\, \hat{H}' \,|\, a \rangle}{\omega' + \omega_a - E_b/\hbar}\,. \tag{9.8}$$

It follows from Eq. (7.44) that $\sum_{\boldsymbol{k}}$ can be replaced by a factor that is proportional to ω^2, and which also involves a frequency integration. We then get for the differential scattering cross section $d\sigma$ per unit solid angle and unit frequency interval

$$d\sigma \sim \sum_{f_{\text{atom}}} \omega^2\, |w|^2\,. \tag{9.9}$$

The δ function of Eq. (9.7) is not present in Eq. (9.9), since it disappeared when making the frequency integration of $\sum_{\boldsymbol{k}}$.

Since $\hat{H}'$ appears twice in Eq. (9.8), the scattering amplitude w will contain bilinear combinations of creation and annihilation operators, as follows from Eq. (7.63). As the photon number is conserved in the scattering process, the bilinear combinations have to contain one each of the creation and annihilation operators, but not two of the same kind. The order in which the two operators occur determines the nature of the physical process as well as the value of E_b. Figure 9.1 gives a graphical illustration of the two basic contributions to the scattering amplitude.

The upper diagram of Fig. 9.1 represents the resonant process of absorption followed by emission. At the right vertex a photon with wave number vector $\boldsymbol{k}'$ is destroyed (operator $\hat{a}_{\boldsymbol{k}'}$), and the atomic state a is destroyed (operator $\langle a \,|$), while the atomic state b is created (operator $|\, b \rangle$). At the left vertex a photon with wave number vector $\boldsymbol{k}$ is created (operator $\hat{a}^\dagger_{\boldsymbol{k}}$), atomic state b is destroyed (operator $\langle b \,|$), while atomic state f is created (operator $|\, f \rangle$). In Eq. (9.8) this means that the surviving contribution of $\hat{H}'$ to the right matrix element of w has the form $|\, b \rangle \langle a \,|\, \hat{a}_{\boldsymbol{k}'}$, while for the left matrix element the contribution is $|\, f \rangle \langle b \,|\, \hat{a}^\dagger_{\boldsymbol{k}}$. In this case the intermediate state contains only an atomic but no photon contribution, which means that $E_b = \hbar\omega_b$. Then the denominator in Eq. (9.8) is $\omega' + \omega_a - E_b/\hbar = \omega' - \omega_{ba}$, where by definition $\omega_{ba} = \omega_b - \omega_a$. Since the energy conservation dictated by the δ function in Eq. (9.7) implies that

$$\omega' - \omega + \omega_{af} = 0\,, \tag{9.10}$$

the denominator of the resonant term becomes

$$\omega' - \omega_{ba} = \omega - \omega_{bf}\,. \tag{9.11}$$

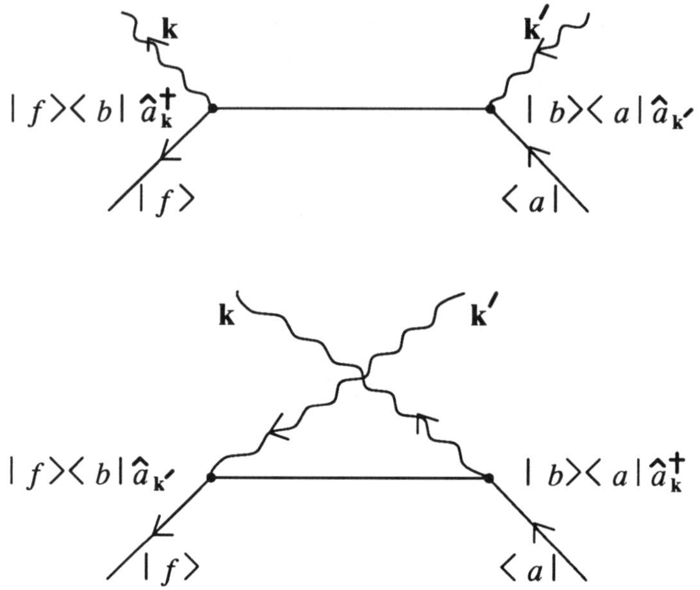

Fig. 9.1. Illustration of the two contributions to the scattering amplitude w (see text).

The lower diagram of Fig. 9.1 differs from the upper one only in the ordering of the creation and annihilation operators. This implies reversal of the time ordering, so that the diagram represents the non-resonant process of emission followed by absorption. In this case the intermediate state contains *two* photons, so that

$$E_b/\hbar = \omega_b + \omega' + \omega . \tag{9.12}$$

The denominator of the non-resonant term then becomes

$$\omega' + \omega_a - E_b/\hbar = -(\omega + \omega_{ba}) . \tag{9.13}$$

Combining these results and using Eq. (7.63) where the coefficients $d_{j\ell\mathbf{k}}$ were introduced, Eq. (9.8) for the scattering amplitude can be written

$$w = \frac{1}{\hbar} \sum_b \left\{ \frac{d^\dagger_{fb\mathbf{k}}\, d_{ba\mathbf{k}'}}{\omega - \omega_{bf}} - \frac{d_{fb\mathbf{k}'}\, d^\dagger_{ba\mathbf{k}}}{\omega + \omega_{ba}} \right\} . \tag{9.14}$$

The non-resonant term is almost never of practical importance, so we will disregard it in the following. For the resonant term our simplified second-order treatment leads to an undamped singularity at the resonant frequency, but because the lifetime of the intermediate state is finite, damping will of course always be present. As in the previous chapters we introduce damping heuristically, and refer to the discussion and arguments of Sects. 7.9, 8.5, and 10.1. Further, as ω^2 in Eq. (9.9) is $\approx \omega_{bf}^2$, it can usually be regarded as a fixed constant of proportionality for the scattering cross section. For the time being we will omit the proportionality constant but reclaim it again in Sect. 9.14, where we will give the expression for the absolute magnitude of the scattering cross section.

Using expressions (7.64) for $d_{j\ell k}$ and letting the indices $\boldsymbol{k}$ and $\boldsymbol{k}'$ be represented by the photon polarization indices α and β as we did in the previous two chapters, we obtain for the scattering amplitude $w = w_{\alpha\beta}$:

$$w_{\alpha\beta} \sim \sum_b \frac{\langle f\,|\,\hat{\boldsymbol{r}}\cdot\boldsymbol{e}_\alpha\,|\,b\rangle\,\langle b\,|\,\hat{\boldsymbol{r}}\cdot\boldsymbol{e}_\beta\,|\,a\rangle}{\omega_{bf}-\omega-i\gamma/2}\,. \tag{9.15}$$

Since $w_{\alpha\beta}$ is a tensor and not a scalar, and since the general polarization states of the incident and scattered photons are mixed states involving products of creation and annihilation operators representing different polarization states, as shown by the radiation coherency matrix (7.53), we need to generalize the expression for the differential scattering cross section $\mathrm{d}\sigma$ such that $|w|^2 = ww^*$ is replaced by the tensor product $\boldsymbol{w}\otimes\boldsymbol{w}^*$ as in Eq. (8.114). We thus get

$$\mathrm{d}\sigma \sim \sum_{f_{\mathrm{atom}}} \boldsymbol{w}\otimes\boldsymbol{w}^*\,, \tag{9.16}$$

where we have summed over all the final atomic substates. This expression is identical to that for the $\boldsymbol{W}$ matrix in Eq. (8.114), since we have implicitly assumed that we have complete redistribution in the initial state (the diagonal density matrix elements $\rho_{\mu_i\mu_i}$ are independent of μ_i), and the summation label μ_f corresponds to our summation label f_{atom}.

9.4. Symmetry Properties of the Phase Matrix

The matrix $\boldsymbol{W} = \boldsymbol{w}\otimes\boldsymbol{w}^*$ describes scattering of the radiation coherency matrix. In a Stokes vector formalism the counterpart is the phase matrix $\boldsymbol{P} \sim \boldsymbol{TWT}^{-1}$, as shown by Eq. (2.48) and in Sects. 5.9, 5.10, and 8.10. In Eq. (5.68) we extended the classical expression for $\boldsymbol{P}$ to the general quantum-mechanical case by introducing the coefficients W_1 and W_2, but did not show how this expression was obtained or that such a symmetry property of $\boldsymbol{P}$ at all applies to all the various scattering cases (including Raman scattering) that can occur. This demonstration will now be done.

The factors W_1 and W_2 enter naturally in our present theoretical framework, and we are in a position to derive explicit expressions for them in terms of the quantum numbers of the transitions. W_1 is a scaling factor for the scattering cross section for circular polarization (Stokes V), which equals unity in the classical case, whereas W_2 represents the fraction of the scattering processes that affect the linear polarization in the same way as classical dipole scattering. The fraction $1 - W_2$ represents isotropic, unpolarized scattering. Note that the parameters $W_{1,2}$ should not be confused with the elements W_{ij} of the $\boldsymbol{W}$ matrix. The notation happens to be similar due to the limited choice offered by the alphabet.

Before proving the symmetry properties of $\boldsymbol{P}$ represented by Eq. (5.68), we rewrite this equation into a form that will be more useful here. In the limit of vanishing magnetic field we obtain from Eqs. (5.68) and (5.58)–(5.64) for the special case when the scattering angles $\theta' = \theta = \pi/2$ (i.e., when the scattering takes place in the equatorial plane, cf. Fig. 3.3)

$$\begin{aligned} \boldsymbol{P} = \boldsymbol{E}_{11} &+ \tfrac{3}{4} W_2 (\boldsymbol{P}_0^2 + \boldsymbol{P}_2^2 \cos 2\phi) \\ &+ \tfrac{3}{2} (W_2 \boldsymbol{E}_{33} + W_1 \boldsymbol{E}_{44}) \cos\phi \end{aligned} \tag{9.17}$$

with

$$\begin{aligned} \boldsymbol{P}_0^2 &= \frac{1}{2} \begin{pmatrix} \frac{1}{3} & 1 & 0 & 0 \\ 1 & 3 & 0 & 0 \\ 0 & 0 & 0 & 0 \\ 0 & 0 & 0 & 0 \end{pmatrix}, \\ \boldsymbol{P}_2^2 &= \frac{1}{2} \begin{pmatrix} 1 & -1 & 0 & 0 \\ -1 & 1 & 0 & 0 \\ 0 & 0 & 0 & 0 \\ 0 & 0 & 0 & 0 \end{pmatrix}. \end{aligned} \tag{9.18}$$

$\boldsymbol{E}_{jj}$ is a matrix that has its diagonal element at position jj equal to unity, while all the remaining elements are zero. Inserting the expressions of Eq. (9.18) in Eq. (9.17), we get

$$\boldsymbol{P} = W_2 \boldsymbol{P}_R + (1 - W_2)\boldsymbol{E}_{11} + W_1 \boldsymbol{E}_{44} \tfrac{3}{2} \cos\phi\,, \tag{9.19}$$

where

$$\boldsymbol{P}_R = \frac{3}{4} \begin{pmatrix} 1 + \cos^2\phi & \sin^2\phi & 0 & 0 \\ \sin^2\phi & 1 + \cos^2\phi & 0 & 0 \\ 0 & 0 & 2\cos\phi & 0 \\ 0 & 0 & 0 & 0 \end{pmatrix} \tag{9.20}$$

represents the classical dipole-type Rayleigh scattering matrix for the linear polarization. The second term on the right hand side of Eq. (9.19) represents isotropic, unpolarized scattering, the third term refers only to scattering of Stokes V (which is decoupled from the other Stokes parameters in the scattering process), while the first term represents dipole-type scattering scaled by the factor W_2.

When trying to reproduce the form represented by Eq. (9.19) from quantum mechanics, it is natural to start by exploring the symmetry properties of the matrix $\boldsymbol{W}$, whose elements are governed by the geometrical ε factors and the 3-j symbols according to Eqs. (8.115) and (8.117) when lower-state coherences are disregarded. These symmetry properties are found most easily with the coordinate system of Fig. 3.3, which was used for Eqs. (9.17)–(9.20). Thus we assume that the incident radiation is travelling along the x-axis, the scattering takes place in the xy plane, and the scattering angle is ϕ. Let us further mark incident polarization states with a prime, which means that index β of Eq. (9.15) should be replaced by $1'$ or $2'$, while $\alpha = 1$ or 2. With our choice of scattering geometry it follows from Eq. (3.86) that

$$\begin{aligned} \varepsilon_0^1 &= \varepsilon_0^{1'} = -1\,, \\ \varepsilon_0^2 &= \varepsilon_0^{2'} = \varepsilon_\pm^1 = \varepsilon_\pm^{1'} = 0\,, \\ \varepsilon_\pm^2 &= -i\, e^{\pm i\phi}\,/\sqrt{2}\,, \\ \varepsilon_\pm^{2'} &= -i/\sqrt{2}\,. \end{aligned} \tag{9.21}$$

In the following we will consider the general case of Raman scattering, which includes Rayleigh scattering as the special case when the initial and final states are the same. Inserting Eq. (9.21) in Eq. (8.117) and using Eq. (2.39), we find that

$$\begin{aligned} W_{12} &= W_{21} = W_{13} = W_{31} = W_{24} = W_{42} = W_{34} = W_{43} = 0\,, \\ W_{22} &= W_{33}\,, \\ W_{23} &= W_{32}\,, \\ W_{14} &= W_{41}\,. \end{aligned} \tag{9.22}$$

The validity of these relations is readily seen by direct inspection of the expressions for the different elements, except for the last relation, $W_{14} = W_{41}$, which is not so obvious in the Raman scattering case, when $J_{\mu_i} \neq J_{\mu_f}$ (see the explicit expression for W_{14} below). It can be shown, however, that this symmetry property like all the other relations in Eq. (9.22) is valid for all Raman scattering cases.

As a consequence of Eq. (9.22) $\boldsymbol{W}$ has the form

$$\boldsymbol{W} = \begin{pmatrix} W_{11} & 0 & 0 & W_{14} \\ 0 & W_{22} & W_{23} & 0 \\ 0 & W_{23} & W_{22} & 0 \\ W_{14} & 0 & 0 & W_{44} \end{pmatrix}. \tag{9.23}$$

It follows that the phase matrix $\boldsymbol{P}$ can be written as

$$\boldsymbol{P} \sim \boldsymbol{T}\boldsymbol{W}\boldsymbol{T}^{-1} = \frac{1}{2}\begin{pmatrix} a & b & 0 & 0 \\ b & c & 0 & 0 \\ 0 & 0 & d & 0 \\ 0 & 0 & 0 & e \end{pmatrix}, \tag{9.24}$$

where

$$\begin{aligned} a &= W_{11} + 2W_{14} + W_{44}\,, \\ b &= W_{11} - W_{44}\,, \\ c &= W_{11} - 2W_{14} + W_{44}\,, \\ d &= 2(W_{22} + W_{23})\,, \\ e &= 2(W_{22} - W_{23})\,. \end{aligned} \tag{9.25}$$

Next we will write down the explicit expressions for the non-zero elements of the $\boldsymbol{W}$ matrix and show how the phase matrix $\boldsymbol{P}$ in Eq. (9.24) can be brought to a form that is identical to that of Eq. (9.19). As a byproduct we will obtain the algebraic expressions for the polarizability coefficients $W_{1,2}$ in terms of the J_μ and J_m quantum numbers.

The derivation will first be made for the case when there is only one upper-level quantum number J_m, since then $W_{1,2}$ turn out to be well-defined constants, expressible in terms of J_m and J_μ only. When there are interferences between states of different total angular momentum J_m, one cannot factorize out the profile function Φ in Eq. (8.115) to form a frequency-independent phase matrix, unless the differences between the resonant frequencies of the different profile functions are small in comparison with the Doppler width. The resulting frequency dependence of the phase matrix can however be described using the algebraic expressions for $W_{1,2}$ for the individual levels as parameters of the frequency-dependent function (cf. Stenflo, 1980), as will be shown below in Sects. 9.8, 9.9, and 9.12.

From Eqs. (8.114), (8.117), and (9.21) we obtain in the case of complete redistribution for the lower level (when $\rho_{\mu_i\mu_i}$ is independent of μ_i)

$$\begin{aligned} W_{11} &\sim \sum_k \begin{pmatrix} J_m & J_{\mu_f} & 1 \\ -k & k & 0 \end{pmatrix}^2 \begin{pmatrix} J_m & J_{\mu_i} & 1 \\ -k & k & 0 \end{pmatrix}^2 , \\ W_{14} &\sim \sum_k \begin{pmatrix} J_m & J_{\mu_f} & 1 \\ -k & k & 0 \end{pmatrix}^2 \begin{pmatrix} J_m & J_{\mu_i} & 1 \\ -k & k-1 & 1 \end{pmatrix}^2 . \end{aligned} \tag{9.26}$$

The summation over k arises from the summation in Eq. (8.114) over μ_i and μ_f.

If J_{μ_f} and J_{μ_i} in W_{14} are exchanged, we obtain W_{41}. Although the equivalence of the expressions for W_{14} and W_{41} may not be obvious when $J_{\mu_f} \neq J_{\mu_i}$, it can be shown that they are the same.

In all our expressions for the matrix elements W_{ij} the not explicitly given proportionality constant contains the product $f_{J_{\mu_i} J_m} f_{J_{\mu_f} J_m}$ between the two oscillator strengths. When we later allow for interference between the different intermediate J_m states, the situation gets more complex, since we have to deal with products of $\pm\sqrt{f_{J_{\mu_{i,f}} J_m}}$, where the sign is determined in a fairly complicated way by the L, S, and J quantum numbers involved. We will treat this interference case in Sects. 9.7, 9.9, and 9.11 below.

Like Eqs. (9.26) we obtain

$$\begin{aligned} W_{22} &= W'_{22} \cos\phi , \\ W_{23} &= W'_{23} \cos\phi , \end{aligned} \tag{9.27}$$

where

$$\begin{aligned} W'_{22} \sim & \sum_k \begin{pmatrix} J_m & J_{\mu_f} & 1 \\ -k & k & 0 \end{pmatrix} \begin{pmatrix} J_m & J_{\mu_i} & 1 \\ -k & k & 0 \end{pmatrix} \\ & \qquad \begin{pmatrix} J_m & J_{\mu_f} & 1 \\ -k-1 & k & 1 \end{pmatrix} \begin{pmatrix} J_m & J_{\mu_i} & 1 \\ -k-1 & k & 1 \end{pmatrix} , \\ W'_{23} \sim & -\sum_k \begin{pmatrix} J_m & J_{\mu_f} & 1 \\ -k & k & 0 \end{pmatrix} \begin{pmatrix} J_m & J_{\mu_i} & 1 \\ -k+1 & k-1 & 0 \end{pmatrix} \\ & \qquad \begin{pmatrix} J_m & J_{\mu_f} & 1 \\ -k+1 & k & -1 \end{pmatrix} \begin{pmatrix} J_m & J_{\mu_i} & 1 \\ -k & k-1 & 1 \end{pmatrix} . \end{aligned} \tag{9.28}$$

The following two relations can also be derived from the properties of the 3-j symbols:

$$
\begin{aligned}
W'_{22} + W'_{23} &= W_{11} - W_{14}\,, \\
W_{44} - W_{14} &= (W_{11} - W_{14})\cos^2\phi\,.
\end{aligned}
\tag{9.29}
$$

Using relations (9.27) and (9.29) in Eq. (9.25), we now obtain

$$
\begin{aligned}
a &= (W_{11} - W_{14})(1+\cos^2\phi) + 4W_{14}\,, \\
b &= (W_{11} - W_{14})\sin^2\phi\,, \\
c &= (W_{11} - W_{14})(1+\cos^2\phi)\,, \\
d &= 2(W_{11} - W_{14})\cos\phi\,, \\
e &= 2(W'_{22} - W'_{23})\cos\phi\,.
\end{aligned}
\tag{9.30}
$$

Comparison between Eq. (9.24) with expressions (9.30) inserted and Eq. (9.19) shows that the two expressions for the phase matrix become identical if the parameters $W_{1,2}$ satisfy the relations

$$
\begin{aligned}
\tfrac{3}{4}W_2 + (1 - W_2) &= k(W_{11} + 3W_{14})\,, \\
\tfrac{3}{4}W_2 &= k(W_{11} - W_{14})\,, \\
\tfrac{3}{4}W_1 &= k(W'_{22} - W'_{23})\,,
\end{aligned}
\tag{9.31}
$$

where k is a proportionality (or normalization) constant. The solution of Eqs. (9.31) is

$$
\begin{aligned}
W_1 &= \frac{W'_{22} - W'_{23}}{W_{11} + 2W_{14}}\,, \\
W_2 &= \frac{W_{11} - W_{14}}{W_{11} + 2W_{14}}\,.
\end{aligned}
\tag{9.32}
$$

9.5. Expressions for $W_{1,2}$

When Eqs. (9.26) and (9.28) are inserted in Eqs. (9.32), it is possible to compute the values of $W_{1,2}$ for any combination of the J_m, J_{μ_i}, and J_{μ_f} quantum numbers. It is however not necessary to evaluate the sums over the 3-j symbols each time, since simple algebraic expressions for these sums in terms of J_m, J_{μ_i}, and J_{μ_f} can be derived and used, as will be seen below.

The most compact way of writing the solutions for W_K, $K = 1,2$, has been given by Landi Degl'Innocenti (1984):

$$
W_K = w^{(K)}_{J_m J_{\mu_i}}\, w^{(K)}_{J_m J_{\mu_f}}\,,
\tag{9.33}
$$

where $w^{(K)}_{J_m J_\mu}$ should not be confused with the scattering amplitudes $w_{\alpha\beta}$. Instead they are algebraic expressions in terms of 6-j symbols:

$$w^{(K)}_{J_m J_\mu} = \begin{Bmatrix} 1 & 1 & K \\ J_m & J_m & J_\mu \end{Bmatrix} \Big/ \begin{Bmatrix} 1 & 1 & 0 \\ J_m & J_m & J_\mu \end{Bmatrix} . \tag{9.34}$$

As the general algebraic expression for 6-j symbols is not so transparent, and as all the various scattering possibilities can be represented by a small number of cases corresponding to different possible combinations of J quantum numbers, it is much more convenient to write down and work with the explicit solutions for these possible cases. These solutions will now be listed in the form of simple algebraic expressions.

In the Rayleigh scattering case, when $J_{\mu_f} = J_{\mu_i} = J_\mu$, only the three cases when $J_m - J_\mu = 0, \pm 1$ can occur (but $J_m = J_\mu = 0$ is forbidden). The simple and compact algebraic expressions for these three cases have previously been given by Chandrasekhar (1950). We list them in the order of increasing value of J_m (for a given value of J_μ).

CASE I. $J_m = J_\mu - 1 = J - 1$

$$\begin{aligned} W_1 &= \frac{J-1}{2J} , \\ W_2 &= \frac{(J-1)(2J-3)}{10J(2J+1)} . \end{aligned} \tag{9.35}$$

CASE II. $J_m = J_\mu = J$

$$\begin{aligned} W_1 &= \frac{1}{2J(J+1)} , \\ W_2 &= \frac{(2J-1)(2J+3)}{10J(J+1)} . \end{aligned} \tag{9.36}$$

CASE III. $J_m = J_\mu + 1 = J + 1$

$$\begin{aligned} W_1 &= \frac{J+2}{2(J+1)} , \\ W_2 &= \frac{(J+2)(2J+5)}{10(J+1)(2J+1)} . \end{aligned} \tag{9.37}$$

Note that Cases I–III also apply to the case of Raman scattering when $J_{\mu_f} = J_{\mu_i}$ while $L_f \neq L_i$.

The symmetry of W_K with respect to an exchange of J_{μ_f} and J_{μ_i}, e.g. as shown by Eq. (9.33), implies invariance of W_K with respect to reversal of the order of the scattering (time reversal). This reduces the number of additional cases to be listed to three, to fully cover all the Raman scattering cases that can occur. We have used the explicit algebraic expressions for the 3-j symbols in Eqs. (9.26) and (9.28) to derive simple algebraic expressions for these additional cases, analogous to the

Chandrasekhar expressions for the Rayleigh scattering case. These solutions will now be listed. In Cases IV and V $J_{\mu_f} - J_{\mu_i} = 1$ (the two cases are listed in the order of increasing value of J_m for fixed values of J_{μ_f} and J_{μ_i}), in Case VI $J_{\mu_f} - J_{\mu_i} = 2$.

CASE IV. $J_m = J_{\mu_f} - 1 = J_{\mu_i} = J$

$$\begin{aligned} W_1 &= -\frac{1}{2(J+1)}, \\ W_2 &= -\frac{2J-1}{10(J+1)}. \end{aligned} \tag{9.38}$$

CASE V. $J_m = J_{\mu_f} = J_{\mu_i} + 1 = J + 1$

$$\begin{aligned} W_1 &= \frac{1}{2(J+1)}, \\ W_2 &= -\frac{2J+5}{10(J+1)}. \end{aligned} \tag{9.39}$$

CASE VI. $J_m = J_{\mu_f} - 1 = J_{\mu_i} + 1$

$$\begin{aligned} W_1 &= -\frac{1}{2}, \\ W_2 &= \frac{1}{10}. \end{aligned} \tag{9.40}$$

We notice that for all the cases (9.35)–(9.39) the symbol J that is used in the algebraic expressions is always $= J_{\mu_i}$. In Cases IV–VI J_{μ_i} is always $< J_{\mu_f}$, but all the cases when $J_{\mu_i} > J_{\mu_f}$ can always be converted to one of the Cases IV–VI by reversing the direction of scattering.

We also notice the negative signs for W_K that appear in Eqs. (9.38)–(9.40). For Rayleigh scattering (Cases I–III) W_K can never be negative, in contrast to the Raman scattering case. A negative value of W_2 means for instance that incoming light that is linearly polarized perpendicular to the scattering plane becomes polarized *parallel* to the scattering plane when emerging from the scattering event. Similarly, a negative value of W_1 means that left-handed circular polarization is converted to right-handed circular polarization.

9.6. Example of Negative Polarizability

To illuminate the physics of negative polarizability, let us consider a concrete case with a scattering angle of 90°. To explicitly distinguish between the different magnetic substates m we let them be Zeeman split by a magnetic field that is perpendicular to the scattering plane. Later we let the field strength go to zero to retrieve the non-magnetic scattering case (principle of spectroscopic stability).

Let us define the positive Stokes Q direction to be perpendicular to the scattering plane (and thus parallel to the field vector $\boldsymbol{B}$). Using Eq. (4.48) with $\gamma = \pi/2$ and $\chi = 0$ we obtain the profile functions

$$\begin{aligned} H_I &= \tfrac{1}{2}[\,H_0 + \tfrac{1}{2}(H_+ + H_-)\,]\,, \\ H_Q &= \tfrac{1}{2}[\,H_0 - \tfrac{1}{2}(H_+ + H_-)\,]\,, \\ H_U &= H_V = 0 \end{aligned} \tag{9.41}$$

for a normal Zeeman triplet. We can directly extend this to the general case of an anomalous splitting pattern if we replace H_q in Eq. (9.41) with the ϕ_q functions that were defined by Eq. (6.37). In the case of optically thin radiation the Stokes parameters of the light beam are proportional to $\phi_{I,Q,U,V}$, so we get

$$\begin{aligned} I &\sim \phi_0 + \tfrac{1}{2}(\phi_+ + \phi_-)\,, \\ Q &\sim \phi_0 - \tfrac{1}{2}(\phi_+ + \phi_-) \end{aligned} \tag{9.42}$$

(cf. Sect. 6.5).

As was done in Eq. (6.52) we can further generalize the expressions for ϕ_q to the case of non-LTE populations of the excited, upper magnetic substates M_u:

$$\phi_q \sim \sum_{M_\ell, M_u} \rho_{M_u M_u}\; S_q(M_\ell, M_u)\, H_q\,, \tag{9.43}$$

where the diagonal density matrix elements $\rho_{M_u M_u}$ represent the relative population of the M_u states. S_q is the transition strength between the states represented by M_u and M_ℓ and is given by Eq. (6.33). $q = M_\ell - M_u$.

In the limit of vanishing field strength $H_+ = H_- = H_0$, which means that Eqs. (9.42) and (9.43) give us

$$\begin{aligned} I &\sim \sum_{M_\ell, M_u} \rho_{M_u M_u}\; [\,S_0 + \tfrac{1}{2}(S_+ + S_-)\,]\,, \\ Q &\sim \sum_{M_\ell, M_u} \rho_{M_u M_u}\; [\,S_0 - \tfrac{1}{2}(S_+ + S_-)\,]\,. \end{aligned} \tag{9.44}$$

The degree of linear polarization of the scattered radiation is $p = Q/I$.

The above expressions describe the emitted radiation without reference to the absorption, which is the other essential half of the scattering process. The effect of absorption is however implicitly included in Eq. (9.44), since it is responsible for the selective population of the M_u states and thus for the values of the density matrix elements of the excited level.

Let us now consider the special case that the incident radiation is linearly polarized perpendicular to the scattering plane (parallel to the magnetic field), i.e., the incident Stokes vector is $\boldsymbol{I}' = I'(1,1,0,0)^\dagger$. Then only absorption transitions with $\Delta M = 0$ are possible. In this case the relative populations $\rho_{M_u M_u}$ of the excited states will only be determined by the populations $\rho_{M_\ell M_\ell}$ of the initial sublevels and the transition strength $S_0(M_\ell, M_u)$. If the initial substates are uncorrelated, as they usually are due to long lifetimes and/or collisions, then the excited M_u substates will also be uncorrelated.

For clarity and illustrative purposes let us now consider the special case when $J_{\mu_i} = 0$, for which there is only one initial substate with $M_\ell = 0$. Then, with our assumption for the polarization of the incident light, only the $M_u = 0$ state will be excited. Eq. (9.44) then simplifies to

$$\begin{aligned} I &\sim S_0(0,0) + \tfrac{1}{2}[\,S_+(1,0) + S_-(-1,0)\,]\,, \\ Q &\sim S_0(0,0) - \tfrac{1}{2}[\,S_+(1,0) + S_-(-1,0)\,]\,, \end{aligned} \tag{9.45}$$

provided that $J_{\mu_f} \geq 1$, so that the $M_\ell = \pm 1$ sublevels exist for the final state. When $J_{\mu_f} = 0$ these substates do not exist, with the consequence that $S_\pm = 0$ and therefore $p = Q/I = 1$. This represents the well-known Rayleigh scattering case $J = 0 \to 1 \to 0$, which corresponds to classical dipole-type scattering.

When $J_{\mu_f} \geq 1$ we have two Raman scattering possibilities: $J = 0 \to 1 \to 1$, representing Case V of the preceding section, and $J = 0 \to 1 \to 2$, representing Case VI. In the former case $S_0(0,0) = 0$ and $S_+(1,0) = S_-(-1,0)$, which gives $p = -1$, i.e., the polarization of the scattered radiation is 100 % linear, *parallel* to the scattering plane. The reason why the polarization plane has been "rotated" by 90° in the scattering process is that emission in the π component is forbidden (mathematically a consequence of the properties of the 3-j symbols). This forces all the radiation to be emitted exclusively as σ components, which are polarized perpendicular to the magnetic field vector and thus parallel to the plane of scattering.

In the second Raman scattering case ($J = 0 \to 1 \to 2$), $S_0(0,0) = 0.40$ and $S_\pm(\pm 1, 0) = 0.30$, which according to Eq. (9.45) gives $p = 1/7 \approx 0.14$.

These results are to be compared with those obtained from our algebraic expressions (9.39) and (9.40) for W_2. According to the phase matrix, Eq. (9.19), we get for the scattered radiation when the incident Stokes vector is $\sim (1,1,0,0)^\dagger$

$$\begin{aligned} I &\sim 1 + \tfrac{1}{2} W_2\,, \\ Q &\sim \tfrac{3}{2} W_2\,, \end{aligned} \tag{9.46}$$

and thus

$$p = \frac{3W_2}{2 + W_2}\,. \tag{9.47}$$

$p = -1$ thus implies that $W_2 = -0.50$, while $p = 1/7$ implies $W_2 = 0.10$, in agreement with Cases V and VI, Eqs. (9.39) (when inserting $J = 0$) and (9.40).

9.7. Upper J-state Interference

In our previous discussion (Sect. 9.4) of the elements W_{ij} of the scattering matrix $\boldsymbol{W}$ and of the expressions that led to the polarizability coefficients W_K, $K = 1, 2$, we disregarded the reduced matrix elements and the profile factors by hiding them in the omitted proportionality factor. This was allowed, since the proportionality factor was the same for all the W_{ij} elements and therefore divided out when forming W_K. The validity of this procedure is however limited to the considered case when only one excited J_m state is involved in the scattering transition. In the case when we need to account for interferences not only between the excited magnetic sublevels m but also between states of different total angular momentum J_m, the reduced matrix elements and profile factors however do not divide out but must be retained throughout the calculations. This means in particular that W_K will vary with wavelength (in contrast to the wavelength-independent expressions of Eqs. (9.35)–(9.40)).

Let us first address the problem of calculating the reduced matrix elements. As shown by Eq. (8.51) the square of the reduced matrix element is proportional to the oscillator strength. This implies that the reduced matrix element alone is proportional to the square root of the oscillator strength times some phase factor. Expressions for this phase factor will be given later in Sect. 9.11, but let us first give some background discussion of the physical situation.

With the Wigner-Eckart theorem of Eq. (7.95) the normal quantum-mechanical matrix element can be factorized into two parts, one reduced matrix element that does not contain any reference to the M quantum number, and a real, M-dependent factor (that contains the 3-j symbol). It is possible to go one step further and do a second reduction of the reduced matrix element, whereby it becomes factorized into two parts, one "doubly reduced" matrix element that only depends on the L and S quantum numbers without reference to J, and one real, J-dependent factor. As the doubly reduced matrix element is common to all the terms in the expressions for W_{ij} (even in the case of J_m-state interference), it can be omitted (or "hidden" in the common proportionality factor). What is then left from the ordinary (singly) reduced matrix element is a real factor, whose magnitude is proportional to the square of the oscillator strength, while its sign depends on the combination of L, S, and J quantum numbers involved. We may thus write

$$\langle J_\mu \,||\, \hat{\boldsymbol{r}} \,||\, J_m \rangle \sim (-1)^{r_{J_\mu J_m}} \sqrt{f_{J_\mu J_m}}\,, \tag{9.48}$$

where $r_{J_\mu J_m}$ is an even or odd integer depending on the values of the L, S, and J quantum numbers. This dependence will be explicitly given in Sect. 9.11.

When we in the following derive explicit expressions for the polarizability in the presence of J-state interference, we will limit ourselves to doing it for the W_2 coefficient. It is straightforward to derive the corresponding expressions for W_1, but this is rarely needed in practice, since the dominating scattering polarization effects are in the linear polarization, governed exclusively by the W_2 coefficient.

For W_2 only the elements W_{11} and W_{14} of the scattering matrix need to be considered, according to Eq. (9.32). To simplify the index notation we will let index ie represent $J_{\mu_i} J_m$, while index fe represents $J_{\mu_f} J_m$ (the letters i, e, and

f thus refer to the initial, excited, and final states, respectively). The previous expressions of Eq. (9.26) can then be generalized (cf. also Eq. (8.117)), to become

$$\begin{aligned}
W_{11} \sim \sum_k \Big| \sum_{J_m} (-1)^{r_{ie}+r_{fe}} \sqrt{f_{ie} f_{fe}}\, \Phi_{ef} \\
\begin{pmatrix} J_m & J_{\mu_f} & 1 \\ -k & k & 0 \end{pmatrix} \begin{pmatrix} J_m & J_{\mu_i} & 1 \\ -k & k & 0 \end{pmatrix} \Big|^2 , \\
W_{14} \sim \sum_k \Big| \sum_{J_m} (-1)^{r_{ie}+r_{fe}} \sqrt{f_{ie} f_{fe}}\, \Phi_{ef} \\
\begin{pmatrix} J_m & J_{\mu_f} & 1 \\ -k & k & 0 \end{pmatrix} \begin{pmatrix} J_m & J_{\mu_i} & 1 \\ -k & k-1 & 1 \end{pmatrix} \Big|^2 .
\end{aligned} \tag{9.49}$$

The same symmetry arguments that we used for the case without upper J-state interference and which led to Eq. (9.32) for W_K also apply to the present case with J-state interference. Thus Eq. (9.32) is still valid, but W_{11} and W_{14} are given by Eq. (9.49).

In the special case of Rayleigh scattering, when $J_{\mu_f} = J_{\mu_i} = J_\mu$ and $L_f = L_i$,

$$\Big[(-1)^{r_{ie}+r_{fe}} \sqrt{f_{ie} f_{fe}} \Big]_{\text{Rayleigh}} = f_{J_\mu J_m} . \tag{9.50}$$

The profile factor Φ is given by Eq. (7.90). To formally simplify the expressions we will here omit the damping constant γ, which is often unimportant for problems where the aim is to explore the effects of J_m-state interference. The final expressions that we obtain for W_2 can readily be generalized afterwards by inserting the damping constant again, so the physics is not really restricted by this formal simplification. Omitting γ for the time being means according to Eq. (7.90) that the profile factor can be written as

$$\Phi_{ef} \sim (\nu_{ef} - \nu)^{-1} , \tag{9.51}$$

where ν_{ef} is the resonant frequency of the respective emission transition.

Eq. (9.49) describes the coherent superposition and mixing of states of different J_m number. For the further development of the theory it is convenient to introduce the notation $W_{ij}^{(n)}$ for the W_{ij} that only accounts for a single J_m state $J_m = J_n$ without the phase, oscillator strength, and profile factors. Thus $W_{ij}^{(n)}$ are identical to the W_{ij} that were given by Eq. (9.26) with $J_m = J_n$. They account for the squared, incoherent portions of Eq. (9.49). For the cross (interference) terms that arise from Eq. (9.49) we introduce the notation $W_{ij}^{(u,v)}$, defined by

$$
\begin{aligned}
W_{11}^{(u,v)} &= \sum_k \begin{pmatrix} J_u & J_{\mu_f} & 1 \\ -k & k & 0 \end{pmatrix} \begin{pmatrix} J_u & J_{\mu_i} & 1 \\ -k & k & 0 \end{pmatrix} \\
&\quad \begin{pmatrix} J_v & J_{\mu_f} & 1 \\ -k & k & 0 \end{pmatrix} \begin{pmatrix} J_v & J_{\mu_i} & 1 \\ -k & k & 0 \end{pmatrix}, \\
W_{14}^{(u,v)} &= \sum_k \begin{pmatrix} J_u & J_{\mu_f} & 1 \\ -k & k & 0 \end{pmatrix} \begin{pmatrix} J_u & J_{\mu_i} & 1 \\ -k & k-1 & 1 \end{pmatrix} \\
&\quad \begin{pmatrix} J_v & J_{\mu_f} & 1 \\ -k & k & 0 \end{pmatrix} \begin{pmatrix} J_v & J_{\mu_i} & 1 \\ -k & k-1 & 1 \end{pmatrix}.
\end{aligned} \tag{9.52}
$$

When calculating W_2 we form according to Eq. (9.32) $W_{11} - W_{14}$ and $W_{11} + 2W_{14}$. These are expressions of the same form as W_{ij} alone, but with the incoherent $W_{ij}^{(n)}$ coefficients replaced by $W_{11}^{(n)} - W_{14}^{(n)}$ and $W_{11}^{(n)} + 2W_{14}^{(n)}$, and correspondingly for the interference coefficients $W_{ij}^{(u,v)}$. It can be shown through algebraic calculations that

$$W_{11}^{(u,v)} + 2W_{14}^{(u,v)} = 0\,, \tag{9.53}$$

which means that the interference terms in the denominator of W_2 and W_1 vanish, while the interference terms in the nominator survive. Eq. (9.53) implies that these interference terms depend on $W_{11}^{(u,v)}$ alone:

$$W_{11}^{(u,v)} - W_{14}^{(u,v)} = 1.5W_{11}^{(u,v)}\,. \tag{9.54}$$

Since $\Delta J = 0, \pm 1$ for allowed electric dipole transitions, there can be interference between at most three J_m states. The maximum case of three-state interference can only occur when $J_{\mu_f} = J_{\mu_i}$ and corresponds to interference between Cases I, II, and III of Eqs. (9.35)–(9.37). In the Raman scattering case J_m-state interference may also occur between Cases IV and V (when $|J_{\mu_f} - J_{\mu_i}| = 1$). Case VI is never involved in any J_m-state interference, since only one J_m state is allowed when $|J_{\mu_f} - J_{\mu_i}| = 2$.

9.8. J-state Interference in the Rayleigh Scattering Case

In this section we will formally include the maximum number of three J_m states that may interfere, and use the notation $J_m = J_n$, $n = 1, 2, 3$, to label the interfering states. $n = 1, 2, 3$ correspond to Cases I, II, III of Eqs. (9.35)–(9.37), respectively. If a transition to a given J_m state is forbidden, its contribution to the expressions below will automatically be zero since the corresponding oscillator strength is zero.

For the Rayleigh scattering case ($i = f$) we get from Eqs. (9.49) and (9.50) with the new notations of $W_{ij}^{(n)}$ and $W_{ij}^{(u,v)}$ that were introduced in the preceding section

$$W_{ij} = \sum_{n=1}^{3} f_{fn}^2 \,(\nu_n - \nu)^{-2}\, W_{ij}^{(n)} + 2 f_{f1} f_{f2} \,(\nu_1 - \nu)^{-1} (\nu_2 - \nu)^{-1}\, W_{ij}^{(1,2)}$$
$$+ 2 f_{f1} f_{f3} \,(\nu_1 - \nu)^{-1} (\nu_3 - \nu)^{-1}\, W_{ij}^{(1,3)} + 2 f_{f2} f_{f3} \,(\nu_2 - \nu)^{-1} (\nu_3 - \nu)^{-1}\, W_{ij}^{(2,3)} . \tag{9.55}$$

With Eqs. (9.32) and (9.54) we may thus write W_2 as

$$W_2 = \Big[\sum_{n=1}^{3} f_{fn}^2\, c_{nn}\,(\nu_n - \nu)^{-2} + f_{f1} f_{f2}\, c_{12}\,(\nu_1 - \nu)^{-1} (\nu_2 - \nu)^{-1}$$
$$+ f_{f1} f_{f3}\, c_{13}\,(\nu_1 - \nu)^{-1} (\nu_3 - \nu)^{-1} + f_{f2} f_{f3}\, c_{23}\,(\nu_2 - \nu)^{-1} (\nu_3 - \nu)^{-1} \Big]$$
$$\Big/ \sum_{n=1}^{3} f_{fn}^2\, d_{nn}\,(\nu_n - \nu)^{-2} , \tag{9.56}$$

where the coefficients c_{ij} and d_{ii} are formed from combinations of the $W_{ij}^{(n)}$ and $W_{ij}^{(u,v)}$ elements and exclusively depend on the J quantum numbers involved.

$n = 1, 2, 3$ represents an increasing sequence of $J_m = J_n$. Let us as in Eqs. (9.35)–(9.37) use the notation $J_{\mu_f} = J_{\mu_i} = J$. After some fairly extensive algebraic calculations we then get for the coefficients of the squared (incoherent) terms (the diagonal elements of the c_{ij} and d_{ij} matrices)

$$d_{11} = \frac{1}{3(2J-1)} , \quad d_{22} = \frac{1}{3(2J+1)} , \quad d_{33} = \frac{1}{3(2J+3)} , \tag{9.57}$$

and

$$c_{11} = \frac{(J-1)(2J-3)}{30J(2J-1)(2J+1)} , \quad c_{22} = \frac{(2J-1)(2J+3)}{30J(J+1)(2J+1)} , \quad c_{33} = \frac{(J+2)(2J+5)}{30(J+1)(2J+1)(2J+3)} , \tag{9.58}$$

while the coefficients for the interference terms are

$$c_{12} = \frac{J-1}{5J(2J+1)} , \quad c_{13} = \frac{2}{5(2J+1)} , \quad c_{23} = \frac{J+2}{5(J+1)(2J+1)} . \tag{9.59}$$

When $\nu = \nu_n$ the influence of the transitions not having the resonant frequency ν_n vanishes, which means that

$$W_2 = c_{nn}/d_{nn} \quad \text{for} \quad \nu = \nu_n \,. \tag{9.60}$$

Inserting the values for c_{nn} and d_{nn} from Eqs. (9.58) and (9.57), we see that the algebraic expressions of Eqs. (9.35)–(9.37) for Cases I, II, III are retrieved when $n = 1, 2, 3$, as expected.

9.9. J-state Interference in the Raman Scattering Case

We need to distinguish between two different situations for Raman scattering: (A) $J_{\mu_f} = J_{\mu_i}$ while $L_f \neq L_i$ (scattering into a different multiplet). Three J_m states may interfere, corresponding to interference between Cases I, II, and III as for Rayleigh scattering. (B) $|J_{\mu_f} - J_{\mu_i}| = 1$. Two J_m states may interfere, corresponding to interference between Cases IV and V. As Case VI corresponds to $|J_{\mu_f} - J_{\mu_i}| = 2$, it cannot be involved in any upper J-state interference.

Let us first turn to case (A) ($J_{\mu_f} = J_{\mu_i}$). In this case the expression for W_2 is the same as Eq. (9.56) for the Rayleigh scattering case, with the exception that the sign factors with parameters r_{ie} and r_{fe} have to be attached to the $W_{ij}^{(u,v)}$ elements and thus also to the interference coefficients c_{uv}, and that the difference between the absorption and emission oscillator strengths has to be accounted for. We then get

$$\begin{aligned} W_2 = \Big[\sum_{n=1}^{3} & f_{in} f_{fn} \, c_{nn} \, (\nu_n - \nu)^{-2} + g_{12} \, c_{12} \, (\nu_1 - \nu)^{-1} \, (\nu_2 - \nu)^{-1} \\ & + g_{13} \, c_{13} \, (\nu_1 - \nu)^{-1} \, (\nu_3 - \nu)^{-1} + g_{23} \, c_{23} \, (\nu_2 - \nu)^{-1} \, (\nu_3 - \nu)^{-1} \Big] \\ & \Big/ \sum_{n=1}^{3} f_{in} f_{fn} \, d_{nn} (\nu_n - \nu)^{-2} \,, \end{aligned} \tag{9.61}$$

where

$$g_{uv} = (-1)^{r_{iu} + r_{fu} + r_{iv} + r_{fv}} \sqrt{f_{iu} \, f_{fu} \, f_{iv} \, f_{fv}} \,, \tag{9.62}$$

and where d_{nn}, c_{nn}, c_{uv} are given by Eqs. (9.57)–(9.59) as in the Rayleigh scattering case.

Let us next turn to case (B) ($|J_{\mu_f} - J_{\mu_i}| = 1$). If we let $n = 4, 5$ represent Cases IV, V, we get

$$\begin{aligned} W_2 = & \big[f_{i4} f_{f4} \, c_{44} \, (\nu_4 - \nu)^{-2} + f_{i5} f_{f5} \, c_{55} \, (\nu_5 - \nu)^{-2} \\ & + g_{45} \, c_{45} \, (\nu_4 - \nu)^{-1} \, (\nu_5 - \nu)^{-1} \big] \\ & / \big[f_{i4} f_{f4} \, d_{44} \, (\nu_4 - \nu)^{-2} + f_{i5} f_{f5} \, d_{55} \, (\nu_5 - \nu)^{-2} \big] \,. \end{aligned} \tag{9.63}$$

This expression may represent two different Raman scattering subcases, corresponding to $J_{\mu_f} - J_{\mu_i} = \pm 1$. If we however let

$$J = \min(J_{\mu_i}, J_{\mu_f}), \tag{9.64}$$

then we can write down the following explicit algebraic expressions that are valid for both these subcases:

$$\begin{aligned} d_{44} &= \frac{1}{3(2J+1)}, \\ d_{55} &= \frac{1}{3(2J+3)}, \\ c_{44} &= -\frac{2J-1}{30(J+1)(2J+1)}, \\ c_{55} &= -\frac{2J+5}{30(J+1)(2J+3)}, \\ c_{45} &= -\frac{1}{5(J+1)}\sqrt{\frac{J(J+2)}{(2J+1)(2J+3)}}. \end{aligned} \tag{9.65}$$

In analogy with Eq. (9.60) for the Rayleigh scattering case we retrieve W_2 of Eqs. (9.38) and (9.39) for Cases IV and V if we in Eq. (9.63) insert $\nu = \nu_4$ or ν_5, which gives $W_2 = c_{44}/d_{44}$ and $W_2 = c_{55}/d_{55}$, respectively.

Finally, for the case when $|J_{\mu_f} - J_{\mu_i}| = 2$, there is as already mentioned no upper J-state interference and therefore no off-diagonal terms of c_{ij} and d_{ij}. Nevertheless we need the single, non-zero diagonal terms of c_{ij} and d_{ij}, i.e., c_{66} and d_{66}, for our later computation of W_2 for complete multiplets (Sect. 9.12 below). d_{66} equals d_{55} of Eq. (9.65), while $c_{66} = 0.1 d_{66}$.

9.10. Expressions for the Relative Oscillator Strengths

To derive the polarizability W_2 from Eqs. (9.56), (9.61), and (9.63) we need to calculate the oscillator strengths, which serve as weights for the various terms. However, we do not have to compute the absolute but only the relative oscillator strengths within a multiplet, since the two common scale factors for the oscillator strengths (one scale factor for the multiplet responsible for the absorption process, another scale factor for the multiplet of the emission process) divide out when forming W_2. We may therefore give the relative oscillator strengths as functions of the L, S, and J quantum numbers alone, since the effects of the radial atomic structure are part of the common factor that divides out. For reasons of symmetry we give the relative oscillator strengths in terms of the line strength $S_{ij} = S_{ji}$, from which the relative oscillator strength is obtained through

$$f_{ij} \sim \frac{S_{ij}}{\lambda_{ij}(2J_i + 1)}, \tag{9.66}$$

where λ_{ij} is the wavelength of the transition. As it does not matter to S_{ij} (in contrast to f_{ij}) which is the initial or final level, we may in the expressions below reverse the direction of all the arrows without affecting the form of the expressions.

Thus $L \to L-1$ can be used for both of the cases $\Delta L = \pm 1$. Because of this symmetry property all the cases that may occur can be covered by giving the expressions for S_{ij} for the two main cases $L \to L-1$ and $L \to L$, each of which has the three subcases $J \to J+1$, $J \to J$, and $J \to J-1$. These expressions for the relative line strength S_{ij}, from Condon and Shortley (1970), are listed below.

$L \to L-1, \quad J \to J+1$:

$$\frac{(J-L-S)(J-L-S+1)(J-L+S+1)(J-L+S+2)}{J+1}. \tag{9.67}$$

$L \to L-1, \quad J \to J$:

$$\frac{(2J+1)(L+S-J)(J+L-S)(J+L+S+1)(J-L+S+1)}{J(J+1)}. \tag{9.68}$$

$L \to L-1, \quad J \to J-1$:

$$\frac{(J+L+S)(J+L+S+1)(J+L-S)(J+L-S-1)}{J}. \tag{9.69}$$

$L \to L, \quad J \to J+1$:

$$\frac{(L+S-J)(J+L+S+2)(J+L-S+1)(J-L+S+1)}{J+1}. \tag{9.70}$$

$L \to L, \quad J \to J$:

$$\frac{(2J+1)\left[J(J+1)+L(L+1)-S(S+1)\right]^2}{J(J+1)}. \tag{9.71}$$

$L \to L, \quad J \to J-1$:

$$\frac{(L+S-J+1)(J+L-S)(J-L+S)(J+L+S+1)}{J}. \tag{9.72}$$

Only combinations of the L, S, and J quantum numbers for which the triangular condition

$$|L-S| \le J \le |L+S| \tag{9.73}$$

is satisfied are allowed.

9.11. Sign of the Interference Terms

With the Wigner-Eckart theorem, Eq. (7.95), the matrix element $\langle JM\,|\,\hat{r}_q\,|\,J'M'\rangle$ can be factorized into two parts, a reduced matrix element $\langle J\,||\,\hat{\boldsymbol{r}}\,||\,J'\rangle$ that is free from any M dependence, and another part (including a 3-j symbol) that accounts for this M dependence. The total angular momentum that is represented by the quantum number J is formed by vector addition of the orbital angular momentum, represented by L, and the electron spin, represented by S. It is possible to factorize $\langle J\,||\,\hat{\boldsymbol{r}}\,||\,J'\rangle$ such that the spin (S) dependence gets separated from the radial dependence, which is accounted for by the S-independent "doubly reduced" matrix element $\langle L\,||\,\hat{\boldsymbol{r}}\,||\,L'\rangle$. According to Brink and Satchler (1968) we may thus write

$$\begin{aligned}\langle J\,||\,\hat{\boldsymbol{r}}\,||\,J'\rangle &= \langle LSJ\,||\,\hat{\boldsymbol{r}}\,||\,L'S'J'\rangle \\ &= (-1)^{L+S+J'+1}\sqrt{2J'+1}\,\sqrt{2L+1}\begin{Bmatrix} L & L' & 1 \\ J' & J & S\end{Bmatrix}\langle L\,||\,\hat{\boldsymbol{r}}\,||\,L'\rangle\,.\end{aligned} \tag{9.74}$$

As $\langle L\,||\,\hat{\boldsymbol{r}}\,||\,L'\rangle$ divides out when forming W_2, and as the magnitude of the preceding factor on the right hand side of Eq. (9.74) is implicitly accounted for by expressions (9.67)–(9.72) of the preceding section, our only remaining concern is the sign of this factor. As seen by Eq. (9.74) this sign is determined by the first sign factor (the power of -1) and the sign of the 6-j symbol.

Since the general expressions for the sign are messy, it is more convenient to list the results separately for each case that may occur. This is done below, with index u representing the upper level, index ℓ the lower level. $r_{J_\ell J_u}$ is the sign parameter defined in Eq. (9.48).

CASE A. $L_u = L_\ell - 1$

$$r_{J_\ell J_u} = J_u - J_\ell + 1\,. \tag{9.75}$$

CASE B.1. $L_u = L_\ell\,,\quad J_u - J_\ell = -1$

$$r_{J_\ell J_u} = 0\,. \tag{9.76}$$

CASE B.2. $L_u = L_\ell\,,\quad J_u - J_\ell = 0$

$$\begin{aligned} r_{J_\ell J_u} &= 0 \;\text{ if }\; J(J+1) \geq S(S+1) - L(L+1)\,, \\ r_{J_\ell J_u} &= 1 \;\text{ if }\; J(J+1) < S(S+1) - L(L+1)\,. \end{aligned} \tag{9.77}$$

CASE B.3. $L_u = L_\ell\,,\quad J_u - J_\ell = 1$

$$r_{J_\ell J_u} = 1\,. \tag{9.78}$$

CASE C. $L_u = L_\ell + 1$

$$r_{J_\ell J_u} = 0\,. \tag{9.79}$$

9.12. W_2 for Complete Multiplets

Let us by indices i, e, and f denote the initial, intermediate (excited), and final quantum numbers. W_2, as expressed by Eqs. (9.56), (9.61), and (9.63) above, is then a function of the six quantum numbers $L_{i,e,f}$, S, and $J_{i,f}$ (S does not change within a multiplet, and J_e has already been summed over in our expressions for W_2). We may summarize our previous three expressions for the Rayleigh (Eq. (9.56)) and Raman (Eqs. (9.61) and (9.63)) scattering cases in the single equation

$$W_2(L_{i,e,f}\,, S\,; J_i, J_f) = C/D\,, \tag{9.80}$$

where

$$\begin{aligned} C &= \sum_{u=1}^{6}\sum_{v=u}^{6} g_{uv}\, c_{uv}\, (\nu_u - \nu)^{-1}\, (\nu_v - \nu)^{-1}\,, \\ D &= \sum_{n=1}^{6} f_{in} f_{fn}\; d_{nn}\, (\nu_n - \nu)^{-2}\,. \end{aligned} \tag{9.81}$$

The summations in Eq. (9.81) can only have contributions from the allowed diagonal terms and the allowed off-diagonal terms with (u, v) equal to $(1,2)$, $(1,3)$, $(2,3)$, or $(4,5)$. C and D are implicitly functions of $L_{i,e,f}$, S, and $J_{i,f}$.

To obtain W_2 when accounting for all the contributing transitions within the multiplet of the emission process while keeping the initial J_i number fixed, we have to sum C and D over all possible final states. Since each final J_f state has a multiplicity of $2J_f + 1$ (a factor that is part of the Wigner-Eckart theorem of Eq. (7.95) and which represents the number of magnetic substates),

$$W_2(L_{i,e,f}\,, S\,; J_i) = \frac{\sum_{J_f}(2J_f+1)C(L_{i,e,f}\,, S\,; J_i, J_f)}{\sum_{J_f}(2J_f+1)D(L_{i,e,f}\,, S\,; J_i, J_f)} = \frac{C(L_{i,e,f}\,, S\,; J_i)}{D(L_{i,e,f}\,, S\,; J_i)}\,. \tag{9.82}$$

To account for all the possible contributions in the absorption multiplet we need to sum over all the possible initial J_i numbers as well while attaching the statistical weights $2J_i + 1$ (which are proportional to the diagonal density matrix elements $\rho_{J_i J_i}$, cf. Sect. 8.10):

$$W_2(L_{i,e,f}\,, S) = \frac{\sum_{J_i}(2J_i+1)\sum_{J_f}(2J_f+1)C(L_{i,e,f}\,, S\,; J_i, J_f)}{\sum_{J_i}(2J_i+1)\sum_{J_f}(2J_f+1)D(L_{i,e,f}\,, S\,; J_i, J_f)} = \frac{C(L_{i,e,f}\,, S)}{D(L_{i,e,f}\,, S)}\,. \tag{9.83}$$

This summation procedure represents a generalization of Eq. (8.114), where only summation over magnetic substates was discussed. The nominator and denominator have to be summed separately, since they each represent linear combinations of W_{ij} matrix elements.

9.13. Principle of Spectroscopic Stability

The correctness of this jungle of algebraic expressions can be verified by invoking the principle of spectroscopic stability in the limit of vanishing fine-structure splitting. We have previously (Sect. 9.6) made use of this principle by introducing a magnetic field to selectively populate the excited Zeeman sublevels, requiring that non-magnetic scattering should be retrieved in the limit of vanishing magnetic field strength. In the present section the electron spin assumes the previous role of the magnetic field.

The question that we are asking is what value W_2 approaches in the limit of vanishing fine-structure splitting, i.e., when all the resonant frequencies ν_n are made to coincide for all n, or, equivalently, when we consider frequencies ν in the dispersion wings at distances $|\nu_n - \nu|$ from the line centers that are large in comparison with the separation $|\nu_u - \nu_v|$ between the resonant frequencies involved.

In this limiting case all the profile factors $(\nu_n - \nu)^{-2}$ and $(\nu_u - \nu)^{-1}(\nu_v - \nu)^{-1}$ divide out, and we are left with a frequency-independent expression with each term in the nominator and denominator depending in a complex algebraic way on all three quantum numbers L, S, and J, since the oscillator strengths and sign factors have such complicated dependencies. From a mathematical point of view we would therefore expect W_2 to depend on all three numbers L, S, and J.

Physically, however, vanishing fine-structure splitting implies vanishing influence of the electron spin, which, in terms of the corresponding quantum numbers, implies $S \to 0$. In this limit we thus have $S = 0$ and $L = J$. This means that when we form W_2 for the whole multiplet, $W_2(L_{i,e,f}\,, S)$ of Eq. (9.83), the limiting value should be independent of the value of S used. We may therefore write for arbitrary values of S

$$\lim_{|\nu_u - \nu_v| \to 0} W_2(L_{i,e,f}\,, S) = W_2(J_{i,e,f},\, S = 0)\,, \tag{9.84}$$

where $|\nu_u - \nu_v| \to 0$ for all u and v. When $S = 0$, $J_{i,e,f} = L_{i,e,f}$. The limiting value $W_2(J_{i,e,f},\, S = 0)$ is obtained directly from one of the expressions (9.35)–(9.40) (depending on the particular combination of the $J_{i,e,f}$ numbers) if we let $J = \min(L_i, L_f)$ in the respective expression (L assumes the role of J).

As an example, the principle of spectroscopic stability demands that a Raman scattering transition with $(L, S) = (3, 4.5) \to (2, 4.5) \to (1, 4.5)$ must approach $W_1 = -0.50$, $W_2 = 0.10$ in the far dispersion wings, since this follows directly from Case VI, Eq. (9.40), as the transition is of type $J = 1 \to 2 \to 3$ (the values of $W_{1,2}$ are invariant with respect to a reversal of the scattering direction).

As the algebraic expressions that are involved in Eq. (9.83) are very complex, it seems like an "algebraic miracle" that the above principle of spectroscopic stability can really be satisfied, and it would be quite hard to prove it by purely algebraic methods. Instead we have let a computer program that calculates W_2 scan through all possible combinations of L and S quantum numbers, and have thereby verified that the principle of spectroscopic stability is indeed always satisfied. Such a verification in fact constitutes a very powerful test of the correctness of the individual algebraic expressions for the coefficients c_n, c_{uv}, d_n (Eqs. (9.57)–(9.59) and (9.65)), the oscillator strengths $f_{J_\mu J_m}$ (Eqs. (9.67)–(9.72)), and the sign parameters $r_{J_\mu J_m}$ (Eqs. (9.75)–(9.79)). If any single one of these various parameters were to be computed with the incorrect algorithm, this would immediately be revealed by the test of spectroscopic stability. The interference terms, containing the c_{uv} coefficients, are of course essential. Without any one of them, there would be no spectroscopic stability.

It turns out that the principle of spectroscopic stability has general validity also for scattering transitions from each initial J_i state separately (before we sum over J_i, but after having summed over J_f). We therefore have an even more restrictive condition for spectroscopic stability that always has to be satisfied for both W_1 and W_2:

$$\lim_{|\nu_u - \nu_v| \to 0} W_{1,2}(L_{i,e,f}\,, S\,; J_i) = W_{1,2}(J_{i,e,f},\, S = 0)\,, \tag{9.85}$$

where $W_{1,2}(L_{i,e,f}\,, S\,; J_i)$ is given by Eq. (9.82).

As a very special case of all this we note that for a transition of the type $S \to P \to S$ $(L = 0 \to 1 \to 0)$, W_1 and W_2 always approach the value 1.0 at large distances from the resonant frequencies of the multiplet, i.e., the scattering in the far dispersion wings behaves like classical dipole-type scattering.

9.14. Role of a Background Continuum

The polarization of the background continuous spectrum was treated in Sect. 5.16. In the visible part of the spectrum this polarization, which is mainly due to Thomson scattering at free electrons and Rayleigh scattering at neutral hydrogen, is usually quite small as compared with the polarization of some of the lines. It increases in importance when we go to the blue and UV parts of the spectrum, due to the steep wavelength dependence (λ^{-4}) of Rayleigh scattering at hydrogen (cf. Eq. (5.83)). As shown in Sect. 5.16 both the Rayleigh and Thomson scattering processes are describable as classical dipole-type scattering with frequency-independent $W_{1,2}$ coefficients of unity.

According to Eq. (5.86) the continuous emission vector can be written

$$\boldsymbol{j}_c = \sigma_c \boldsymbol{J}_{\nu,c} + \kappa_c B\nu\, \mathbf{1}\,, \tag{9.86}$$

where σ_c represents the sum of the Thomson and Rayleigh scattering coefficients as shown by Eq. (5.84). $\boldsymbol{J}_\nu$ is the "average Stokes vector $\boldsymbol{I}_\nu$" as defined by Eq. (5.85). Index c for $\boldsymbol{J}_\nu$ indicates that the phase matrix that is valid for the continuum and is identical to the classical dipole-type phase matrix should be used for the calculation of $\boldsymbol{J}_\nu$.

As seen from Eqs. (6.104), (6.105), (6.94), (5.95), and (5.96), the line emission vector in the line wings can be written as

$$\boldsymbol{j}_L = \sigma_L\, \varphi_\nu\, [\, k_c \boldsymbol{J}_{\nu,L} + (\alpha^{-1} - k_c) J_\nu\, \mathbf{1}\,]\,, \tag{9.87}$$

where α, given by Eq. (6.90), is the fraction of the line emission events that are part of scattering processes. k_c, given by Eq. (5.67), is the fraction of the scattering processes that occur as coherent scattering. It represents a collisional depolarization factor. σ_L is the line scattering coefficient, which as in Eq. (6.94) is found to be

$$\sigma_L = \alpha \frac{h\nu}{4\pi} N_u A_{ul} \Big/ \int \varphi_\nu J_\nu \,\mathrm{d}\nu \,, \tag{9.88}$$

where indices u and ℓ refer to the upper and lower levels, respectively. Inserting expression (6.90) for α we obtain

$$\sigma_L = \frac{h\nu}{4\pi} N_\ell B_{\ell u} A_{ul} \Big/ \sum_k R_{uk} \,, \tag{9.89}$$

where we have used the statistical equilibrium relation

$$\sum_k N_k R_{ku} = N_u \sum_k R_{uk} \tag{9.90}$$

for the transition rate coefficients. The left hand side of Eq. (9.90) represents the transitions *to* level u, the right hand side the transitions *from* level u.

To arrive at Eqs. (9.87) and (9.88) we have for the parameter e, which occurs in Eqs. (6.105) and (5.95) and is explicitly given by Eq. (5.34), assumed that the incident radiation is unpolarized. We have further replaced $\int \varphi_\nu J_\nu \,\mathrm{d}\nu$ by J_ν, which is a very good approximation in the far line wings. For clarity we have written Eqs. (9.88) and (9.89) for the case of a single line transition only — when several line transitions overlap we simply have to sum over their respective contributions to the line emission vector.

For the average Stokes vector $\boldsymbol{J}_{\nu,L}$ that is based on the phase matrix for the line and is a function of the $W_{1,2}$ polarizability coefficients, we compute $\boldsymbol{J}_\nu$ as defined by Eq. (5.85) with the general phase matrix given by Eq. (9.19). If we omit the $\boldsymbol{E}_{44}$ term in the phase matrix we get the simple relation

$$\boldsymbol{J}_{\nu,L} \approx W_2 \,\boldsymbol{J}_{\nu,c} + (1 - W_2)\, J_\nu \,\mathbf{1} \,. \tag{9.91}$$

Omission of the $\boldsymbol{E}_{44}$ term is usually permissible, since scattering of Stokes V is generally unimportant (in comparison with scattering of the linear polarization). Inserting Eq. (9.91) in Eq. (9.87) and using Eq. (9.86), we obtain the total emission vector

$$\begin{aligned} \boldsymbol{j} = \boldsymbol{j}_L + \boldsymbol{j}_c &= (k_c\, W_2\, \sigma_L\, \varphi_\nu + \sigma_c)\, \boldsymbol{J}_{\nu,c} \\ &\quad + [\, \sigma_L\, \varphi_\nu\, (\alpha^{-1} - k_c W_2)\, J_\nu + \kappa_c\, B_\nu \,]\, \mathbf{1} \,. \end{aligned} \tag{9.92}$$

The degree of linear polarization in the Stokes Q direction of the emitted radiation is given by

$$p = j_Q / j_I \,, \tag{9.93}$$

where $j_{I,Q}$ are the first two components of the emission vector $\boldsymbol{j}$. To calculate this degree of polarization we have to evaluate the first two components of the average Stokes vector $\boldsymbol{J}_{\nu,c}$, which we denote $J_{\nu,I}$ and $J_{\nu,Q}$. Using Eqs. (5.68) and (5.60) and assuming that the incident radiation is unpolarized (so that only the first column of the phase matrix needs to be used — the full expressions without this limiting assumption have been given in Stenflo (1976b)), we get

$$\begin{aligned} J_{\nu,I} &= J_\nu + \tfrac{1}{8}(1 - 3\mu^2) P_\nu \,, \\ J_{\nu,Q} &= \tfrac{3}{8}(1 - \mu^2) P_\nu \,, \end{aligned} \tag{9.94}$$

where

$$P_\nu = J_\nu - 3K_\nu\,, \tag{9.95}$$

with

$$K_\nu = \tfrac{1}{2}\int_{-1}^{1} \mu^2 I_\nu \,\mathrm{d}\mu \tag{9.96}$$

being the second order moment of I_ν (while J_ν is the zero order moment). For an isotropic radiation field $P_\nu = 0$. Inserting Eqs. (9.94) in Eq. (9.92) we get

$$\begin{aligned} j_I =& (\alpha^{-1}\,\sigma_L\,\varphi_\nu + \sigma_c)\,J_\nu + \kappa_c B_\nu \\ &+ \tfrac{1}{8}(1-3\mu^2)(k_c\,W_2\,\sigma_L\,\varphi_\nu + \sigma_c)\,P_\nu\,, \\ j_Q =& \ \tfrac{3}{8}(1-\mu^2)(k_c\,W_2\,\sigma_L\,\varphi_\nu + \sigma_c)\,P_\nu\,. \end{aligned} \tag{9.97}$$

Usually the last term in j_I may be disregarded, since it is similar in magnitude to j_Q (at the limb, where $\mu = 0$, it is $= \frac{1}{3}j_Q$). It is small, since P_ν is generally $\ll J_\nu$. In this case we may write

$$p \approx k_G\,[\,\alpha\,k_c\,W_2\,r_L + \alpha_c\,(1-r_L)\,]\,, \tag{9.98}$$

where

$$k_G = \tfrac{3}{8}(1-\mu^2)\,P_\nu/J_\nu \tag{9.99}$$

represents a "geometrical depolarization factor" (introduced in Stenflo (1982)) that accounts for the polarizability due to the degree of anisotropy of the radiation field,

$$r_L = \frac{\alpha^{-1}\sigma_L\,\varphi_\nu}{\alpha^{-1}\sigma_L\,\varphi_\nu + \sigma_c + \kappa_c B_\nu/J_\nu} \tag{9.100}$$

represents the fraction of the total opacity that is due to line radiation, and

$$\alpha_c = \frac{\sigma_c\,J_\nu}{\sigma_c\,J_\nu + \kappa_c B_\nu} \tag{9.101}$$

represents the fraction of the continuum photons that are due to Rayleigh or Thomson scattering.

So far we have written the explicit expressions only for the case of one line transition (plus background continuum) with a profile function φ_ν. When several transitions (e.g. within a multiplet) contribute, we have to sum over them. Thus, when the polarizability W_2 is calculated, the denominator, which is proportional to the number of line photons, can be expressed as (cf. Eq. (9.32))

$$W_{11} + 2W_{14} \sim \sum_n \alpha_n^{-1}\,\sigma_{L,n}\,\varphi_{\nu,n}\,, \tag{9.102}$$

where the different line contributions in the multiplet are marked by index n. We may now recast Eq. (9.100) in the more general form

$$r_L = \frac{W_{11} + 2W_{14}}{W_{11} + 2W_{14} + a_c}\,, \tag{9.103}$$

where the parameter a_c is independent of wavelength and represents the contribution of the continuum.

It is convenient to introduce

$$W_{2,\,\text{eff}} = W_2\, r_L + b_c(1 - r_L) \tag{9.104}$$

as an "effective" scattering polarizability W_2, which accounts for the effect of the continuum through the wavelength-independent parameters a_c (contained in r_L) and b_c, where

$$b_c = \frac{\alpha_c}{\alpha\, k_c} \tag{9.105}$$

represents the continuum value of $W_{2,\,\text{eff}}$. Then

$$p = \alpha\, k_c\, k_G\, W_{2,\,\text{eff}}\,, \tag{9.106}$$

i.e., the degree of linear polarization p scales in proportion to $W_{2,\,\text{eff}}$. To explore the effect of the continuum on the line polarization one may thus compute $W_{2,\,\text{eff}}$ for different values of the two free parameters a_c and b_c.

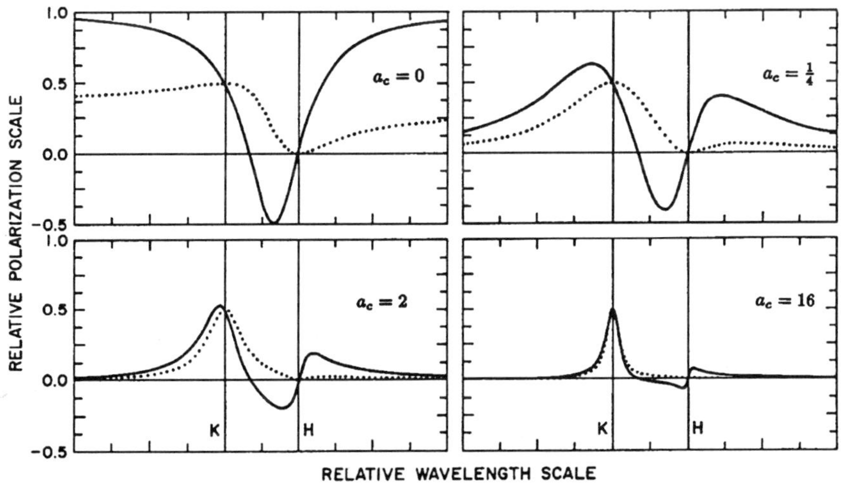

Fig. 9.2. Wavelength variation of $W_{2,\,\text{eff}}$ for a scattering process of the type $S_{\frac{1}{2}} \rightarrow P_{\frac{1}{2},\frac{3}{2}} \rightarrow S_{\frac{1}{2}}$, for different values of the continuum opacity parameter a_c (normalized as described in the text), keeping the continuum polarization parameter $b_c = 0$. The thin vertical lines mark the locations of the two resonant frequencies. The thick solid curve represents $W_{2,\,\text{eff}}$ with J_m-state interference taken into account, while the dotted curve represents the case when this interference has been omitted (incoherent superposition of the two scattering transitions). From Stenflo (1980).

An example of such calculations of $W_{2,\,\text{eff}}$ is given in Figs. 9.2 and 9.3, from Stenflo (1980). The diagrams represent Rayleigh scattering of the type $S_{\frac{1}{2}} \rightarrow P_{\frac{1}{2},\frac{3}{2}} \rightarrow S_{\frac{1}{2}}$ with quantum-mechanical interference between the two $J_m = \frac{1}{2}$ and $\frac{3}{2}$ states. This type of coherent scattering transition is represented by the Ca II

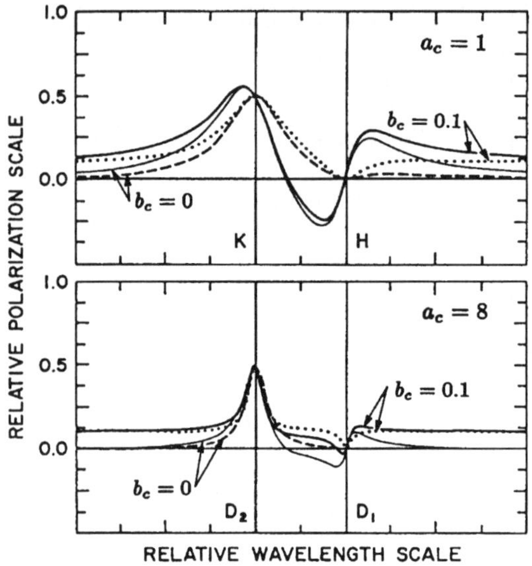

Fig. 9.3. Effect on $W_{2,\mathrm{eff}}$ (for the same scattering case as in Fig. 9.2) from varying the continuum polarization parameter b_c. The dotted and dashed curves represent the case when J_m-state interference has been omitted. The choices 1 and 8 for the continuum opacity parameter a_c lead to profile shapes that closely resemble the observed polarization profiles of the Ca II K and H and the Na I D_2 and D_1 lines, respectively. From Stenflo (1980).

K and H lines at 3933 and 3968 Å and by the Na I D_2 and D_1 lines at 5890 and 5896 Å. The normalization used for the continuum parameter a_c is such that for $a_c = 1$ the continuum and line opacities are equal at a frequency that is the average of the two resonant frequencies (i.e., halfways between the two lines).

In Fig. 9.2 the continuum polarization parameter b_c is set equal to zero, while the opacity parameter a_c is varied. For $a_c = 0$, $W_{2,\mathrm{eff}}$ approaches unity in the far line wings, as expected from our discussion of the principle of spectroscopic stability (Sect. 9.13). We see in the figure that this asymptotic limit is not approached when the J_m-state interference term is omitted. When an unpolarized continuum is added, $W_{2,\mathrm{eff}}$ asymptotically approaches zero in the distant line wings.

The effect of varying the amount of continuum polarization is illustrated in Fig. 9.3. The asymptotic value of $W_{2,\mathrm{eff}}$ in the far wings equals the continuum polarization parameter b_c.

The simple model provided by Eq. (9.106) is able to represent the observations remarkably well. Figure 9.4 shows the model fit (dotted curve) to the observations (solid curve) for the Ca II K and H resonant scattering transition. The parameter values used for the fit are $a_c = 1$ (in terms of the same units as used for Figs. 9.2 and 9.3), $b_c = 0.057$, and $\alpha\, k_c\, k_G = 0.0176$.

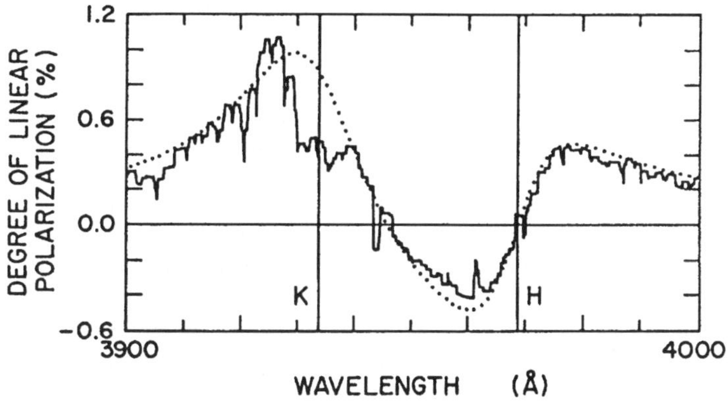

Fig. 9.4. Theoretical fit (dotted curve) to the linear polarization observed at the National Solar Observatory (Kitt Peak) 10 arcsec inside the solar limb at the heliographic north pole (solid curve). The observations have been treated with a 1 Å running window to crudely remove the influence of depolarizing blend lines. Positive values correspond to linear polarization parallel to the solar limb, negative values to polarization perpendicular to the limb. The theoretical curve is based on Eq. (9.106) with the values of the model parameters being $a_c = 1$, $b_c = 0.057$, and $\alpha\, k_c\, k_G = 0.0176$. From Stenflo (1980).

9.15. Clarifying Remarks Concerning the Dependence of the Polarizability on Oscillator Strength

According to the semi-classical theory for radiative scattering, Eq. (5.20), the line scattering coefficient σ_L should be proportional to the oscillator strength f_{ij}, not to f_{ij}^2, since the scattering probability is governed by the radiative absorption rate, and the absorbed photons are simply reradiated in the scattering process. We have further seen, through Eq. (9.102), that the denominator of W_2 (cf. Eq. (9.32)), $W_{11} + 2W_{14}$, is proportional to σ_L and thus to f_{ie}, the absorption oscillator strength. This appears to be in contradiction with our expressions (9.56), (9.61), and (9.63) for W_2, where f_{ij} appears *twice* (both as absorption and emission oscillator strength) in each of the various terms in the nominator and denominator. Let us now clarify this situation to show that everything remains consistent, and that the classical expression is indeed reproduced by the quantum-mechanical theory.

In Sects. 8.4 and 8.9 we showed how the correct Einstein transition rates could be obtained from quantum field theory. The same methods can be applied to show how the absolute magnitude of σ_L can be expressed in terms of the Einstein transition rates, starting from the expression (9.4) for the transition rate $1/\tau$ in second-order time-dependent perturbation theory, or, equivalently, from the expressions (8.110) for the scattering amplitudes obtained from quantum-field theory. When inspecting Eqs. (8.111) and (9.15) we see that the scattering amplitude $w_{\alpha\beta}$ contains two matrix elements, one for absorption, the other for emission, each containing a reduced matrix element that is proportional to $\sqrt{f_{ij}}$. When constructing the tensor product $\boldsymbol{w} \otimes \boldsymbol{w}^*$ to get the elements W_{ij} of the $\boldsymbol{W}$ scattering matrix, the oscillator strength will therefore appear *twice* in each element of $\boldsymbol{W}$, although the classical expression (5.20) for σ_L would lead us to believe that the oscillator strength should appear only once.

The source of this apparent confusion lies in the proportionality constant that was omitted in Eqs. (8.111) and (9.15). From Eq. (8.110) we see that this proportionality constant is $\sim a^{-1}$, where a is the normalized damping parameter that is proportional to γ, the inverse lifetime of the excited state. This lifetime is determined by the decay due to spontaneous radiative transitions to any allowed lower level as well as to collisional transitions, as described by Eq. (8.55). In our previous treatment we have implicitly assumed that this damping parameter or inverse lifetime is the same for all terms appearing in W_2, which is justified as all the lower levels are involved in forming the radiative decay rate $\sum_i A_{ji}$. This has allowed us to separate γ^{-1} as a common proportionality constant for the scattering amplitudes $w_{\alpha\beta}$.

In Eq. (9.89) we gave an explicit expression for σ_L in terms of the transition rates, obtained from the non-LTE theory of Chapter 6. The denominator, $\sum_k R_{uk}$, represents the total transition rate away from the upper (excited) level u, and is in fact identical to the inverse lifetime γ. Eq. (9.89) can therefore be rewritten as

$$\sigma_L = \frac{h\nu}{4\pi} N_{J_{\mu_i}} B_{J_{\mu_i} J_m} A_{J_m J_{\mu_f}} / \gamma . \tag{9.107}$$

This is indeed the expression we get when deriving the absolute value of σ_L from Eq. (8.110), using the methods of Sects. 8.4 and 8.9.

Since the B and A coefficients are each proportional to an oscillator strength, Eq. (9.107) seems to imply that the oscillator strength appears twice in σ_L. However, the terms that make up γ are also proportional to oscillator strengths. In the special case of a two-level atom with only one radiative decay route, the emission oscillator strength appears both in the nominator and denominator and thus divides out, so that σ_L becomes proportional to $f_{J_{\mu_i} J_m}$ alone, in complete agreement with the semi-classical treatment that led to Eq. (5.20).

9.16. Origin of Partial Polarization in the Scattered Radiation

Any coherent superposition of electromagnetic waves represents radiation that is 100 % polarized. Mathematically such radiation can be represented by the Jones vector, and as was noted in Sects. 2.6.1 and 4.1, the Jones vector is unable to represent anything that is less than 100 % polarized. The polarization may be elliptical with arbitrary shape and orientation of the polarization ellipse. In terms of the Stokes parameters, 100 % polarization means that $Q^2 + U^2 + V^2 = I^2$. In quantum terminology we may say that each photon is always 100 % polarized.

For a scattering phase matrix $\boldsymbol{P}$ given by Eq. (9.19) with $W_{1,2} = 1$, the scattered radiation is always 100 % elliptically polarized whenever the incoming radiation is 100 % elliptically polarized. When $W_2 < 1$ the scattered radiation is however only partially polarized. If for instance $W_2 = 0$ the scattered radiation is unpolarized even when the incoming radiation is 100 % linearly polarized. On the other hand, when we consider each individual scattering event, both the incoming and scattered photons are always 100 % polarized. How can then a scattering matrix with $W_2 < 1$ at all exist?

The source of the partial polarization can be found in the *incoherent* sum in Eq. (8.114) of the scattering matrices (the tensor products of the scattering amplitudes) over all the initial and final magnetic substates μ_i and μ_f. As was shown in Sects. 2.6 and 4.1 the Jones formalism cannot account for partially polarized light, but we need the coherency matrix or Stokes vector formalisms to accomodate an incoherent superposition of stochastically uncorrelated wave trains. While a coherent superposition would still leave the radiation 100 % polarized, an incoherent superposition of an ensemble of uncorrelated photon states can result in arbitrary degrees of partial polarization.

If we calculate the scattering phase matrix from the tensor product $\boldsymbol{w} \otimes \boldsymbol{w}$ before any incoherent summation is done, then this phase matrix that represents scattering from an initial magnetic substate μ_i to a final substate μ_f produces scattered radiation that is always 100 % elliptically polarized when the incoming radiation is also 100 % polarized. The different $\mu_i - \mu_f$ combinations however correspond to different phase matrices. When these matrices are added (which corresponds to an incoherent superposition of uncorrelated scattering events), an average phase matrix that may have $W_{1,2} < 1$ results, which leads to partially polarized scattered radiation.

Scattering from different initial substates (into all the possible μ_f states) produces entirely different amounts of polarization. Let us as an example consider the Rayleigh scattering transition $J = 6 \rightarrow 5 \rightarrow 6$. According to Eq. (9.35) $W_2 = 0.0577$. If the incident radiation is unpolarized and the scattering angle is 90°, the polarization of the scattered radiation is according to Eqs. (9.19) and (9.20) $Q/I = P_{21}/P_{11} = \frac{3}{4}W_2/(1-\frac{1}{4}W_2) = 4.4\,\%$, with the plane of polarization oriented perpendicular to the scattering plane. If we now consider scattering of unpolarized radiation from only the initial substate $\mu_i = +$ or -6, then $Q/I = -50.6\,\%$, with the plane of linear polarization oriented *parallel* to the scattering plane. (For the computation of this number we do not have any simple algebraic expressions at our disposal, so we have instead evaluated the 3-j symbols in the scattering amplitudes and then formed P_{21} and P_{11} from products of scattering amplitudes.) Scattering from the initial state with $\mu_i = 0$ on the other hand gives $+30.7\,\%$ linear polarization oriented perpendicular to the scattering plane. Since the polarization of the scattered radiation is such a sensitive function of μ_i, small deviations from complete redistribution between the initial magnetic substates, i.e., a variation with μ_i of the population $\rho_{\mu_i \mu_i}$, may have considerable effects on the incoherent scattering superposition of the contributions from the various μ_i states.

As a second example we consider scattering for the Rayleigh transition $J = 4 \rightarrow 4 \rightarrow 4$. While the average Q/I for the complete redistribution case is $+32.0\,\%$ as follows from Eq. (9.36) and the expression for P_{21}/P_{11} above, it is $-63.6\,\%$ for scattering from the $\mu_i = 0$ substate and $+72.9\,\%$ for scattering from $\mu_i = \pm 4$. Such examples may serve as a warning to watch out for the possibility of non-LTE populations of the initial sublevels (while the phase relations between the initial substates or the non-diagonal terms of the density matrix may be neglected).

CHAPTER 10

COLLISIONS, PARTIAL REDISTRIBUTION, AND TURBULENT MAGNETIC FIELDS

10.1. Introduction

In the preceding chapters we have developed the foundations for treating general problems involving radiative scattering in a magnetic field. The theory however needs to be completed by including the effects of collisions and partial redistribution before it is ready for general use on the sun.

It would take us too far to make a comprehensive treatment of collisional physics here. Instead we start with a simplified classical approach to gain some physical insight about collisional effects and then treat the quantum-mechanical theory in a phenomenological way. Thereby we will take over from literature the numerical values of the two main parameters of collision theory, the rates of collisional destruction of the atomic orientation ($\gamma_c^{(1)}$) and alignment ($\gamma_c^{(2)}$), expressed in units of the total elastic collision rate γ_c that is responsible for line broadening.

A preliminary overview of partial redistribution (the relation between the incoming and outgoing frequencies and the scattering directions in the *observer's* coordinate system) in terms of the Mueller matrix formalism was given in Chapter 5 for classical scattering. This theory will now be completed and integrated with the collisional theory. The resulting redistribution matrix incorporates the results from the best available detailed quantum-mechanical calculations. Analytical expressions will be given at various levels of approximation. Particularly useful are the approximate forms in the line core and wings, respectively. In the general radiative transfer equation (e.g. Eq. (6.104)) all the physics of the coherence phenomena, collisions, and partial frequency redistribution resides in the redistribution matrix $\boldsymbol{R}$.

After these theoretical tools have been established we turn to a discussion of practical applications of the Hanle effect, in particular for the diagnostics of the magnetic field structure in prominences, in the chromosphere and above, and of turbulent magnetic fields at various levels of the atmosphere. Since the elusive turbulent fields are of fundamental importance for our understanding of solar magnetism, we will describe how they may be constrained in different ways by observations of Hanle depolarization, line broadening, and the transverse Zeeman effect, although they remain spatially unresolved. These joint constraints provide information on both the turbulent field strength and the angular distribution of the field vectors.

10.2. Classical Collision Theory

The spectral properties of a fluctuating component $E_q(t)$ of the radiation field is found by forming the Fourier transform $\widetilde{E}_q(\omega)$ and the power spectrum $\widetilde{E}_q\widetilde{E}_q^*$. Thus we may Fourier transform the Jones vector $\boldsymbol{J}$ of Eq. (2.28) to get $\widetilde{\boldsymbol{J}}$. Since in the Jones calculus all polarizing filters are represented by linear operators (cf. Eq. (2.29)), and since the Fourier transform is itself a linear operator, the whole Jones calculus remains valid if we replace all electric fields with their Fourier transforms. We may then form a coherency matrix $\boldsymbol{D}_{\rm FT} = \widetilde{\boldsymbol{J}}\widetilde{\boldsymbol{J}}^\dagger$ from the Fourier transforms as in Eq. (2.33) as well as a Stokes vector $S_{k,{\rm FT}} = {\rm Tr}\,(\boldsymbol{\sigma}_k \boldsymbol{D}_{\rm FT})$ as in Eq. (2.43). The whole Mueller calculus then applies to $S_{k,{\rm FT}}$. In particular $S_{0,{\rm FT}}$ is the power spectrum that is recorded with spectrographs and detectors. The spectral behavior of the polarization effects (including the Hanle effect) is therefore contained in the bilinear products formed by $\widetilde{\boldsymbol{J}}\widetilde{\boldsymbol{J}}^\dagger$. In terms of the spherical vector components we have to explore the products $\widetilde{E}_q\widetilde{E}_{q'}^*$.

As we have seen in Sects. 3.2 and 3.12, the spherical vector component E_q of the electric field of the scattered radiation is proportional to the corresponding component d_q of the electric dipole moment of a classical oscillator. It is thus sufficient to discuss the properties of d_q and regard them as representative of E_q.

A classical oscillator that has been excited will perform damped oscillations governed by Eq. (3.33) without the driving term on the right-hand side after the exciting radiation has been turned off (the photon absorption has taken place). The solution of this equation gives the time dependence of the oscillating electric dipole moment. To first order in the classical damping constant γ_N of Eq. (3.33) and the Larmor precession frequency ω_L of Eq. (3.44), the damped oscillations of the spherical vector component d_q of the electric dipole moment are described by

$$d_q = d_{0q}\, e^{i(\omega_0 + q\omega_L)t}\, e^{-(\gamma_N/2)t}\,. \tag{10.1}$$

These oscillations can be viewed as an antenna that generates the emitted (scattered) radiation.

To get the spectral properties of the scattered radiation we need to calculate the Fourier transform $\widetilde{d}_q(\omega)$ of $d_q(t)$. The Fourier integral however cannot extend over infinite time, since the damped oscillation of Eq. (10.1) only lasts as long as the excited atom is undisturbed by collisions. Accordingly it only exists between time $t = 0$ (time of excitation) and $t = t_c$ (time of a collision, which is assumed to instantly scramble the phase of the oscillation). In the presence of collisions $\widetilde{d}_q$ is therefore given by

$$\widetilde{d}_q \sim \int_0^{t_c} e^{i(\omega_0 - \omega + q\omega_L + i\gamma_N/2)t}\,{\rm d}t = \frac{i\,[\,1 - e^{i(\omega_0 - \omega + q\omega_L)t_c}\, e^{-(\gamma_N/2)t_c}\,]}{\omega_0 - \omega + q\omega_L + i\gamma_N/2}\,. \tag{10.2}$$

The spectral properties of the coherency matrix and the Stokes parameters depend on the bilinear products $\widetilde{d}_q\widetilde{d}_{q'}^*$. Let τ_c be the average time between two collisions, which are assumed to occur randomly according to Poisson statistics.

The times t_c then have the probability distribution $\exp(-t_c/\tau_c)$, over which $\widetilde{d}_q\widetilde{d}^*_{q'}$ has to be averaged. Let us *define* the collisional damping constant γ_c as

$$\gamma_c = 2/\tau_c\,. \tag{10.3}$$

The factor 2 is used in this definition to make γ_c agree with the conventional definition of the collisional broadening parameter when it is combined with the radiative broadening (see below). The needed average is then given by

$$\langle\, \widetilde{d}_q\widetilde{d}^*_{q'}\,\rangle = \int_0^\infty \widetilde{d}_q\widetilde{d}^*_{q'}\, e^{-(\gamma_c/2)t_c}\,\mathrm{d}t_c \Big/ \int_0^\infty e^{-(\gamma_c/2)t_c}\,\mathrm{d}t_c\,. \tag{10.4}$$

The denominator equals $2/\gamma_c$.

In the special case with a vanishing magnetic field ($\omega_L = 0$) we find after some calculations that

$$\langle\, |\widetilde{d}_0|^2\,\rangle \sim k'_c\,\frac{1}{\pi}\,\frac{\gamma_{\rm tot}/2}{(\omega_0-\omega)^2+(\gamma_{\rm tot}/2)^2}\,, \tag{10.5}$$

where

$$\begin{aligned} k'_c &= \frac{\gamma_N}{\gamma_N+\gamma_c/2}\,,\\ \gamma_{\rm tot} &= \gamma_N+\gamma_c\,. \end{aligned} \tag{10.6}$$

k'_c represents the fraction of scattering processes that are coherent (undisturbed by collisions) and corresponds to k_c of Eq. (5.67). We use a prime for k_c to mark that the expression of Eq. (10.6) is based on the special, simplified treatment of the present section and will be replaced by more correct expressions later. $\gamma_{\rm tot}$ is the total damping constant representing combined radiative and collisional broadening.

In the case of non-zero magnetic fields the complicated expressions may be greatly simplified if we limit ourselves to the weak-field regime where $\omega_L \ll \Delta\omega_D$, i.e., the Zeeman splitting is much smaller than the Doppler width. Then we can perform frequency averaging (integration over ω) as in Sect. 5.11. Physically such averaging is done by the Doppler motions of the scattering atoms. We give the final result of these calculations in normalized form, such that the expression becomes unity in the absence of collisions and magnetic fields. Using $\langle\langle\,\rangle\rangle$ to denote averaging over both t_c and ω, we get

$$\langle\langle\, \widetilde{d}_q\widetilde{d}^*_{q'}\,\rangle\rangle \sim \frac{k'_c}{1+i\tan\alpha_{q-q'}}\,, \tag{10.7}$$

where the "Hanle mixing angle" $\alpha_{q-q'}$ is given by

$$\tan\alpha_{q-q'} = \frac{(q-q')\,\omega_L}{\gamma_N+\gamma_c/2} \tag{10.8}$$

(cf. Eqs. (5.53) or (8.80)).

Note that the denominator in $\tan\alpha_{q-q'}$ and in k'_c is *not* $\gamma_{\rm tot}$, since only half the value of γ_c enters to combine with γ_N, in contrast to the line broadening parameter. The reason for this difference is that $\alpha_{q-q'}$ and k'_c are determined by coherence phenomena, which directly involve the unsquared amplitudes, while line broadening is not related to coherence.

In the quantum-mechanical theory that we will discuss in the following sections, we need to replace $\gamma_c/2$ with the collisional destruction rate $\gamma_c^{(K)}$ of the $2K$-multipole of the excited state, where $K = |q - q'|$. This gives us the collisional branching ratio

$$k_c^{(K)} = \frac{\gamma_N}{\gamma_N + \gamma_c^{(K)}}, \tag{10.9}$$

while the Hanle mixing angle is determined by

$$\tan \alpha_K = \frac{K g_e \, \omega_L}{\gamma_N + \gamma_c^{(K)}}. \tag{10.10}$$

As in Eq. (5.53) we have here attached the Landé factor g_e of the upper, excited level. The expression of Eq. (10.6) for $\gamma_{\rm tot}$ remains unchanged.

As we will see later, detailed quantum-mechanical collision theory gives $\gamma_c^{(K)}/\gamma_c = 0.43$ and 0.38 for $K = 1$ and 2, respectively, while our simple classical treatment gives 0.50, which is a good result in view of the idealizations made.

10.3. Scalar Frequency Redistribution Function

In the absence of collisions the probability $p(\xi, \xi')$ that in the *rest frame* of the scattering atom an incident photon of frequency ξ' scatters such that the emitted photon has frequency ξ is according to Eqs. (9.4) and (9.10) determined by the delta function $\delta(\xi' - \xi + \nu_{af})$, where $\nu_{af} = E_{af}/h$ and E_{af} is the energy difference between the initial (a) and final (f) states of any given Raman scattering transition. Note that in our notation the parameters representing the incident radiation always are placed to the right of those for the scattered radiation, in contrast to commonly used notations in previous literature (e.g. Mihalas, 1978). Our convention directly corresponds to the index conventions for matrices, which are used throughout the present monograph.

The rest frame absorption profile $f(\xi)$ that determines the probability for photon absorption is a Lorentz function of the form

$$f(\xi) = \frac{1}{\pi} \frac{\gamma/4\pi}{(\nu_0 - \xi)^2 + (\gamma/4\pi)^2}, \tag{10.11}$$

where ν_0 is the resonance frequency, and $\gamma = \gamma_N + \gamma_c$ is the sum of the natural (radiative) and collisional damping constants. $f(\xi)$ is the result of a convolution of two Lorentz functions, one describing radiative broadening alone (with damping constant γ_N), the other describing broadening by elastic collisions (damping constant γ_c).

$f(\xi)$ and $p(\xi, \xi')$ are area normalized such that integration over all frequencies ξ gives unity. In addition, integration of $p(\xi, \xi')$ over all ξ' gives unity as well.

Elastic collisions during the excited state scramble the phase and erase the atomic memory of the excitation process. The correlation between the incident and scattered frequencies ξ' and ξ is thereby destroyed. Let k_c denote the fraction of the scattering events that are undisturbed by elastic collisions. If we disregard the unimportant inelastic collisions,

$$k_c = \frac{\gamma_N}{\gamma_N + \gamma_c} . \tag{10.12}$$

When polarization effects are disregarded, the probability function $p(\xi, \xi')$ then becomes

$$p(\xi, \xi') = k_c \, \delta(\xi' - \xi + \nu_{af}) + (1 - k_c) \, f(\xi) \tag{10.13}$$

(cf. Mihalas, 1978). The *scalar* frequency redistribution function for the *rest frame* can now be written

$$R_0(\xi, \boldsymbol{n} \, ; \, \xi', \boldsymbol{n}') = f(\xi') \, p(\xi, \xi') \, g(\boldsymbol{n}, \boldsymbol{n}') , \tag{10.14}$$

where $g(\boldsymbol{n}, \boldsymbol{n}')$ is the *angular phase function* accounting for the directional probability distribution. It is normalized such that averaging over all solid angles Ω or Ω' gives unity. In polarization problems g is replaced by the *phase matrix* $\boldsymbol{P}(\boldsymbol{n}, \boldsymbol{n}')$, and R_0 becomes the 4×4 matrix $\boldsymbol{R}_0$ (cf. Eq. (5.11)).

It is convenient to introduce the notation

$$R_0 / g(\boldsymbol{n}, \boldsymbol{n}') = k_c \, R_{\mathrm{II},\,0} + (1 - k_c) \, R_{\mathrm{III},\,0} \, , \tag{10.15}$$

where

$$\begin{aligned} R_{\mathrm{II},\,0} &= f(\xi') \, \delta(\xi' - \xi + \nu_{af}) , \\ R_{\mathrm{III},\,0} &= f(\xi') \, f(\xi) . \end{aligned} \tag{10.16}$$

When converting to the observer's frame in which the incident and scattered frequencies are ν' and ν, we use relation (5.9) and integrate over the Maxwellian velocity distribution (5.10). Thereby $R_{\mathrm{II},\,0}$ and $R_{\mathrm{III},\,0}$ get transformed into $R_{\mathrm{II}}(\nu, \boldsymbol{n} \, ; \, \nu', \boldsymbol{n}')$ and $R_{\mathrm{III}}(\nu, \boldsymbol{n} \, ; \, \nu', \boldsymbol{n}')$, respectively. The scalar frequency redistribution function then becomes

$$R(\nu, \boldsymbol{n} \, ; \, \nu', \boldsymbol{n}') / g(\boldsymbol{n}, \boldsymbol{n}') = k_c \, R_{\mathrm{II}} + (1 - k_c) R_{\mathrm{III}} \, . \tag{10.17}$$

Note that through our definition (10.16) R_{II} and R_{III} do not contain the phase function $g(\boldsymbol{n}, \boldsymbol{n}')$, in contrast to the definition in previous literature (e.g. Mihalas, 1978). This allows us to retain the same functions and keep them scalar when we go to the polarized case, where $g(\boldsymbol{n}, \boldsymbol{n}')$ is replaced by $\boldsymbol{P}(\boldsymbol{n}, \boldsymbol{n}')$. In Sects. (10.7) and (10.8) we will give explicit expressions for R_{II} and R_{III} at various levels of approximation.

10.4. Branching Ratios for the Polarized Case

When an atomic state gets polarized, e.g. due to anisotropic radiative excitation, it acquires orientation (magnetic dipole moment) and alignment (electric quadrupole moment), which represent the two $2K$-multipoles, with $K = 1$ and 2. We then have to distinguish between three different effects of elastic collisions represented by three different rates: The previously discussed elastic collision rate γ_c that is responsible for line broadening and decorrelation between frequencies ξ' and ξ, and the rates $\gamma_c^{(K)}$ for the collisional destruction of the $2K$-multipoles with $K = 1$ and 2. The angular redistribution is also determined by the multipolar state of the excited level (cf. Eq. 10.22)). We therefore need to decompose the redistribution problem in its multipolar components.

Let $p(\xi, \xi')^{(K)}$ be the rest frame probability that the atom scatters from incoming frequency ξ' to emitted frequency ξ without collisional destruction of the $2K$-multipole. The special case of $K = 0$ corresponds to the unpolarized (unindexed) case of the previous section. The fraction k_c of the scattering atoms is unaffected by collisions. It will therefore scatter coherently in the rest frame while preserving the multipoles. The incoherent part relating to the $2K$-multipole (when $K \neq 0$) however does not have the branching ratio $1 - k_c$ in contrast to the unpolarized case. The ratio is instead $\beta^{(K)}$, which represents the fraction of radiative transitions for which the excited state has been subject to a collision that does *not* destroy the $2K$-multipole. The rate of such elastic collisions is $\gamma_c - \gamma_c^{(K)}$. We then get

$$p(\xi, \xi')^{(K)} = k_c\, \delta(\xi' - \xi + \nu_{af}) + \beta^{(K)} f(\xi)\,, \tag{10.18}$$

where

$$\beta^{(K)} = \left(\frac{\gamma_N}{\gamma_N + \gamma_c} \right) \left(\frac{\gamma_c - \gamma_c^{(K)}}{\gamma_N + \gamma_c^{(K)}} \right). \tag{10.19}$$

The second factor in Eq. (10.19) represents the probability that a collision that does not destroy the $2K$-multipole occurs during the lifetime of this multipole. The first factor, which equals k_c, represents the probability that spontaneous emission occurs before the atom is disturbed by any elastic collisions (cf. Omont et al., 1972; Domke and Hubeny, 1988; Faurobert-Scholl, 1992, 1993).

Eq. (10.13) for the unpolarized case can be retrieved from Eq. (10.18) for the special case when $K = 0$. As $\gamma_c^{(0)}$ represents the rate of quenching collisions (inelastic collisions corresponding to collisional excitation, ionization, or deexcitation), which is generally $\ll \gamma_c$ and γ_N, we get from Eqs. (10.19) and (10.12)

$$\beta^{(0)} \approx 1 - k_c\,. \tag{10.20}$$

This shows that $p(\xi, \xi')^{(0)}$ indeed equals $p(\xi, \xi')$ of the scalar problem. Eq. (10.18) is therefore valid for all three cases $K = 0, 1, 2$.

The fractions k_c and $\beta^{(K)}$ do not add up to unity (except when $K = 0$). The remaining fraction,

$$1 - k_c - \beta^{(K)} = \frac{\gamma_c^{(K)}}{\gamma_N + \gamma_c^{(K)}}\,, \tag{10.21}$$

represents the probability that the $2K$-multipole will be collisionally destroyed during the scattering process. This destruction of the $2K$-multipole not only leads to complete frequency redistribution but also destroys the directional dependence of the phase matrix. $\boldsymbol{P}(\boldsymbol{n},\boldsymbol{n}')$ therefore becomes isotropic and replaced by $\boldsymbol{E}_{11}$, a 4×4 matrix that has its first component equal to unity, while all the remaining components are zero. $\boldsymbol{E}_{11}$ is the result of directional averaging of $\boldsymbol{P}$. In contrast, the processes represented by $p(\xi,\xi')^{(K)}$ preserve the directional dependence of the polarization, since the $2K$-multipole is not collisionally destroyed.

To find the rest-frame redistribution matrix we need to decompose the phase matrix $\boldsymbol{P}$ into three components that relate to each of the three multipoles with $K=0,1,2$:

$$\boldsymbol{P}=\boldsymbol{P}_0+\boldsymbol{P}_1+\boldsymbol{P}_2\,. \tag{10.22}$$

While $\boldsymbol{P}_1$ and $\boldsymbol{P}_2$ vanish when an angular average is performed, $\boldsymbol{P}_0$ is isotropic and simply represented by $\boldsymbol{E}_{11}$. $\boldsymbol{P}_2$ relates exclusively to the linear polarization, $\boldsymbol{P}_1$ to the circular polarization. Such a decomposition is necessary, since the branching ratios with index (K) cannot be applied to the full phase matrix $\boldsymbol{P}$ but only to component $\boldsymbol{P}_K$, as will be seen explicitly in the next section.

According to Eq. (5.68)

$$\begin{aligned}\boldsymbol{P}_0&=\boldsymbol{E}_{11}\,,\\ \boldsymbol{P}_1&=\tfrac{3}{4}W_1\boldsymbol{P}^1\,,\\ \boldsymbol{P}_2&=\tfrac{3}{4}W_2\boldsymbol{P}^2\,.\end{aligned} \tag{10.23}$$

Explicit expressions for $\boldsymbol{P}^1$ and $\boldsymbol{P}^2$ for an arbitrary scattering geometry and a weak magnetic field along the polar axis have been given in Eqs. (5.58)–(5.64). In the absence of a magnetic field and for the special scattering geometry of Fig. 3.3 with $\theta'=\theta=\pi/2$ (scattering in the equatorial plane) we have according to Eq. (9.19)

$$\begin{aligned}\boldsymbol{P}^1&=2\boldsymbol{E}_{44}\cos\phi\,,\\ \boldsymbol{P}^2&=\tfrac{4}{3}(\boldsymbol{P}_R-\boldsymbol{E}_{11})\,.\end{aligned} \tag{10.24}$$

$\boldsymbol{P}_R$ is the classical Rayleigh scattering phase matrix for the linear polarization.

10.5. Frequency Redistribution Matrix in the Rest Frame

The frequency redistribution matrix $\boldsymbol{R}_0$ for polarized radiation is obtained if we in expression (10.14) for the scalar redistribution function R_0 replace the angular phase function $g(\boldsymbol{n},\boldsymbol{n}')$ with the phase matrix in the component form (10.22) and apply the probability functions $p^{(K)}$ of Eq. (10.18) to each phase matrix component $\boldsymbol{P}_K$:

$$\boldsymbol{R}_0(\xi,\boldsymbol{n}\,;\,\xi',\boldsymbol{n}')/f(\xi')=\sum_{K=0}^{2}p(\xi,\xi')^{(K)}\boldsymbol{P}_K\,, \tag{10.25}$$

or, with Eq. (10.18) and the functions defined by Eq. (10.16),

$$\boldsymbol{R}_0(\xi,\boldsymbol{n}\,;\,\xi',\boldsymbol{n}') = k_c\,R_{\mathrm{II},\,0}\,\boldsymbol{P} + R_{\mathrm{III},\,0}\sum_{K=0}^{2}\beta^{(K)}\boldsymbol{P}_K\,. \tag{10.26}$$

10.6. Frequency Redistribution Matrix in the Observer's Frame

Exactly as in the scalar redistribution case, the functions $R_{\mathrm{II},\,0}$ and $R_{\mathrm{III},\,0}$ are transformed to R_{II} and R_{III} when we do the conversion to the observer's frame. We therefore obtain the frequency redistribution matrix $\boldsymbol{R}(\nu,\boldsymbol{n}\,;\,\nu',\boldsymbol{n}')$ from Eqs. (10.25) and (10.26) simply by removing index 0 everywhere and replacing frequency ξ with ν. Somewhat more explicitly the frequency redistribution matrix can be written in terms of the classical Rayleigh phase matrix $\boldsymbol{P}_R$ as

$$\begin{aligned}\boldsymbol{R}(\nu,\boldsymbol{n}\,;\,\nu',\boldsymbol{n}') = k_c\,R_{\mathrm{II}}\,[\,&W_2\boldsymbol{P}_R + (1-W_2)\boldsymbol{E}_{11} + \tfrac{3}{4}W_1\boldsymbol{P}^1\,]\\ &\beta^{(2)}R_{\mathrm{III}}\,[\,W_2\boldsymbol{P}_R + (1-W_2)\boldsymbol{E}_{11}\,] + \beta^{(1)}R_{\mathrm{III}}\,\tfrac{3}{4}W_1\boldsymbol{P}^1\\ &+R_{\mathrm{III}}\,(1-k_c-\beta^{(2)})\boldsymbol{E}_{11}\,.\end{aligned} \tag{10.27}$$

This form is most similar to the one used by Domke and Hubeny (1988), who based their treatment of the redistribution problem on the results of Omont et al. (1972). To keep our expressions simple we have used the assumption that the initial state has an infinite radiative lifetime, and quenching collisions have been disregarded.

The explicit expressions for $\boldsymbol{P}_R$ in a magnetic field with a general scattering geometry can be found directly by expressing $\boldsymbol{P}_R$ in terms of $\boldsymbol{P}^2$ via Eq. (10.24) and using the general expressions for $\boldsymbol{P}^2$ that were given in Eqs. (5.59)–(5.64).

We will later find it convenient to use the branching ratio defined by Eq. (10.9), in terms of which

$$\beta^{(K)} = k_c^{(K)} - k_c\,. \tag{10.28}$$

Using Eqs. (10.23) and (10.20) we then obtain from Eq. (10.26)

$$\boldsymbol{R}(\nu,\boldsymbol{n}\,;\,\nu',\boldsymbol{n}') = k_cR_{\mathrm{II}}\boldsymbol{P} + R_{\mathrm{III}}\Big[\frac{3}{4}\sum_{K=1}^{2}(k_c^{(K)}-k_c)\,W_K\boldsymbol{P}^K + (1-k_c)\boldsymbol{E}_{11}\Big]\,. \tag{10.29}$$

Through a detailed and extensive quantum-mechanical treatment of the collisional effects, Berman and Lamb (1969) find that

$$\begin{aligned}\gamma_c^{(1)}/\gamma_c &= 0.43\,,\\ \gamma_c^{(2)}/\gamma_c &= 0.38\,.\end{aligned} \tag{10.30}$$

We thus only need to know the line-broadening parameters γ_N and γ_c, since the values of all the other collision-related parameters, k_c, $k_c^{(K)}$, $\beta^{(K)}$, automatically follow from Eq. (10.30).

10.7. Expressions for $R_{\rm II}$ and $R_{\rm III}$

It is convenient for numerical calculations to express the redistribution functions in terms of the dimensionless frequency parameter

$$x = \Delta\nu/\Delta\nu_D\,, \tag{10.31}$$

where $\Delta\nu_D$ is the Doppler width. As

$$\boldsymbol{R}(\nu, \boldsymbol{n}\,;\,\nu', \boldsymbol{n}')\,\mathrm{d}\nu\,\mathrm{d}\nu' = \boldsymbol{R}(x, \boldsymbol{n}\,;\,x', \boldsymbol{n}')\,\mathrm{d}x\,\mathrm{d}x' \tag{10.32}$$

we have

$$\boldsymbol{R}(\nu, \boldsymbol{n}\,;\,\nu', \boldsymbol{n}') = \boldsymbol{R}(x, \boldsymbol{n}\,;\,x', \boldsymbol{n}')/(\Delta\nu_D)^2\,. \tag{10.33}$$

The corresponding relations are of course valid for the scalar redistribution functions $R_{\rm II}$ and $R_{\rm III}$. The explicit expressions for them are the same as in ordinary scalar (unpolarized) redistribution theory and have been given in previous literature (cf. Mihalas, 1978; Saliba, 1985). Notice that in contrast to these previous authors we define $R_{\rm II}$ and $R_{\rm III}$ without the angular phase function or the phase matrix included, since the phase matrix contains different multipole components with different branching ratios as we have seen.

Let Θ be the angle between the incident and scattered directions $\boldsymbol{n}'$ and $\boldsymbol{n}$. We then get

$$R_{\rm II}(x, \boldsymbol{n}\,;\,x', \boldsymbol{n}') = \frac{1}{\pi\,|\sin\Theta|}\,\exp[\,-\tfrac{1}{2}(x-x')^2/\sin^2\tfrac{1}{2}\Theta\,]\;H(a_{\rm II},\,v_{\rm II})\,, \tag{10.34}$$

where H is the Voigt function of Eq. (4.15) with the Voigt parameters

$$\begin{aligned} a_{\rm II} &= a/\cos\tfrac{1}{2}\Theta\,,\\ v_{\rm II} &= \tfrac{1}{2}(x+x')/\cos\tfrac{1}{2}\Theta\,. \end{aligned} \tag{10.35}$$

a is the usual damping parameter of Eq. (3.53) with the damping rate $\gamma = \gamma_N + \gamma_c$, the total line-broadening rate.

Similarly

$$R_{\rm III}(x, \boldsymbol{n}\,;\,x', \boldsymbol{n}') = \frac{a_{\rm III}}{\pi^2}\int_{-\infty}^{+\infty}\frac{e^{-u^2}H(a_{\rm III},\,x/\sin\Theta\,-u/\tan\Theta)}{(x'-u)^2+a^2}\,\mathrm{d}u\,, \tag{10.36}$$

where

$$a_{\rm III} = a/\sin\Theta\,. \tag{10.37}$$

In unpolarized redistribution analysis it has been found useful to work with angular averages defined as in Eq. (5.31), since they provide a good approximation for most practical purposes while greatly reducing the computational complexity. This is also the case for polarized problems if we make the angular average over the scalar redistribution functions $R_{\rm II}$ and $R_{\rm III}$ separately, while keeping the phase matrix components intact, since their directional properties are the cause of the polarization effects.

We may thus replace $R_{\rm II}$ and $R_{\rm III}$ in the expressions of the preceding section with their angular averages

$$R_{\rm II}(x,x') = \frac{1}{\pi^{3/2}} \int_{|x'-x|/2}^{\infty} e^{-u^2} \left[\tan^{-1}\left(\frac{\underline{x}+u}{a}\right) - \tan^{-1}\left(\frac{\overline{x}-u}{a}\right)\right] du\,, \quad (10.38)$$

where

$$\begin{aligned} \overline{x} &\equiv \max(x,x')\,, \\ \underline{x} &\equiv \min(x,x')\,, \end{aligned} \quad (10.39)$$

and

$$\begin{aligned} R_{\rm III}(x,x') = \frac{1}{\pi^{5/2}} \int_0^{\infty} e^{-u^2} &\left[\tan^{-1}\left(\frac{x'+u}{a}\right) - \tan^{-1}\left(\frac{x'-u}{a}\right)\right] \\ &\left[\tan^{-1}\left(\frac{x+u}{a}\right) - \tan^{-1}\left(\frac{x-u}{a}\right)\right] du\,. \end{aligned} \quad (10.40)$$

According to Jefferies and White (1960) one may write

$$R_{\rm II}(\nu,\nu') \approx a_\nu\, \varphi_{\nu'}\, \delta(\nu-\nu') + (1-a_\nu)\,\varphi_\nu \varphi_{\nu'} \quad (10.41)$$

(cf. Eq. (5.96)), where $a_\nu \approx 0$ in the line core ($|\Delta\nu| \lesssim 3\Delta\nu_D$) and ≈ 1 in the wings. The transition between core and wings however is not well defined by this approximate expression. While the first term in Eq. (10.41) represents frequency coherence, the second represents complete frequency redistribution. This means that we have frequency coherence in the line wings, complete redistribution in the line core, a property that was discussed in Sects. 5.7–5.10.

For $R_{\rm III}$ it is a good approximation to use

$$R_{\rm III}(\nu,\nu') \approx \varphi_\nu \varphi_{\nu'} \quad (10.42)$$

everywhere (cf. Mihalas, 1978). $R_{\rm III}$ thus approximates complete frequency redistribution at all frequencies.

10.8. Core and Wing Redistribution Matrices

It is illuminating for the physical understanding to write down the explicit expressions for the redistribution matrix in the line core and wings. With the approximate expressions (10.41) and (10.42) this can now readily be done.

10.8.1. Line Core

As

$$R_{\rm II} \approx R_{\rm III} \quad (10.43)$$

in the line core according to Eqs. (10.41) and (10.42), we immediately get from Eq. (10.29)

$$\boldsymbol{R}_{\rm core}(\nu, \boldsymbol{n}\,;\,\nu', \boldsymbol{n}') \approx R_{\rm III}\left[\boldsymbol{E}_{11} + \tfrac{3}{4}k_c^{(1)}\,W_1\,\boldsymbol{P}^1 + \tfrac{3}{4}k_c^{(2)}\,W_2\,\boldsymbol{P}^2\right]. \tag{10.44}$$

For $R_{\rm III}$ Eq. (10.42) may be used. Eq. (10.44) agrees with our classically derived expressions (5.46) and (5.68) in the absence of collisions.

Note that the matrices $\boldsymbol{P}^1$ and $\boldsymbol{P}^2$ that were given by Eqs. (5.58)–(5.64) include the Hanle effect for weak magnetic fields. The "Hanle mixing angles" α_1 and α_2 appearing in expressions (5.58) and (5.59) for $\boldsymbol{P}^1$ and $\boldsymbol{P}^2$ are in the presence of collisions given by Eq. (10.10) with $K = q - q'$ for $K = 1, 2$. Inserting the numerical values of Eq. (10.30) we explicitly get

$$\begin{aligned} \tan\alpha_1 &= \frac{g_e\,\omega_L}{\gamma_N + 0.43\gamma_c}\,, \\ \tan\alpha_2 &= \frac{2g_e\,\omega_L}{\gamma_N + 0.38\gamma_c}\,. \end{aligned} \tag{10.45}$$

The denominator represents the inverse lifetime of the $2K$-multipole. Comparing Eq. (10.45) with Eq. (10.8) of our simplified classical treatment, we see that agreement with the classical theory would have been reached if $\gamma_c^{(K)}/\gamma_c$ had been 0.50 instead of 0.43 and 0.38 according to the elaborate quantum theory of Berman and Lamb (1969). Also the classically derived branching ratio k_c' of Eq. (10.6) would then have agreed with the branching ratios in Eq. (10.44).

10.8.2. Line wings

As

$$R_{\rm III} \ll R_{\rm II} \tag{10.46}$$

in the wings, we immediately get from Eqs. (10.29) and (10.41)

$$\boldsymbol{R}_{\rm wings}(\nu, \boldsymbol{n}\,;\,\nu', \boldsymbol{n}') \approx k_c\,\varphi_{\nu'}\,\delta(\nu - \nu')\,\boldsymbol{P}_{\rm non-mag}\,. \tag{10.47}$$

$\boldsymbol{P}_{\rm non-mag}$ represents the ordinary, collision-free, *non-magnetic* phase matrix, which is given by Eqs. (5.68) and (5.58)–(5.64) if we let the Hanle mixing angles $\alpha_{1,2}$ be zero in the expressions for $\boldsymbol{P}^1$ and $\boldsymbol{P}^2$. Eq. (10.47) agrees with the classically derived expression (5.45) in the absence of collisions. The reason why the magnetic-field effects vanish in the line wings has been explained in Sect. 5.8.

10.9. Continuum Polarization and Depolarizing Blend Lines

In the blue and ultraviolet portion of the solar spectrum, for $\lambda \lesssim 4200$ Å, the continuum polarization is comparable in magnitude to or even dominates the line polarization. This part of the spectrum is very crowded with overlapping lines. The majority of these lines is not formed by scattering. Most lines instead depolarize the spectrum by removing (absorbing) the polarized continuum photons or diluting them with the less polarized line photons.

If one makes a scatter plot of the observed polarization $p(\lambda)$ vs. intensity $I(\lambda)$ for a spectral region with insignificant line polarization from coherent scattering, then one finds a rather well-defined linear regression relation indicating a proportionality between p and I (Stenflo et al., 1983a). Let us denote the continuum values of p and I by p_c and I_c. Since for strong depolarizing lines $p/p_c \to 0$ when $I/I_c \to 0$, we may write the linear regression relation in the form

$$\frac{p_c - p}{p_c} \approx \frac{I_c - I}{I_c} \,. \tag{10.48}$$

This form implies that the *shapes* of the linear polarization profiles of depolarizing lines are similar to their intensity profiles. There is of course considerable scatter around relation (10.48), since $(p_c - p)/p_c$ depends on the details of the line formation, but the relation is useful as a statistical approximation.

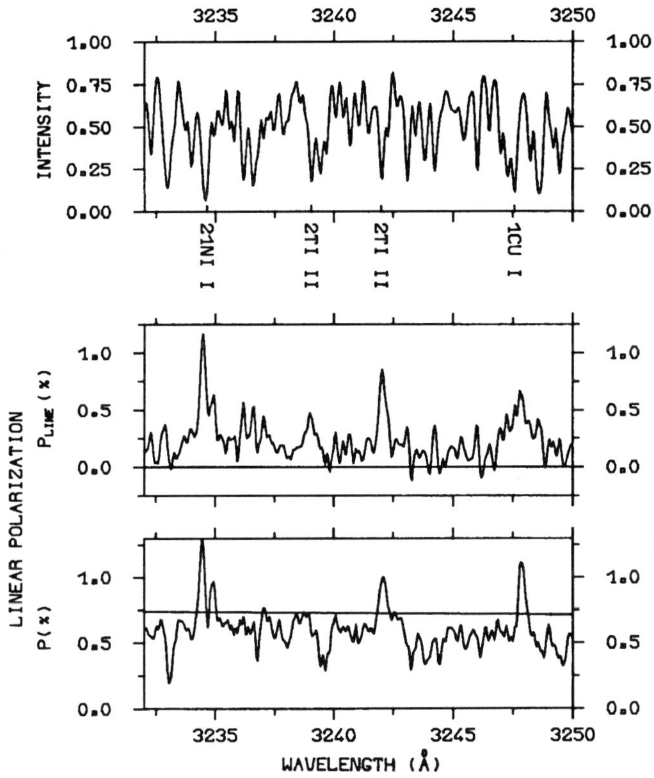

Fig. 10.1. Portion of a spectral recording 10 arcsec inside the solar limb showing intensity I (top diagram), degree of linear polarization p (bottom diagram, where the horizontal line marks the continuum level p_c), and line polarization p_{line} (middle diagram) after statistical removal of the depolarizing lines (see the text). The strong polarization signals of the nickel, titanium, and copper lines stand out much more clearly in the p_{line} diagram. From Stenflo et al. (1983a).

Eq. (10.48) allows us to make a crude removal of the often confusing effects of depolarizing blend lines, so that the few intrinsically polarized lines in a crowded

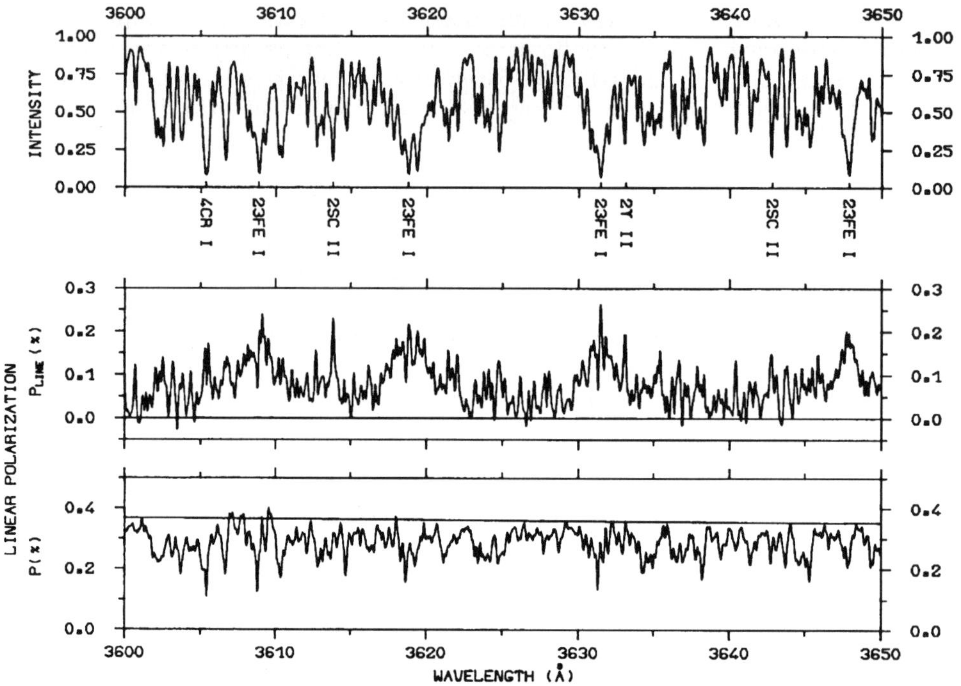

Fig. 10.2. Same as Fig. 10.1, but for a region with broad and strongly polarizing lines due to multiplet no. 23 of neutral iron, and a couple of polarizing lines of ionized scandium. From Stenflo et al. (1983a).

region of the spectrum can be brought out. This is done by defining the "line polarization" p_{line} as

$$p_{\text{line}} = p - p_c I/I_c\,. \tag{10.49}$$

We see that with this definition blend lines obeying Eq. (10.48) as well as the continuum portions of the spectrum will have $p_{\text{line}} = 0$. Only lines that behave differently will stand out with a $p_{\text{line}} \neq 0$. The method of course requires that we can identify the continuum levels p_c and I_c properly, which is generally possible with the help of the above-mentioned scatter plots.

In strong, polarizing lines with extended damping wings, like the Ca II H and K lines or the Ca I 4227 Å line (cf. Fig. 5.5), we may define a quasi-continuum represented by the line-wing intensity and polarization to remove the depolarizing blend lines as in Eq. (10.49).

This procedure is good for survey purposes but should not be used for detailed quantitative analysis, since the continuum and line opacities enter jointly in the radiative-transfer problem in a non-trivial way. Examples of I, p, and p_{line}, from Stenflo et al. (1983a), are given in Figs. 10.1–3. The recordings have been made 10 arcsec inside the solar limb near the heliographic north pole. p represents the degree of linear polarization Q/I in the direction parallel to the solar limb. The horizontal line in the diagrams for p represents p_c.

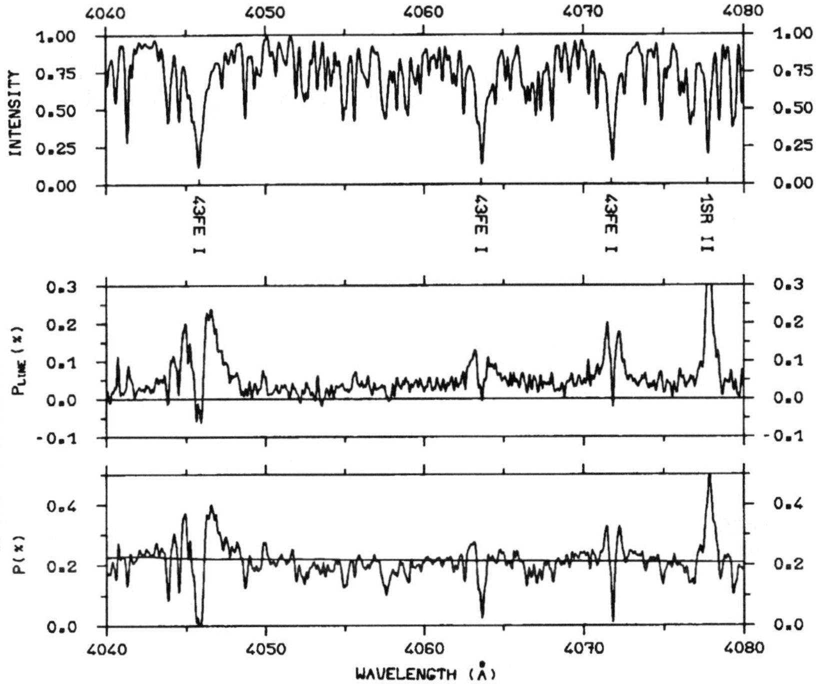

Fig. 10.3. Same as Fig. 10.1, but for a region that includes strongly polarizing lines of ionized strontium and multiplet no. 43 of neutral iron. From Stenflo et al. (1983a).

10.10. Hanle Effect in Solar Prominences

Prominences are arch-like clouds or condensations in the solar corona, permeated by a supporting magnetic field. They have in the past served as a testing ground for the Hanle effect for the following main reasons: (1) They are optically thin in interesting scattering lines like the He I D_3 5876 Å line, which means that radiative-transfer effects disappear and we only need to consider single scattering processes. (2) Due to the low number density of helium atoms, collisional effects can be disregarded. (3) The incident radiation field can be considered as known, since the illumination comes from the underlying photosphere. The anisotropy of the radiation field is determined by the known limb-darkening function and the height of the scattering particle above the solar surface. The anisotropy increases with height and becomes less dependent on the form of the limb-darkening function. (4) The typical field strengths in prominences fall within the range of optimum sensitivity for the Hanle effect. Because the fields are rather weak, the polarization due to the ordinary, longitudinal Zeeman effect is small and hard to measure.

The emphasis of both the observational and theoretical work has been on the He I D_3 line (e.g. Leroy et al., 1977; Sahal-Bréchot et al., 1977; Bommier, 1980; Landi Degl'Innocenti, 1982; Querfeld et al., 1985; cf. the review of Stenflo, 1991a, in which more references can be found). An example of diagnostic diagrams used to interpret the observations was given in Fig. 8.2, which represented more elaborate versions of the idealized diagrams of Fig. 5.4. There are two observables, the amount of linear polarization p and the angle β of rotation of the plane of polarization, while there are three unknowns, the three spatial components of the vector magnetic field. The resulting non-uniqueness in the interpretation can be resolved via additional observational constraints. One way is to use spectral lines with different sensitivities to the Hanle effect. Another is to use the symmetry-breaking due to multiple scattering in lines of finite optical thickness, like the hydrogen Hα line, but then radiative transfer effects have to be taken into account. The observed circular polarization due to the longitudinal Zeeman effect provides an additional observable. It has further been possible to measure collisional depolarization in the hydrogen Hβ line (Bommier et al., 1986), which provides information on the electron density.

The anisotropy of the unpolarized, incident radiation field that is the actual source of the polarization of the scattered radiation can be conveniently described (for the purposes of polarization calculations) in terms of a mathematical decomposition of the incident radiation in two components at the location of the scattering particle: (1) A component for which the electric vectors of the electromagnetic radiation have an isotropic distribution. (2) A component for which the electric vectors exclusively lie in the horizontal plane.

The second component acts as if it were only coming from disk center and therefore scatters at 90°. It is responsible for the polarization of the scattered light. The first component on the other hand only produces unpolarized scattered photons, which dilute the polarized photons of the second component. The amount of dilution is represented by the "geometric depolarization factor" k_G, which equals the ratio between the second component and the sum of the first and second components (the total intensity). For 90° scattering k_G would be unity (maximum anisotropy). It decreases towards zero as the incident radiation field becomes increasingly isotropic.

This geometric depolarization factor is the same as the k_G that was introduced in Eqs. (9.98), (9.99), and (9.106). In the absence of limb darkening of the solar disk k_G is only a function of height h of the scattering particle above the photosphere. It is determined by the cone angle that the solar disk occupies as seen from the vantage point of the particle in the prominence and is given by

$$k_G = \tfrac{1}{2}(1 + \cos\gamma)\cos\gamma\,, \tag{10.50}$$

where the angle γ is defined by

$$\sin\gamma = R_\odot/(R_\odot + h)\,. \tag{10.51}$$

$R_\odot$ is the radius of the sun.

The measured values of the observables p and β are found to fall practically all over the allowed region of diagnostic diagrams like Fig. 8.2. It is however of particular importance to note that observed points can fall near the outer contours of these diagrams only if the magnetic field is *both* oriented almost along the line of sight *and* is rather homogeneous within the spatial resolution element (5000–10,000 km). An unresolved, inhomogeneous field that varies across the spatial resolution element or along the line of sight within the prominence will generally lead to a decrease of β as compared with a homogeneous field that carries the same magnetic flux.

For prominences that are observed edge-on, with the line of sight almost parallel to the length axis of the prominence, the likelihood for the magnetic field to be nearly parallel to the line of sight is high. Observations of such prominences show that about $\frac{1}{4}$ of all measurements give points that fall near the outer contours of the diagnostic diagrams in spite of the long line-of-sight integration through the prominence when it is viewed edge-on (Leroy, 1985). This is a remarkable result, indicating that the prominence magnetic field is largely horizontal and more homogeneous than the threadlike emission structures, which show large fluctuations on scales much smaller than those resolved in the polarimetric observations. Accordingly the prominence magnetic field is largely decoupled from the small-scale fluctuations of the thermodynamic structure (which is expected if the field is force-free).

10.11. Hanle Effect in the Upper Solar Atmosphere

It is attractive to use the Hanle effect for diagnostics of magnetic fields in the upper solar atmosphere, since there we approach the idealized conditions that we have encountered in prominences (optically rather thin medium, negligeable collision rate, large anisotropies in the radiation field). At the same time it gets harder to use the ordinary Zeeman effect when we go higher in the atmosphere. In the chromosphere-corona transition region, where the corona is being heated to a million degrees, the radiation used for diagnostics is primarily in the form of emission lines in the vacuum ultraviolet region of the spectrum, below about 1700 Å. Since the Zeeman splitting $\Delta\lambda_H \sim \lambda^2$, the polarization signal due to the ordinary Zeeman effect is very small in these lines. The signal is further reduced by the circumstance that the magnetic field diverges with height due to the declining gas pressure, which determines the degree of field confinement. We are therefore dealing with considerably weaker fields than in the photosphere (from where the visible spectrum emanates).

Most strong emission lines in the vacuum ultraviolet region are resonance lines, for which coherent scattering plays an important role. Since the radiation field at these wavelengths is highly anisotropic (as shown by the high contrast of the emission structures recorded in the light of these lines), the expected polarization effects are large. Polarization in the vacuum ultraviolet is however still an almost unexplored territory, since the observational technology is hard to master. Initial attempts on the satellites Intercosmos 16 (Stenflo et al., 1976, 1980b) and SMM (Henze et al., 1982; Henze and Stenflo, 1987) have not yet been followed up. Theoretical explorations of scattering polarization in the transition region and corona have been made by Stenflo and Stenholm (1976), Bommier and Sahal-Bréchot (1982), and Sahal-Bréchot et al. (1986).

When going deeper in the atmosphere the observations get easier, but the full radiative-transfer theory with partial redistribution has to be used for the interpretation. This has long appeared to be a formidable obstacle, but with the development of the radiative-transfer theory and computational techniques the diagnostic use of the Hanle effect in these deeper layers has become increasingly attractive. One example is the application of the Hanle effect in the Ca I 4227 Å line, which has the largest scattering polarization in the visible solar spectrum (cf. Fig. 5.5). Faurobert-Scholl (1992) interpreted observations of Stenflo et al. (1980a) and Stenflo (1982) (cf. Fig. 5.6) with detailed radiative-transfer calculations and concluded that the observed Hanle depolarization in the line core was consistent with the presence of a magnetic canopy in the low chromosphere with the canopy boundary at a height of about 950 km above $\tau_{5000} = 1$ (the level of continuum formation for $\lambda = 5000$ Å). Such canopies with a largely horizontal field overlying a field-free atmosphere are expected to form as a result of the rapid flaring out of the field lines of the concentrated kG photospheric magnetic flux tubes, since these flux tubes are embedded in a cool atmosphere that cannot maintain pressure balance as the flux emerges at chromospheric levels. The existence of extended and low-lying canopies has been predicted by Giovanelli (1980), Jones and Giovanelli (1983), and Solanki and Steiner (1990).

Another increasingly important area for the application of the Hanle effect is the diagnostics of turbulent magnetic fields, which will be discussed next. The theoretical tools for treating this difficult problem have only recently been developed.

10.12. Hanle Depolarization for Microturbulent Magnetic Fields

The solar atmosphere is in a turbulent state, and a major portion of the turbulent spectrum has scales that are too small to be spatially resolved. When the scale gets smaller than the photon mean free path, which is about 100 km in the photosphere, the turbulent elements cannot be spatially resolved even in principle, due to the spatial smoothing along the line of sight. Still the photon mean free path is orders of magnitude larger than the viscous damping scale, at which the turbulent spectrum ends. Although the turbulent elements may be unresolved, they reveal their existence by their Doppler-broadening effect on the spectral lines.

As the solar plasma has a very high electrical conductivity, any weak magnetic field will be tangled up and amplified by the turbulence, until some equilibrium state at small scales is reached. The theoretical treatment of such magnetoturbulence is highly complicated by the circumstance that the solar atmosphere is strongly gravitationally stratified. Nevertheless we expect every volume element of the solar atmosphere to be filled with a ubiquitous, tangled magnetic field. Since it is tangled the magnetic polarities are mixed on small scales, such that the magnetic fluxes cancel within the spatial resolution element, leaving no or an insignificant amount of net magnetic flux to be measured. The longitudinal Zeeman effect is practically "blind" to this type of field, since the circular polarization is proportional to the net magnetic flux.

The Hanle depolarization of the scattering polarization on the other hand has only one sign (reduction of the polarization), so there are no cancellation effects. The amount of scattering polarization in the line core depends on the strength of the unresolved turbulent field, although it may carry no net flux. The expected turbulent field strengths lie in the range where most scattering lines of interest have their highest sensitivity to the Hanle effect. For these reasons the Hanle effect emerges as an attractive diagnostic tool to explore and constrain the properties of the turbulent field.

The simplest approach to treat the depolarizing effects of a turbulent field is to do it in the microturbulent limit, i.e., by assuming that the turbulent elements are optically thin. The statistical effects of the turbulent elements can then be incorporated by averaging the magnetic-field dependent terms in the radiative transfer equation before solving it. In our case with the Hanle effect in the weak-field limit, we simply need to average the phase matrix $\boldsymbol{P}$ over the distribution of turbulent properties. As seen by Eqs. (10.22) and (10.23), the problem boils down to an averaging of the two matrices $\boldsymbol{P}^1$ and $\boldsymbol{P}^2$. These averaged matrices replace the unaveraged ones in our radiative-transfer formalism. The solution of this averaged radiative-transfer equation gives the polarized line profiles in the presence of the turbulent magnetic field.

The matrices $\boldsymbol{P}^1$ and $\boldsymbol{P}^2$ in the form given by Eqs. (5.58)–(5.64) refer to the special coordinate system in which the magnetic field vector $\boldsymbol{B}$ is along the vertical axis. To obtain these two matrices for the general case of an arbitrary field direction we have to apply the Mueller rotation matrices as in Eq. (5.66). If we insert these rotated matrices in the redistribution expressions of Sects. 10.6 and 10.8, we can compute the Hanle effect for any variation of the magnetic field vector throughout the solar atmosphere.

To carry out the averaging over the fluctuating physical parameters in the microturbulent limit (when the optical length scale of the turbulent elements is much smaller than unity), we need to use a statistical description of the turbulent magnetic field in terms of a distribution function, which specifies the frequency of occurrence of magnetic fluxes as a function of field strength and field direction. For each given field strength we may carry out the angular integration of $\boldsymbol{P}^1$ and $\boldsymbol{P}^2$ over the field directions. This has been done by Stenflo (1982) and Landi Degl'Innocenti (1985b, 1988) for an isotropic distribution of field vectors and by Stenflo (1987) for a more general angular distribution. Since the trigonometric derivations are rather lengthy, we will only quote the results here. As in Stenflo (1987) we consider an angular distribution that is axisymmetric around the vertical direction, with which the field vectors make inclination angles ϑ_B. We use the notation $\mu_B = \cos\vartheta_B$ and assume that the distribution $f(\mu_B)\,\mathrm{d}\mu_B$ of the inclination angles is given by

$$f(\mu_B) \sim \mu_B^a\,, \tag{10.52}$$

characterized by the single, free parameter a. The special case of $a = 0$ corresponds to an isotropic distribution of field vectors, while $a \to \infty$ implies that all field vectors get aligned along the vertical direction. The parameter a thus specifies how peaked the distribution is around the vertical direction.

If we assume that the incident radiation field is symmetrical with respect to the vertical axis and denote the angular averages of $\boldsymbol{P}^1$ and $\boldsymbol{P}^2$ by $\boldsymbol{P}^1_H$ and $\boldsymbol{P}^2_H$ (index H for "Hanle"), then the results can be written in the form

$$\boldsymbol{P}^K_H = k_H^{(K)}\,\boldsymbol{P}^K\,, \quad K = 1,2\,, \tag{10.53}$$

where

$$k_H^{(1)} = 1 - \frac{2}{a+3}\sin^2\alpha_1\,,$$
$$k_H^{(2)} = 1 - \frac{6}{(a+3)(a+5)}[\,(a+1)\sin^2\alpha_1 + \sin^2\alpha_2\,]\,. \tag{10.54}$$

The Hanle mixing angles α_1 and α_2 are determined by Eq. (10.45) if we for the field that determines the Larmor frequency ω_L insert the value B_t for the strength of the turbulent field. The matrices $\boldsymbol{P}_H^K$ replace $\boldsymbol{P}^K$ in the redistribution matrices, Eqs. (10.29) and (10.44).

Another special type of angular distribution is the case when the field vectors are confined to the horizontal plane but are randomly distributed in azimuth. The results for this case can also be described by Eqs. (10.53) and (10.54), if we simply let $a = -1$.

In practice only $k_H^{(K)}$ for the case $K = 2$ is of importance, since the really observable effect is the depolarization of the linear polarization, while $K = 1$ only refers to the circular polarization.

As in Stenflo (1982) we define the "characteristic field strength" B_0 for the Hanle effect by

$$B_0 = \frac{2m\,\gamma_N}{e\,g_e}\,, \tag{10.55}$$

where g_e is the Landé factor of the excited state. The sensitivity range of the Hanle effect is approximately given by $0.2B_0 \lesssim B \lesssim 5B_0$. From Eq. (3.44) we see that B_0 is the field strength for which the Larmor precession rate (including the Landé factor) equals the inverse radiative lifetime γ_N of the excited state.

According to Eqs. (10.10), (10.9), (10.55), and (3.44) the Hanle mixing angle α_K for a magnetic field of strength B_t is determined by

$$\tan\alpha_K = K\,k_c^{(K)} B/B_0\,. \tag{10.56}$$

In Eq. (5.71) we introduced the dimensionless parameter Ω, defined by

$$\tan\alpha_2 = \Omega\,, \tag{5.71}$$

to describe the Hanle effect, and used it in Figs. 5.3 and 5.4. With Eq. (10.56) we have

$$\Omega = \frac{2B_t}{B_0/k_c^{(2)}}\,. \tag{10.57}$$

Figure 10.4 shows a plot of the magnetic depolarization $k_H^{(2)}$ vs. $2/\Omega$ for the two cases $a = 0$ (isotropic distribution) and $a = -1$ (random distribution confined to the horizontal plane). We note in particular that $k_H^{(2)}$ does not go to zero when $B_t \to \infty$ but instead approaches 0.20 (for $a = 0$) and 0.25 (for $a = -1$). The reason is that the Hanle effect vanishes for a vertical field (when the incident radiation field is symmetric around the field vector) as well as for a horizontal field that is

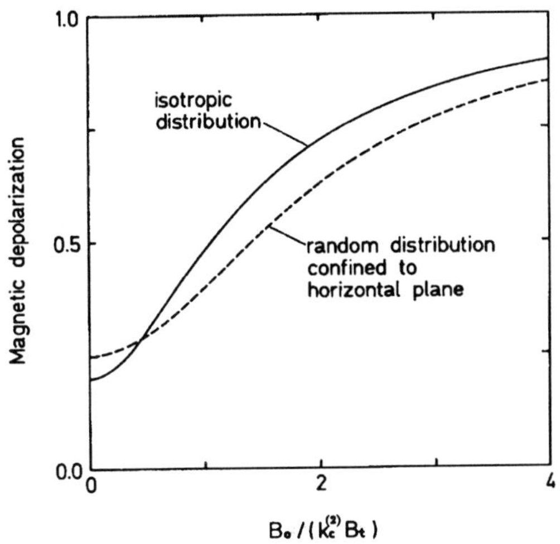

Fig. 10.4. Magnetic depolarization $k_H^{(2)}$ vs. $B_0/(k_c^{(2)} B_t)$ for a turbulent field of strength B_t with an isotropic distribution of field vectors (solid line) and a randomly oriented field that is confined to the horizontal plane (dashed line). B_0 represents a field strength for which the Larmor precession rate equals the inverse radiative lifetime of the excited state. $k_c^{(2)}$ is the branching ratio representing the probability that the alignment of the excited state will not be destroyed during the scattering process. From Stenflo (1982).

perpendicular to the line of sight. If the field were strictly longitudinal, then $k_H^{(2)}$ would go to zero as B_t goes to infinity.

10.13. Properties of the Turbulent Magnetic Field According to the Hanle Effect

It is possible to obtain empirical estimates of the Hanle depolarization parameter $k_H^{(2)}$ if we can reliably predict what the polarization amplitude in the line core should be in the absence of magnetic fields. The discrepancy between this theoretical prediction and the observed polarization amplitude then gives us the amount of Hanle depolarization. Using such a procedure and assuming a given value of the angular distribution parameter a of Eq. (10.52) (e.g. $a = 0$ for an isotropic distribution) we may thus arrive at a determination of the strength B_t of the turbulent magnetic field. If the parameter a is also considered as an unknown, then we get a two-parameter family of solutions to the equation

$$k_H^{(2)}(a, B_t) = k_{H,\,\mathrm{empirical}}^{(2)}\,, \tag{10.58}$$

where the right-hand side contains the fixed, empirical constraint determined by the observations. Using Eqs. (10.54), (10.56), and (10.57) we can express Eq. (10.58) in the more explicit form

$$\frac{6}{(a+3)(a+5)}\left(\frac{a+1}{1+4/\Omega^2}+\frac{1}{1+1/\Omega^2}\right)=1-k^{(2)}_{H,\mathrm{empirical}}\,. \tag{10.59}$$

The solutions are then obtained as a function of the two parameters a and Ω.

The translation of Ω to B_t depends on the "Hanle sensitivity" $B_0/k_c^{(2)}$ of the spectral line used for the observational constraint, which in turn depends on the lifetime, Landé factor, and collisional properties of the excited state. Thus each spectral line gives a different two-parameter $(a,\, B_t)$ solution that constrains the properties of the turbulent field. It is therefore of great advantage to do the observations for a set of spectral lines that spans as large a range of $B_0/k_c^{(2)}$ values as possible.

This multi-line approach was used by Stenflo (1982) in combination with the "last-scattering approximation" to estimate the value of B_t from the observational data of Stenflo et al. (1980a). Although of limited accuracy the last-scattering approximation has the enormous advantage that it bypasses the radiative-transfer calculations. It implies according to Sect. 9.14 and Eq. (9.106) that the polarization amplitude p_0 in the line core can be written in the form

$$p_0=k_H^{(2)}\alpha\, W_2\, k_c^{(2)}\, k_G\,. \tag{10.60}$$

Here we have used the approximation $W_{2,\mathrm{eff}}\approx W_2$, since the continuum opacity can generally be neglected in the core. We have further replaced the line-wing collisional depolarization factor k_c in Eq. (9.106) by $k_c^{(2)}$, which is the correct branching ratio for the line core according to our redistribution theory, Eq. (10.44). In addition, the polarization amplitude is scaled with the Hanle depolarization factor $k_H^{(2)}$. As before, α is the fraction of spontaneous emission transitions that are part of scattering events, while k_G is the "geometric depolarization" factor that is determined primarily by the limb-darkening function for the wavelengths in the line core.

If we for a given spectral line are able to estimate the values of α, $k_c^{(2)}$, and k_G, then the Hanle depolarization factor $k_H^{(2)}$ can be obtained from the observed polarization amplitude p_0 in that line via Eq. (10.60). For strong resonance lines near the limb it is often safe to assume that $\alpha\approx 1$ (implying that scattering dominates over collisions), while $k_c^{(2)}$ is expected to be close to unity (since the density and therefore also the collision rate decreases exponentially with height). k_G can be estimated from the limb-darkening function that is directly observed in the chosen spectral line.

This method was applied to a number of lines of different Hanle sensitivity $B_0/k_c^{(2)}$, which all gave mutually consistent results indicating the existence of Hanle depolarization due to a turbulent magnetic field with a strength B_t in excess of 10 G assuming an isotropic distribution of field vectors ($a=0$). For other values of the angular-distribution parameter a we get solutions for B_t as a function of a that depend on the spectral line used (the value of $B_0/k_c^{(2)}$). Figure 10.5, from Stenflo (1987), shows two such curves of B_t vs. a for the two typical cases $B_0/k_c^{(2)}=10\,\mathrm{G}$ and 30 G. Both curves are constrained to have $B_t=10\,\mathrm{G}$ for $a=0$.

The observationally allowed parameter range for the turbulent field lies *above* these two curves. The meaning of the other curves will be explained in subsequent sections. In particular, the other constraints indicate that the angular distribution of field vectors is indeed not very far from isotropic, which means that the assumption of $a = 0$ should in general be a good one.

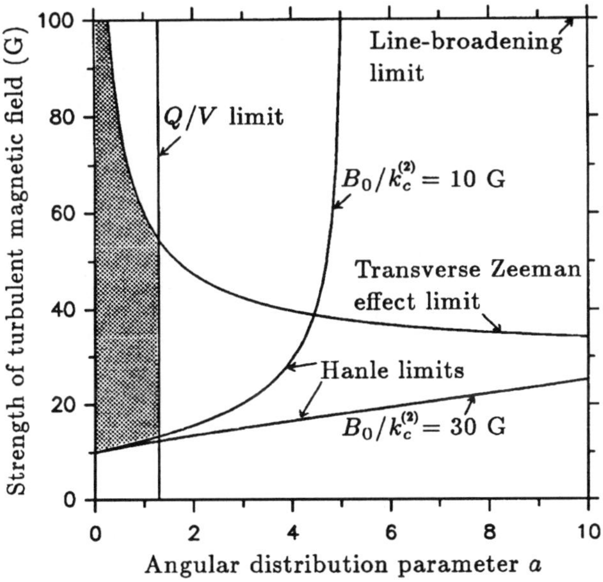

Fig. 10.5. Joint observational constraints on the strength B_t and angular distribution parameter a of a ubiquitous, turbulent magnetic field on the sun (see text). The observationally allowed region is marked by the shaded area. From Stenflo (1987).

The last-scattering approximation is useful for exploratory purposes to illuminate the physics that is involved and to obtain crude estimates of the effects. Reliable quantitative determinations of the properties of the turbulent magnetic field however cannot be obtained with such short-cuts but require full numerical solutions of the polarized transfer equations, taking partial redistribution into account. This has been done by Faurobert-Scholl (1993) through careful theoretical analysis of the center-to-limb variation of the polarization in the Sr I 4607 Å line (cf. Fig. 5.1). The theoretical polarization amplitudes have been compared with the ones observed by Stenflo et al. (1980a) to determine the Hanle depolarization factor. Since the radiation comes from higher layers when observing closer to the solar limb, the observed center-to-limb variation of the Sr I polarization contains height information on the strength B_t of the turbulent field. Assuming an isotropic distribution of field vectors ($a = 0$) it is found that the field strength B_t decreases from 30–60 G at a height of $z = 150$ km above $\tau_{5000} = 1$ (level of continuum formation at $\lambda = 5000$ Å) to 10–30 G at $z = 250$ km. Such a decrease of B_t may be understood as a result of the corresponding decrease of the kinetic energy of the turbulent motions, due to the rapid decrease of the density with height.

10.14. Magneto-turbulent Line Broadening

Although a turbulent magnetic field may give no visible signatures in solar magnetograms (maps of the longitudinal field component), the Zeeman splitting produces a potentially observable effect by broadening the spectral lines, similar to the turbulent Doppler broadening. As this Zeeman broadening component is correlated with the Landé factors of the lines, it may be statistically separated from the other line broadening effects. This approach was first tried by Unno (1959).

The success of such a statistical separation depends on the degree to which the line width can be accuractely modelled by a regression equation describing its dependence on line strength, excitation potential, wavelength, and Landé factor. Such a statistical modelling has been done for 402 unblended Fe I lines by Stenflo and Lindegren (1977). Figure 10.6 illustrates the adopted parametrization of the spectral lines. The main parameters are the relative line depth d, the line strength parameter S defined as the area of the part of the profile that is below the half-level chord (the shaded region of the diagram), and the length of the half-level chord expressed in velocity units as the formal Doppler width v_D of a Gaussian profile $\exp(-v^2/v_D^2)$ that has the same chord length as the spectral line under consideration.

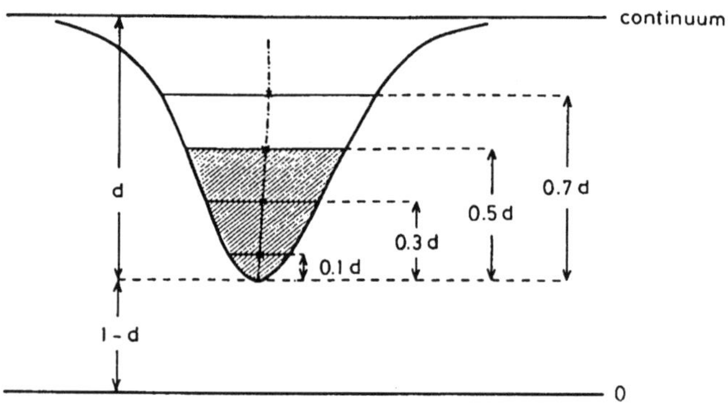

Fig. 10.6. Illustration of the adopted line parameters. From Stenflo and Lindegren (1977).

Empirical tests of the dependence of v_D on S show that the quadratic relation $x_1 + x_2 S + x_3 S^2$ approximates v_D well (cf. the left diagram of Fig. 10.7). The relative deviations from this relation, $[v_D - (x_1 + x_2 S + x_3 S^2)]/v_D$, are found to depend linearly on excitation potential χ_e of the lower level of the line transition. We therefore need to add a term $x_4 v_D \chi_e$ to the regression equation for v_D, but to avoid a coupling between the left- and right-hand sides of this equation we write the new term as $x_4\, v_0\, \chi_e$, where $v_0 = 2.51 + 0.0434 S^2$ represents an approximation of v_D (the exact form of which is unimportant, since x_4 is rather small). The right diagram of Fig. 10.7, giving $v_D - x_4\, v_0\, \chi_e$ vs. S, illustrates how much the scatter is reduced when the effect of the excitation potential on the line broadening is taken into account.

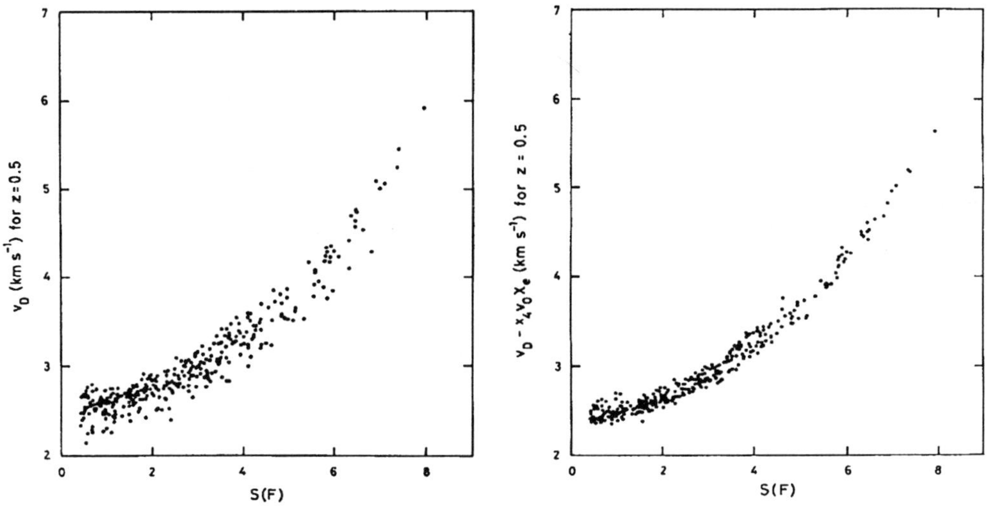

Fig. 10.7. The diagram to the left shows the line-width parameter v_D as a function of line-strength parameter S for a sample of 402 unblended Fe I lines. The diagram to the right illustrates how the scatter is reduced when a linear dependence on excitation potential is taken into account. It represents the relation between v_D and S in the limit of zero excitation potential. From Stenflo and Lindegren (1977).

The magneto-turbulent, Zeeman line broadening acts approximately like turbulent Doppler broadening, since we are generally dealing with a random field distribution with a corresponding distribution of Zeeman splittings. For the unpolarized Stokes I line profiles these Zeeman shifts combine in a way similar to the line-broadening Doppler shifts. Let us by v_B denote the Zeeman displacement $\Delta\lambda_H$ in velocity units due to a magnetic field of strength B. According to Eq. (6.47)

$$v_B = z\, c\, g_{\text{eff}} \lambda\, B\,, \tag{10.61}$$

where

$$z = 4.67 \times 10^{-13}\ \text{Å}^{-1}\ \text{G}^{-1}\,, \tag{10.62}$$

c is the speed of light, and g_{eff} the effective Landé factor (given by Eq. (6.44)). Then in a first approximation

$$v_D^2 = v_{D_0}^2 + v_B^2\,, \tag{10.63}$$

where v_{D_0} is the non-magnetic line width. When $v_B \ll v_{D_0}^2$, which is the case for the large majority of lines in the visible solar spectrum (since they have intermediately large Landé factors, and the turbulent field strengths are much smaller than 1 kG),

$$v_D - v_{D_0} \approx v_B^2/(2v_D)\,. \tag{10.64}$$

With Eqs. (10.64) and (10.61) we can now write the magneto-turbulent contribution to the line width in the form $x_5\, g_{\text{eff}}^2 \lambda^2/v_0$, where the coefficient x_5 is proportional to B_t^2. The final form of the regression equation for v_D then becomes

$$\begin{aligned} v_D = x_1 + x_2 S + x_3 S^2 + x_4\, v_0\, \chi_e \\ + x_5\, g_{\text{eff}}^2 \lambda^2 / v_0 + x_6 \langle\, g_{\text{eff}}^2 \,\rangle \lambda^2 / v_0 \,. \end{aligned} \tag{10.65}$$

The last term is necessary to absorb the wavelength dependence of v_D (or, more precisely, the dependence on λ^2/v_0). Without it a regression analysis would give a non-zero x_5 coefficient even in the absence of a magnetic field, since g_{eff}^2 has a non-zero average.

According to Eqs. (10.61) and (10.64) the value of the x_5 coefficient determined from the regression fit to the observational data can be converted to a magnetic field strength

$$B_{\text{obs}} = \frac{\sqrt{2x_5}}{z\, c}\,, \tag{10.66}$$

where B_{obs} represents the "observed" or "characteristic" field strength of the turbulent field distribution. In order to relate B_{obs} in greater quantitative detail to a particular angular distribution of field vectors of strength B_t, one has to calculate the broadening contributions as functions of field direction and strength for a given model atmosphere. As an additional parameter we may introduce the magnetic filling factor α_m, which is the fraction of the solar surface that is occupied by the particular field distribution, while the remaining fraction $(1 - \alpha_m)$ is field free. Then we may write

$$B_t = \gamma_t\, B_{\text{obs}} / \sqrt{\alpha_m}\,, \tag{10.67}$$

where γ_t is a dimensionless factor that usually does not differ much from unity, although its value depends on the choice of angular distribution of field vectors, the model atmosphere, and, as it turns out, to a small degree on the line strength.

Our usual picture of the turbulent field is that it is ubiquitous and space filling, which implies that $\alpha_m = 1$. If the angular distribution is isotropic, then γ_t is 0.8–0.9, which means that B_{obs} may be regarded as almost equivalent to B_t.

Regression equations of type (10.65) can be formulated not only for the width of the half-level chord, but also for the other chords illustrated in Fig. 10.6. From the error bars for the fitted values of the x_5 coefficients when using a data set of 402 unblended Fe I lines, Stenflo and Lindegren (1977) could set a 1-σ upper limit of 100 G to B_t. This implies a turbulent magnetic energy that is substantially smaller than the kinetic energy of the turbulent motions in the line-forming layers. The derived 100 G upper limit has been marked in the constraint diagram of Fig. 10.5.

10.15. Constraints on the Turbulent Magnetic Field from the Transverse Zeeman Effect

One may make use of the 180° ambiguity of the transverse Zeeman effect to constrain the properties of the turbulent magnetic field. As seen already from Eq. (4.48) for the angular dependence of the terms contributing to the different Stokes parameters, Stokes V changes sign when the field direction is reversed, while Stokes Q and U are invariant with respect to such a field reversal. If we take the example of a vertical magnetic field observed at some distance from the center of the solar disk, its transverse component is directed along the radius vector (the line connecting the disk center with the point of observation). The signs and magnitudes of Q and U are independent of whether the field lines point into the sun or out of it, while the sign of Stokes V depends on this direction.

Let us now consider the case that the photosphere is filled with a small-scale turbulent field having an angular distribution that is preferentially along the vertical direction but with a random mixture of the opposite magnetic polarities. Then the net contribution to Stokes V of this field will be zero, since the plus and minus signs are equally frequent. Since however the azimuth distribution of the transverse components is preferentially along the radius vector, the contributions from the various turbulent elements will add up constructively to give a net non-zero average Stokes Q or U signal.

Let us in the following choose our Stokes coordinate system such that the Stokes Q direction is parallel or perpendicular to the radius vector. Then Stokes $U = 0$ (disregarding magneto-optical effects, which can be neglected in the present context, since we are discussing small effects), and we only need to consider Stokes Q as representative of the transverse Zeeman effect. From the above considerations we see that the observed magnitude of Stokes Q in apparently "non-magnetic" regions depends on the two parameters B_t and a characterizing the turbulent field distribution. The effects on Stokes Q for a given field strength B_t are largest for a vertical field ($a \to \infty$), zero for an isotropic field distribution.

In regions with a net, observable Stokes V signal the Q/V ratio can be used to constrain the angular-distribution parameter a for the unresolved fields within the spatial resolution element, regardless of the turbulent or "non-turbulent" (like discrete flux tube) nature of these unresolved fields. Let us now consider more explicitly how these constraints on B_t and a are obtained from the observed values of Stokes Q and the Q/V ratio.

The Stokes Q and V signals observed in a given spectral line may be formally written as

$$\begin{aligned} Q &= \alpha_m \left(\frac{B}{B_Q}\right)^2 \sin^2\gamma\, \cos 2\chi\,, \\ V &= \alpha_m\, \frac{B}{B_V}\, \cos\gamma\,, \end{aligned} \tag{10.68}$$

where α_m is the magnetic filling factor, B the field strength, γ the field inclination (with respect to the line of sight), χ the field azimuth (in a plane perpendicular to the line of sight). The radiative-transfer physics is hidden in the parameters B_Q and B_V, which are only weakly dependent on field direction (vanishing dependence for weak spectral lines) and field strength (vanishing dependence in the weak-field limit).

For the turbulent field we may assume that $\alpha_m = 1$. Let the center-to-limb distance of the observed region be given by $\mu = \cos\vartheta$, where ϑ is the heliocentric angle (the angle between the line of sight and the vertical direction). Then the observed Stokes Q and V, averaged over the angular distribution $f(\mu_B)$ of Eq. (10.52), may be written

$$\begin{aligned} \langle Q\rangle &= g_Q(\mu)\,(B_t/B_Q)^2\,, \\ \langle V\rangle &= g_V(\mu)\,B_t/B_V\,, \end{aligned} \tag{10.69}$$

where

$$g_Q(\mu) = \int_0^{2\pi} \mathrm{d}\varphi \int_0^1 \mathrm{d}\mu_B \, f(\mu_B) \sin^2\gamma \, \cos 2\chi \, ,$$
$$g_V(\mu) = \int_0^{2\pi} \mathrm{d}\varphi \int_0^1 \mathrm{d}\mu_B \, f(\mu_B) \cos\gamma \tag{10.70}$$

in the case when there is only one magnetic polarity. In the more general case when there is a mixture of polarities, g_Q will remain unchanged, while g_V will be reduced by cancellation. While μ_B is the cosine of the inclination angle of the field with respect to the vertical direction, φ is the azimuthal field angle in the horizontal plane. $f(\mu_B)$ is normalized such that the double integration over it is unity.

With spherical trigonometry we can express γ and χ in terms of μ_B, φ, and μ. We can then perform the integrations in Eq. (10.70) to obtain

$$g_Q(\mu) = \frac{a}{a+3}\,(1-\mu^2)\, ,$$
$$g_V(\mu) = \frac{a+1}{a+2}\,\mu\, . \tag{10.71}$$

Observations with very high polarimetric accuracy (about 10^{-4}) in regions at various center-to-limb distances μ in as quiet or "non-magnetic" regions as possible have provided an upper limit to $|Q|/(1-\mu^2)$. In combination with Eqs. (10.69) and (10.71) this limit can be used to derive the following constraint relation for the turbulent magnetic field:

$$B_t \lesssim 30\sqrt{\frac{a+3}{a}}\ \mathrm{G} \tag{10.72}$$

(Stenflo, 1987). This relation, labeled "transverse Zeeman effect limit", has been plotted in Fig. 10.5. The observationally allowed region lies below this curve.

An additional but considerably more indirect constraint can be obtained from the observed Q/V ratio. For a unipolar angular distribution (10.52) of field vectors within the resolution element this ratio is according to Eqs. (10.69) and (10.71)

$$\frac{\langle Q\rangle}{\langle V\rangle} \sim \frac{a(a+2)}{(a+1)(a+3)}\,\frac{1-\mu^2}{\mu}\,B\, . \tag{10.73}$$

Here we have written B instead of B_t, since the field may have a significant contribution from a "non-turbulent" magnetic field when it is unipolar, although it is still modelled with the angular distribution function of Eq. (10.52). The field may also be intermittent with a small filling factor α_m. The constant of proportionality in Eq. (10.73) equals B_V/B_Q^2.

For a more general field of mixed polarities only $\langle V\rangle$ but not $\langle Q\rangle$ will be reduced due to cancellation. The unipolar case of Eq. (10.73) then represents a *lower* limit to the actual $\langle Q\rangle/\langle V\rangle$ ratio for such a field. In contrast, however, observations with very high sensitivity always showed a significant $\langle V\rangle$ signal even in the most quiet regions selected, while in these regions only upper limits to $\langle Q \rangle$ (of about 3×10^{-5} in units of the intensity of the continous spectrum) were found (Stenflo, 1987). No region on the sun without a significant $\langle V \rangle$ signal could be found. This means that the observations give us an *upper* limit to $|\langle Q \rangle / \langle V \rangle|$.

Application of the "line-ratio technique" to the observed $\langle V \rangle$ profiles of the Fe I 5250.22 and 5247.06 Å lines (see Chapter 12), a technique that allows us to estimate the intrinsic field strength B, indicates that even for the weakest fluxes with a $\langle V \rangle$ amplitude of about 5×10^{-4} in units of the continuum intensity, the fields are highly intermittent with field strengths B of close to 1 kG. If we adopt this value in Eq. (10.73), then the observed upper limit to $|\langle Q \rangle / \langle V \rangle|$ implies that

$$a \lesssim 1.3 \tag{10.74}$$

(Stenflo, 1987). This quite large spread in field direction for the kG flux tubes may be understood in terms of the models of Solanki and Steiner (1990), according to which the magnetic flux tubes are embedded in a cool environment while their interiors are heated. This leads to a rapid flaring out of the field lines as we move up in the atmosphere. Thereby largely horizontal canopies form, as discussed in connection with the use of the Hanle effect to diagnose magnetic fields in the upper solar atmosphere. These canopies are particularly pronounced and extended in regions of small flux (small $\langle V \rangle$), i.e., where the number density of flux tubes is small, since there the field may expand rather freely before it bumps into the flux of other expanding flux tubes. Therefore the spread of field directions within the higher layers of isolated flux tubes can be expected to be large.

Buoyancy forces try to keep the kG flux tubes aligned with the vertical direction. As strong fields are more buoyant than weak fields, the vertical buoyancy forces are smaller for the continually changing, weaker, turbulent fields than for the quasi-stable flux tubes. Therefore we expect the angular distribution of the turbulent field to be more isotropic (with a smaller value of the parameter a) than that of the flux tube field.

Using such quite indirect arguments one may conclude that the constraint of Eq. (10.74) is also valid for the turbulent field. In this sense this constraint has been added to Fig. 10.5 with the label "Q/V limit".

10.16. Concluding Remarks on the Turbulent Field

In summary, the various constraints of Fig. 10.5 are very complementary to each other. While the Hanle effect is insensitive to vertical magnetic fields but provides an observational signature for horizontal fields, the transverse Zeeman effect is insensitive to a horizontal, random distribution, but responds to predominantly vertical fields (of mixed polarity). Therefore the constraints in Fig. 10.5 from the Hanle and the transverse Zeeman effects curve in opposite directions. As they represent lower and upper limits, respectively, the allowed region is squeezed into the rather narrow region between these curves. The line-broadening limit puts a general cap on the field strength. Finally, indirect arguments using Q/V data can be used to put a cap on the magnitude of the angular distribution parameter a.

The recent radiative-transfer calculations with partial redistribution of the polarized radiation in the Sr I 4607 Å line by Faurobert-Scholl (1993) not only constrain but allow actual determinations of the turbulent field strength B_t. Assuming an isotropic distribution of field vectors and fitting the observed center-to-limb variation of the polarization amplitude in the strontium line, B_t was found to vary with height h from 30–60 G at $h \approx 150$ km to 10–30 G at $h \approx 250$ km (the zero point for h is the level of formation of the 5000 Å continuum). It is particularly gratifying that this range falls within the narrow, allowed region that is left in Fig. 10.5, and that all the various independent methods used to constrain this elusive field give mutually consistent results.

CHAPTER 11

SOLUTIONS OF THE POLARIZED TRANSFER EQUATION

11.1. Formulations of the Transfer Equation

We will here make the common assumption that atomic coherences in the absorption and stimulated emission terms may be neglected, and that the continuous spectrum may be treated in LTE. According to Eqs. (4.50), (5.95), and (6.104), the radiative transfer equation for the Stokes vector $\boldsymbol{I}_\nu$ can then be written as

$$\frac{\mathrm{d}\boldsymbol{I}_\nu}{\mathrm{d}\tau_c} = (\boldsymbol{\eta} + \boldsymbol{E})\boldsymbol{I}_\nu - \boldsymbol{S}_\nu \,, \tag{11.1}$$

where τ_c represents the continuum optical depth. Note that Eq. (6.104) was given in terms of the line optical depth. The choice of depth scale influences the form of the source function, which in the case of Eq. (11.1) can be expressed as

$$\boldsymbol{S}_\nu = S_L\boldsymbol{\eta}\mathbf{1} + S_c\mathbf{1} + \boldsymbol{j}_{\mathrm{coh}}/\kappa_c \,. \tag{11.2}$$

The three terms represent line emission excluding coherent scattering, continuous emission, and coherent scattering, respectively. If the continuum is formed in LTE, $S_c = B_\nu$, the Planck function. We recall our notation that $\boldsymbol{E}$ is the 4×4 unity matrix (the diagonal elements being unity, the off-diagonal zero), while $\mathbf{1}$ is the 4-vector with the first position being unity, the remaining zero.

The line absorption matrix is

$$\boldsymbol{\eta} = \begin{pmatrix} \eta_I & \eta_Q & \eta_U & \eta_V \\ \eta_Q & \eta_I & \rho_V & -\rho_U \\ \eta_U & -\rho_V & \eta_I & \rho_Q \\ \eta_V & \rho_U & -\rho_Q & \eta_I \end{pmatrix} . \tag{11.3}$$

For a normal Zeeman triplet

$$\begin{aligned} \eta_{I,Q,U,V} &= \eta_0 H_{I,Q,U,V} \,, \\ \rho_{Q,U,V} &= 2\eta_0 F_{Q,U,V} \,. \end{aligned} \tag{11.4}$$

$H_{I,Q,U,V}$ and $F_{Q,U,V}$ have been defined by Eqs. (4.48) and (4.49), η_0 by Eq. (4.46). To cover the case of anomalous Zeeman splitting patterns we write

$$\boldsymbol{\eta} = \eta_0\sqrt{\pi}\Delta\nu_D\boldsymbol{\Phi} \,, \tag{11.5}$$

where the matrix $\boldsymbol{\Phi}$ has been given by Eqs. (6.57)–(6.59) and (6.52). Note that Eq. (6.52) accounts not only for anomalous splitting but also for deviations from complete redistribution among the Zeeman sublevels. In the common case of complete redistribution, Eqs. (6.37)–(6.40) should be used to define the components of $\boldsymbol{\Phi}$.

The corresponding radiative transfer equation for the coherency matrix $\boldsymbol{D}$ is according to Eqs. (4.41) and (6.110)

$$\frac{\mathrm{d}\boldsymbol{D}}{\mathrm{d}\tau_c} = \boldsymbol{A}\boldsymbol{D} + \boldsymbol{D}\boldsymbol{A}^\dagger - \boldsymbol{F}\,. \tag{11.6}$$

We recall that

$$\boldsymbol{D} = \frac{1}{2}\begin{pmatrix} I+Q & U+iV \\ U-iV & I-Q \end{pmatrix}. \tag{2.47}$$

The complex absorption matrix $\boldsymbol{A}$ can be written in the form

$$\boldsymbol{A} = \tfrac{1}{2}(\boldsymbol{\sigma}_0 + \boldsymbol{\eta}_D - i\boldsymbol{\rho}_D)\,, \tag{11.7}$$

where the Pauli spin matrix $\boldsymbol{\sigma}_0$ (unity matrix) represents continuous absorption, $\boldsymbol{\eta}_D$ line absorption, and $\boldsymbol{\rho}_D$ anomalous dispersion (magneto-optical effects in the line).

$$\begin{aligned} \boldsymbol{\eta}_D &= \begin{pmatrix} \eta_I + \eta_Q & \eta_U + i\eta_V \\ \eta_U - i\eta_V & \eta_I - \eta_Q \end{pmatrix}, \\ \boldsymbol{\rho}_D &= \begin{pmatrix} \rho_Q & \rho_U + i\rho_V \\ \rho_U - i\rho_V & -\rho_Q \end{pmatrix}. \end{aligned} \tag{11.8}$$

Note that ρ_I does not need to be used, since its contribution to $\boldsymbol{A}$ is $i\rho_I\boldsymbol{\sigma}_0/2$, and this contribution is canceled by the second term in the transfer equation, since $i\boldsymbol{\sigma}_0\boldsymbol{D} + \boldsymbol{D}(i\boldsymbol{\sigma}_0)^\dagger = 0$.

If we disregard coherent scattering, we get from Eqs. (4.42), (4.44), and (4.47) for the source matrix

$$\boldsymbol{F} = \tfrac{1}{2}(S_L\boldsymbol{\eta}_D + B_\nu\boldsymbol{\sigma}_0)\,. \tag{11.9}$$

11.2. Formal Solution of the Stokes Transfer Equation

To solve the Stokes transfer equation (11.1) we introduce the 4×4 matrix $\boldsymbol{L}$ that obeys the corresponding homogeneous differential equation:

$$\frac{\mathrm{d}\boldsymbol{L}}{\mathrm{d}\tau_c} = (\boldsymbol{\eta} + \boldsymbol{E})\boldsymbol{L} \tag{11.10}$$

(cf. Landi Degl'Innocenti and Landi Degl'Innocenti, 1985). Expanding $\mathrm{d}(\boldsymbol{L}\boldsymbol{L}^{-1})/\mathrm{d}\tau_c = \mathrm{d}\boldsymbol{E}/\mathrm{d}\tau_c = 0$ and using Eq. (11.10), we see that

$$\frac{\mathrm{d}\boldsymbol{L}^{-1}}{\mathrm{d}\tau_c} = -\boldsymbol{L}^{-1}(\boldsymbol{\eta} + \boldsymbol{E})\,. \tag{11.11}$$

It then follows from Eq. (11.1) that

$$\frac{\mathrm{d}}{\mathrm{d}\tau_c}(\boldsymbol{L}^{-1}\boldsymbol{I}_\nu) = -\boldsymbol{L}^{-1}\boldsymbol{S}_\nu\,. \tag{11.12}$$

This equation may be integrated to give

$$\boldsymbol{I}_\nu(\tau_c) = \boldsymbol{L}(\tau_c)\int_{\tau_c}^{\tau_0}\boldsymbol{L}^{-1}(\tau)\boldsymbol{S}_\nu(\tau)\,\mathrm{d}\tau \;+\boldsymbol{L}(\tau_c)\boldsymbol{L}^{-1}(\tau_0)\boldsymbol{I}_\nu(\tau_0)\,. \tag{11.13}$$

When $\tau_0 \to \infty$, $\boldsymbol{L}^{-1}(\tau_0)\boldsymbol{I}_\nu(\tau_0) \to 0$, since $\boldsymbol{L}^{-1}$ represents an exponential attenuation (as shown explicitly below) that increases faster than $\boldsymbol{I}_\nu\tau_0$. Therefore

$$\boldsymbol{I}_\nu(\tau_c) = \boldsymbol{L}(\tau_c)\int_{\tau_c}^{\infty}\boldsymbol{L}^{-1}(\tau)\boldsymbol{S}_\nu(\tau)\,\mathrm{d}\tau\,. \tag{11.14}$$

The natural boundary condition for $\boldsymbol{L}$ at the top of the atmosphere ($\tau_c = 0$) is that it equals the unity matrix there, i.e.

$$\boldsymbol{L}(0) = \boldsymbol{E}\,. \tag{11.15}$$

The solution for the emergent Stokes vector is then

$$\boldsymbol{I}_\nu(0) = \int_0^\infty \boldsymbol{L}^{-1}\boldsymbol{S}_\nu\,\mathrm{d}\tau\,. \tag{11.16}$$

Let us define the exponential of a matrix $\boldsymbol{M}$ in terms of its Taylor expansion:

$$e^{\boldsymbol{M}} = \sum_{n=0}^{\infty}\frac{\boldsymbol{M}^n}{n!}\,. \tag{11.17}$$

It can be manipulated as a scalar exponential function as long as no other non-commuting matrix is involved in the algebra. In particular, we can solve Eq. (11.10) to obtain

$$\boldsymbol{L}(\tau_c) = \exp\Big[\int_0^{\tau_c}(\boldsymbol{\eta}+\boldsymbol{E})\,\mathrm{d}\tau\Big]\,, \tag{11.18}$$

where we have applied the boundary condition (11.15) to fix the scale factor. The solution (11.16) then becomes

$$\boldsymbol{I}_\nu(0) = \int_0^\infty \exp\Big[-\int_0^\tau(\boldsymbol{\eta}(\tau')+\boldsymbol{E})\,\mathrm{d}\tau'\Big]\,\boldsymbol{S}_\nu(\tau)\,\mathrm{d}\tau\,. \tag{11.19}$$

11.3. Formal Solution of the Coherency Matrix Transfer Equation

We proceed in a similar way as before, and introduce the complex 2×2 matrix $\boldsymbol{L}_D$ that obeys the homogeneous differential equation

$$\frac{\mathrm{d}\boldsymbol{L}_D}{\mathrm{d}\tau_c} = \boldsymbol{A}\boldsymbol{L}_D\,. \tag{11.20}$$

The Hermitian conjugate of this equation is

$$\frac{\mathrm{d}\boldsymbol{L}_D^\dagger}{\mathrm{d}\tau_c} = \boldsymbol{L}_D^\dagger\boldsymbol{A}^\dagger\,. \tag{11.21}$$

In analogy with Eq. (11.15), the natural surface boundary condition for $\boldsymbol{L}_D$ is

$$\boldsymbol{L}_D(0) = \boldsymbol{L}_D^\dagger(0) = \boldsymbol{\sigma}_0\,. \tag{11.22}$$

Let us next define a 2×2 matrix $\boldsymbol{G}$ by

$$\boldsymbol{D} = \boldsymbol{L}_D\boldsymbol{G}\,\boldsymbol{L}_D^\dagger\,. \tag{11.23}$$

If we take the derivative of this expression and identify the result with the transfer equation (11.6) while making use of Eqs. (11.20) and (11.21), we obtain

$$\frac{\mathrm{d}\boldsymbol{G}}{\mathrm{d}\tau_c} = -\boldsymbol{L}_D^{-1}\boldsymbol{F}(\boldsymbol{L}_D^\dagger)^{-1}\,. \tag{11.24}$$

When solving Eq. (11.24) we can use the same arguments for the boundary conditions at $\tau_c = \infty$ as for Eqs. (11.12) and (11.13). Further, as $\boldsymbol{G}(0) = \boldsymbol{D}(0)$, we obtain the solution for the coherency matrix emerging from the atmosphere:

$$\boldsymbol{D}(0) = \int_0^\infty \boldsymbol{L}_D^{-1}\,\boldsymbol{F}\,(\boldsymbol{L}_D^\dagger)^{-1}\,\mathrm{d}\tau\,. \tag{11.25}$$

As the solution to the homogeneous equation is

$$\boldsymbol{L}_D(\tau_c) = \exp\Big[\int_0^{\tau_c} \boldsymbol{A}(\tau)\,\mathrm{d}\tau\Big]\,, \tag{11.26}$$

we obtain

$$\boldsymbol{D}(0) = \int_0^\infty \exp\Big[-\int_0^\tau \boldsymbol{A}(\tau')\,\mathrm{d}\tau'\Big]\,\boldsymbol{F}(\tau)\,\exp\Big[-\int_0^\tau \boldsymbol{A}^\dagger(\tau')\,\mathrm{d}\tau'\Big]\,\mathrm{d}\tau\,. \tag{11.27}$$

11.4. Contribution Functions

For most diagnostic work it is important to know where in the solar atmosphere a given spectral line is formed. This information is provided by the contribution functions $\boldsymbol{C_I}$ and $\boldsymbol{C_D}$ defined by

$$\boldsymbol{I}(0) = \int_0^\infty \boldsymbol{C_I}(\tau_c)\,\mathrm{d}\tau_c \tag{11.28}$$

for the Stokes vector $\boldsymbol{I}$,

$$\boldsymbol{D}(0) = \int_0^\infty \boldsymbol{C_D}(\tau_c)\,\mathrm{d}\tau_c \tag{11.29}$$

for the coherency matrix. While $\boldsymbol{C_I}$ is a 4×1 vector, $\boldsymbol{C_D}$ is a 2×2 complex matrix (cf. van Ballegooijen, 1985).

For practical purposes a more useful definition of the contribution functions is in terms of $x = \log \tau_c$ instead of τ_c, since x is more closely related to the geometrical height scale and is the variable normally used for numerical integrations of the transfer equations and for tabulations of model atmospheres. Thus we define

$$\begin{aligned} \boldsymbol{I}(0) &= \int_{-\infty}^{+\infty} \boldsymbol{C_I}(x)\,\mathrm{d}x\,, \\ \boldsymbol{D}(0) &= \int_{-\infty}^{+\infty} \boldsymbol{C_D}(x)\,\mathrm{d}x\,. \end{aligned} \tag{11.30}$$

Since $\mathrm{d}\tau_c = \tau_c\,(\ln 10)\,\mathrm{d}x$,

$$\boldsymbol{C}(x) = (\ln 10)\,\tau_c\,\boldsymbol{C}(\tau_c)\,. \tag{11.31}$$

$\boldsymbol{C}(x)$ represents the contribution per unit interval in x to the emergent Stokes vector or coherency matrix.

When diagrams of the contribution functions are shown, it is always $\boldsymbol{C}(x)$ that is plotted, not $\boldsymbol{C}(\tau_c)$. When analytical expressions for the contribution functions are given in the following, we will however use $\boldsymbol{C}(\tau_c)$ for convenience, remembering that one has to make the simple conversion (11.31) before plotting.

According to Eqs. (11.16) and (11.25),

$$\begin{aligned} \boldsymbol{C_I}(\tau_c) &= \boldsymbol{L}^{-1}\boldsymbol{S}_\nu\,, \\ \boldsymbol{C_D}(\tau_c) &= \boldsymbol{L}_D^{-1}\boldsymbol{F}\,(\boldsymbol{L}_D^\dagger)^{-1}\,. \end{aligned} \tag{11.32}$$

It follows from the relations (2.43), (2.46), and (2.47) between $\boldsymbol{I}$ and $\boldsymbol{D}$ that $\boldsymbol{C_I}$ and $\boldsymbol{C_D}$ are connected according to

$$\begin{aligned} C_{I,Q,U,V} &= \mathrm{Tr}\,(\sigma_{0,1,2,3} C_D)\,, \\ C_D &= \frac{1}{2}\begin{pmatrix} C_I + C_Q & C_U + iC_V \\ C_U - iC_V & C_I - C_Q \end{pmatrix}. \end{aligned} \tag{11.33}$$

C_I and C_D however do not provide "clean" information on the height of line formation, since the contributions from the line and the continuous spectrum are mixed. In the far line wings these contribution functions mainly tell us where the continuum is formed, not where the line is formed. To obtain something that is of practical use and which exclusively tells us where the line Stokes parameters are formed we need to define the *line depression* contribution functions (Grossmann-Doerth et al., 1988a)

$$\begin{aligned} \boldsymbol{r}(0) &= \int_0^\infty C_{\boldsymbol{r}}(\tau_c)\,\mathrm{d}\tau_c\,, \\ \boldsymbol{r}_D(0) &= \int_0^\infty C_{\boldsymbol{r}_D}(\tau_c)\,\mathrm{d}\tau_c\,, \end{aligned} \tag{11.34}$$

where the line depressions are related to $\boldsymbol{I}_\nu$ and $\boldsymbol{D}$ by

$$\begin{aligned} \boldsymbol{r} &= \boldsymbol{1} - \boldsymbol{I}_\nu / I_c\,, \\ \boldsymbol{r}_D &= \tfrac{1}{2}\boldsymbol{\sigma}_0 - \boldsymbol{D}/I_c\,. \end{aligned} \tag{11.35}$$

I_c is the continuum intensity.

If we now take the derivative of $\boldsymbol{I}_\nu = I_c(\boldsymbol{1} - \boldsymbol{r})$ and use the tranfer equation (11.1) as well as the special version of it for I_c ($\mathrm{d}I_c/\mathrm{d}\tau_c = I_c - B_\nu$), we obtain a transfer equation for $\boldsymbol{r}$ that has the same form as Eq. (11.1):

$$\frac{\mathrm{d}\boldsymbol{r}}{\mathrm{d}\tau_c} = \boldsymbol{\eta}_r \boldsymbol{r} - \boldsymbol{S}_r\,, \tag{11.36}$$

where

$$\begin{aligned} \boldsymbol{\eta}_r(\tau_c) &= \boldsymbol{\eta}(\tau_c) + \boldsymbol{E} B_\nu / I_c(\tau_c)\,, \\ \boldsymbol{S}_r(\tau_c) &= \boldsymbol{\eta}_r(\tau_c)\boldsymbol{1} - \boldsymbol{S}_\nu(\tau_c)/I_c(\tau_c)\,. \end{aligned} \tag{11.37}$$

In analogy with Eqs. (11.32), (11.10), and (11.18) we then get

$$C_{\boldsymbol{r}}(\tau_c) = \boldsymbol{L}_r^{-1}(\tau_c)\boldsymbol{S}_r(\tau_c)\,, \tag{11.38}$$

where

$$\boldsymbol{L}_r(\tau_c) = \exp\Big[\int_0^{\tau_c} \boldsymbol{\eta}_r(\tau)\,\mathrm{d}\tau\Big] \tag{11.39}$$

is the solution of

$$\frac{\mathrm{d}\boldsymbol{L}_r}{\mathrm{d}\tau} = \boldsymbol{\eta}_r \boldsymbol{L}_r \tag{11.40}$$

with the boundary condition $\boldsymbol{L}_r(0) = \boldsymbol{E}$.

$\boldsymbol{C}_{\boldsymbol{r}_D}$ may be obtained in a similar way, or may be obtained directly from $\boldsymbol{C}_{\boldsymbol{r}}$ via Eq. (11.33).

Some examples of line depression contribution functions are illustrated in Fig. 11.1, from Sánchez Almeida (1992). The diagram shows $\boldsymbol{C}_{\boldsymbol{r}}(x)$ (although the axes are labeled $C_{I,Q,U,V}$) per unit interval of $x = \log \tau_5$ (the continuum optical depth at $\lambda = 5000$ Å), for the Fe I 5250.2 Å line at a wavelength $\Delta\lambda = 50$ mÅ from the line center (the approximate location of the Stokes V maximum). The magnetic field parameters are $B = 1200$ G, $\gamma = 45°$, and $\chi = 0$.

11.5. Response Functions

Let $a = a(\tau_c)$ be a scalar physical parameter like temperature, density, velocity, field strength, field angle, or microturbulent velocity. If we perturb a to $a + \delta a$, where δa is a function of τ_c, and restrict ourselves to small perturbations so that only first-order terms in δa, $\delta \boldsymbol{I}_\nu$, $\delta \boldsymbol{\eta}$, and $\delta \boldsymbol{S}_\nu$ need to be retained, the emergent Stokes vector gets changed to $\boldsymbol{I}_\nu(0) + \delta \boldsymbol{I}_\nu(0)$, where

$$\delta \boldsymbol{I}_\nu(0) = \int_0^\infty \boldsymbol{R}_a(\tau_c)\, \delta a(\tau_c)\, \mathrm{d}\tau_c \,. \tag{11.41}$$

This defines the *response function* $\boldsymbol{R}_a$ for the Stokes vector $\boldsymbol{I}_\nu$ with respect to the parameter a (Landi Degl'Innocenti and Landi Degl'Innocenti, 1977). Response functions are useful when one wants to explore the sensitivity of Stokes profiles of different spectral lines to changes in the various physical parameters at different heights in the atmosphere.

Differentiation of Eq. (11.1) gives

$$\frac{\mathrm{d}}{\mathrm{d}\tau_c} \delta \boldsymbol{I}_\nu = (\boldsymbol{\eta} + \boldsymbol{E}) \delta \boldsymbol{I}_\nu - \delta \boldsymbol{S}_a \,, \tag{11.42}$$

where

$$\delta \boldsymbol{S}_a(\tau_c) = \left(\frac{\partial \boldsymbol{S}_\nu}{\partial a} - \frac{\partial \boldsymbol{\eta}}{\partial a} \boldsymbol{I}_\nu \right) \delta a(\tau_c) \,. \tag{11.43}$$

As Eq. (11.42) has the same form as Eq. (11.1), the formal solution can be obtained in the same way as before. We therefore only have to make a symbol substitution in the expression (11.32) for the contribution function to obtain the solution for the response function:

$$\boldsymbol{R}_a = \boldsymbol{L}^{-1} \left(\frac{\partial \boldsymbol{S}_\nu}{\partial a} - \frac{\partial \boldsymbol{\eta}}{\partial a} \boldsymbol{I}_\nu \right) . \tag{11.44}$$

We can likewise define a response function $\boldsymbol{R}_{\boldsymbol{r}a}$ for the line depression function $\boldsymbol{r}$ by

$$\delta \boldsymbol{r}(0) = \int_0^\infty \boldsymbol{R}_{\boldsymbol{r}a}\, \delta a\, \mathrm{d}\tau_c \tag{11.45}$$

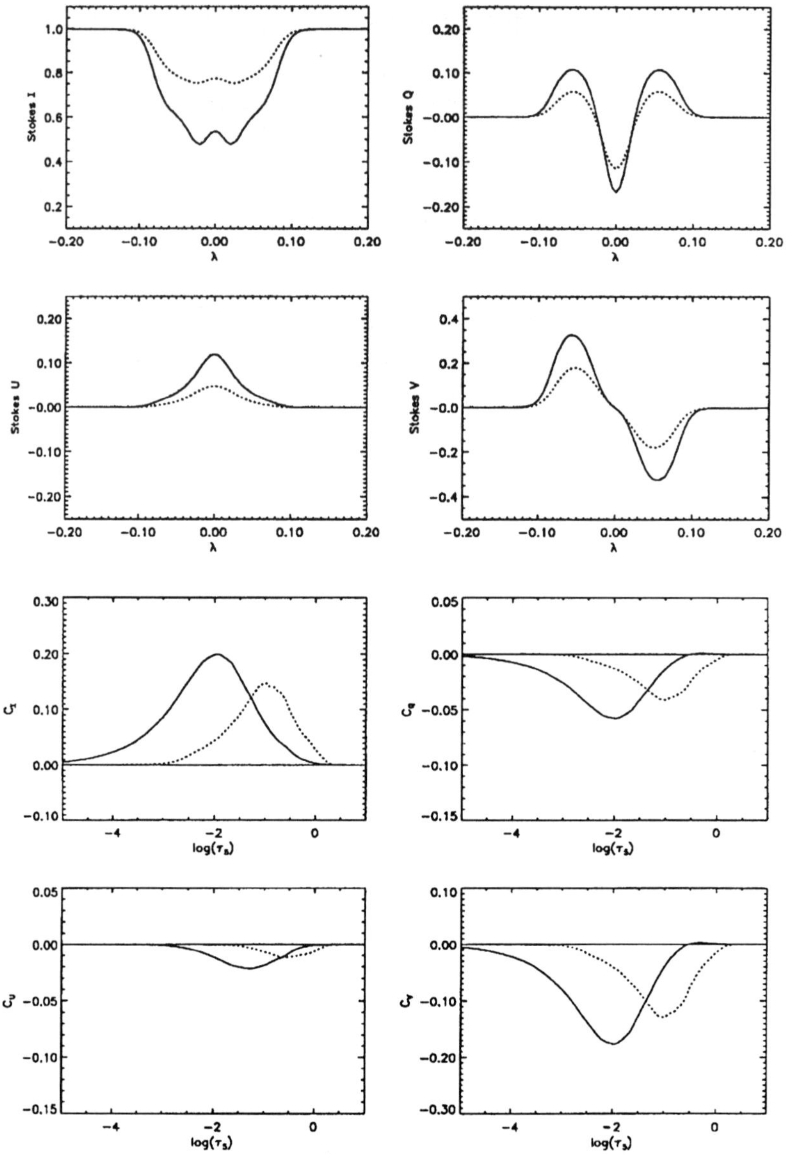

Fig. 11.1. Upper set of four diagrams: Stokes profiles of the Fe I 5250.2 Å line for a field strength $B = 1200\,\mathrm{G}$, field inclination $\gamma = 45°$, and field azimuth $\chi = 0$. The solid curves have been obtained for the thermodynamic properties of a quiet sun model atmosphere, the dotted curves for a facular atmosphere. The wavelengths are given in Å relative to the line center. Lower set of diagrams: Line depression contribution functions $\boldsymbol{C_r}(x)$ (here labeled as $C_{I,Q,U,V}$) for $\Delta\lambda = 50\,\mathrm{m}$Å, in units of the intensity $I_c(0)$ of the local continuum, as functions of $x = \log \tau_5$, where τ_5 is the continuum optical depth at 5000 Å. The solid and dotted curves represent the same model atmospheres and field parameters as the curves in the upper set of diagrams. From Sánchez Almeida (1992).

(cf. Grossmann-Doerth et al., 1988a). Differentiation of Eq. (11.36) gives

$$\frac{\mathrm{d}}{\mathrm{d}\tau_c}\delta\boldsymbol{r} = \boldsymbol{\eta}_r\delta\boldsymbol{r} - \left(\frac{\partial \boldsymbol{S}_r}{\partial a} - \frac{\partial \boldsymbol{\eta}_r}{\partial a}\boldsymbol{r}\right)\delta a\,, \tag{11.46}$$

from which we get

$$\boldsymbol{R}_{ra} = \boldsymbol{L}_r^{-1}\left(\frac{\partial \boldsymbol{S}_r}{\partial a} - \frac{\partial \boldsymbol{\eta}_r}{\partial a}\boldsymbol{r}\right), \tag{11.47}$$

where $\boldsymbol{L}_r$ is given by Eq. (11.39).

It is straightforward to define and derive corresponding expressions for the response functions of the coherency matrix $\boldsymbol{D}$, but we refrain from doing this here.

As in the case of the contribution functions, graphical representations in terms of $x = \log\tau_c$ are more useful than representations in terms of τ_c. Like Eq. (11.31) the relation between the two representations is

$$\boldsymbol{R}(x) = (\ln 10)\,\tau_c\,\boldsymbol{R}(\tau_c)\,. \tag{11.48}$$

In Fig. 11.2, from Ruiz Cobo and del Toro Iniesta (1992), we give an example of normalized response functions $\boldsymbol{R}_a(x)$ for the Fe I 6302.5 Å line at $\Delta\lambda = 120$ mÅ from line center. A model atmosphere representing a sunspot umbra has been used.

11.6. Analytical Solutions

We will here only consider the Stokes version (11.1) of the transfer equation, not the coherency matrix formulation. The formal solution (11.16) of Eq. (11.1) may be expressed in explicit analytic form under various simplifying assumptions. These analytical solutions are often of considerable use in spite of their limited validity range, since general numerical solutions are both computer intensive and less physically transparent.

The first simplifying assumption that we will introduce is that the absorption matrix $\boldsymbol{\eta}$ is depth independent, which implies that the magnetic field is assumed to be homogeneous, and that the strength η_0 and shapes of the absorption-dispersion profiles are independent of depth. This idealization is called a ***Milne-Eddington atmosphere***.

Next let us assume a plane-parallel stratification of the atmosphere and redefine the optical depth scale τ to be measured in the vertical direction instead of along the line of sight. If θ is the angle between the line of sight and the vertical direction, and $\mu = \cos\theta$, we then need to replace τ_c in our previous expressions by τ/μ. (We now omit index c, both for convenience and to distinguish τ from our previous τ_c scale along the line of sight.)

With these assumptions we obtain from Eq. (11.18)

$$\boldsymbol{L}^{-1}(\tau) = \exp[-(\boldsymbol{\eta} + \boldsymbol{E})\tau/\mu]\,, \tag{11.49}$$

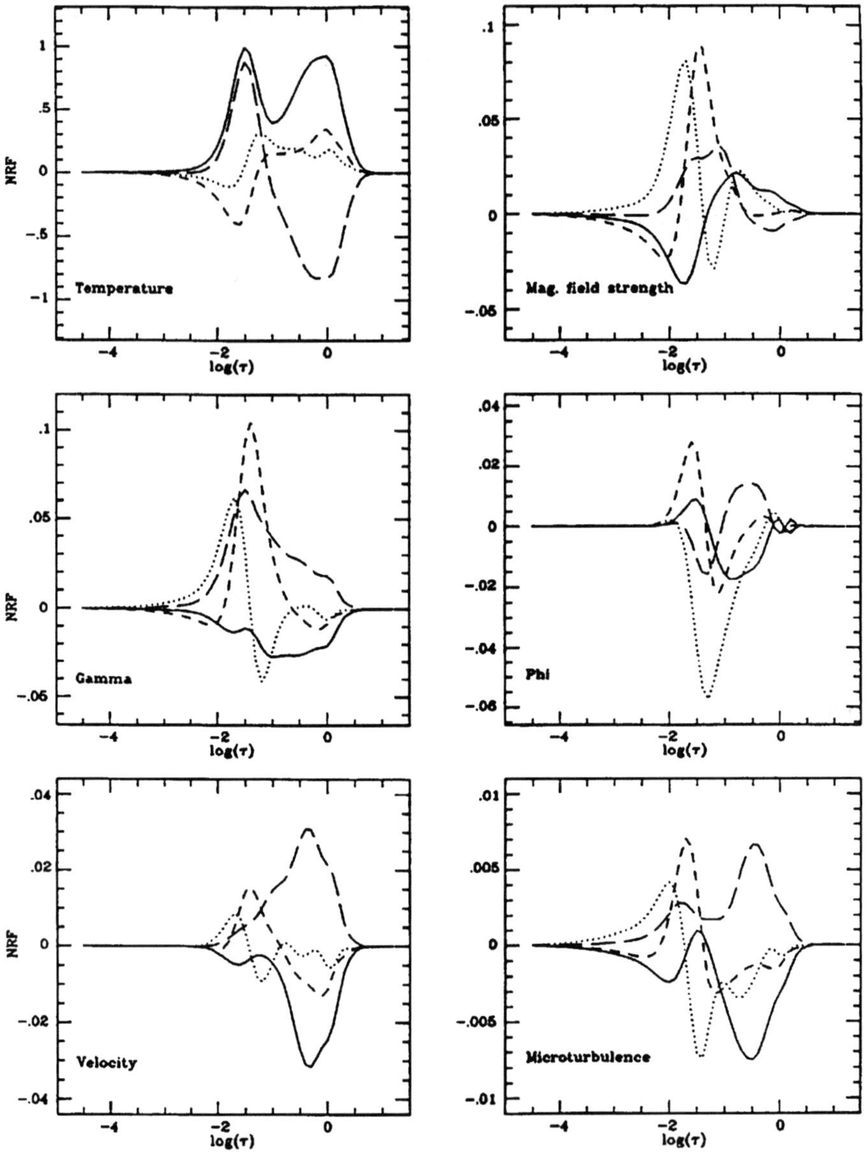

Fig. 11.2. Normalized response functions $\boldsymbol{R}_a(x)$ for the Fe I 6302.5 Å line at $\Delta\lambda = 120$ mÅ from line center, calculated with a sunspot umbral model atmosphere. The different diagrams give the six cases when the perturbed parameter a represents temperature, field strength, field inclination γ, field azimuth, flow velocity, and microturbulent velocity. The response functions for Stokes I are given as solid lines, for Stokes Q as dotted lines, for Stokes U as short-dashed lines, and for Stokes V as long-dashed lines. The normalization of the curves is with respect to the maximum value of $\boldsymbol{R}_a$ for Stokes I with respect to temperature. From Ruiz Cobo and del Toro Iniesta (1992).

which we insert in the solution (11.16). Partial integration of Eq. (11.16) then gives

$$\boldsymbol{I}_\nu(0) = [\boldsymbol{\eta}(0) + \boldsymbol{E}]^{-1}\boldsymbol{S}_\nu(0) + \mu \int_0^\infty \exp[-(\boldsymbol{\eta} + \boldsymbol{E})\tau/\mu]\,(\boldsymbol{\eta} + \boldsymbol{E})^{-1}\frac{\mathrm{d}\boldsymbol{S}_\nu}{\mathrm{d}\tau}\,\mathrm{d}\tau/\mu\,, \tag{11.50}$$

where, if coherent scattering $\boldsymbol{j}_{\mathrm{coh}}$ is neglected,

$$\frac{\mathrm{d}\boldsymbol{S}_\nu}{\mathrm{d}\tau} = \left(\frac{\mathrm{d}S_L}{\mathrm{d}\tau}\boldsymbol{\eta} + \frac{\mathrm{d}S_c}{\mathrm{d}\tau}\boldsymbol{E}\right)\mathbf{1} \tag{11.51}$$

according to Eq. (11.2) and our assumption that $\boldsymbol{\eta}$ is depth independent.

Let us at this point introduce the assumption of LTE, which implies that $S_L = S_c = B_\nu$ (the Planck function), so that Eq. (11.2) becomes

$$\boldsymbol{S}_\nu = B_\nu(\boldsymbol{\eta} + \boldsymbol{E})\mathbf{1}\,. \tag{11.52}$$

Inserting this in Eq. (11.50) gives us the solution

$$\boldsymbol{I}_\nu(0) = B_\nu(0)\mathbf{1} + \mu \int_0^\infty \frac{\mathrm{d}B_\nu}{\mathrm{d}\tau}\exp[-(\boldsymbol{\eta} + \boldsymbol{E})\tau/\mu]\,\mathbf{1}\,\mathrm{d}\tau/\mu\,. \tag{11.53}$$

To solve this equation analytically we have to make some assumption about the depth dependence of the Planck function B_ν. The simplest assumption is that of a linear τ dependence:

$$B_\nu(\tau) = B_0(1 + \beta_0\tau)\,. \tag{11.54}$$

The solution of Eq. (11.53) then becomes

$$\boldsymbol{I}_\nu(0) = B_0\big[\boldsymbol{E} + \beta_0\mu(\boldsymbol{\eta} + \boldsymbol{E})^{-1}\big]\mathbf{1}\,. \tag{11.55}$$

As a special case we obtain the intensity of the unpolarized continuum:

$$I_c(0) = B_0(1 + \beta_0\mu)\,. \tag{11.56}$$

It is convenient to introduce a symbol β for the "limb-darkening factor":

$$\beta = \frac{\beta_0\mu}{1 + \beta_0\mu}\,. \tag{11.57}$$

The solution for the emergent Stokes vector in units of the intensity of the local continuum is then, according to Eqs. (11.55)–(11.57),

$$\boldsymbol{I}_\nu(0)/I_c(0) = \big[(1-\beta)\boldsymbol{E} + \beta(\boldsymbol{\eta} + \boldsymbol{E})^{-1}\big]\mathbf{1}\,. \tag{11.58}$$

Explicit expressions are obtained by straightforward inversion of the 4×4 matrix $\boldsymbol{\eta} + \boldsymbol{E}$, which leads to the following solution, first derived by Unno (1956) without magneto-optical effects (anomalous dispersion), and later by Rachkovsky (1962a,b) when magneto-optical effects were fully accounted for:

$$\begin{aligned} I/I_c &= 1+\beta+\beta(\eta_I+1)\big[(\eta_I+1)^2+d\big]/\Delta\,, \\ Q/I_c &= -\beta\big[(\eta_I+1)^2\eta_Q+(\eta_I+1)(\rho_U\eta_V-\rho_V\eta_U)+s\rho_Q\big]/\Delta\,, \\ U/I_c &= -\beta\big[(\eta_I+1)^2\eta_U+(\eta_I+1)(\rho_V\eta_Q-\rho_Q\eta_V)+s\rho_U\big]/\Delta\,, \\ V/I_c &= -\beta\big[(\eta_I+1)^2\eta_V+(\eta_I+1)(\rho_Q\eta_U-\rho_U\eta_Q)+s\rho_V\big]/\Delta\,, \end{aligned} \tag{11.59}$$

where

$$\begin{aligned} \Delta &= (\eta_I+1)^2\big[(\eta_I+1)^2-a+d\big]-s^2\,, \\ a &= \eta_Q^2+\eta_U^2+\eta_V^2\,, \\ d &= \rho_Q^2+\rho_U^2+\rho_V^2\,, \\ s &= \eta_Q\rho_Q+\eta_U\rho_U+\eta_V\rho_V\,. \end{aligned} \tag{11.60}$$

a represents absorption, d dispersion effects, while s is a kind of "scalar product" between absorption and dispersion. Δ is the common denominator of all the expressions in Eq. (11.59).

This solution can easily be generalized by adding an extra exponential term to the B_ν variation of Eq. (11.54), for instance to simulate a chromospheric temperature rise:

$$B_\nu = B_0(1+\beta_0\tau)+Ae^{-\alpha\tau}\,. \tag{11.61}$$

$\mathrm{d}B_\nu/\mathrm{d}\tau$ in the solution (11.53) then gets the additional term $-\alpha A\exp(-\alpha\tau)$. As the two exponential functions in the integrand can be combined, we can obtain the explicit solution directly as before, with the result

$$\boldsymbol{I}_\nu(0)/I_c(0) = \big[(1-\beta)\boldsymbol{E}+\beta(\boldsymbol{\eta}+\boldsymbol{E})^{-1}-\beta'(\boldsymbol{\eta}'+\boldsymbol{E})^{-1}\big]\mathbf{1}\,, \tag{11.62}$$

where

$$\begin{aligned} \beta' &= \frac{A\alpha\mu}{B_0(1+\beta_0\mu)}\,, \\ \boldsymbol{\eta}' &= \boldsymbol{\eta}+\alpha\mu\boldsymbol{E}\,. \end{aligned} \tag{11.63}$$

$(\boldsymbol{\eta}'+\boldsymbol{E})^{-1}$ is obtained simply by making the substitution $\eta_I \to \eta_I+\alpha\mu$ in $(\boldsymbol{\eta}+\boldsymbol{E})^{-1}$ (cf. Landi Degl'Innocenti, 1992). We therefore do not need to write down the explicit expressions here.

11.7. Diagonalization of the Transfer Equation

The differential matrix equation (11.1) would be greatly simplified if the absorption matrix $\boldsymbol{\eta}$ were diagonalized, since then the four component equations would decouple from each other, and we would get to deal with four independent scalar equations instead.

Since magneto-optical effects ($\rho_{Q,U,V}$) break the symmetry of $\boldsymbol{\eta}$, a direct, mathematical diagonalization, as carried out by Šidlichovsky (1976) and Landi

Degl'Innocenti and Landi Degl'Innocenti (1985) leads to quite complicated and not very transparent algebraic expressions. To enhance the physical transparency we will approach the problem differently here, using concepts introduced by Stepanov (1958a) when he first diagonalized $\boldsymbol{\eta}$ in the absence of magneto-optical effects. His idea was to transform the transfer equation to a new polarization system, in which the complex basis vectors represent mutually orthogonal elliptical polarization states (instead of as before using a linear polarization basis). We enhance the transparency of the formulation and simplify the algebra by focusing our attention on the 2×2 matrix $\boldsymbol{A}$ that was given by Eq. (11.7) and describes the absorption-dispersion effects for the coherency matrix, instead of trying to diagonalize the 4×4 matrix $\boldsymbol{\eta}$ directly. After $\boldsymbol{A}$ has been diagonalized, the results are transformed to the Stokes vector formalism.

A polarization basis of linear and real unit vectors $\boldsymbol{e}_1$ and $\boldsymbol{e}_2$ as used in Eq. (2.26) may be transformed into an orthogonal basis of complex unit vectors $\boldsymbol{e}'_1$ and $\boldsymbol{e}'_2$ representing elliptical polarization via

$$\begin{aligned} \boldsymbol{e}'_1 &= \boldsymbol{e}_1 \cos\varphi + i\boldsymbol{e}_2 \sin\varphi , \\ i\boldsymbol{e}'_2 &= -\boldsymbol{e}_1 \sin\varphi + i\boldsymbol{e}_2 \cos\varphi . \end{aligned} \tag{11.64}$$

They can be represented in the complex plane as in Fig. 11.3. Note that we use $i\boldsymbol{e}'_2$ instead of $\boldsymbol{e}'_2$ in Eq. (11.64) for reasons of symmetry (see below).

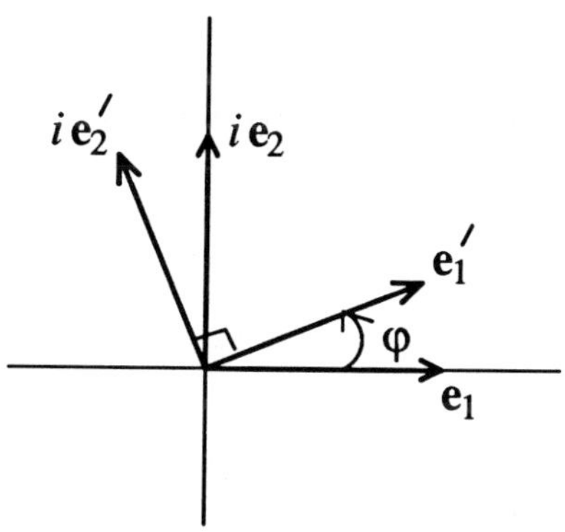

Fig. 11.3. Geometrical relation between the basis vectors according to the transformation (11.64).

When $\varphi = 5\pi/4$ we retrieve as a special case the spherical unit vectors $\boldsymbol{e}_\pm$ defined by Eq. (3.34), since then $\boldsymbol{e}'_1 = \boldsymbol{e}_+$ and $i\boldsymbol{e}'_2 = \boldsymbol{e}_-$. This case represents a circular polarization basis.

The Jones matrix $\boldsymbol{w}$ that transforms $\boldsymbol{e}_{1,2}$ into $\boldsymbol{e}'_{1,2}$ is the same as the one that transforms the components $E_{1,2}$ of the electric vector into $E'_{1,2}$. According to Eq. (11.64)

$$\boldsymbol{w} = \begin{pmatrix} \cos\varphi & i\sin\varphi \\ i\sin\varphi & \cos\varphi \end{pmatrix} = \boldsymbol{\sigma}_0 \cos\varphi + i\boldsymbol{\sigma}_2 \sin\varphi , \tag{11.65}$$

where $\boldsymbol{\sigma}_{0,2}$ are the Pauli spin matrices given by Eq. (2.31). Note that if $\boldsymbol{e}_2'$ had been used instead of $i\boldsymbol{e}_2'$ in the definition (11.64) of the transformation, $\boldsymbol{w}$ would not have been symmetric and could not have been expressed in terms of such a simple combination of Pauli spin matrices. The algebra can now be greatly simplified by making use of the relations (4.31) for bilinear products of the spin matrices.

It is easy to verify that the inverse of $\boldsymbol{w}$ is

$$\boldsymbol{w}^{-1} = \boldsymbol{\sigma}_0 \cos\varphi - i\boldsymbol{\sigma}_2 \sin\varphi\,. \tag{11.66}$$

Since the term $\boldsymbol{AD}$ in the transfer equation (11.6) transforms according to Eq. (2.36) as $\boldsymbol{D}$ alone, we get in the new system

$$\boldsymbol{A}'\boldsymbol{D}' = \boldsymbol{w}(\boldsymbol{AD})\boldsymbol{w}^\dagger = (\boldsymbol{wAw}^{-1})(\boldsymbol{wDw}^\dagger)\,, \tag{11.67}$$

and accordingly

$$\boldsymbol{A}' = \boldsymbol{wAw}^{-1}\,. \tag{11.68}$$

Note that it is important to make a distinction between $\boldsymbol{w}^{-1}$ and $\boldsymbol{w}^\dagger$, as we will see later.

In the case of a normal Zeeman triplet,

$$\boldsymbol{A} = \tfrac{1}{2}\eta_0 \sum_{k=0}^{3} \mathcal{H}^k \boldsymbol{\sigma}_k + \tfrac{1}{2}\boldsymbol{\sigma}_0 \tag{11.69}$$

according to Eqs. (4.23) and (4.42)–(4.45). The treatment can easily be generalized to the case of arbitrary anomalous splitting patterns by replacing H_q and F_q that are part of $\mathcal{H}^k$ with ϕ_q and ψ_q as in Eqs. (6.37), (6.52), and (6.57). We will however limit ourselves in the present treatment to the normal triplet case to keep the notations and the algebra as transparent as possible.

If now expression (11.69) is inserted in Eq. (11.68), we only have to apply the rules (4.31) for the spin matrix products to obtain the transformed matrix $\boldsymbol{A}'$. The results is

$$\begin{aligned} 2\boldsymbol{A}' &= (1 + \eta_0 \mathcal{H}_I)\boldsymbol{\sigma}_0 + \eta_0(\mathcal{H}_Q \cos 2\varphi + \mathcal{H}_V \sin 2\varphi)\boldsymbol{\sigma}_1 \\ &\quad + \eta_0 \mathcal{H}_U \boldsymbol{\sigma}_2 + \eta_0(\mathcal{H}_V \cos 2\varphi - \mathcal{H}_Q \sin 2\varphi)\boldsymbol{\sigma}_3\,. \end{aligned} \tag{11.70}$$

For $\boldsymbol{A}'$ to be diagonal, the coefficients before $\boldsymbol{\sigma}_2$ and $\boldsymbol{\sigma}_3$ have to be zero, since $\boldsymbol{\sigma}_2$ and $\boldsymbol{\sigma}_3$ contain the off-diagonal terms. This condition is satisfied if

$$\begin{aligned} \mathcal{H}_U &= 0\,, \\ \cot 2\varphi &= \mathcal{H}_Q/\mathcal{H}_V\,. \end{aligned} \tag{11.71}$$

$\mathcal{H}_U$ becomes zero if the coordinate system is chosen such that the Stokes Q direction is parallel to the projected field direction ($\chi = 0$, cf. Eqs. (3.74) and (4.48)). The second condition however cannot be satisfied by real trigonometric functions, since magneto-optical effects make the ratio $\mathcal{H}_Q/\mathcal{H}_V$ complex. Our analysis can however be mathematically extended to the case of complex angles φ by using the definition of the trigonometric functions in terms of exponential functions. The algebra is the same in the complex case, and it follows from Eq. (11.71) that

$$\begin{aligned} \cos 2\varphi &= \mathcal{H}_Q / \sqrt{\mathcal{H}_Q^2 + \mathcal{H}_V^2}\,, \\ \sin 2\varphi &= \mathcal{H}_V / \sqrt{\mathcal{H}_Q^2 + \mathcal{H}_V^2}\,. \end{aligned} \tag{11.72}$$

We can now write Eq. (11.70) in the form

$$\boldsymbol{A}' = \tfrac{1}{2}(1+\eta_I)\boldsymbol{\sigma}_0 + \tfrac{1}{2}(\eta_Q' - i\rho_Q')\boldsymbol{\sigma}_1\,, \tag{11.73}$$

where according to Eq. (11.72) when inserted in Eq. (11.70)

$$\begin{aligned} \eta_Q' - i\rho_Q' &= \eta_0 \sqrt{\mathcal{H}_Q^2 + \mathcal{H}_V^2} \\ &= \left[(\eta_Q - i\rho_Q)^2 + (\eta_V - i\rho_V)^2\right]^{1/2}. \end{aligned} \tag{11.74}$$

The term $i\rho_I\boldsymbol{\sigma}_0$ in Eq. (11.70) has been disregarded, since it subtracts out when forming $\boldsymbol{A}'\boldsymbol{D}' + \boldsymbol{D}'\boldsymbol{A}'^{\dagger}$ in the transfer equation. Squaring Eq. (11.74) we get

$$(\eta_Q' - i\rho_Q')^2 = a - d - 2is\,, \tag{11.75}$$

where a, d, and s have been defined in Eq. (11.60) (and we recall that $\eta_U = \rho_U = 0$ in our coordinate system).

Let us define the angle ψ by

$$\tan 2\psi = \frac{2s}{a-d}\,. \tag{11.76}$$

Then we can write

$$\begin{aligned} \eta_Q' &= b\cos\psi\,, \\ \rho_Q' &= b\sin\psi\,, \end{aligned} \tag{11.77}$$

where

$$b = \left[(a-d)^2 + 4s^2\right]^{1/4}. \tag{11.78}$$

Notice that in the absence of magneto-optical effects $\psi = \psi_0$ and $b = b_0$, where

$$\begin{aligned} \psi_0 &= 0\,, \\ b_0 &= \sqrt{a} = \sqrt{\eta_Q^2 + \eta_V^2}\,. \end{aligned} \tag{11.79}$$

If we define

$$\eta_\pm = \eta_I \pm \eta_Q'\,, \tag{11.80}$$

the expression (11.73) for $\boldsymbol{A}'$ can be written

$$2\boldsymbol{A}' = \begin{pmatrix} 1+\eta_+ & 0 \\ 0 & 1+\eta_- \end{pmatrix} - i \begin{pmatrix} \rho_Q' & 0 \\ 0 & -\rho_Q' \end{pmatrix}. \tag{11.81}$$

Although we have thus succeeded in diagonalizing the Jones absorption-dispersion matrix $\boldsymbol{A}$, the goal of diagonalizing the corresponding Mueller matrix has not been reached. From a comparison between the Jones and Mueller matrices of Eqs. (4.45) and (4.35), we can immediately write down the Mueller absorption-dispersion matrix in our transformed system, and find

$$\boldsymbol{\eta}' = \begin{pmatrix} \eta_I & \eta'_Q & 0 & 0 \\ \eta'_Q & \eta_I & 0 & 0 \\ 0 & 0 & \eta_I & \rho'_Q \\ 0 & 0 & -\rho'_Q & \eta_I \end{pmatrix} . \tag{11.82}$$

The Stokes parameters in this system are denoted I', Q', U', and V'. As Eq. (11.82) is not diagonal, we have to make a transformation to still another system, in which the new Stokes parameters I'', Q'', U'', and V'' are defined by

$$\begin{aligned} I'' &\equiv I_+ = \tfrac{1}{2}(I' + Q'), \\ Q'' &\equiv I_- = \tfrac{1}{2}(I' - Q'), \\ U'' &= \tfrac{1}{2}(U' + iV'), \\ V'' &= \tfrac{1}{2}(iU' + V'). \end{aligned} \tag{11.83}$$

Here we have for convenience introduced the notation $I_\pm$, since these parameters transfer independently through the atmosphere and are governed by the absorption coefficients $\eta_\pm$ of Eq. (11.80), as we will see below.

The transformation matrix defined by Eq. (11.83) is

$$\boldsymbol{M} = \frac{1}{2} \begin{pmatrix} 1 & 1 & 0 & 0 \\ 1 & -1 & 0 & 0 \\ 0 & 0 & 1 & i \\ 0 & 0 & i & 1 \end{pmatrix} . \tag{11.84}$$

From a simple direct inversion of Eq. (11.83) we see that

$$\boldsymbol{M}^{-1} = 2\boldsymbol{M}^\dagger . \tag{11.85}$$

The absorption-dispersion matrix in the new system then becomes

$$\boldsymbol{\eta}'' = \boldsymbol{M}\boldsymbol{\eta}'\boldsymbol{M}^{-1} = \begin{pmatrix} \eta_+ & 0 & 0 & 0 \\ 0 & \eta_- & 0 & 0 \\ 0 & 0 & c_- & 0 \\ 0 & 0 & 0 & c_+ \end{pmatrix} , \tag{11.86}$$

where $\eta_\pm$ have been given by Eq. (11.80), and

$$c_\pm = \eta_I \pm i\rho'_Q . \tag{11.87}$$

The complete diagonalization implies that the transfer equations of all four Stokes parameters are now entirely decoupled from each other, even in the presence of magneto-optical effects.

For a full formulation of the Stokes transfer equation for $\boldsymbol{I}''$ we need to relate it to Stokes $\boldsymbol{I}$ of the original system, so that we can give the correct expression for the source function and formulate the boundary conditions in the double-primed system. After having solved the four independent scalar equations for $\boldsymbol{I}''$, we need to transform back to the Stokes $\boldsymbol{I}$ system. Eq. (11.83) gives the relation between $\boldsymbol{I}''$ and $\boldsymbol{I}'$, but what is still missing is the relation between $\boldsymbol{I}'$ and $\boldsymbol{I}$. This relation is determined by the Jones matrix (11.65) that defines the connection between coherency matrices $\boldsymbol{D}$ and $\boldsymbol{D}'$, with φ being determined by Eq. (11.71).

When transforming $\boldsymbol{D}$ to $\boldsymbol{D}'$ we however have to be extra careful, since the Jones matrix $\boldsymbol{A}$ and the coherency matrix $\boldsymbol{D}$ do not transform in the same way when magneto-optical effects are present. While $\boldsymbol{D}$ transforms according to Eq. (2.36), the transformation of $\boldsymbol{A}$ is given by Eq. (11.68). A difference between these two transformations occurs when $\boldsymbol{w}^\dagger \neq \boldsymbol{w}^{-1}$, which happens when the angle φ governing $\boldsymbol{w}$ is complex. The imaginary part of φ is exclusively due to magneto-optical effects. Thus from Eqs. (11.65) and (11.66) it follows that

$$\boldsymbol{w}^\dagger = \boldsymbol{\sigma}_0(\cos\varphi)^* - i\boldsymbol{\sigma}_2(\sin\varphi)^* \neq \boldsymbol{w}^{-1}\,, \tag{11.88}$$

where $*$ means complex conjugation.

Let us for convenience denote the Stokes parameters I, Q, U, and V by $S_{0,1,2,3}$. Then the relation between the transformed coherency matrix $\boldsymbol{D}'$ and the Stokes parameters are obtained from Eqs. (2.46) and (2.36):

$$\boldsymbol{D}' = \tfrac{1}{2}\sum_{k=0}^{3} S'_k\boldsymbol{\sigma}_k = \tfrac{1}{2}\sum_{k=0}^{3} S_k\boldsymbol{w}\boldsymbol{\sigma}_k\boldsymbol{w}^\dagger\,. \tag{11.89}$$

If we insert Eqs. (11.65) and (11.88), we obtain the relation between S_k and S'_k by identifying the coefficients in front of $\boldsymbol{\sigma}_k$ in Eq. (11.89). Once we have achieved the transition to the primed system, we get to the double-primed one via Eq. (11.83). The result is

$$\begin{aligned}
I_\pm &= \tfrac{1}{2}[\;\; I\cos(\varphi-\varphi^*) \;\pm Q\cos(\varphi+\varphi^*) \\
&\qquad +Ui\sin(\varphi-\varphi^*) \;\pm V\sin(\varphi+\varphi^*)]\,,\\
U'' &= \tfrac{1}{2}[\;\; Ii\sin(\varphi-\varphi^*) - Qi\sin(\varphi+\varphi^*) \\
&\qquad +U\cos(\varphi-\varphi^*) + Vi\cos(\varphi+\varphi^*)]\,,\\
V'' &= \tfrac{1}{2}[-I\;\sin(\varphi-\varphi^*) \;- Q\sin(\varphi+\varphi^*) \\
&\qquad +Ui\cos(\varphi-\varphi^*) \;+ V\cos(\varphi+\varphi^*)]\,.
\end{aligned} \tag{11.90}$$

From these relations it follows that the LTE source function $\boldsymbol{S}_\nu = B_\nu\boldsymbol{1}$ gets transformed into

$$\boldsymbol{S}''_\nu = \tfrac{1}{2}B_\nu\begin{pmatrix}\cos(\varphi-\varphi^*)\\ \cos(\varphi-\varphi^*)\\ i\sin(\varphi-\varphi^*)\\ -\sin(\varphi-\varphi^*)\end{pmatrix}. \tag{11.91}$$

From Eq. (11.91) we see that when magneto-optical effects are present, so that $\varphi^* \neq \varphi$, there are weak source terms in U'' and V''. Even when the boundary conditions for U'' and V'' are zero (sufficiently deep down in the atmosphere η_0 and therefore also the magneto-optical effects vanish), U'' and V'' signals will be built up from these source terms. In the absence of magneto-optical effects, however, U'' and V'' remain zero throughout the atmosphere, and only the independent transfers of I_+ and I_- need to be calculated.

After the decoupled radiative transfer equations have been solved in terms of the parameters $I_\pm$, U'', and V'', the matrix that describes the relations (11.90) has to be inverted to retrieve the Stokes vector $\boldsymbol{I}_\nu$ of the original system. Here we will do this inversion explicitly only for the special case when magneto-optical effects are disregarded. Then $\varphi^* = \varphi$, and $U'' = V'' = 0$ throughout the atmosphere. In this case Eq. (11.90) can readily be inverted to give

$$\begin{aligned} I &= I_+ + I_- \,, \\ Q &= (I_+ - I_-)\cos 2\varphi \,, \\ U &= 0 \,, \\ V &= (I_+ - I_-)\sin 2\varphi \,. \end{aligned} \tag{11.92}$$

This solution, which refers to the special case when the field azimuth $\chi = 0$, can by simple Mueller matrix rotation be transformed into a solution for an arbitrary value of χ. While Stokes I and V are unaffected by an azimuth rotation, Stokes Q and V transform as H_Q and H_U in Eq. (4.48), so instead of Eq. (11.92) we get

$$\begin{aligned} Q &= (I_+ - I_-)\cos 2\varphi\cos 2\chi \,, \\ U &= (I_+ - I_-)\cos 2\varphi\sin 2\chi \,. \end{aligned} \tag{11.93}$$

According to Eq. (11.72)

$$\begin{aligned} \cos 2\varphi &= \eta_Q/\sqrt{\eta_Q^2 + \eta_V^2} \,, \\ \sin 2\varphi &= \eta_V/\sqrt{\eta_Q^2 + \eta_V^2} \end{aligned} \tag{11.94}$$

in the absence of magneto-optical effects.

11.8. Special Cases: Longitudinal and Transverse Fields

In the longitudinal case the field inclination $\gamma = 0$, which implies that $\eta_Q = \eta_U = \rho_Q = \rho_U = 0$. According to Eq. (11.71) we then have $\cot 2\varphi = 0$, which means that $\varphi = \pi/4$. The diagonalization thus refers to a circular polarization basis, and as $\varphi^* = \varphi$, the source function (11.91) becomes

$$\boldsymbol{S}''_\nu = \tfrac{1}{2} B_\nu \begin{pmatrix} 1 \\ 1 \\ 0 \\ 0 \end{pmatrix} . \tag{11.95}$$

U'', V'', and all magneto-optical effects then vanish, and we only have to deal with the two independent parameters $I_\pm$ with absorption coefficients $\eta_\pm$, which we denote $\eta_{\|\pm}$ to mark that they refer to the longitudinal case.

From Eq. (11.74) follows for the longitudinal case that $\eta'_Q = \eta_V$. According to Eqs. (11.80), (4.48), and (4.49) (with $\gamma = 0$) we then get

$$\eta_{\|\pm} = \eta_0(H_I \pm H_V) = \eta_0 H_\pm \,, \tag{11.96}$$

where $H_\pm$ is defined (as H_q) by Eq. (4.49). In the absence of magnetic fields Eq. (11.96) reduces to $\eta_{\|\pm} = \eta_0 H_0$.

Let $I_0(\Delta\lambda)$ be the solution of the *non-magnetic* scalar transfer equation with absorption coefficient $1 + \eta_0 H_0$. It then follows from Eq. (11.96) that the solution for $I_\pm$ is

$$I_\pm(\Delta\lambda) = I_0(\Delta\lambda \mp \Delta\lambda_H)\,. \tag{11.97}$$

Eq. (11.92) allows us to retrieve the original Stokes parameters, and we get

$$\begin{aligned} I &= I_+ + I_- \,, \\ V &= I_+ - I_- \,, \end{aligned} \tag{11.98}$$

while $Q = U = 0$. We want to stress that this is an exact result valid for all line strengths and field strengths in the case of a normal Zeeman triplet.

In the case of anomalous splitting patterns we have more than one σ component for each of the two polarization states. As the different components have different Zeeman displacements, the line profiles of the resulting $I_\pm$ parameters will differ in shape from the I_0 profile. In addition we have the effect that the different components of the same polarization state interact with each other in the transfer process (causing line saturation effects). For practical purposes it is however generally a good approximation to use an effective Landé factor $g_{\rm eff}$ determined by Eq. (6.44), and thereby treat the splitting pattern in terms of an equivalent normal Zeeman triplet with the same effective splitting as the anomalous pattern. In this section we therefore limit the discussion to the case of normal Zeeman triplets.

In the transverse case, when the field inclination $\gamma = \pi/2$, we have $\eta_V = \rho_V = \mathcal{H}_V = 0$, which according to Eq. (11.71) means that $\tan 2\varphi = 0$ and thus that $\varphi = 0$. The diagonalization of $\boldsymbol{\eta}$ then refers to a linear polarization basis, corresponding to the orthogonal polarization states of the π and σ components of the Zeeman splitting pattern.

The transformed source function will in the transverse case also be given by Eq. (11.95). The magneto-optical effects vanish, $U'' = V'' = 0$ throughout the atmosphere, and the decoupled transfer of $I_\pm$, representing the π and σ component radiation, is governed by the absorption coefficients $\eta_{\perp\pm}$. In the same way as we derived Eq. (11.96), we now find that

$$\begin{aligned} \eta_{\perp+} &= \eta_0 H_0 \,, \\ \eta_{\perp-} &= \tfrac{1}{2}\eta_0(H_+ + H_-)\,. \end{aligned} \tag{11.99}$$

With the arguments that led to Eq. (11.97) the solution for the emergent π component radiation can be given immediately as

$$I_+ = I_0(\Delta\lambda)\,. \tag{11.100}$$

For the σ component radiation a general solution in terms of the non-magnetic profile I_0 cannot be given in this way, since there are two σ components with different Zeeman displacements but identical polarization states. They interact with each other, which leads to saturation effects influencing the line profile. In the limiting case of *weak lines*, however, the saturation effects vanish, and Eq. (11.99) implies that

$$I_- = \tfrac{1}{2}\left[I_0(\Delta\lambda - \Delta\lambda_H) + I_0(\Delta\lambda + \Delta\lambda_H)\right]. \tag{11.101}$$

Note that although Eq. (11.101) is only valid for weak spectral lines, there is no limitation on the strength of the magnetic field. The longitudinal solution (11.97) is valid for any line *and* field strength.

From Eq. (11.92) we get for the transverse case (as $\varphi = 0$)

$$\begin{aligned} I &= I_+ + I_-\,,\\ Q &= I_+ - I_-\,. \end{aligned} \tag{11.102}$$

11.9. Weak Field Approximation

For weak fields, when the Zeeman splitting v_H expressed in units of the Doppler width is $\ll 1$, we can make a Taylor expansion of H_q of Eq. (4.49) and only retain terms up to second order in v_H:

$$H_\pm \approx H_0 \mp v_H \frac{\partial H_0}{\partial v} + \tfrac{1}{2} v_H^2 \frac{\partial^2 H_0}{\partial v^2}\,. \tag{11.103}$$

Replacing H with F, we obtain the corresponding expression for $F_\pm$. We recall that

$$\begin{aligned} v &= \Delta\lambda/\Delta\lambda_D\,,\\ \Delta\lambda &= \lambda - \lambda_0\,,\\ v_H &= \Delta\lambda_H/\Delta\lambda_D = \Delta\nu_H/\Delta\nu_D\,, \end{aligned} \tag{11.104}$$

as in Eq. (6.39).

According to Eqs. (3.44) and (3.89), extended by the use of an effective Landé factor g_{eff} as in Eqs. (6.43) and (6.44), the Zeeman splitting is

$$\Delta\nu_H = \frac{e}{4\pi m} g_{\text{eff}}\, B = 1.401 g_{\text{eff}}\, B \text{ MHz} \tag{11.105}$$

if B is given in G. In wavelength units

$$\Delta\lambda_H = 4.67 \times 10^{-13} g_{\text{eff}}\, \lambda^2 B\,, \tag{11.106}$$

where B should be given in G, λ and $\Delta\lambda_H$ in Å. Note that in contrast to the definition (6.47) we have in Eqs. (11.105) and (11.106) incorporated g_{eff} in the definitions of $\Delta\nu_H$ and $\Delta\lambda_H$ to be able to cover the general splitting case while using the expressions (4.48) and (4.49) for $H_{I,Q,U,V}$ and $F_{I,Q,U,V}$, which were developed in a classical context.

It is convenient to introduce the definitions

$$\begin{aligned} h_1 &= -v_H \frac{\partial H_0}{\partial v}\,, \\ f_1 &= -v_H \frac{\partial F_0}{\partial v}\,, \\ h_2 &= -\tfrac{1}{4} v_H^2 \frac{\partial^2 H_0}{\partial v^2}\,, \\ f_2 &= -\tfrac{1}{4} v_H^2 \frac{\partial^2 F_0}{\partial v^2}\,. \end{aligned} \tag{11.107}$$

Indices 1 and 2 thus correspond to the first and second order in v_H (as well as to the first and second derivative). With these definitions and Eq. (11.103) we obtain from Eqs. (4.48) and (4.49)

$$\begin{aligned} H_I &\approx H_0 - h_2(1+\cos^2\gamma)\,, \\ H_Q &\approx h_2 \sin^2\gamma \cos 2\chi\,, \\ H_U &\approx h_2 \sin^2\gamma \sin 2\chi\,, \\ H_V &\approx h_1 \cos\gamma\,, \end{aligned} \tag{11.108}$$

with corresponding expressions for $F_{I,Q,U,V}$.

It is convenient as we have done before to choose the coordinate system such that the field azimuth $\chi = 0$. After having solved the problem for this special geometry, we then transform the solution back to represent an arbitrary value of χ (cf. the transformation of Eq. (11.92) into (11.93)). In the $\chi = 0$ system the Mueller absorption-dispersion matrix of Eq. (11.3) becomes, according to Eqs. (11.4) and (11.108),

$$\begin{aligned} \boldsymbol{\eta}/\eta_0 \approx\ & \left[H_0 - h_2(1+\cos^2\gamma)\right] \boldsymbol{E} \\ & + \sin^2\gamma \begin{pmatrix} 0 & h_2 & 0 & 0 \\ h_2 & 0 & 0 & 0 \\ 0 & 0 & 0 & 2f_2 \\ 0 & 0 & -2f_2 & 0 \end{pmatrix} \\ & + \cos\gamma \begin{pmatrix} 0 & 0 & 0 & h_1 \\ 0 & 0 & 2f_1 & 0 \\ 0 & -2f_1 & 0 & 0 \\ h_1 & 0 & 0 & 0 \end{pmatrix}. \end{aligned} \tag{11.109}$$

$h_2(1+\cos^2\gamma)$ in the first term describes the magnetic line broadening of the Stokes I profile. This Zeeman broadening is of second order, proportional to B^2, and is also somewhat angular dependent. The second term describes the transverse Zeeman effect. It is also of second order, proportional to $B_\perp^2$, wherer $B_\perp = B\sin\gamma$. The third and last term describes the longitudinal Zeeman effect and is of first order, proportional to $B_\| = B\cos\gamma$.

Since the longitudinal Zeeman effect is of first order while the transverse one is of second order, $|Q|$, $|U| \ll |V|$ when $v_H \ll 1$ (unless $\gamma \approx \pi/2$). Furthermore, since $|f_2| \ll |f_1|$, and $|f_1|$ is already small, f_2 may be neglected. The magneto-optical effects then only appear in the last term, where they represent Faraday rotation of the plane of linear polarization (conversion between Stokes Q and U). Since the linear polarization is only produced by the second term in second order (since only the first column of $\boldsymbol{\eta}$ contributes to the source function $\boldsymbol{\eta}\mathbf{1}$), the magneto-optical effects vanish relative to the other effects in the third term as $v_H \to 0$. Therefore Eq. (11.109) may be simplified to give

$$\begin{aligned}\boldsymbol{\eta}/\eta_0 \approx \left[H_0 - h_2(1+\cos^2\gamma)\right]\boldsymbol{E}\\ +h_2\sin^2\gamma \begin{pmatrix} 0 & 1 & 0 & 0\\ 1 & 0 & 0 & 0\\ 0 & 0 & 0 & 0\\ 0 & 0 & 0 & 0\end{pmatrix}\\ +h_1\cos\gamma \begin{pmatrix} 0 & 0 & 0 & 1\\ 0 & 0 & 0 & 0\\ 0 & 0 & 0 & 0\\ 1 & 0 & 0 & 0\end{pmatrix}.\end{aligned} \tag{11.110}$$

As the magneto-optical effects can thus be neglected in the weak-field limit, the diagonalized matrix $\boldsymbol{\eta}''$ of Eq. (11.86) has no imaginary parts in it. The transformation angle φ is real and therefore given by Eq. (11.94). $\eta_\pm$, defined by Eq. (11.80) and used in Eq. (11.86), now becomes

$$\eta_\pm \approx \eta_I \pm \sqrt{\eta_Q^2 + \eta_V^2}\,. \tag{11.111}$$

As according to Eq. (11.108)

$$\begin{aligned}\eta_Q &\approx \eta_0 h_2 \sin^2\gamma\,,\\ \eta_V &\approx \eta_0 h_1 \cos\gamma\end{aligned} \tag{11.112}$$

for $\chi = 0$, and as $|\eta_Q| \ll |\eta_V|$ in the weak-field limit,

$$\eta_\pm \approx \eta_I \pm \eta_V\,, \tag{11.113}$$

and $\sin 2\varphi \approx 1$ according to Eq. (11.94).

Using Eqs. (11.108) and (11.107) in Eq. (11.113), we get

$$\eta_\pm \approx \eta_0\left[H_0 \mp v_H\cos\gamma\frac{\partial H_0}{\partial v}\right] \approx \eta_0 H(a, v\mp v_H\cos\gamma)\,. \tag{11.114}$$

The last part of Eq. (11.114) implies that the emergent intensities can be written as

$$I_\pm(\Delta\lambda) \approx I_0(\Delta\lambda \mp \Delta\lambda_H\cos\gamma)\,. \tag{11.115}$$

As in Eq. (11.97) I_0 represents the solution of the non-magnetic, scalar transfer equation.

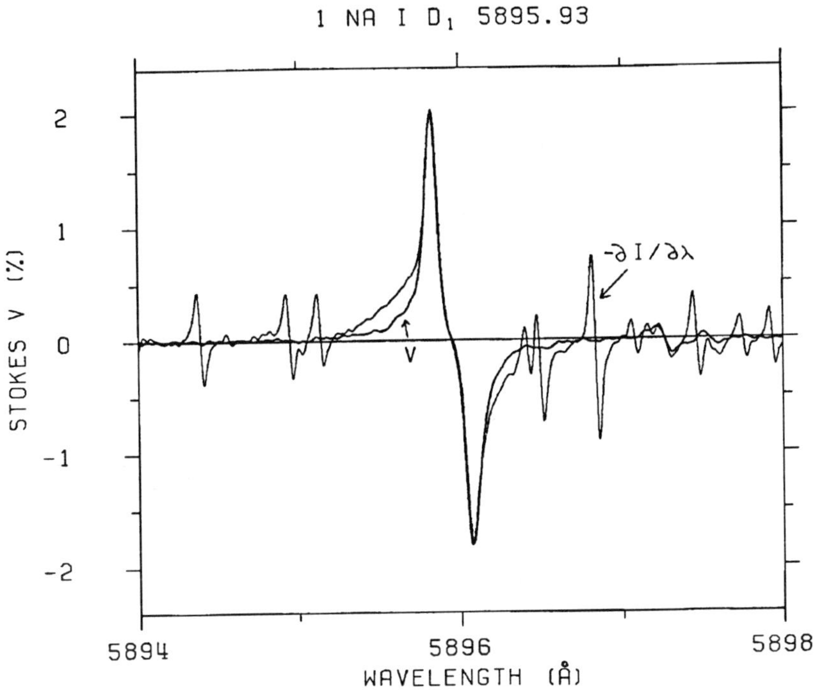

Fig. 11.4. Stokes V (thick curve) plotted in units of the continuum intensity together with $-\partial I/\partial\lambda$ (thin curve), which has been scaled to agree with the Stokes V amplitude of the Na I D_1 line. The recording has been made with the NSO/Kitt Peak Fourier Transform Spectrometer. From Stenflo et al. (1984).

According to Eqs. (11.92) and (11.115), with $\sin 2\varphi \approx 1$,

$$\begin{aligned} I &= I_+ + I_- \approx 2I_0\,, \\ V &\approx I_+ - I_- \approx -v_H \cos\gamma \,\frac{\partial I}{\partial v}\,. \end{aligned} \tag{11.116}$$

Examples illustrating the validity of this weak-field approximation are given in Fig. 11.4 for the Na I D_1 5895.93 Å line and in Fig. 11.5 for the hydrogen Hβ 4861.34 Å line, from Stenflo et al. (1984). Stokes V, recorded with a Fourier Transform Spectrometer in a plage region, is plotted together with the intensity gradient $\partial I/\partial\lambda$, which has been scaled so that its amplitude agrees with the amplitude of the Na I or the Hβ line. The many telluric lines surrounding the Na I D_1 line are conspicuous in $\partial I/\partial\lambda$ but are entirely absent in Stokes V (since Stokes V is exclusively of solar origin, in contrast to Stokes I). Since the factor of proportionality between V and $\partial I/\partial\lambda$ includes the Landé factor, the amplitudes of Stokes V and the normalized $\partial I/\partial\lambda$ do not agree for the lines surrounding the Hβ line, since they have different Landé factors. We note however how well the shapes of the Stokes V profiles of the Na I D_1 and Hβ lines are reproduced by $\partial I/\partial\lambda$. This is evidence for the usefulness of the weak-field approximation (11.116).

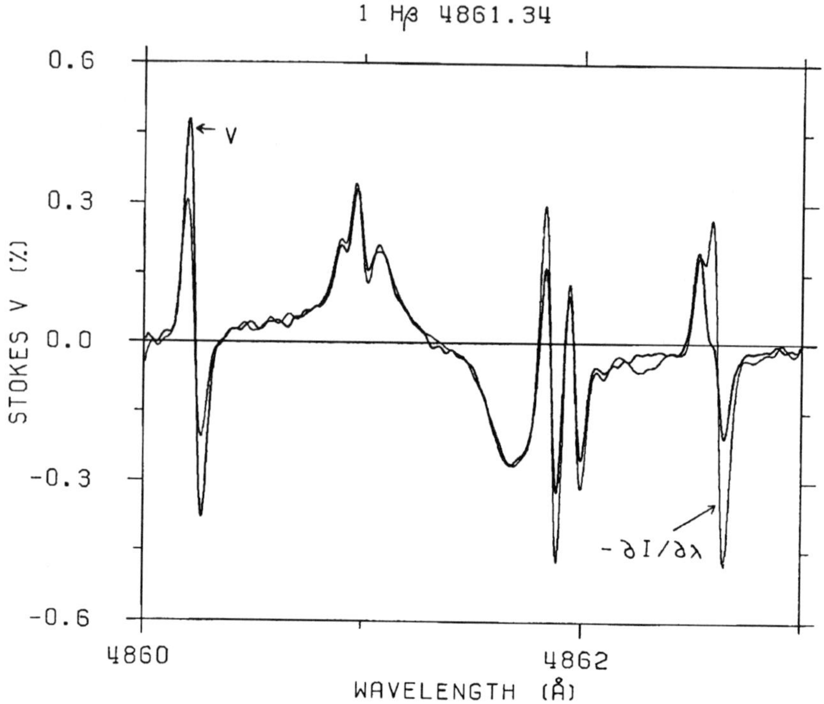

Fig. 11.5. Stokes V and $-\partial I/\partial\lambda$ as in Fig. 11.4, but for the region around the hydrogen Hβ line. $\partial I/\partial\lambda$ has been scaled so that its amplitude in Hβ agrees with that of Stokes V. From Stenflo et al. (1984).

According to Eqs. (11.92) and (11.94),

$$Q = (I_+ - I_-)\cos 2\varphi \approx V\eta_Q/\eta_V \,. \tag{11.117}$$

Using Eqs. (11.112) and (11.116) we then get

$$Q \approx -v_H \sin^2\!\gamma\,\frac{h_2}{h_1}\,\frac{\partial I}{\partial v}\,. \tag{11.118}$$

With the notation

$$\begin{aligned} H' &= \partial H_0/\partial v\,,\\ H'' &= \partial^2 H_0/\partial v^2\,, \end{aligned} \tag{11.119}$$

we can write

$$Q \approx -\Big(\frac{v_H \sin\gamma}{2}\Big)^2\,\frac{H''}{H'}\,\frac{\partial I}{\partial v}\,. \tag{11.120}$$

To transform this solution to a system of arbitrary field azimuth χ, we simply multiply the expression of Eq. (11.120) by $\cos 2\chi$ to get the new Q, by $\sin 2\chi$ to get the new U.

In the limit of *weak lines*, $\partial I/\partial v \sim H'$ and $\partial^2 I/\partial v^2 \sim H''$, so that

$$Q \approx -\left(\frac{v_H \sin\gamma}{2}\right)^2 \frac{\partial^2 I}{\partial v^2}. \tag{11.121}$$

The full expression (11.120) that is valid for any line strength has the drawback that it uses not only the observable emergent radiation I, but also H'' and H', which contain the unknown Doppler width $\Delta\lambda_D$. It is however possible to obtain a good value for the Doppler width by profile fitting or from flux tube models.

11.10. Approximate Determination of Vector Fields

11.10.1. Weak field case

The three unknowns that we want to determine to get the vector magnetic field are the field strength B, which is contained in the dimensionless parameter v_H (cf. Eqs. (11.104)–(11.106)), the field inclination γ, and the field azimuth χ. Summarizing our results from the previous section, Eqs. (11.116) and (11.120), the three parameters v_H, γ, and χ are determined by the three relations

$$\begin{aligned} v_H \cos\gamma &= -V\Big/\left(\frac{\partial I}{\partial v}\right), \\ \left(\frac{v_H \sin\gamma}{2}\right)^2 &= \left|\frac{H'}{H''}\sqrt{Q^2+U^2}\Big/\left(\frac{\partial I}{\partial v}\right)\right|, \\ \tan 2\chi &= U/Q \end{aligned} \tag{11.122}$$

in the weak-field limit. These relations have previously been given by Jefferies et al. (1989) and Jefferies and Mickey (1991).

11.10.2. General case: calibration curves

If we now let the field strength be arbitrary but restrict ourselves to *weak lines*, then only the first column of the absorption-dispersion matrix $\boldsymbol{\eta}$ is of importance, and $I, Q, U, V \sim \eta_{I,Q,U,V}$. This is so since the radiation field incident at the lower boundary is unpolarized, and only the first column of $\boldsymbol{\eta}$ can act on unpolarized radiation. In this case it follows from Eq. (4.48) that

$$\begin{aligned} Q &= q(B)\sin^2\gamma\cos 2\chi\,, \\ U &= u(B)\sin^2\gamma\sin 2\chi\,, \\ V &= v(B)\cos\gamma\,, \end{aligned} \tag{11.123}$$

where $u(B) = q(B)$. In this weak-line case, q, u, and v are functions of B alone, without any γ or χ dependence.

From our previous Eqs. (11.116) and (11.120) we see that Eq. (11.123) also applies to lines of arbitrary strength as long as the fields are weak. In the *weak-field limit*, when $v_H \ll 1$,

$$\begin{aligned} v(B) &\sim B\,, \\ q(B) &= u(B) \sim B^2\,. \end{aligned} \tag{11.124}$$

The functions $v(B)/B$ and $q(B)/B^2$ thus describe the Zeeman saturation that occurs when we depart from the weak-field regime.

In the general case, when both the spectral line and the magnetic field may be strong, q may be somewhat different from u, and q, u, and v may depend on γ as well. This γ dependence arises both from line saturation and magneto-optical effects, as can be seen from the Milne-Eddington solution (11.59). For survey purposes it is however often a reasonable assumption and considerable simplification to disregard these effects and assume that q, u, and v are given by

$$\begin{aligned} q(B) &\approx u(B) \approx Q(B;\, \gamma = \pi/2,\, \chi = 0)\,, \\ v(B) &\approx V(B;\, \gamma = 0)\,, \end{aligned} \tag{11.125}$$

and use these functions as *calibration curves* to determine the vector magnetic field. For a somewhat more sophisticated analysis, a set of calibration curves for different values of γ may be used.

For each position $\Delta\lambda$ within the line profile, the field strength B can be determined from

$$\frac{\sqrt{Q^2+U^2}}{q(B)} + \left[\frac{V}{v(B)}\right]^2 = 1 \tag{11.126}$$

according to Eq. (11.123). Knowing B, the inclination γ can then be obtained from

$$\frac{\sin^2\gamma}{\cos\gamma} = \frac{\sqrt{Q^2+U^2}}{V}\,\frac{v(B)}{q(B)} \tag{11.127}$$

and the azimuth χ from

$$\tan 2\chi = U/Q\,. \tag{11.128}$$

These relations are useful for "first-look" conversions of observations performed with a filter magnetograph, when only one position $\Delta\lambda$ (integrated over a certain band width) is used, but measurements at thousands of different spatial points have to be converted to obtain a map of the vector field. The neglect of magneto-optical effects may however lead to serious errors in the determined azimuth χ, and the influence of spatially unresolved inhomogeneities in the magnetic field may lead to large overestimates of the inclination γ and underestimates of the field strength B. The diagnostics of spatially unresolved structures will be treated in the next chapter. It is however clear that for a reliable quantitative analysis of vector magnetic fields filter magnetographs are not so suitable, but line profile or multi-line information is needed.

Fig. 11.6, from Stenflo (1985), shows examples of normalized calibration curves, to illustrate the different Zeeman saturations in q and v and their dependence on $\Delta\lambda$. A filter with a $1/8$ Å pass band has been assumed. The curves represent the spectral line Fe I 5250.2 Å that has a Landé factor of 3.0, and have been calculated for a Milne-Eddington atmosphere with $\eta_0 = 1.72$, damping parameter

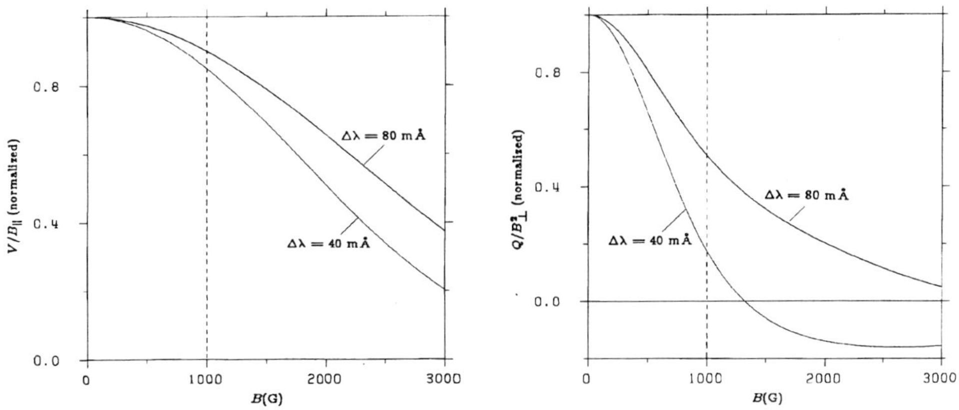

Fig. 11.6. $v(B)/B$ (left diagram) and $q(B)/B^2$ (right diagram), calculated according to the prescription (11.125) for a Milne-Eddington model atmosphere, the Fe I 5250.2 Å line, $\Delta\lambda = 40$ and 80 mÅ, and a filter band width of 1/8 Å. The curves have been normalized to unity for $B = 0$. While $v(B)/B \sim V/B_{\|}$, $q(B)/B^2 \sim Q/B_{\perp}^2$. The vertical dashed lines indicate the typical field strength in solar magnetic flux tubes (1 kG). From Stenflo (1985).

$a = 0.25$, and Doppler width $\Delta\lambda_D = 25.25$ mÅ. In this case $v_H = 1$ corresponds to $B = 654$ G.

11.11. Numerical Solutions

As the previous analytical and approximate methods to solve the transfer equation (11.1) have a limited range of validity, most quantitative work on solar magnetic fields requires numerical solutions based on some numerically given model atmosphere. In the past the most commonly used numerical method to solve Eq. (11.1) has been the classical fourth-order Runge-Kutta technique (Abramowitz and Stegun, 1970). One starts at the lower boundary and integrates the four coupled differential equations for the four Stokes parameters outwards. The lower boundary should be chosen below the level where the continuous spectrum is formed. At this boundary, $\tau_c = \tau_0$, it is often sufficient to assume that the radiation field is unpolarized and given by the local Planck function, i.e., that $\boldsymbol{I}_\nu(\tau_0) = B_\nu(\tau_0)\mathbf{1}$. A more elaborate boundary condition will be given by Eq. (11.137) below.

The Runge-Kutta method is however not so accurate, is not very stable numerically, and may also be slow. Instead, a superior numerical scheme called DELO (**D**iagonal **E**lement **L**ambda **O**perator) has recently become the method of choice (Murphy, 1990). It has the additional advantage of computing the contribution functions at the same time as solving the transfer equations, in contrast to the Runge-Kutta scheme. We will therefore in the following give a brief outline of the DELO method.

A main problem of solving Eq. (11.1) is that $\boldsymbol{\eta}$ in front of $\boldsymbol{I}_\nu$ is a non-diagonal matrix. To formally circumvent this we move $\boldsymbol{\eta}\boldsymbol{I}_\nu$ to let it be incorporated inside the source term. Thus we rewrite Eq. (11.1) in the form

$$\frac{\mathrm{d}\boldsymbol{I}_\nu}{\mathrm{d}\tau_c} = \boldsymbol{I}_\nu - \boldsymbol{X}\,, \tag{11.129}$$

where the new source term is

$$\boldsymbol{X} = \boldsymbol{S}_\nu - \boldsymbol{\eta}\boldsymbol{I}_\nu\,. \tag{11.130}$$

The simple form (11.129) of the transfer equation allows us to write down an explicit analytical solution when $\boldsymbol{X}$ is linearly interpolated between grid points. This leads to a recursive relation for $\boldsymbol{I}_\nu$, which is used to derive the emergent Stokes vector $\boldsymbol{I}_\nu(0)$ and the contribution functions.

Let us introduce a grid of depth points $1, 2, \ldots, N$ along the line of sight (or in the vertical direction, if a plane-parallel stratification is assumed). Point 1 represents the surface, point N the lower boundary. At depth point n we use the notations $\boldsymbol{I}_\nu = \boldsymbol{I}_n$, $\boldsymbol{S}_\nu = \boldsymbol{S}_n$, $\boldsymbol{\eta} = \boldsymbol{\eta}_n$, $\boldsymbol{X} = \boldsymbol{X}_n$, and $\tau_c = \tau_n$.

If we in our formal solution (11.13) make the substitution $\tau_c = \tau_n$ and $\tau_0 = \tau_{n+1}$, and we remove $\boldsymbol{\eta}$ in the expression (11.18) for $\boldsymbol{L}$ (since $\boldsymbol{\eta}$ is formally absent in Eq. (11.129)), then we obtain an integral solution relating $\boldsymbol{I}_n$ and $\boldsymbol{I}_{n+1}$ to each other:

$$\boldsymbol{I}_n = \int_{\tau_n}^{\tau_{n+1}} e^{-(\tau-\tau_n)}\boldsymbol{X}(\tau)\,\mathrm{d}\tau \;+ e^{-\Delta\tau_n}\boldsymbol{I}_{n+1}\,, \tag{11.131}$$

where

$$\Delta\tau_n = \tau_{n+1} - \tau_n\,. \tag{11.132}$$

As the integral is over a small optical depth interval $\Delta\tau_n$, the source function $\boldsymbol{X}$ can be linearly interpolated within this interval:

$$\boldsymbol{X}(\tau) = \boldsymbol{X}_n + (\boldsymbol{X}_{n+1} - \boldsymbol{X}_n)(\tau - \tau_n)/\Delta\tau_n\,. \tag{11.133}$$

If we insert this in Eq. (11.131), we see from Eq. (11.130) that the integral in Eq. (11.131) produces terms in $\boldsymbol{I}_n$ and $\boldsymbol{I}_{n+1}$. Solving the integral analytically and grouping the $\boldsymbol{I}_n$ and $\boldsymbol{I}_{n+1}$ terms, we obtain after some algebra

$$\boldsymbol{I}_n = \boldsymbol{P}_n + \boldsymbol{Q}_n\boldsymbol{I}_{n+1}\,, \tag{11.134}$$

where

$$\begin{aligned} \boldsymbol{P}_n &= \boldsymbol{R}_n(a_n\boldsymbol{S}_n + b_n\boldsymbol{S}_{n+1})\,,\\ \boldsymbol{Q}_n &= \boldsymbol{R}_n(e^{-\Delta\tau_n}\boldsymbol{E} - b_n\boldsymbol{\eta}_{n+1})\,, \end{aligned} \tag{11.135}$$

and

$$\begin{aligned} \boldsymbol{R}_n &= (\boldsymbol{E} + a_n\boldsymbol{\eta}_n)^{-1}\,,\\ a_n &= 1 - \frac{1}{\Delta\tau_n}(1 - e^{-\Delta\tau_n})\,,\\ b_n &= \frac{1}{\Delta\tau_n}(1 - e^{-\Delta\tau_n}) - e^{-\Delta\tau_n}\,. \end{aligned} \tag{11.136}$$

From these expressions we see that only one matrix inversion ($\boldsymbol{R}_n$) per depth point is involved.

Usually a logarithmic τ scale $x = \log \tau$ is used with the depth points being equidistant in x, but for the present discussion we do not concern ourselves with the choice of the depth scale.

The recursive relation (11.134) is used by starting with some Stokes vector $\boldsymbol{I}_N$ at the lower boundary and iterating outwards to the surface. For $\boldsymbol{I}_N$ one may use an asymptotic expression obtained from Eq. (11.53), if we in that equation let the "surface" be at depth point N, and assume that the source function gradient $\mathrm{d}B_\nu/\mathrm{d}\tau_c$ and matrix $\boldsymbol{\eta}$ are asymptotically constant (Milne-Eddington assumption). The integral in Eq. (11.53) can then be solved, and we get

$$\boldsymbol{I}_N = B_{\nu,N}\mathbf{1} + \left(\frac{\mathrm{d}B_\nu}{\mathrm{d}\tau_c}\right)_N (\boldsymbol{\eta}_N + \boldsymbol{E})^{-1} . \tag{11.137}$$

Note that in all the expressions of this section we have used a τ_c scale along the line of sight. The special case of a plane-parallel stratification, when the optical depth scale is in the vertical direction, can readily be obtained by making the replacement $\tau_c \rightarrow \tau_c/\mu$ everywhere.

The Stokes contribution function $\boldsymbol{C_I}(\tau_c)$ that was defined by Eq. (11.28) represents the contribution to $\boldsymbol{I}_\nu(0)$ from the sources $\boldsymbol{S}_\nu$ at depth τ_c, per unit interval in τ_c. We can obtain $\boldsymbol{C}_{\boldsymbol{I},n}$ at depth point n by calculating the emergent intensity $\boldsymbol{I}_1$ with the recursive expression (11.134), if we only allow for source contributions at depth n, setting $\boldsymbol{S}_m = 0$ for all $m \neq n$. In this case $\boldsymbol{P}_m$ vanishes for $m < n$, and repeated use of Eq. (11.134) gives

$$\Delta \boldsymbol{I}_1 = \prod_{k=1}^{n-1} \boldsymbol{Q}_k \boldsymbol{I}_n . \tag{11.138}$$

Here we have put a Δ in front of $\boldsymbol{I}_1$ to mark that it is only the part of $\boldsymbol{I}_1$ due to the differential layer $\Delta\tau_n$ at depth point n that is expressed by Eq. (11.138).

As we due to our definition of $\Delta\boldsymbol{I}_1$ also have to discard the contributions from all the layers below the interval $\Delta\tau_n$, the only contribution to $\boldsymbol{I}_n$ arises from the sources $\boldsymbol{S}_n$, which according to Eq. (11.1) means that we can replace $\boldsymbol{I}_n$ in Eq. (11.138) by $\boldsymbol{S}_n\Delta\tau_n$. According to the definition of the contribution function, we then get

$$\Delta \boldsymbol{I}_1 = \boldsymbol{C}_{\boldsymbol{I},n} \Delta\tau_n , \tag{11.139}$$

where according to Eq. (11.138)

$$\boldsymbol{C}_{\boldsymbol{I},n} = \prod_{k=1}^{n-1} \boldsymbol{Q}_k \boldsymbol{S}_n . \tag{11.140}$$

If we compare this expression with Eq. (11.32) we can make the identification

$$\boldsymbol{L}_n^{-1} = \prod_{k=1}^{n-1} \boldsymbol{Q}_k . \tag{11.141}$$

Since $\boldsymbol{Q}_k$ are computed to obtain the solution for $\boldsymbol{I}_\nu(0)$, the contribution functions $\boldsymbol{C}_{\boldsymbol{I}}(\tau_c)$ are obtained as byproducts when solving for the emergent radiation.

11.12. Model Atmospheres and Multi-Dimensional Radiative Transfer

The temperature-density structure of magnetic-field regions is generally very different from the atmospheric structure in non-magnetic regions. The distribution of magnetic flux is observed to be extremely intermittent, with the magnetic field lines concentrated in bundles, which, because of the buoyancy forces, try to orient themselves in the vertical direction. These flux tubes have to be partially evacuated to establish pressure balance between the outside gas pressure and the inside magnetic plus gas pressure. This makes the flux tubes more transparent, which means that the surfaces of equal optical depth become depressed (the so-called "Wilson depression") inside the flux regions. The flux tube boundary diverges rapidly with height to adjust to the exponentially decreasing outside gas pressure, and a magnetic "canopy" develops above the underlying "non-magnetic" regions. All this implies that the atmospheric thermodynamic stratification is very far from plane parallel, the usual assumption in one-dimensional (1-D) radiative transfer.

In model calculations one normally makes the reasonable idealization that the outside atmosphere, assumed to be non-magnetic, becomes asymptotically plane parallel at large distances from the flux concentration. Therefore a plane-parallel reference atmosphere is of great use to represent the average, non-magnetic part of the atmosphere. A frequently used reference atmosphere is HSRASP, which is the so-called Harvard-Smithsonian Reference Atmosphere HSRA (Gingerich et al., 1971), extended downwards into the convection zone by connecting it with the convection-zone model of Spruit (1974). This extension is not needed for purely non-magnetic problems but is necessary for flux tubes, since due to the Wilson depression the radiation in the flux tubes comes from larger geometrical depths.

Since the fundamental constituents (flux tubes) of magnetic regions are not optically thick in the horizontal direction, plane-parallel models are inadequate for detailed quantitative analysis of Zeeman-effect observations. The geometry needs to be modelled as realistically as possible. As the magnetic field of individual flux tubes expands with height, it encounters and merges with the field from other flux tubes at some height in the atmosphere, so that the magnetic filling factor of the corona is always 100 % . When viewing a magnetic-field region obliquely (near the limb), the line of sight may sample Stokes profile contributions from more than one flux tube. An example of a model of an array of cylindrical flux tubes used to compute synthetic Stokes profiles near the solar limb is shown in Fig. 11.7.

When radiative scattering is ignored, the source function depends only on the *local* physical parameters. The transfer equation can be solved in the same way as in a 1-D or plane-parallel problem, with the difference that each line of sight has its own, different 1-D model atmosphere, which is deduced by tracing the line of sight through the 2 or 3-D model. The emergent Stokes spectrum is thus different for each line of sight. To compare with observations, which are always subject to finite spatial resolution, the Stokes spectra representing the different lines of

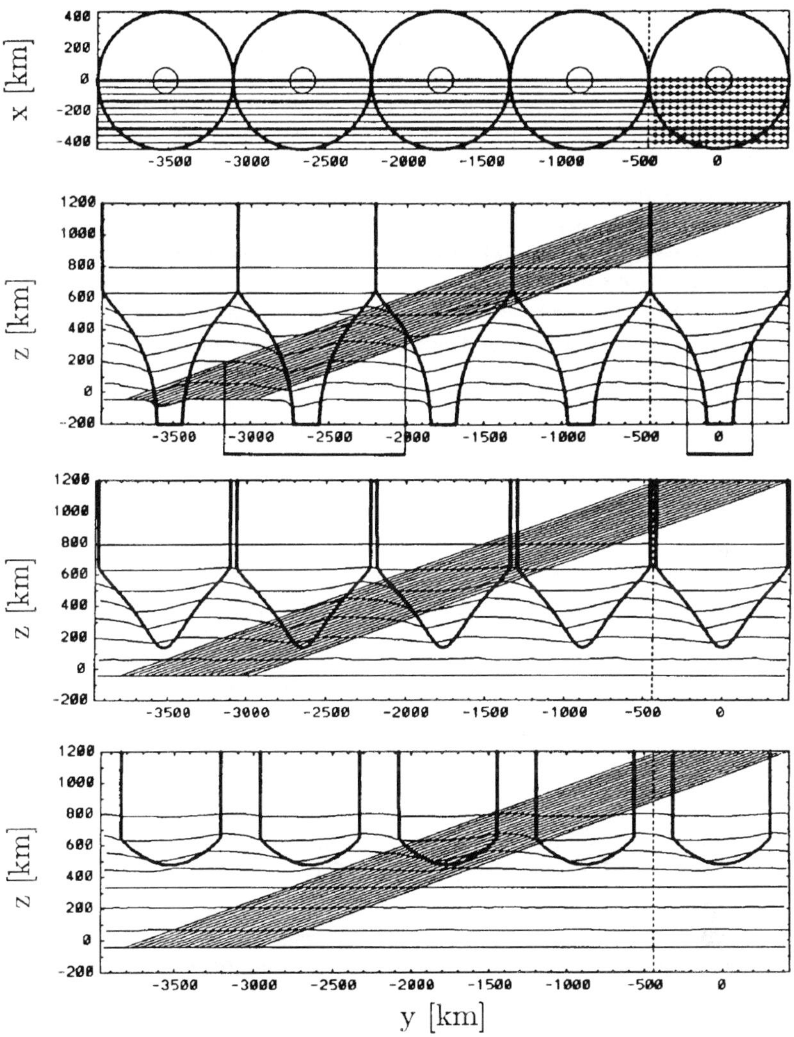

Fig. 11.7. Example of model geometry for an array of flux tubes. The first diagram gives a view from above. The thick circles represent the flux tube boundary at the merging height of +600 km, the thin circles the boundary at −200 km. The horizontal lines represent projections of vertical planes cutting through the model. The three thick horizontal lines correspond to the vertical cuts illustrated in the following three diagrams. The slanted straight lines represent different lines of sight, which enter the upper boundary of the model at the points to the right in the upper diagram. The lines of sight make an angle of 70° to the vertical, corresponding to a solar disk position of $\mu = 0.34$. The thin contours in the vertical planes represent surfaces of equal optical depth, equidistant in $\log \tau_c$ from $\log \tau_c = -6$ to $+1$. From Bünte et al. (1993b).

sight have to be averaged. This type of analysis is usually called 1.5-D radiative transfer, since the physics involves locally determined source functions along lines of sight embedded in a multi-dimensional atmosphere.

The name multi-dimensional, 2-D, or 3-D radiative transfer is reserved for *non-local* problems, when radiative scattering is involved and couples the source function to physical conditions at other spatial points. An example of multi-dimensional radiative transfer in a flux tube geometry is illustrated in Fig. 11.8. Due to the partial evacuation of the flux tube and the temperature increase with depth in the outside atmosphere, the interior will be surrounded and illuminated by hot flux tube walls. The radiation field from the non-local "wall photons" will enhance the line source function in the flux tube interior. The scattered photons can more easily escape in the vertical direction, in which the optical thickness of the overlaying atmosphere is the smallest. The scattered "wall photons" raise the core intensity of the emergent spectral lines, thereby mimicking a higher flux tube temperature that would be derived with a 1.5-D analysis.

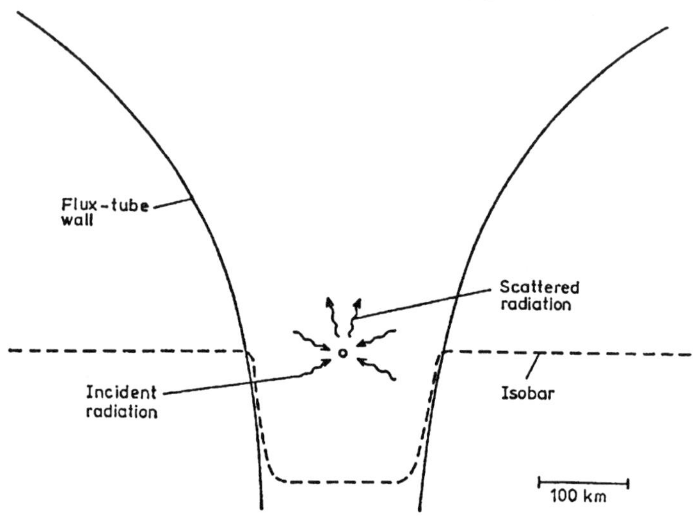

Fig. 11.8. Example of multi-dimensional transfer effects. The atoms in the flux tube interior scatter photons from the hot walls of the flux tube. The scattered radiation escapes the atmosphere preferentially in the vertical direction, contributing to line weakenings that may be mistaken for temperature effects. From Stenflo (1978a).

Although multi-dimensional radiative transfer of the Stokes vector is a tractable problem, as has been demonstrated by Stenholm and Stenflo (1978), the numerical complications are so great in comparison with the magnitude of the effects that one wants to avoid it whenever possible. The multi-dimensional effects appear to be small enough for normal Zeeman-effect analysis in the photosphere and the temperature minimum region, but may be important in the chromosphere. The situation is of course entirely different in the case of the Hanle effect, even deep in the photosphere, since the Hanle effect is fundamentally a coherent scattering problem and thus of non-local nature. In the present chapter we have however disregarded the Hanle effect.

11.13. Stokes Inversion

So far we have only discussed the *direct problem*, how to calculate the emergent Stokes spectrum from a given model atmosphere. The problem that one really wants to solve is however the *inverse* one, since it is the emergent Stokes spectrum that is observed, while the model atmosphere is the unknown to be deduced from the Stokes observations.

The relation between the observables (the Stokes line profiles) and the model parameters (the magnetic field and thermodynamic quantities as functions of height) is generally highly non-linear without any analytical formulation. The inverse problem however becomes tractable if it is linearized, which is commonly done in the context of a least-squares fitting procedure, as will be described next.

Let y_i, $i = 1, 2, \ldots, n$ be the observables with errors σ_i, while Y_i are the corresponding synthetic observables that are computed from the model. While there are n observables, the model is characterized by m free parameters. The goodness of the model fit to the observations is then governed by

$$\chi^2 = \frac{1}{n-m}\sum_{i=1}^{n}\frac{1}{\sigma_i^2}(y_i - Y_i)^2\,. \tag{11.142}$$

$n - m$ represents the number of degrees of freedom of the problem. If the model is good, we expect that χ^2 is of order unity.

Often different additional weights are applied to the observables, but for the present discussion we will assume that these weights have already been included in σ_i.

The least-squares fitting procedure consists of finding the parameter values for which χ^2 has its global minimum. Let the free model parameters be a_k, $k = 1, 2, \ldots, m$. They may represent any physical parameter at any location, e.g. the temperature gradient at $\tau_c = 10^{-3}$, or the magnetic-field angle at selected depth points. Ideally the a_k parameters, which may be described in terms of an m-vector $\boldsymbol{a}$, should represent for each relevant depth point all the physical parameters, i.e., $\boldsymbol{a}$ should fully describe the model atmosphere. In practice, however, one has to severely limit the number of free parameters and select the set of observables with utmost carc, to ensure stability and uniqueness of the numerical inversion.

χ^2 is a function of $\boldsymbol{a}$ and is represented by a surface in the m-dimensional parameter space. Our task is to find the minimum value of χ^2 on this surface. χ^2 has a local minimum when

$$\frac{\partial\chi^2}{\partial a_k} = 0 \tag{11.143}$$

for all allowed values of k. To find the location of this minimum we linearize the problem by making a Taylor expansion to first order of the synthetic observables:

$$Y(\boldsymbol{a} + \delta\boldsymbol{a}) = Y_0(\boldsymbol{a}) + \sum_{k=1}^{m}\frac{\partial Y_0}{\partial a_k}\delta a_k\,. \tag{11.144}$$

Inserting this expansion in Eq. (11.142) and applying the minimum condition (11.143), we obtain the matrix equation

$$\boldsymbol{b} = \boldsymbol{A}\,\delta\boldsymbol{a}\,, \tag{11.145}$$

where

$$A_{jk} = \sum_{i=1}^{n} \frac{1}{\sigma_i^2} \frac{\partial Y_{0i}}{\partial a_j} \frac{\partial Y_{0i}}{\partial a_k} \tag{11.146}$$

is the so-called curvature matrix, and

$$b_k = \sum_{i=1}^{n} \frac{1}{\sigma_i^2} (y_i - Y_{0i}) \frac{\partial Y_{0i}}{\partial a_k}\,. \tag{11.147}$$

The location $\boldsymbol{a}$ of the χ^2 minimum is determined iteratively, starting with an initial "guess" $\boldsymbol{a}_0$. From this starting point on the m-dimensional hypersurface we compute the corrections $\delta\boldsymbol{a}$ that should be applied to find the correct location. According to Eq. (11.145),

$$\delta\boldsymbol{a} = \boldsymbol{A}^{-1}\boldsymbol{b}\,. \tag{11.148}$$

The new, improved value for the location of the minimum is $\boldsymbol{a}_1 = \boldsymbol{a}_0 + \delta\boldsymbol{a}$, which serves as the starting point for the next iteration. The iterations continue until convergence has been obtained.

This procedure for localizing the minimum is however efficient only in the vicinity of the minimum, where the χ^2 surface is parabolic. To speed up the search process when the starting value is far away, an algorithm due to Marquardt (1963), described in Bevington (1969), uses a factor that interpolates the behavior of the algorithm between that of a gradient search, which is efficient at large distances from the true minimum, and that of the linearization method described above.

The observables have to be chosen very carefully so that they respond in the most "orthogonal" way to the perturbations of the model free parameters $\boldsymbol{a}$, otherwise the inversion of the matrix $\boldsymbol{A}$ will be numerically unstable. It is therefore essential to first explore the usefulness of different types of observables by test calculations, e.g. by using the response functions of Eqs. (11.41) and (11.44). One also has to make sure that the local minimum of the χ^2 hypersurface is also a global minimum (e.g. by repeating the computation with widely different starting parameters $\boldsymbol{a}_0$ for the iteration), otherwise the derived solution may not be unique.

As one has to limit the number of free parameters to make the problem well posed, the number of depth points at which the parameters can be kept free is usually small. To obtain the values of these parameters at the other depth points in the dense grid needed for the numerical solution of the transfer equations (for the calulation of the synthetic observables Y_i), one has to employ interpolation methods, like cubic splines.

Inversion of Stokes data to derive the values of free model parameters have been done by Harvey et al. (1972), Stenflo (1975), Auer et al. (1977), Frazier and Stenflo (1978), Skumanich and Lites (1987), Keller et al. (1990), and Ruiz Cobo and del Toro Iniesta (1992). They have all used the least-squares fitting approach outlined above.

CHAPTER 12

DIAGNOSTICS OF SMALL-SCALE MAGNETIC FIELDS

12.1. Introduction

Since Babcock (1953) and Kiepenheuer (1953) invented the photoelectric magnetograph, which maps the distribution of longitudinal magnetic flux across the solar surface by recording the circular polarization in a Zeeman-sensitive spectral line, the spatial resolution of the magnetic-field observations has been gradually increased from about one arcmin to less than one arcsec. Even the seeing limit determined by the earth's atmosphere can now be overcome by image reconstruction techniques like speckle polarimetry (see next chapter). Each new advance in spatial resolution has revealed new, smaller flux structures not seen before, and we have reasons to believe that this structuring continues well beyond the presently achievable spatial resolution.

If the observations were really able to resolve the fields, then we could apply the line-formation theory developed in the preceding chapters directly to diagnose the field, and the present chapter would not be needed. As it is, however, all magnetic fields outside sunspots are spatially unresolved. A new set of diagnostic methods therefore has to be developed to allow us to extract information about the spatially unresolved field.

This situation is typical in stellar astrophysics. Although the stars are spatially unresolved point objects, it has been possible to determine the stratification of temperatures, densities, velocity fields, and element abundances by extracting the information stored in different parts of the stellar spectrum. Similarly, the spatially unresolved magnetic flux elements on the sun have left their "finger prints" in the polarized profiles of many different types of spectral lines. The height variation of the intrinsic thermodynamic and magnetic structure of the unresolved flux elements may be derived through the combined analysis of carefully selected sets of polarized spectral lines. As the diagnostic methods that we develop for the sun concern spatially unresolved structures, they can be applied to other stars as well.

Since all the magnetic flux elements outside sunspots are spatially unresolved, the diagnostic methods presented in the present chapter are fundamental not just to exploring small-scale fields but to solar magnetism in general. The lack of spatial information is compensated for by using additional information in the spectral domain, e.g. combinations of two or more simultaneously observed spectral lines, spectrally resolved polarized line profiles, or resolved Zeeman splittings in the infrared. It may seem that this indirect approach could not take us very far, but a fortunate circumstance provides exceptional help: A two-component model, with one magnetic and one non-magnetic component, turns out to be a good approximation in most practical situations. This circumstance greatly reduces the number of free parameters needed to fit the data in comparison with the case of a multi-component atmosphere or a continuous distribution of the unresolved physical parameters. After solutions for the two-component model have been established as new references for further refinements, one may build ever more sophisticated models with additional components or continuous distributions added, thereby introducing more physical realism.

Applications of the spectral techniques outlined in the present chapter have revealed that the circular-polarization signals of which magnetograms are made up have their sources in spatially unresolved, discrete patches of magnetic flux that only occupy a small fraction, typically 1 % (varying between a fraction of a percent in very "quiet" regions to tens of percent in active-region plages), with no discernible flux in between (except for the "hidden", turbulent field that was discussed in Chapter 10). We call these flux patches flux tubes, although they could sometimes be more slab-like than tube-like. While the unpolarized spectrum (Stokes I) receives contributions from both the flux tubes and their "non-magnetic" surroundings, with the surroundings dominating due to their larger relative area, the polarized spectrum has its sources exclusively inside the flux tubes.

The Stokes Q, U, and V spectra thus provide diagnostic information on the flux tubes that is uncontaminated by the surrounding radiation, stray light, or limited spatial resolution, in contrast to the unpolarized spectra. This is a huge advantage for polarimetric observations. It is as if we would observe stars at daylight and had the means of completely filtering out the background radiation from the bright sky. In polarized light the flux tubes are similar to point-like objects against an entirely "black" background.

In the present chapter we will show how this polarimetric information can be exploited to build increasingly sophisticated empirical models of the magnetic and thermodynamic structure of the flux tubes and their immediate surroundings. The free parameters of these models have to satisfy not only the observational but also the theoretical constraints, since the requirements of magnetohydrodynamic (MHD) force balance (between gravitational and dynamical forces, gas pressure, magnetic pressure and tension) in all spatial directions have to be satisfied. The resulting empirically determined thermodynamic structure can be compared with the structure that emerges from numerical MHD simulations of collapsing flux tubes. The discrepancies between the empirical and theoretical models may result from simplifying assumptions made when deriving the empirical models but may also indicate what physics is missing in the theoretical description (e.g. the wave dissipation processes that contribute to the heating of the flux tube interiors, and which may be of critical importance for the overall energy balance of the outer solar atmosphere).

12.2. Diagnostic Contents of Stokes V Spectra: An Overview

The most powerful existing instrument in terms of spectral resolution, wavelength coverage, and straylight level is the Fourier Transform Spectrometer (FTS) at the McMath-Pierce facility of the National Solar Observatory (Kitt Peak). It was converted into a polarimeter in the end of the 1970's, which allowed fully resolved Stokes spectra with high signal-to-noise ratio to be recorded over 1000 Å portions of the solar spectrum. This has made it possible to use many hundreds of simultaneously recorded polarized line profiles for magnetic-field diagnostics.

Figure Eq. 12.1, from Stenflo et al. (1984), gives an example of a 6 Å wide segment of one of the first Stokes V recordings with the FTS polarimeter, as an illustration of the diagnostic contents of such spectra. Stokes V is given in units of the continuum intensity I_c. To achieve such high signal-to-noise ratio as in this recording a trade-off with spatial and time resolution had to be made. Thus a circular aperture of 10 arcsec diameter and an integration time of about 35 min were used.

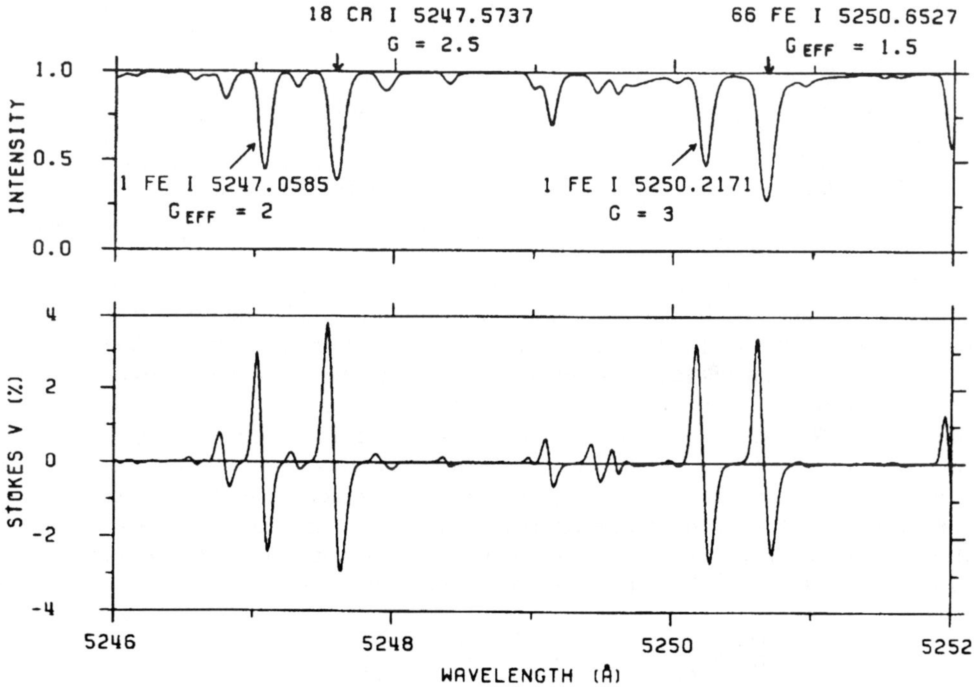

Fig. 12.1. Stokes I and V spectra recorded in a strong plage near disk center with the FTS polarimeter of the National Solar Observatory. From Stenflo et al. (1984).

To aid the understanding of the meaning of the Stokes V spectrum, let us consider the special case of a height-independent longitudinal magnetic field that covers the fraction α_m (the magnetic filling factor) of the spatial resolution element. Then a Taylor expansion gives according to Eqs. (11.97) and (11.98)

$$
\begin{aligned}
I &= \alpha_m I_m + (1-\alpha_m) I_{nm} \,, \\
I_m &\approx I_{m,0} + \tfrac{1}{2}(\Delta\lambda_H)^2 \frac{\partial^2 I_{m,0}}{\partial\lambda^2} + \dots \,, \\
V &\approx -\alpha_m\, \Delta\lambda_H \left[\frac{\partial I_{m,0}}{\partial\lambda} + \tfrac{1}{6}(\Delta\lambda_H)^2 \frac{\partial^3 I_{m,0}}{\partial\lambda^3} + \dots \right] .
\end{aligned}
\tag{12.1}
$$

Index m refers to the magnetic component, nm to the non-magnetic component, while index 0 refers to the case of zero Landé factor.

$$\Delta\lambda_H = z\, g \lambda^2 B \,, \tag{12.2}$$

where z is a fundamental, numerical constant with a value given by Eq. (10.62).

For most lines in the visible part of the solar spectrum the higher-order terms in the Taylor expansion are unimportant. In this case Stokes V is simply proportional to the Landé factor g, the field strength B, and the intensity gradient $\partial I_{m,0}/\partial\lambda$ *of the magnetic component.* To properly calibrate a magnetograph such that polarization V may be converted to field strength B one would need to know $I_{m,0}$ and α_m, which however are not observable quantities. The observables are Stokes I and V, and Stokes I is dominated by the contribution I_{nm} from the non-magnetic component. The anti-symmetric shape of the Stokes V profiles is due to $\partial I_{m,0}/\partial\lambda$, since all $\partial I/\partial\lambda$ profiles have qualitatively this shape, as was illustrated in Figs. 11.4 and 11.5.

The *average* field strength over the resolution element is

$$\langle B \rangle = \alpha_m B \,, \tag{12.3}$$

which would be obtained from the observed Stokes V if the calibration problem could be solved (i.e., if I_m were known). The value of $\langle B \rangle$ however does not tell us what the intrinsic field strength B may be as long as the filling factor α_m is unknown. We need additional information to untangle α_m and B from each other.

This additional information can be obtained by using a combination of two lines, which we refer to by indices 1 and 2. Let us assume that the two lines have $\lambda_1 \approx \lambda_2$ and $I_{m,1} \approx I_{m,2}$, which is generally the case when they belong to the same atomic multiplet and have approximately the same oscillator strengths. In this case the value of the field strength B to be used in Eq. (12.2) is the same for the two lines even when B is strongly height dependent, since the lines are formed in the same atmospheric layers. The Stokes V ratio then becomes

$$
\begin{aligned}
\frac{V_1}{V_2} = \frac{g_1}{g_2} &\left(\frac{\partial I_{m,0}}{\partial\lambda} + \frac{z^2\lambda^4}{6}\, g_1^2\, B^2 \frac{\partial^3 I_{m,0}}{\partial\lambda^3} + \dots \right) \\
&\left(\frac{\partial I_{m,0}}{\partial\lambda} + \frac{z^2\lambda^4}{6}\, g_2^2\, B^2 \frac{\partial^3 I_{m,0}}{\partial\lambda^3} + \dots \right)^{-1} .
\end{aligned}
\tag{12.4}
$$

Note that by forming this line ratio we have managed to eliminate the magnetic filling factor α_m. In the weak-field limit Eq. (12.4) reduces to

$$V_1/V_2 \approx g_1/g_2 \,. \tag{12.5}$$

Let us now check how the Stokes V line ratios of Eqs. (12.4) and (12.5) compare with what we actually see in Fig. 12.1. The two lines Fe I 5247.06 and 5250.22 Å represent an ideal combination for this purpose, since they belong to the same multiplet (no. 1) of iron, and have practically the same line strengths, excitation potentials, and wavelengths. As they are thus formed in the same way in the atmosphere, the conditions $I_{m,1} \approx I_{m,2}$ and $\lambda_1 \approx \lambda_2$ for Eq. (12.4) are fulfilled. The only difference of importance is due to the effective Landé factors, 2.0 and 3.0, respectively, for the two lines. If the magnetic fields were intrinsically weak ($\lesssim$ 0.5 kG), then the Stokes V amplitudes should be in the proportion 2 : 3 according to Eq. (12.5), whereas Fig. 12.1 shows that the ratio is much closer to 1 : 1. The reason for this is "differential Zeeman saturation", meaning that the higher order terms in the Taylor expansion are important for the line with the larger Landé factor while being rather small for the other line. This measured deviation from the weak-field relation (12.5) is a function of field strength B and therefore allows the value of B to be determined, regardless of what the value of the filling factor α_m is.

The conversion of the ratio V_1/V_2 to field strength B however depends on I_m according to Eq. (12.4), and as this is not an observable, some model dependence of the derived field strengths is introduced. This model dependence is however quite weak, since only differential line broadening but not line weakening may have an effect. If there is only differential line weakening by a factor w but no differential line broadening, then the *shapes* of the line profiles inside and outside the flux element are the same. When the filling factor is small $\ll 1$, the rest intensities may be written

$$(1 - I_{m,0}/I_{m,0_c}) \approx w\,(1 - I/I_c)\,. \tag{12.6}$$

Index c refers to the continuous spectrum and I is the observed Stokes I of Eq. (12.1). In this case the unobserved $I_{m,0}$ in Eq. (12.4) may be replaced by the observable I. Then the field strength B is the only unknown that is left in the equation and may be solved for in an unambiguous way. As the field strength varies with height we have to remember that the derived value of B refers to the height of formation of the lines used. As we will see later the geometrical height of formation is a sensitive function of the opacity structure and thus also of the temperature of the flux tube interior.

Let us next compare the two iron lines 5247.06 and 5250.65 Å in Fig. 12.1. Although the latter line has a smaller Landé factor it has a larger Stokes V amplitude. This behavior cannot be explained in terms of Zeeman saturation, which is small for both lines, but is the result of different temperature sensitivities. The line weakening w is more pronounced for the 5247.06 Å line because of its lower excitation potential (0.09 eV) as compared with that (2.20 eV) of the 5250.65 Å line.

For two lines with differential line weakening factors w_1 and w_2 but with no differential line broadening (between flux tube interior and exterior — the widths of the two lines may be different from each other) we have in the absence of Zeeman saturation according to Eqs. (12.4) and (12.6)

$$\frac{V_1}{V_2} = \frac{w_1}{w_2}\,\frac{g_1}{g_2}\,\frac{\partial I_1}{\partial \lambda} \Big/ \frac{\partial I_2}{\partial \lambda}\,. \tag{12.7}$$

Comparison of the Stokes V amplitudes of lines of different temperature sensitivities thus provides information on the temperature difference between flux tube interior and exterior.

The reason why the line weakening is larger for lines with low excitation potential is due to the temperature dependence of the level populations. If $n_{0,s}$ is the population number density for level s of the neutral atom, and χ_i and $\chi_{0,s}$ are the ionization and excitation potentials, then in the limit of small relative temperature changes $\Delta T/T$ we obtain the relative population change by Taylor expansion of the Saha-Boltzmann law:

$$\frac{\Delta n_{0,s}}{n_{0,s}} \approx -\left(\frac{3}{2}+\frac{\chi_i-\chi_{0,s}}{kT}\right)\frac{\Delta T}{T}. \tag{12.8}$$

Here we have ignored the temperature dependence of the electron density n_e and made use of the fact that most metal atoms in the solar photosphere are singly ionized. Eq. (12.8) shows that lines with smaller excitation potentials $\chi_{0,s}$ of their lower levels are more sensitive in their level populations than lines with higher excitation potential. The temperature dependence outside sunspots is always such that the population decreases when the temperature increases. Note however that this situation can change in the cool sunspots, where neutrals become the dominant species.

Another temperature effect that has influence on the line weakening is the difference in the temperature gradient $\mathrm{d}T/\mathrm{d}\tau$ between the inside and the outside of the flux tube. As can be seen from Eq. (11.53) the line depth scales with $\mathrm{d}B_\nu/\mathrm{d}\tau$, where B_ν is the Planck function. Due to the general differential cooling of the lower layers and heating of the upper layers of magnetic elements, $\mathrm{d}T/\mathrm{d}\tau$ is shallower in flux tube interiors, which leads to weakened spectral lines.

The discussion so far can be summarized as follows. If we use the observed intensity profile for calibration to obtain an "observed" longitudinal field strength B_{obs}, assuming the weak-field limit and a homogeneous magnetic field, then we may write

$$B_{\mathrm{obs}} = w_B\, w_T\, \alpha_m\, B\,, \tag{12.9}$$

where w_B is the Zeeman saturation factor that is a function of field strength B and equals unity for $B=0$, while w_T is the temperature weakening factor (same as the w and $w_{1,2}$ factors used in Eqs. (12.6) and (12.7)), which is a function of ΔT (temperature difference between flux tube interior and exterior) and is unity for $\Delta T = 0$. In the weak-field limit in the absence of temperature effects B_{obs} would equal the *average* field strength $\langle B\rangle$ of Eq. (12.3) (net flux divided by the area of the resolution element). A magnetic line ratio would give us the value of w_B and thus also of B, a thermal line ratio would give us w_T and thus ΔT-information, while the Stokes V amplitude gives us B_{obs} and thus, in combination with the other parameters, the magnetic filling factor α_m. Note that the values of w_B and w_T depend on the wavelength within the line profile that is used for the determination of B_{obs}.

Another property that clearly stands out in Fig. 12.1 is the difference between the Stokes V amplitudes in the blue and red line wings. Both the amplitude and area of Stokes V in the blue line wing are systematically larger than their values in the red wing. This contradicts Eq. (12.1), since the area asymmetry of $\partial I_{m,0}/\partial\lambda$ is always zero. If it did not vanish, then the levels of the continuum intensity I_{m_c} on the blue and red sides of the line would be different. Such Stokes V asymmetries can only arise if there are correlations between magnetic and velocity *gradients* in the atmosphere. For a non-zero area asymmetry we need such correlated gradients *along* the line of sight. The asymmetries thus contain information on the *dynamics* of the spatially unresolved flux tubes. As we will see later, they even inform us about the dynamics of the immediate, non-magnetic surroundings of the flux tubes.

Additional dynamical information is obtained from the wavelength position of the Stokes V zero crossing with respect to the position of the Stokes I profile, as well as from the widths of the Stokes V profiles. While the zero crossing tells us about the systematic mass flows inside the flux tubes, the Stokes V line width tells us about the non-thermal Doppler broadening inside the flux elements. The directional properties of the motions may be diagnosed from the observed center-to-limb variations of these effects.

12.3. Evidence for the Validity of a Two-component Model

The interpretation of the observable effects that we have reviewed in the preceding section is enormously simplified by the possibility of applying a simple two-component approach with a magnetic filling factor α_m. It was one of the pleasant surprises in solar exploration that such an approach can be used at all. The prime evidence for the validity of the two-component approach has come from line-ratio observations with magnetographs being able to sample a large number of points on the sun. If one makes a scatter-plot of the fields $B_{\mathrm{obs},1}$ and $B_{\mathrm{obs},2}$ observed simultaneously in two spectral lines, then the points in the diagram would fall along a 45° line if $w_{B,1}\, w_{T,1}$ were equal to $w_{B,2}\, w_{T,2}$. Instead the points fall beside the 45° line because the various w factors deviate differently from unity.

The line ratio may according to Eq. (12.9) be written in the form

$$\frac{B_{\mathrm{obs},1}}{B_{\mathrm{obs},2}} = \frac{w_{B,1}\; w_{T,1}}{w_{B,2}\; w_{T,2}} = f(B, \Delta T)\,, \tag{12.10}$$

since the flux $\alpha_m B$ is the same for both spectral lines (which is true even when the lines are formed at different heights, if the flux within the spatial resolution element is conserved). Since w_B and w_T are functions of B and ΔT, the line ratio is a function f of these two parameters.

The line ratio is sampled at many different solar locations with different filling factors α_m (we recall that B_{obs} is proportional to α_m). If B and ΔT would vary from place to place on the sun, then we would observe a considerable spread in the line-ratio values. This is not the case, as indicated by the scatter plot in Fig. 12.2 of B_{obs} for the line pair at 5250.22 ($g = 3.0$) and 5247.06 Å ($g_{\mathrm{eff}} = 2.0$) that was discussed in the preceding section. These two lines have the same temperature sensitivities and are formed at nearly the same heights, so that $w_{T,1}/w_{T,2} = 1$. The line ratio $B_{\mathrm{obs},1}/B_{\mathrm{obs},2}$ is therefore exclusively a function of the field strength B. The points in the diagram fall along a well-defined regression line with a slope that is less than unity, as expected from differential Zeeman saturation. A field strength of 1 kG follows from the value of the slope (assuming a rectangular cross section, see below). The scatter around the regression line does not exceed the instrumental scatter, which indicates that the scatter of the intrinsic B values is small. There is a slight tendency for the value of the slope to decrease with increasing flux, but to a first approximation the regression relation is a straight line (as drawn in the figure).

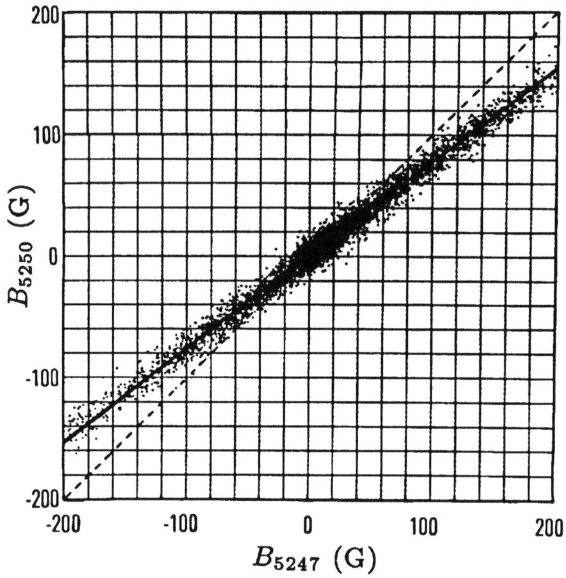

Fig. 12.2. Scatter plot of the apparent field strengths $B_{\rm obs}$ recorded simultaneously in the Fe I 5247.06 and 5250.22 Å lines, for the data used in Frazier and Stenflo (1978). The slope of the straight regression line corresponds to an intrinsic field strength of 1 kG at the level of line formation.

If there were a distribution of intrinsic field strengths between zero and kG values, then the region between the regression line and the 45° line would be populated with points, which is not the case. This has led to the conclusion that it is a good approximation to consider the flux tube properties as being "unique" (Frazier and Stenflo, 1972, 1978). It means that whatever solar region we look at, the difference between apparently strong and weak "observed" fields $B_{\rm obs}$ is not due to variations of the intrinsic properties of these flux regions, but is instead due to the filling factor or number density of flux elements.

To identify the true spread of the intrinsic properties we have to reduce the instrumental contribution to the observed spread. This has been done through Stokesmeter recordings with long integration times using the vertical grating spectrometer and the FTS at the McMath-Pierce facility of the National Solar Observatory. As a trade-off we are not able to sample so many points on the sun when pushing for extremely high polarimetric accuracy. Figure 12.3, from Stenflo and Harvey (1985), gives examples of the results obtained.

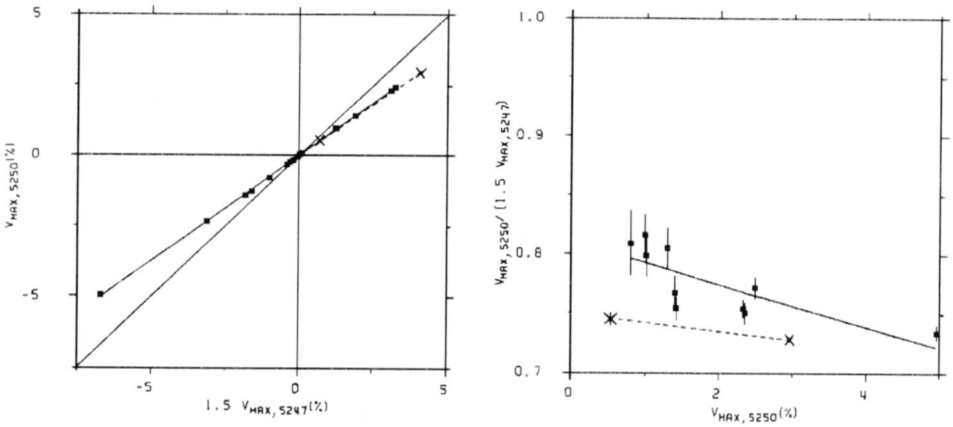

Fig. 12.3. The diagram to the left shows the polarization amplitude $V_{\max}$ in the Fe I 5250.22 Å line vs. 1.5 times the polarization amplitude in the Fe I 5247.06 Å line. The diagram to the right shows the ratio between these two quantities vs. the polarization amplitude of the 5250 Å line (expressed in units of the continuum intensity) for the points with a polarization amplitude in excess of 0.4 %. Filled squares and solid curve: Spectrometer data. Crosses and dashed line: FTS data. From Stenflo and Harvey (1985).

As seen in the left diagram, the spread of the points around the regression line is indeed very small. In the right diagram the Stokes V ratio for the points with a V/I_c amplitude in excess of 0.4 % has been plotted. The difference between the data from the grating spectrometer (filled symbols and solid line) and from the FTS (the two crosses and dashed line) is largely due to the difference in spectral resolution. The FTS data point with the larger V amplitude has been extracted from the recording that was shown in Fig. 12.1.

Figure 12.3 shows that there is a small trend for the field strength to increase with increasing filling factor (increasing deviation of the line ratio from unity with increasing Stokes V amplitude). The most striking feature of the diagram is however the complete absence of any points in the upper 2/3 of the diagram to the right. Although the statistical sample is limited, it suggests an absence of weak fields.

As we will see later, observations in the near infrared have confirmed and extended these conclusions. In the infrared the Zeeman splitting may be measured directly, which reduces the model dependence and increases the sensitivity. This has allowed field strengths for still smaller filling factors than those of Fig. 12.3 to be measured and has made more sophisticated modelling possible, including 3-component models. As we will see, the intrinsic spread of field strengths in the kG range is rather small, but weaker fields (below kG) become more common when we go to very small observed fluxes (or B_{obs} values).

12.4. Line Ratios

The observed polarization at any given wavelength in a Zeeman-sensitive spectral line depends on the detailed line-of-sight variation of the magnetic field, temperature, density, turbulent velocities, etc. The contributions from these various physical parameters are coupled to each other in a highly non-linear way. The basic philosophy of the line-ratio technique is to observe the polarization simultaneously in two spectral lines that respond as similarly as possible to all the various physical parameters except one. The ratio of the $B_{\rm obs}$ values in the two lines is then primarily a function of this parameter alone, which allows it to be untangled from the other parameters when interpreting the data.

A generalization of the line-ratio approach is to use a set of observables (which may represent a combination of amplitudes, asymmetries, widths, and shifts of Q, U, and V polarized line profiles) carefully selected such that each observable responds as differently as possible to the atmospheric parameters as compared with the other observables. We may then apply an inversion approach, as done by Keller et al. (1990), to determine from the set of observables the free physical parameters of the model flux tube, as outlined in Sect. 11.13. As however this is a complex and computer intensive undertaking, the much simpler line-ratio approach is usually the prefered method that has advanced our understanding of the nature of the small-scale magnetic field.

It is important that the line pair is chosen with great care so that a certain effect or parameter may be isolated as cleanly as possible. In the present section we will give examples of three types of line ratios, one that isolates the field strength (magnetic line ratio), one that is dominated by the temperature effects (thermal line ratio), and one (for the infrared) that isolates Zeeman line broadening from Doppler line broadening.

The "cleanest" line ratio that has been found so far is the already mentioned magnetic line ratio based on the line pair Fe I 5247.06 and 5250.22 Å (Stenflo, 1973). These lines, which belong to the same multiplet and have the same line strength, are almost identical in all respects except for the Landé factor and the Zeeman splitting pattern. A $B_{\rm obs,1}/B_{\rm obs,2}$ ratio different from unity can therefore only be produced by Zeeman saturation, which is a function of the intrinsic field strength alone.

In early work in the 1970's this line ratio was used as a $B_{\rm obs}$ ratio derived from scans of the solar disk with a Babcock-type magnetograph (cf. next chapter), whereby the Stokes V signal was recorded within fixed spectral windows in the two spectral lines. With the advent of the FTS polarimeter it became possible to determine the $V_1(\Delta\lambda)/V_2(\Delta\lambda)$ ratio as a function of wavelength position $\Delta\lambda$ with respect to the centers of lines 1 and 2. Such "ratio profiles" provide a verification of the consistency of the interpretation in terms of Zeeman saturation and also introduce additional observational constraints on the field distribution. An example is given in Fig. 12.4, from Stenflo and Harvey (1985), showing in the left diagram Stokes V vs. $-\Delta\lambda$ for the blue wings of the 5250.22 and 5247.06 Å lines. The V signal of the latter line has been multiplied by 1.5 to scale it to the 1.5 times larger Landé factor of the other line. Two cases are shown, labeled strong and

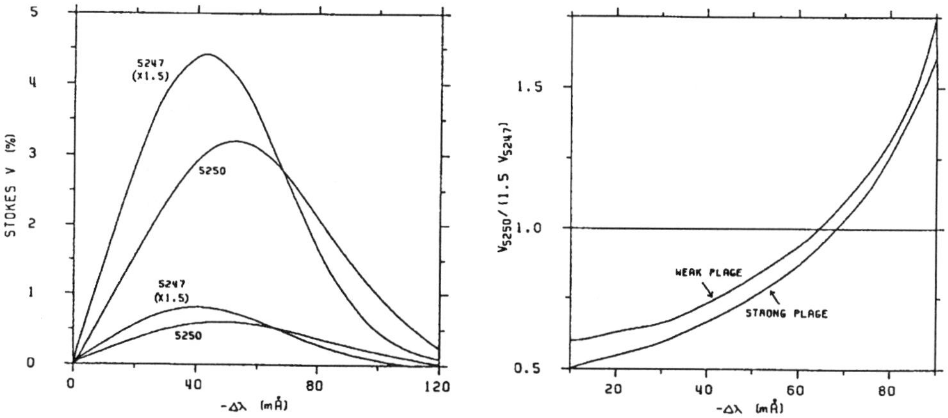

Fig. 12.4. The diagram to the left shows the blue wings of the Stokes V profiles of the Fe I 5250.22 and 5247.06 Å lines for a strong plage (upper pair of curves) and a weak plage (lower pair of curves). The ratio between the respective pair of curves is displayed in the right diagram. From Stenflo and Harvey (1985).

weak plage, which differ in V amplitude by a factor of six. They correspond to the two FTS points in Fig. 12.3. The strong plage spectrum was shown in Fig. 12.1.

There are two main effects caused by the higher-order non-linear terms in the Taylor expansion (12.1): Suppression of the Stokes V amplitude (which we have called Zeeman saturation), and broadening of the V profile. Due to the broadening effect the V ratio becomes larger than unity in the far line wings, but as the signal gets small in that portion of the line, the ratio between the integrated V profiles is always less than unity. It is particularly remarkable that the line-ratio curves (right diagram) are so similar for the two cases that differ in amplitude by as much as a factor of six. This demonstrates that the B-dependence on filling factor is quite minor.

To convert the line ratio to field strength B we have to solve the radiative-transfer problem for V in the two lines. To reproduce the full line-ratio profile one needs a realistic, numerical model atmosphere. An example of such a calculation is given in Fig. 12.5, from Emonet (1992), showing a nearly perfect fit between the synthetic and observed line ratio profiles in the case of the strong plage data of Figs. 12.1 and 12.4. The model uses a magnetic field that varies linearly with $\log \tau_{5000}$, where τ_{5000} is the continuum optical depth at 5000 Å *inside* the flux tube. The data are reproduced for a model that has $B = 1432$ G at $\log \tau_{5000} = -1.46$, a typical height of formation for the wings of the 5250.22 and 5247.06 Å lines. The field decreases with height according to $\mathrm{d}B/\mathrm{d}(\log \tau_{5000}) = 273$ G in this model.

In the discussion so far we have implicitly assumed that the magnetic cross section is rectangular, i.e., that the field is single-valued at each given height. If we have a distribution of field strengths within the resolution element, then we have to spatially integrate the Stokes V signals to obtain the V line ratio. The peak field strength of the distribution then has to be higher than the field strength of the single-valued case to fit a given observed line ratio. Thus, when the magnetic

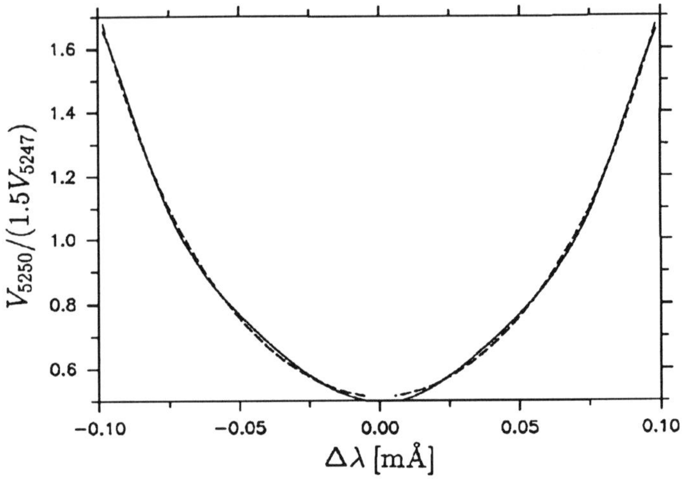

Fig. 12.5. Comparison between observed (solid) and theoretical (dashed) profiles of the magnetic line ratio for the strong plage of Figs. 12.1 and 12.4. The theoretical model has $B = 1432$ G at $\log\tau_{5000} = -1.46$ and a gradient of $dB/d(\log\tau_{5000}) = 273$ G, where τ_{5000} represents the continuum optical depth inside the flux tube. From Emonet (1992).

line ratio was first applied to the quiet sun (Stenflo, 1973), the field strength was found to be 1.1 kG if a rectangular magnetic cross section was assumed, while for a Gaussian cross section the field amplitude had to be 2.3 kG to fit the same line ratio. We will later see that infrared data favor a rectangular cross section.

A "thermal" line ratio cannot be as "clean" as the magnetic line ratio, since the lines need to come from different multiplets to have significantly different excitation potentials, and when the thermal response is different, the heights of formation will also be different. Since an ideal thermal line pair thus cannot be found, one may relax the requirement that the Landé factors and line strengths should be equal. The main requirement should simply be that the line ratio should be dominated by thermal effects (cf. Landi Degl'Innocenti and Landolfi, 1982).

In the discussion of Fig. 12.1 we noticed that the Fe I 5250.65 Å line was much less weakened by temperature effects than the other lines, due to its higher excitation potential. We may therefore form a "thermal line ratio" with this line, but we should not use the neighboring 5250.22 Å line as the other partner of the pair, since it is affected by Zeeman saturation. Figure 12.6 gives an example of the observed center-to-limb variation of the Stokes V amplitude (V_{max}) ratios for the line pair Fe I 5247.06 and 5250.65 Å (filled symbols and thick curve) and the pair Cr I 5247.57 and Fe I 5250.65 Å (crosses and thin curve). μ is the cosine of the heliocentric angle (the angle between the line of sight and the vertical direction).

As the value of the Zeeman saturation factor w_B (cf. Eqs. (12.9) and (12.10)) is almost unity for all of these three lines, the line ratios plotted represent primarily $w_{T,1}/w_{T,2}$ (although the differential effect of the line profile *shapes* may also play a role). The line ratios are considerably smaller than unity as expected from the smaller temperature weakening of the 5250.65 Å line. The conversion of such

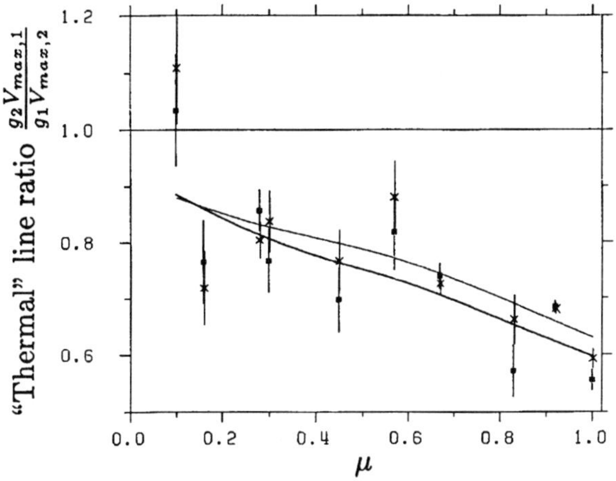

Fig. 12.6. Center-to-limb variation of the "thermal" line ratio $g_2 V_{\rm max,1}/(g_1 V_{\rm max,2})$. The filled squares and thick curve represent the line pair Fe I 5247.06 (line 1) and 5250.65 Å (line 2), the crosses and thin curve the line pair Cr I 5247.57 and Fe I 5250.65 Å. The curves are cubic spline fits to the points. From Stenflo et al. (1987a).

line-ratio data to information on the height variation of ΔT however requires careful numerical radiative-transfer calculations with realistic flux tube models, for instance as done by Zayer et al. (1990).

Since the Zeeman splitting varies with λ^2 according to Eq. (12.2), the infrared region of the spectrum contains lines for which the Zeeman splitting in the kG flux tubes is complete, something that does not occur for any line in the visible part of the spectrum. For such lines the field strength can be measured directly from the wavelength separation of the sigma components, instead of, as in the visible, having to translate polarization ratios to field strength. Still, even in the complete splitting case, a line ratio is of great advantage, since it allows us to untangle the *distribution* of field strengths from the other atmospheric parameters.

Field strength distributions cause significant broadening of the Stokes V profiles when the Zeeman splitting is large. Such Zeeman line broadening, which is correlated with the Landé factors of the lines, is insignificant for lines in the visible part of the spectrum (cf. our search for a turbulent magnetic field in Sect. 10.14). A field-strength distribution will always be present even when the magnetic cross section of an element is single-valued, since the field strength always varies along the line of sight within the line-forming layers due to the height divergence of the field lines. To separate the contributions from the Zeeman broadening and the non-thermal Doppler broadening we need to use a combination of two similar lines with different Landé factors (Zayer et al., 1989).

An example of such an infrared line pair is shown in Fig. 12.7, from Rüedi et al. (1992). The two lines are Fe I 1.5649 μm (the "Zeeman line") with a Landé factor g of 3.0, and Fe I 1.5653 μm with $g = 1.53$. While the solid curves represent the observed V/I_c profiles, the dashed curves have been derived from the two-component model that fits the observations. In the upper diagram the field strength at $z = 0$ (where the continuum optical depth in the external atmosphere $\tau_{5000} = 1$) is 1.52 kG, and the Zeeman splitting is complete for both lines (with the consequence that the Stokes V amplitudes of the two lines are of similar magnitude). The $g = 3.0$ line is however significantly broader than the $g = 1.53$ line. In the lower diagram the field strength is only half as large (0.75 kG), leading to much less Zeeman broadening and a relative suppression of the Stokes V amplitude of the $g = 1.53$ line, since for these weaker fields the Zeeman splitting is not complete for this line.

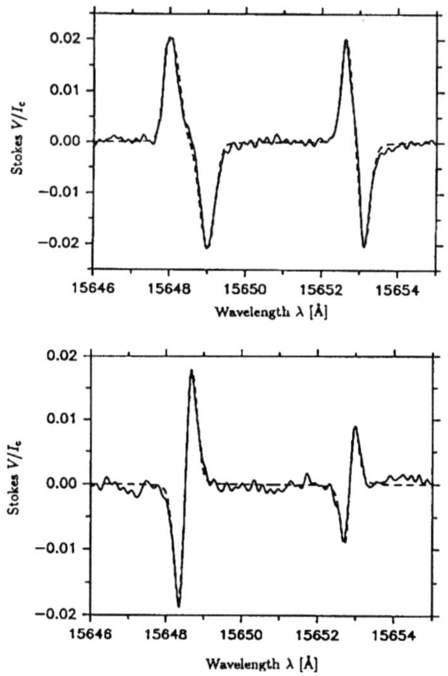

Fig. 12.7. Observed (solid) and synthetic (dashed) Stokes V profiles of an infrared line pair. In the upper diagram the field strength at geometrical height $z = 0$ is 1.52 kG, in the lower diagram it is 0.75 kG. Such a line pair allows the Zeeman line broadening to be separated from the Doppler broadening. From Rüedi et al. (1992).

The lower diagram of Fig. 12.7 in fact represents the first identification of intrinsically weak (below kG) fields in the lower photosphere. The infrared region thus has a great deal of diagnostic potential, but to properly exploit it for Stokes inversion and flux tube model building we need to be able to sufficiently constrain the free physical parameters like the macroturbulent velocities and the height gradients from the observations. This is only feasible by using line combinations like the example of Fig. 12.7.

12.5. Fraction of Magnetic Flux in kG Form

The first determinations of the kG field strengths of the spatially unresolved flux elements were done with the line-ratio method via regression-line analysis of scatter-plot diagrams like that of Fig. 12.2. The observable that contains the information on the intrinsic field strength is the value of the *slope* of the regression line. Using Stokesmeter recordings of polarized line profiles with very high signal-to-noise ratio one can abandon the statistical approach and apply the line ratio to each individual point, as was shown in Fig. 12.3. The results of the statistical approach are thereby confirmed. Still, the observational uncertainties in the line ratio become excessive for small fluxes or filling factors, corresponding to the points near origo in the scatter-plot diagram. This necessitated the introduction of a polarization cut-off of 0.4 % in the right panel of Fig. 12.3, below which the line ratio could not be reliably measured.

In the regression analysis it is the points with the largest filling factors that carry the most weight when the slope is determined, so neither the regression analysis of Fig. 12.2 nor the individual line ratios of Fig. 12.3 provide information on whether the smallest observed fluxes are due to flux elements with intrinsically strong or weak fields.

Still it is these small flux values that are the most numerous when we make a raster scan across an arbitrary area of the sun, as shown by the heavy clustering of the measured points around origo in Fig. 12.2. Since these small fluxes are representative of most of the area of the sun, their accumulated flux may amount to a substantial fraction of the total flux through the solar surface. On the other hand we know from the application of the line ratio to the larger fluxes that at least some fraction of the total flux through the solar surface, both in active and quiet regions, has its sources in kG flux elements. The question that we now want to address is how large this fraction really is.

Let Φ_i be the true vertical magnetic flux through the sampling area A (the resolution element) at position i on the solar disk. The absolute value of the total flux through the scanned region is then

$$\Phi_{\rm tot} = \sum_i |\Phi_i| \,. \tag{12.11}$$

Let us further assume that there are two types of flux within each resolution element: strong-field flux $\Phi_{s,i}$ and weak-field flux $\Phi_{w,i}$. In the strong-field elements the contribution to the magnetograph signal is according to Eq. (12.9) reduced by the factor

$$\delta = w_B \, w_T \tag{12.12}$$

due to Zeeman saturation and temperature line weakening. In the weak-field flux elements, on the other hand, we assume that these effects are absent (which must certainly be true for Zeeman saturation at least), which means that $\delta = 1$ there. Then, according to Eqs. (12.3), (12.9), and (12.12), the observed apparent field at solar position i is

$$B_{{\rm obs}_i} = (\delta \, \Phi_{s,i} + \Phi_{w,i})/A \,. \tag{12.13}$$

The fraction R_s of the total flux that occurs in strong-field form is defined by

$$R_s = \sum_i |\Phi_{s,i}|/\Phi_{\rm tot}\,. \tag{12.14}$$

We now want to determine the value of R_s from observational data. This was done by Howard and Stenflo (1972) and Frazier and Stenflo (1972) through the use of the line pair Fe I 5250.22 (line 1) and Fe I 5232.95 Å (line 2). While line 1, as we have seen, has a small value of δ due to both Zeeman saturation and temperature line weakening, these effects are rather insignificant for line 2, which is much broader, has a smaller Landé factor (1.3), and is much less affected by temperature effects in the line wings. We may therefore assume that $\delta \approx 1$ for line 2.

To deduce the value of R_s we need another observable (besides δ, which is determined from the slope of the regression line) that is sensitive to the effect of the weak-field fluxes Φ_w. Such an observable can be obtained from the data in the scatter-plot diagram by forming the ratio ρ between the total apparent fluxes in the two lines:

$$\rho = \sum_i |B_{\rm obs,1}|_i / \sum_i |B_{\rm obs,2}|_i\,, \tag{12.15}$$

or, using Eq. (12.13),

$$\rho = \sum_i |\,\delta\,\Phi_{s,i} + \Phi_{w,i}\,|/\Phi_{\rm tot}\,, \tag{12.16}$$

since according to Eqs. (12.11) and (12.13)

$$\Phi_{\rm tot} = \sum_i |\,\Phi_{s,i} + \Phi_{w,i}\,|\,. \tag{12.17}$$

Let us now make the reasonable assumption that as soon as there are any strong-field contributions at all for a given solar position i (i.e., when $\Phi_{s,i} \neq 0$), then that contribution dominates over the weak-field one within the same resolution element, which implies that $|\Phi_{w,i}| < \delta\,|\Phi_{s,i}|$. This assumption allows us to write

$$|\,\delta\,\Phi_{s,i} + \Phi_{w,i}\,| = \delta\,|\Phi_{s,i}| + \epsilon_i\,|\Phi_{w,i}|\,, \tag{12.18}$$

where $\epsilon_i = +1$ or -1, depending on whether $\Phi_{s,i}$ and $\Phi_{w,i}$ are of equal or opposite signs. For the special case that $\Phi_{s,i} = 0$, Eq. (12.18) is always valid with $\epsilon_i = +1$.

Using Eqs. (12.14), (12.16), and (12.18),

$$\rho = \delta\,R_s + \sum_i \epsilon_i\,|\Phi_{w,i}|/\Phi_{\rm tot}\,. \tag{12.19}$$

If δ were equal to unity, then ρ would also equal unity according to Eqs. (12.16) and (12.17). Eq. (12.19) then becomes

$$1 = R_s + \sum_i \epsilon_i\,|\Phi_{w,i}|/\Phi_{\rm tot}\,. \tag{12.20}$$

Subtraction of Eq. (12.19) from Eq. (12.20) finally gives

$$R_s = \frac{1-\rho}{1-\delta} \, . \tag{12.21}$$

The determinations of ρ and δ from the Fe I 5250.22 and 5232.95 Å line pair by Howard and Stenflo (1972) and Frazier and Stenflo (1972) resulted in a value for R_s that was ≈ 1 within the uncertainties of the observational data. These uncertainties could be translated to a 10 % uncertainty in R_s. The conclusion then follows that at least 90 % of the total magnetic flux that is recorded by solar magnetographs is in strong, kG form. This figure of 90 % has not yet needed revision, but new diagnostic techniques in the near infrared have now provided us with the tools to make more accurate determinations not only of R_s, but of the full distribution of field strengths in different regions on the sun.

In our statement about the 90 % limit, the qualification "recorded by solar magnetographs" is important. It means that when the true field distribution across the solar disk is smeared by a spatial window that has the same size as the effective spatial resolution element of the magnetograph observations (which is about $4'' \times 4''$ in the case of the observations of Frazier and Stenflo (1972) if we include the seeing effects in addition to the $2.4'' \times 2.4''$ sampling aperture), then at least 90 % of that flux is due to kG fields. Note that this is a resolution-dependent statement. Thus the properties of a hypothetical small-scale turbulent magnetic field of the kind discussed in Chapter 10 are *not* constrained. For such a field the magnetic polarities are mixed on a small scale so that the flux contributions cancel out when averaged with a spatial window of a few arcsec. The turbulent field is therefore not "seen" by the magnetograph and is accordingly unconstrained by the method discussed in the present section, although it may carry total fluxes much in excess of the 10 % weak-field flux limit. For a detailed discussion of how to diagnose this turbulent field we refer to Chapter 10.

12.6. Integrated Stokes *V* Profiles

In connection with the Taylor expansion (12.1) we noted that the Stokes I profile I_m of the flux tube is not directly an observable, since the actually observed Stokes I profile has its main contributions from the surroundings. In the case of weak magnetic fields, however, when the higher-order terms in the expansion can be neglected, I_m may be retrieved directly through integration, since $I_m \sim \int V \, \mathrm{d}\lambda$, although the scale factor $\alpha_m \Delta\lambda_H$ remains an unknown.

For the purpose of a statistical analysis of the Stokes V spectrum it is therefore useful to *define* the integrated Stokes V profile I_V by

$$\frac{I_{V_c} - I_V}{I_c} = -\frac{1}{\alpha_m \Delta\lambda_H} \int_{\lambda_1}^{\lambda} \frac{V(\lambda')}{I_c} \, \mathrm{d}\lambda' \, , \tag{12.22}$$

where I_{V_c} is the continuum intensity in the magnetic element on the blue side of the line (see below), and I_c is as before (cf. Eq. (12.6)) the continuum intensity averaged over the resolution element. The lower integration limit should be chosen to be sufficiently far to the blue side of the line so that there is no significant Stokes V signal for $\lambda < \lambda_1$ (cf. Solanki and Stenflo, 1984).

Under the assumptions used for Eq. (12.1) (height-independent field) I_V is clearly identical to I_m in the weak-field limit, and this limit represents a very good approximation for almost all visible spectral lines except narrow lines with a large Landé factor, like the Fe I 5250.22 Å line. It is nevertheless important to make a distinction between I_m and I_V, since real, observed Stokes V profiles are asymmetric due to correlated gradients of the magnetic and velocity fields, which is not accounted for in Eq. (12.1).

The situation is illustrated in Fig. 12.8, from Solanki and Stenflo (1985). The diagram to the left shows an observed Stokes V profile with definitions of the amplitudes $a_{b,r}$ and areas $A_{b,r}$ of the blue and red line wings. Not only is there an amplitude asymmetry ($a_b \neq a_r$), but an area asymmetry ($A_b \neq A_r$) as well. An area asymmetry of the observed magnitude can only arise if there are correlated magnetic and velocity field gradients *along the line of sight* (cf. Auer and Heasley, 1978). Since $A_b - A_r = \int V \, d\lambda \neq 0$, the continuum levels of the integrated V profile I_V will be different on the blue and red sides of the line, as indicated by the middle diagram of Fig. 12.8. Thus, if as defined by Eq. (12.22), $I_V = I_{m_c}$ on the blue side, then $I_V < I_{m_c}$ on the red side. Such an I_V profile cannot represent a Stokes I_m profile of a real magnetic element.

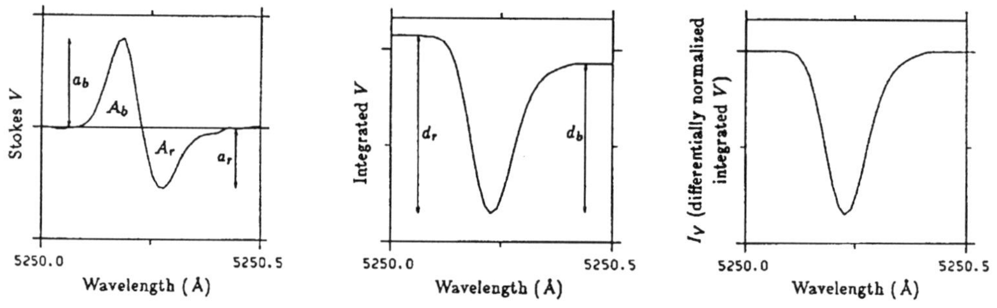

Fig. 12.8. Illustration of the conversion of a Stokes V profile to an I_V profile. Left panel: Observed Stokes V profile of the Fe I 5250.22 Å line. Middle panel: Relative line depth of the I_V profile derived from Stokes V according to Eq. (12.22). Right panel: The I_V profile after differential renormalization, to make the continuum levels on the red and blue sides of the line equal. From Solanki and Stenflo (1985).

For a statistical regression analysis of the I_V profiles on an equal basis with the ordinary Stokes I profiles, I_V needs to be differentially renormalized to force the continuum levels on both sides of the line to be equal. This is done through multiplication of the blue Stokes V wing by $\sqrt{A_r/A_b}$, the red wing by $\sqrt{A_b/A_r}$, before V is inserted for integration in Eq. (12.22). With this procedure the I_V profile emerges with a single-level continuum, as illustrated by the diagram to the right in Fig. 12.8. These differentially renormalized I_V profiles lend themselves to the same parametrization as was used for the Stokes I profiles in Fig. 10.6.

12.7. Regression Analysis of the Stokes V Profiles

We now parametrize the differentially renormalized I_V profiles and the simultaneously recorded Stokes I profiles, using for both the parametrization described by Fig. 10.6. The main parameters are the ones described in Sect. 10.14: the relative line depth d, the area S of the profile below the half-level chord (the shaded area of Fig. 10.6), and the width v_D of the half-level chord expressed in velocity units as the Doppler width of a Gaussian with the same half width. The parameters that have been extracted from the I_V profiles are marked by index V, those extracted from the I profiles by index I. Additional parameters that will be used are the central wavelengths λ_I and λ_V, and for the Stokes V profiles the amplitude and area asymmetries $\Delta a = a_b - a_r$ and $\Delta A = A_b - A_r$. λ_V is the wavelength of the zero crossing of the Stokes V profile. The corresponding quantity for the Stokes I profile would be the wavelength of its minimum, but in practice we use for λ_I the center of gravity of the lower half of the Stokes I profile, since this quantity is less affected by noise.

The power of this type of statistical approach does not lie in the analysis of the Stokes V parameters d_V, v_{D_V}, etc., by themselves, but in the analysis of the *difference* between these Stokes V parameters and the corresponding Stokes I parameters. While it is hard to achieve high accuracy in the absolute values of the derived physical parameters of the atmosphere, the difference between the interior and exterior of the flux tube can be obtained with a high degree of confidence by employing such a *differential* approach.

For the differential line width $v_{D_V} - v_{D_I}$ we may use a regression equation of the same form as the one of Eq. (10.65) for v_{D_I} alone, i.e.,

$$v_{D_V} - v_{D_I} = x_1 + x_2 S_I + x_3 S_I^2 + x_4\, v_0\, \chi_e + x_5\, g_{\rm eff}^2 \lambda^2 / v_0\,, \tag{12.23}$$

where v_0 is a second-order polynomial fit of v_{D_I} to S_I. In comparison with Eq. (10.65) we have omitted the x_6 term, which was only essential when searching for the minute effects of a turbulent magnetic field.

Like the turbulent case the x_5 coefficient may be translated to a field strength, but in the present case the derivation is different, since we assume a longitudinal magnetic field (for observations at the center of the solar disk) and a two-component model with filling factor α_m. We further assume that the line profile is Gaussian and that the Zeeman splitting is small in comparison with the line width. Let us denote by w_c the relative change I_{m_c}/I_{nm_c} of the continuum level in the flux tubes, and let as before w_T represent the thermal line weakening (cf. Eq. (12.9)), assuming that only the line depth but not the line width is affected by the differential thermodynamics. With the numerical constant z given by Eq. (10.62) we can then from Eqs. (12.1) and (12.22) determine the field strength that corresponds to the x_5 coefficient:

$$B_{v_D} = \frac{\sqrt{3x_5}}{zc}\,(1 - 3w_c\, w_T\, \alpha_m\,)^{-1/2}\,. \tag{12.24}$$

Here we have used index v_D for the field strength B to indicate that it has been derived from the differential v_D parameter. The source of the first term (unity) in the square root is the I_V profile, while the second term with the w factors is due to the field-dependent contribution to Stokes I in Eq. (12.1).

For the differential line-depth effects (d_V/d_I) a similar but slightly modified regression equation has been found to work well:

$$\ln(d_V/d_I) = x_1 + x_2 S_I + x_3 S_I^2 + x_4 \chi_e + x_5 \chi_e S_I + x_6 \, g_{\text{eff}}^2 \lambda^2 / v_0^2 \, . \tag{12.25}$$

An interpretation of the magnetic-field dependent coefficient x_6 with the same assumptions as were used for Eq. (12.24) results in the same expression for the corresponding field strength B_d derived from the differential line depth, if we simply replace x_5 in Eq. (12.24) by x_6.

Regression analysis with Eqs. (12.23) and (12.25) has been applied to FTS polarimetric observations of the 402 Fe I lines that were used by Stenflo and Lindegren (1977) for the turbulent magnetic field. The derived field strengths fall in the range 1.4–1.7 kG for both B_{v_D} and B_d (Solanki and Stenflo, 1984), which is consistent with the results of the line-ratio method when one accounts for the circumstance that field strengths of the regression analysis refer to somewhat larger depths in the atmosphere than the line-ratio method. The reason for this is that by accounting for the S_I dependence in the regression equation one effectively reduces the differential line widths or depths to the case of $S_I = 0$, i.e., to the case of the weakest lines, which are the ones that are formed the deepest, where the field strengths are the largest. In contrast the values derived from the line ratio refer to the height of formation of the wings of medium-strong lines.

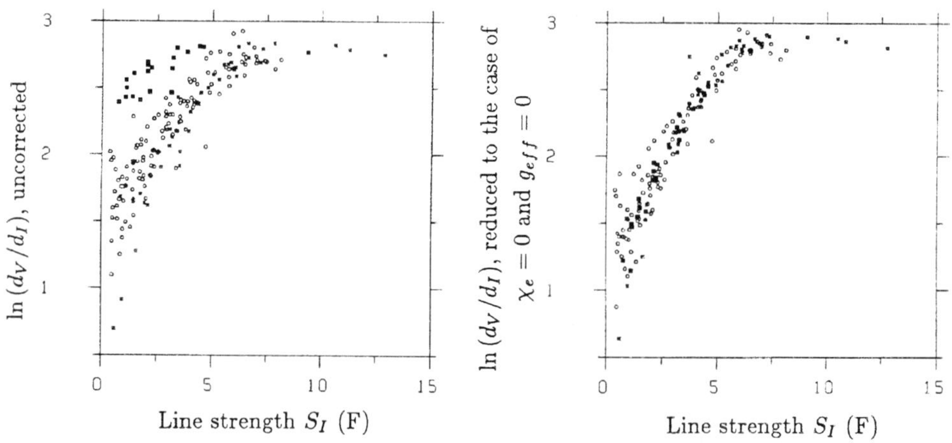

Fig. 12.9. $\ln(d_V/d_I)$ vs. S_I observed in an enhanced network region, for Fe I lines with $\chi_e < 3$ eV (stars), $\chi_e \geq 3$ eV (open circles), and Fe II lines (filled squares). The left panel shows the original data, the right panel the data after they have been reduced to the case of $\chi^\star = 0$ and $g_{\text{eff}} = 0$ via a regression analysis. From Solanki (1987).

Figure 12.9 gives an example of a scatter plot of $\ln(d_V/d_I)$ vs. S_I for a network region at the center of the quiet solar disk. The diagram to the left shows a plot of the original data, where Fe I lines with $\chi_e < 3$ eV are represented by stars, those with $\chi_e \geq 3\,\mathrm{eV}$ by open circles, Fe II lines by filled squares. We may generalize Eq. (12.25) to include the Fe II lines by replacing χ_e by $\chi^\star = \chi_e + \chi_i$, where χ_i is the ionization potential for neutral iron (and is set = 0 for the Fe I lines, 7.87 eV for the Fe II lines). The $\chi^\star$-dependent terms in the regression equations then have to be slightly modified as described in Solanki and Stenflo (1985). If we make the regression fit and then subtract from the $\ln(d_V/d_I)$ values the terms that depend on $\chi^\star$ and $g_{\rm eff}$, then we obtain the diagram to the right of Fig. 12.9, which represents $\ln(d_V/d_I)$ for the case of zero $\chi^\star$ and Landé factor. The greatly reduced scatter in the right diagram is evidence that the regression model used is good.

Since the magnetic filling factor is unknown, the apparent values of d_V/d_I directly evaluated from the observed Stokes I and V spectra deviate from the "true" values by a certain factor. In terms of $\ln(d_V/d_I)$ the filling factor represents an offset, which however is common to all the points of Fig. 12.9 (since all the spectral lines have been recorded strictly simultaneously with the FTS). It is the *shape* of the dependence on S_I and $\chi^\star$ that contains the information on the temperature structure, regardless of where the zero point of the $\ln(d_V/d_I)$ diagram is. The zero point in Fig. 12.9 has been chosen such that it corresponds to $\alpha_m B = 1\,\mathrm{G}$ in Eq. (12.22). By choosing this small value for the average field strength, the values of $\ln(d_V/d_I)$ become correspondingly large.

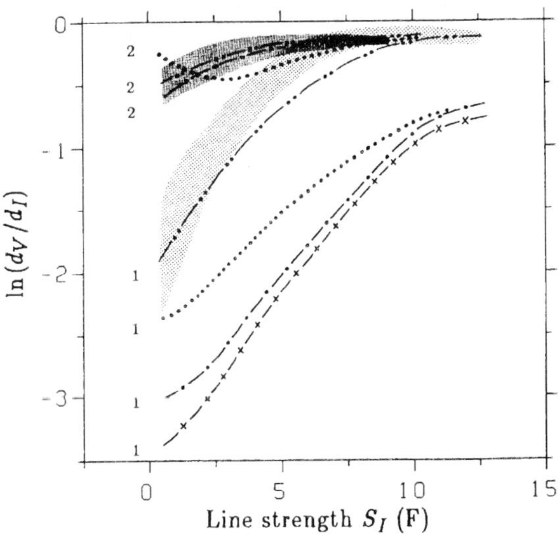

Fig. 12.10. $\ln(d_V/d_I)$ vs. S_I for the same observational data as in the left panel of Fig. 12.9 (light shading for the Fe I lines, dark for the Fe II lines) and for a number of previously published facular models (the various curves, labeled '1' for Fe I, '2' for Fe II). From Solanki (1984).

When we compare with model calculations, which are done for a filling factor of unity, we may shift the zero point of the diagrams such that the results for the Fe II lines, which have little temperature sensitivity, overlap. Figure 12.10, from Solanki (1984), shows such a comparison, using the same observational data as in the left diagram of Fig. 12.9. The region where the Fe II data fall is marked by dark shading, that of the Fe I data by light shading. The various plotted curves represent predictions by a number of previously published facular models (which are now outdated). The label '1' to the left of the curves refer to the predictions for Fe I, the label '2' to the predictions for Fe II. It is obvious from Fig. 12.10 that many of the previous models can immediately be ruled out by such a diagnostic diagram. The d_V/d_I ratio accordingly places quite stringent constraints on the allowed temperature structure of the flux tubes. These constraints have served as a starting point for a flurry of new empirical models of flux tube thermodynamics (reviewed by Solanki, 1993). This model building activity has later been augmented by the introduction of Stokes inversion methods, which may use the d_V/d_I models as an "initial guess" in a perturbation approach to solve the non-linear inversion problem.

12.8. Diagnostics of Flux Tube Dynamics

Since the dynamical state of the flux tube interior greatly differs from that of the external atmosphere, we have to use Stokes V data to isolate the flux tube contribution from its surroundings and apply differential methods as we did in the preceding section for the temperature diagnostics. Only then can the difference between the interior and exterior be accurately determined.

The main diagnostic parameters for the dynamics are: (1) The Stokes V zero crossing wavelength λ_V, which contains information on the systematic mass flows inside the flux tubes. (2) The Stokes V amplitude and area asymmetries, which provide information on the correlations between the magnetic and velocity field gradients. (3) The line width difference $v_{D_V} - v_{D_I}$ between the I_V and I line profiles, which gives information on the difference of the non-thermal line broadening between flux tube interior and exterior. (4) The center-to-limb variation of the above parameters, which tells us about the directional properties of the motions and gives some height information as well.

In Fig. 12.11, from Stenflo et al. (1987a), we illustrate the center-to-limb variation of the Stokes V zero crossing wavelength λ_V relative to the position λ_I of the Stokes I profile, expressed in velocity units by plotting $c(\lambda_V - \lambda_I)/\lambda_I$. The observational data, obtained with the FTS polarimeter, are given for the Fe I 5250.22 (filled squares) and 5247.06 Å (crosses) lines. It is important to note that λ_I does not constitute an absolute wavelength reference, since the average Stokes I profiles are Doppler shifted due to the velocity-brightness correlations in the solar granulation. Balthasar (1984) has determined the *absolute* value of λ_I for the undisturbed, non-magnetic atmosphere with respect to laboratory wavelengths (after subtraction of the gravitational redshift and the relative motion between the observer and the solar region observed). The solid line represents the negative value of the λ_I determined this way. This curve can accordingly be regarded as the true zero level to which $\lambda_V - \lambda_I$ should be referred. It is well supported by absolute wavelength calibrations of Stokes V profiles near disk center (Solanki, 1986).

If we disregard the point at $\mu = 0.1$, which is very close to the solar limb, the observational points hardly deviate from the solid curve by more than their error bars. This sets an upper limit of about 0.2 km s^{-1} to the possible net

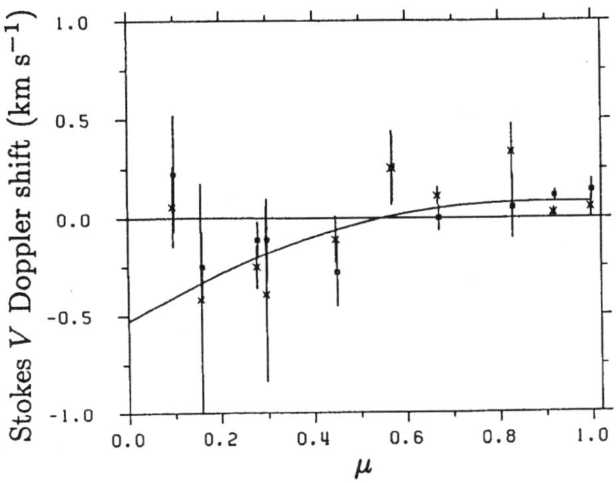

Fig. 12.11. Center-to-limb variation of the position of the Stokes V zero crossing wavelength with respect to the position of the Stokes I profile, $c(\lambda_V - \lambda_I)/\lambda_I$, in velocity units. The filled squares represent the Fe I 5250.22 Å, the crosses the Fe I 5247.06 Å line. The solid curve represents $-c(\lambda_I - \lambda_0)/\lambda_I$, where λ_0 is an absolute line center wavelength for the respective spectral line (the laboratory wavelength, adjusted to include the effects of the gravitational redshift and the motion of the observer). This curve serves as the corrected zero level for the Stokes V Doppler shifts. From Stenflo et al. (1987a).

mass flows inside the flux tube, which is a strong constraint in view of the much larger average downdraft velocities, typically 0.5 km s^{-1} or more, observed at the location of flux concentrations (Simon and Leighton, 1964; Tanenbaum et al., 1969; Frazier, 1970), and the very large velocities of several km s^{-1} needed to explain the Stokes V asymmetries and line widths (see below). This result of essentially zero downdrafts inside the flux tubes is gratifying from a theoretical point of view, since even a small downdraft velocity in the photosphere would as a consequence of mass conservation lead to drainage of the thin corona in a matter of minutes (because diffusion across the field lines from the external atmosphere in the denser layers is too slow to be able to replenish the mass loss).

The left diagram of Fig. 12.12, from Solanki and Stenflo (1985), shows the variation of the absolute amplitude asymmetry $a_b - a_r$ with line strength S_I for the same set of 400 Fe I and 50 Fe II lines that were used for the temperature and magnetic field diagnostics in the preceding section. The portions of the diagram where the lines of different excitation potential or ionization stage fall are marked with different shadings. To theoretically reproduce such Stokes V asymmetries large mass motions with correlated magnetic and velocity field gradients are required. This general interpretation in terms of mass motions is strongly supported by the diagram to the right in Fig. 12.12, from Solanki (1985), which gives the magnitude of the macroturbulent velocity $\xi_{\rm mac}$ that is needed to reproduce the observed widths of the I_V profiles and the $v_{D_V} - v_{D_I}$ data. The qualitative similarity between the asymmetry $a_b - a_r$ and the macroturbulent broadening parameter $\xi_{\rm mac}$ in their dependence on S_I and $\chi^\star$ is evidence that mass motions are a common source of both diagrams.

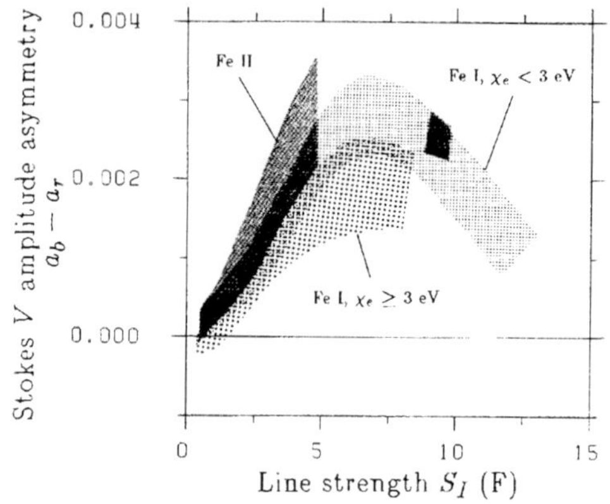

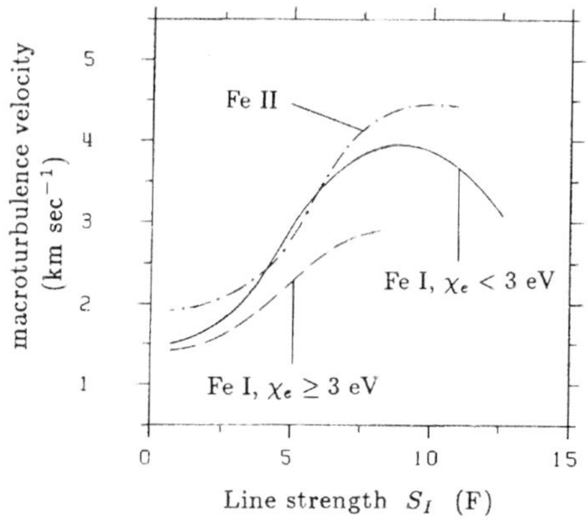

Fig. 12.12. Stokes V amplitude asymmetry $a_b - a_r$ (left diagram, from Solanki and Stenflo, 1985) and macroturbulent velocity $\xi_{\rm mac}$ (right diagram, from Solanki, 1985) as functions of line strength S_I. The amplitudes $a_{b,r}$ are given in units of the continuum intensity.

The high values of several km s^{-1} for $\xi_{\rm mac}$ show that we are dealing with quite large non-thermal velocities inside the flux tubes. Still the Stokes V zero-crossing data of Fig. 12.11 tell us that these large-amplitude motions do not carry any significant *net* mass flow. One type of motion that could reconcile these apparently conflicting requirements is flux tube oscillations, which lead to Doppler line broadening when the FTS polarimeter performs temporal and spatial averaging over flux tubes in different phases of their oscillations (Solanki, 1989; Grossmann-Doerth et al., 1991).

The *area* asymmetries $A_b - A_r$ respond very differently to the flux tube dynamics as compared with the amplitude asymmetries $a_b - a_r$. While the amplitude asymmetry can be due to correlated magnetic and velocity field gradients both parallel and perpendicular to the line of sight in the layer where the Stokes V maxima are formed, the observed area asymmetry can only be caused by correlated line-of-sight gradients, as first noted by Illing et al. (1975). Furthermore, as the area asymmetry is an integral quantity, it is formed over a larger range of heights.

A unique feature of the area asymmetry is its center-to-limb variation, which exhibits a sign reversal at $\mu \approx 0.4$. This is illustrated in Fig. 12.13, from Bünte et al. (1993b), for the Fe I 5250.22 Å line. Thus the red Stokes V area dominates over the blue one on the limbward side of $\mu \approx 0.4$. In contrast, the amplitude asymmetry does not exhibit any such sign change.

Models that try to reproduce the area asymmetry of course also have to satisfy the constraint of no zero-crossing shifts ($\lambda_V - \lambda_I \approx 0$). Using previous ideas of van Ballegooijen (1985), Grossmann-Doerth et al. (1988b, 1989a) have found a way to accomplish this. The condition of no zero-crossing shift of the Stokes V profile is satisfied when there are no mass flows inside the magnetic flux tube. Still a flow in the non-magnetic region immediately *outside* the flux tube can affect the V profile and produce an area asymmetry when the line of sight crosses the boundary between the magnetic and non-magnetic regions. Although only unpolarized

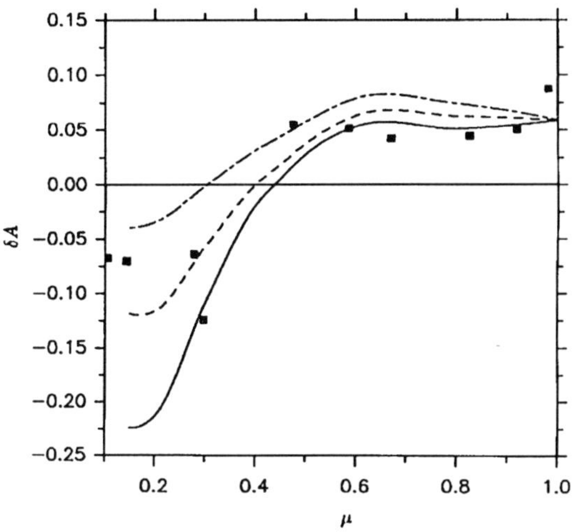

Fig. 12.13. Center-to-limb variation of the relative area asymmetry $\delta A = (A_b - A_r)/(A_b + A_r)$ for the Fe I 5250.22 Å line. The data, from Stenflo et al. (1987a), are represented by the filled squares. The theoretical curves, from Bünte et al. (1993b), differ in the assumed values for the flow speed in the external convection cell responsible for the area asymmetry (cf. Sect. 12.10 and Fig. 12.18).

radiation enters the flux tube from the non-magnetic region, it affects the formation of the polarized profiles and produces an area asymmetry when the incident unpolarized absorption profile is Doppler shifted to an asymmetric position with respect to the two Zeeman-shifted sigma components of the absorption coefficient in the magnetic region. In Sect. 12.10 we will provide an illustration of a model that can reproduce the observed center-to-limb variation of the area asymmetry. The three curves in Fig. 12.13 have been obtained from such a model, which has an external convection cell with a cool downflow near the flux tube boundary and a hot upflow further away. The curves differ in the flow velocities assumed.

The observational data thus lead us to the following general picture of the flux tube dynamics. The field that is of kG strength in the lower photosphere diverges with height, forming "canopies" that overlie the non-magnetic atmosphere. Inside the magnetic region large-amplitude oscillations cause substantial non-thermal line broadening and may be a source of waves that could contribute to the heating of the upper layers of the flux tube. In the external, non-magnetic atmosphere below the canopies a flow pattern develops, with inflow and downdrafts immediately outside the flux tube. The situation is reminiscent of the drain in a bathtub, accompanied by the development of a whirl flow (Stenflo, 1975), which may help to stabilize the flux tube against the fluting or interchange instability (Schüssler, 1984; Bünte et al., 1993a,c). At the same time this picture explains the observed correlations between flux concentrations and downdrafts (Simon and Leighton, 1964; Tanenbaum et al., 1969; Frazier, 1970), without the need for systematic downdrafts *inside* the flux tubes.

12.9. Infrared Diagnostics

While the widths of spectral lines are to a first approximation proportional to wavelength, the Zeeman splitting increases with the square of the wavelength. Therefore the incomplete Zeeman splitting in the visible part of the spectrum can become complete for certain lines in the infrared. When the splitting is complete, the field strength can be read off directly from the wavelength separation of the sigma components, regardless of the magnitude of the magnetic filling factor. Harvey and Hall (1975) first demonstrated this for the Fe I 1.5649 μm line, which has a Landé factor of 3.0.

The near infrared region around 1.6 μm is of special interest, since the solar opacity has a minimum at these wavelengths, which means that the radiation emanates from deeper layers in the atmosphere than in any other part of the solar spectrum. This enhances the Zeeman splitting still further, since the field strength increases with depth.

12.9.1. Weak and Strong Field Regimes

The amount of circular polarization (Stokes V) that we measure in a spectral line is primarily a measure of the magnetic flux through the resolution element (i.e., of the magnetic filling factor), but it does not by itself provide any information on the field strength. It is the shape of the Stokes V profile that carries the field-strength information. The situation may be clarified by distinguishing between three splitting regimes (outside sunspot umbrae): (1) The weak-field regime, which applies to the great majority of lines in the visible part of the solar spectrum, and for which $V \sim \partial I/\partial\lambda$ (cf. Eq. (12.1)). (2) The partial splitting regime, an example of which is the Fe I 5250.22 Å line. (3) The complete splitting regime, represented for kG flux tubes by the Fe I 1.5649 μm line.

Figure 12.14, from Stenflo et al. (1987b), illustrates how these various regimes are connected to each other. The left panel shows half the separation between the centers of gravity λ_r and λ_b of the red and blue Stokes V peaks, $0.5(\lambda_r - \lambda_b)$, expressed in units of $0.425\Delta\lambda_I$, where $\Delta\lambda_I$ is the total half width of the Stokes I profile. On the abscissa the Zeeman splitting $\Delta\lambda_H$ is expressed in the same units. The significance of $0.425\Delta\lambda_I$ is that for a Gaussian profile it represents the distance from the line center to the inflexion point in the wing (where $\partial I/\partial\lambda$ has a maximum). According to Eq. (12.1) the Stokes V peak should be located at this inflexion point in the case of weak fields, implying that half the Stokes V peak separation should be unity in terms of such a normalization if the Stokes I profile were Gaussian. Even for Gaussian profiles there will however be slight deviations from unity in the diagram since the position of the V maximum and the center of gravity of the upper half of the V peak are not entirely identical quantities. The solid and dashed curves correspond to different Voigt profiles used to represent Stokes I.

The complete splitting regime is reached when half the V peak separation directly gives the Zeeman splitting $\Delta\lambda_H$ and thus the field strength B. It corresponds to the dotted 45° line in the left diagram of Fig. 12.14. The weak-field

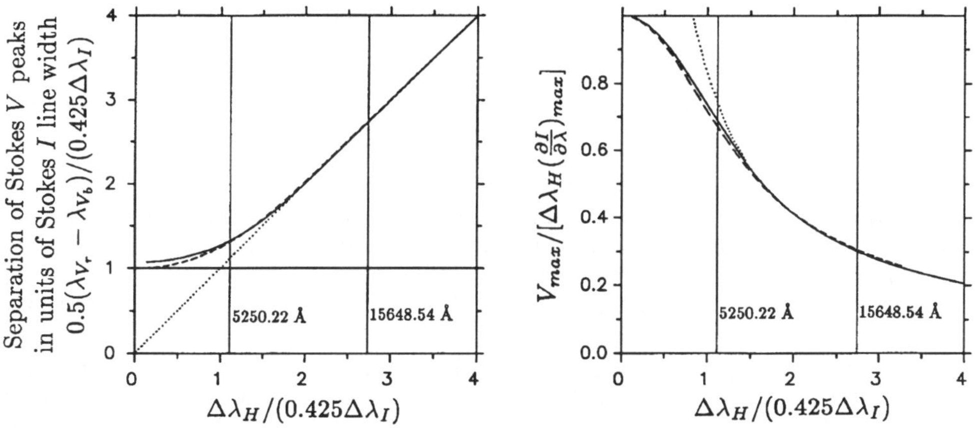

Fig. 12.14. Illustration of the transition from the weak-field to the complete splitting regime, from Stenflo et al. (1987b). The left panel gives half the wavelength separation between the Stokes V peaks vs. Zeeman splitting $\Delta\lambda_H$, both normalized by $0.425\Delta\lambda_I$, the distance from the line center to the inflexion point of a Gaussian Stokes I profile. The right panel gives the Stokes V polarization amplitude $V_{\rm max}$ in units of $\Delta\lambda_H\,(\partial I/\partial\lambda)_{\rm max}$, its value in the weak-field limit. The dotted curves represent the asymptotic complete splitting limits, extrapolated to small values of the Zeeman splitting. The vertical lines mark the splitting values for the disk-center profiles of the Fe I 5250.22 Å and Fe I 1.5649 μm lines in the case of a 1 kG magnetic field.

regime on the other hand represents the case when the V peak separation is merely determined by the inflexion point separation of the Stokes I profile, a quantity that is devoid of field-strength information. The transition between these two regimes, the partial splitting case, occurs over the range $0.5 \lesssim \Delta\lambda_H/(0.425\Delta\lambda_I) \lesssim 1.5$.

The two vertical lines mark the splittings in the case of the disk-center profiles of the Fe I 5250.22 Å and 1.5649 μm lines for a field strength of 1 kG. While the infrared line is located well within the complete splitting regime, the 5250.22 Å line lies in the transition zone of partial splitting. The Fe I 5247.06 Å line has a splitting that is 2/3 that of the 5250.22 Å line and is therefore substantially closer to the weak-field level of unity in the diagram. This difference between these two lines contains field-strength information.

The right panel of Fig. 12.14 shows the Stokes V amplitude $V_{\rm max}$ normalized to its weak-field value $\Delta\lambda_H\,(\partial I/\partial\lambda)_{\rm max}$. In the complete splitting regime $V_{\rm max}$ becomes a constant, independent of the field strength. The functional dependence of the normalized amplitude therefore becomes $\sim 1/\Delta\lambda_H$ in this regime, i.e., a hyperbola (dotted curve). The solid and dashed curves illustrate the transition between the weak-field and complete splitting regimes.

If we disregard temperature line weakenings, then the solid and dashed curves can be seen as representing the ratio between the apparent and true magnetic fluxes, or $B_{\rm obs}/\langle B\rangle$, when the "observed" field $B_{\rm obs}$ is derived from the observed $V_{\rm max}$. This ratio deviates from unity with increasing splitting $\Delta\lambda_H$ due to what we have previously called Zeeman saturation. Thermodynamic effects in the flux tubes (line weakenings) reduce $B_{\rm obs}$ by an additional factor, but since this factor is common to the Fe I 5250.22 and 5247.06 Å lines, the $V_{\rm max}$ ratio, i.e., B_{5250}/B_{5247}, provides direct information on the differential Zeeman saturation and thus on the field strength.

12.9.2. ZEEMAN BROADENING

Due to the large Zeeman splitting in the infrared, a distribution of field strengths within the resolution element will induce significant Zeeman line broadening. In Fig. 12.7 we showed that the separation of the Zeeman broadening from the thermal and non-thermal Doppler broadening requires the use of an "infrared line ratio", for example with the "Zeeman line" Fe I 1.5649 μm and the "Doppler line" Fe I 1.5653 μm that were illustrated in Fig. 12.7.

The Zeeman broadening determined this way may be due to different kinds of field-strength distributions: (1) A lateral variation of the field strength across an unresolved flux element. (2) The existence of different flux elements with different field strengths within the spatial resolution element. (3) A variation of the field strength along the line of sight.

The previously unknown magnetic cross section of a flux element has led to some ambiguity in the determined field strengths, since the field strength amplitude of a flux element determined by any method (line ratio or infrared splittings) depends on the assumed *shape* of the cross section. Usually a rectangular cross section (single-valued field) has been assumed, which then removes the ambiguity, but empirical support for this assumption has only recently become available.

The field-strength variation along the line of sight can be determined by requiring that both the observed infrared line profiles and the 5250-5247 line ratio (which refers to higher atmospheric layers) should be satisfied simultaneously with a self-consistent magnetohydrostatic flux tube model that is used to derive the synthetic observations that fit the real data. The Zeeman line broadening follows from this line-of-sight variation as an automatic byproduct. In this way Zayer et al. (1989) found that the line-of-sight variation alone produced the observed amount of Zeeman broadening of the Fe I 1.5649 μm line. This speaks in favor of unique flux tubes with rectangular magnetic cross sections within the spatial resolution element, since additional lateral field-strength variations would make the Zeeman broadening too large.

12.9.3. MULTI-COMPONENT MODELLING

The majority of the Stokes V infrared spectra look similar to the two cases shown in Fig. 12.7 and can be interpreted in terms of our standard two-component model with no horizontal variations of the field strength within the flux elements or from one flux tube to the next within the same spatial resolution element. A number of spectra however show strange looking Stokes V profiles. The most "pathological" of them that has been analysed so far is shown in Fig. 12.15, from Rüedi et al. (1992). No such profiles have ever been seen in the visible due to the smaller Zeeman splitting there. Obviously two-component models are unable to reproduce spectra like that of Fig. 12.15.

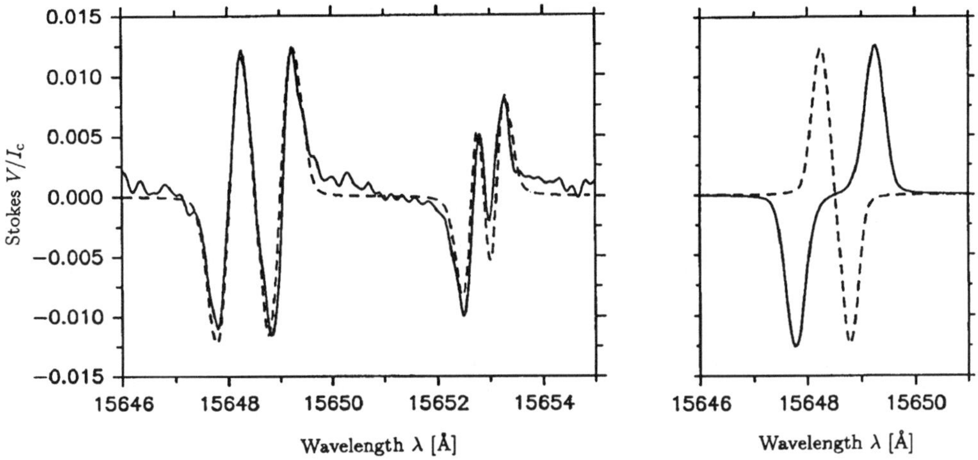

Fig. 12.15. Left panel: Observed (solid) and synthetic (dashed) Stokes V/I_c spectrum with unusually looking profiles. A model with two magnetic components has been used to fit the observations. Right panel: The synthetic profiles of the Fe I 1.5649 μm line for each of the two magnetic components used for the fit. They correspond to a field strength at height $z = 0$ of 1.70 kG (solid curve) and −1.05 kG (dashed curve), respectively. From Rüedi et al. (1992).

It is clear that we have to add at least one additional magnetic component to be able to model such spectra. With each new component one gets more free parameters, so one might fear that the problem could get out of hand with too many free parameters. Fortunately, for all the 27 spectra that have been analysed so far (Rüedi et al., 1992), not a single case has been encountered where more than two magnetic components (plus a third, non-magnetic component) with rectangular cross sections were needed. This intriguing result provides additional support to the concept of discrete flux tubes.

It is important to stress that this model-fitting success is far from being a trivial result. When first attempts to fit the observed spectra were made with a height independent magnetic field (which is unphysical), then a whole distribution of magnetic components were needed, otherwise the Zeeman line broadening could not be fitted. As soon as self-consistent, magnetohydrostatic flux tube models with a diverging field geometry were introduced, however, the Zeeman broadening was automatically taken care of, and no more than two magnetic components were ever needed.

The model fit to the observed "pathological" spectrum in the left panel of Fig. 12.15 is shown by the dashed curve. The contributions to the Stokes V profile of the Fe I 1.5649 μm line from each of the two magnetic components are shown in the right panel. The corresponding field strengths at geometrical height $z = 0$ ($\tau_{5000} = 1$, level of continuum formation in the external atmosphere at 5000 Å) are 1.70 and −1.05 kG, respectively. This implies that we have two discrete flux elements of *opposite* polarities and significantly different field strengths within a few arcsec from each other. A number of other cases requiring two magnetic components have been found, both with the same and with opposite polarities.

12.9.4. FIELD-STRENGTH DISTRIBUTIONS

We have seen how we, from two-component modelling of infrared Stokes V spectra like those of Fig. 12.7 and three-component modelling of spectra like that of Fig. 12.15, can derive the field strengths B at geometrical height $z = 0$ for each magnetic component. The results for all the 27 spectra that have been analysed so far are summarized in Fig. 12.16. The left diagram shows $B(z = 0)$ vs. $\langle B\rangle$, where $\langle B\rangle$ is the field strength of the component when averaged over the spatial resolution element. The open circles represent spectra for which only a single magnetic component was needed for the fit, the pluses the cases where two magnetic components were needed (each such spectrum is thus represented by *two* pluses). A regression line has been drawn through the points above 1.3 kG (which have a plasma β below unity, see below), to indicate the slight filling-factor dependence of B for the "normal" kG flux elements. This weak dependence is consistent with the line-ratio results of Fig. 12.3.

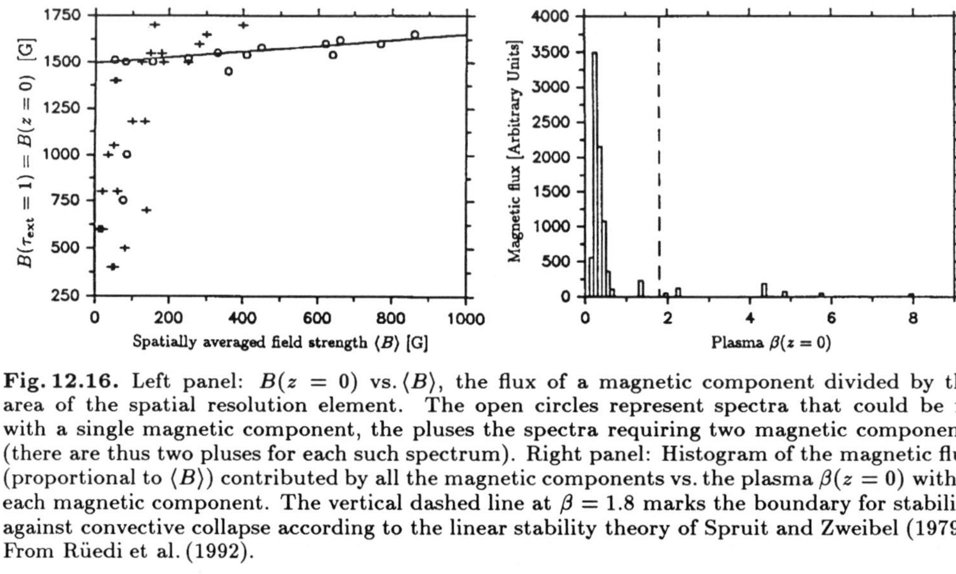

Fig. 12.16. Left panel: $B(z = 0)$ vs. $\langle B\rangle$, the flux of a magnetic component divided by the area of the spatial resolution element. The open circles represent spectra that could be fit with a single magnetic component, the pluses the spectra requiring two magnetic components (there are thus two pluses for each such spectrum). Right panel: Histogram of the magnetic flux (proportional to $\langle B\rangle$) contributed by all the magnetic components vs. the plasma $\beta(z = 0)$ within each magnetic component. The vertical dashed line at $\beta = 1.8$ marks the boundary for stability against convective collapse according to the linear stability theory of Spruit and Zweibel (1979). From Rüedi et al. (1992).

We see from this diagram that for large filling factors ($\langle B\rangle > 140$ G) all field strengths $B(z = 0)$ are > 1.4 kG, while for smaller filling factors the incidence of weaker fields become increasingly abundant, although these weak fields generally correspond to the weaker component in models with two magnetic components within the resolution element, except for two cases (the two open circles that fall significantly below the regression line). As a matter of fact this analysis (Rüedi et al., 1992), independently complemented by the results of Rabin (1992a,b, see below), constituted the first real identification of intrinsically weak flux elements. The lowest field strength found so far is about 0.4 kG for the three-component case (two magnetic and one non-magnetic components) according to Fig. 12.16,

0.75 kG for the two-component case. The 0.75 kG case is the one that was shown in the lower panel of Fig. 12.7, where the Stokes V amplitude of the "Doppler line" became suppressed due to the incomplete Zeeman splitting. In contrast the upper panel of Fig. 12.7 represents the case when the Zeeman splitting is complete for both the Zeeman and Doppler lines.

Each of the determined magnetic components can instead of its field strength $B(z=0)$ also be represented by its plasma $\beta = 2\mu_0 P/B^2$ at height $z=0$, where P is the gas pressure inside the flux tube. β is the ratio between the gas pressure and the magnetic pressure. The right panel of Fig. 12.16 shows a histogram that gives the relative contributions to the magnetic flux from the various magnetic components per unit interval in plasma β. We see that the distribution is peaked around $\beta \approx 0.3$, but it has a low tail that extends to high values of β. The dashed vertical line in the diagram, at $\beta = 1.8$, is the theoretical limit below which flux tubes are stable against convective collapse according to the linear stability analysis of Spruit and Zweibel (1979). According to this theory flux elements with $\beta > 1.8$ spontaneously collapse to a strong-field state. Our empirical values of β are however about a factor of six smaller than this theoretical limit, which implies a very low internal gas pressure and thus highly evacuated flux tubes.

Although weaker intrinsic fields are more abundant when the observed flux values are small, their accumulated contribution to the total magnetic flux still remains small. Thus, if we add up all the absolute flux values $|\langle B\rangle|$ in Fig. 12.16, then only about 10 % of this total flux is due to fields with $B(z=0) < 1.25$ kG. Although the statistical sample is limited, this result is consistent with the old line-ratio result of Howard and Stenflo (1972) and Frazier and Stenflo (1972) that more than 90 % of the total flux is in strong-field form. The diagnostic advantages in the infrared will however allow us to explore this 90 % statement in much greater quantitative detail.

Much improved field-strength statistics has become possible with the Near-Infrared Magnetograph (NIM) of Rabin (1992a,b), with which the polarized spectrum around the Fe I 1.5649 μm line is recorded with a two-dimensional InSb detector array. Thereby it becomes possible to produce 2-D maps of the intrinsic field strength across the solar surface. Initial results with NIM are shown in Fig. 12.17, from Rabin (1992a). The histogram to the right shows the observed distribution of intrinsic field strengths with NIM, while the histogram to the left represents the apparent field strengths $B_{\rm obs}$ (cf. Eq. (12.9)) recorded across the same area of the solar disk with the solar magnetograph of the National Solar Observatory. We see that $B_{\rm obs}$, which is primarily a measure of the magnetic flux through the spatial resolution element, has little to do with the intrinsic field strengths but is more a measure of the filling factor. While all the values of $B_{\rm obs}$ are below 0.5 kG, most values of B occur in the range 1.1–1.5 kG, with only a weak tail of the distribution extending below 1 kG. This also supports the conclusion that most of the magnetic flux in the solar photosphere occurs in kG form.

The field-strength values of Fig. 12.17 have been obtained directly from the Zeeman splitting without any model fitting. They therefore represent the values of B at the height of formation of the sigma components, which is unknown without model fitting. In contrast the B values of Fig. 12.16 refer to geometrical height $z=0$ and have been obtained with a magnetohydrostatic model that is able to

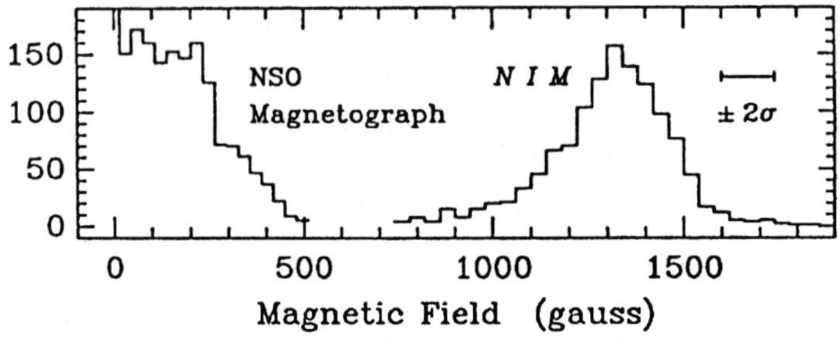

Fig. 12.17. Initial results with the Near-Infrared Magnetograph (Rabin, 1992a). The two histograms represent the apparent flux densities B_{obs} (left) and the intrinsic field strengths B (right).

reproduce the full observed Stokes V profiles. As $z = 0$ lies below the level where the sigma components are formed, the B values of Fig. 12.16 are systematically larger than those of Fig. 12.17. The two sets of results are thus fully consistent with each other.

12.9.5. The 12 μm Region

The spectral region around a wavelength of 12 μm contains some emission lines that have been identified as due to Mg I and that exhibit strikingly large Zeeman splittings, as first noted by Brault and Noyes (1983). Of particular interest is the Mg I 12.32 μm line, which has been explored polarimetrically by Hewagama et al. (1993). Most 12 μm work has been done in sunspots, but some recordings in active plage regions indicate Zeeman splittings corresponding to intrinsically weak fields there, about 300–500 G.

At a first glance these small values may seem to contradict the kG field strengths that have been found with other methods, but they are in fact consistent with the empirical flux tube models when one takes into account the difference in height of formation of the spectral lines used. According to Carlsson et al. (1992) the Mg I 12.32 μm emission line is formed at a height in the upper photosphere where the field strength in a magnetohydrostatic flux tube with $B(z = 0) \approx 1.6$ kG has fallen to a value of 300–500 G due to the expansion of the flux tube as the external gas pressure exponentially decreases. The 12 μm data are thus consistent with the results of other diagnostic methods, as shown by detailed non-LTE radiative-transfer calculations of Mg I line formation in a flux tube atmosphere (Bruls and Solanki, 1994). Although this region of the spectrum is technologically difficult to use for high-resolution observations, in particular due to the diffraction limit of normal-size solar telescopes, it has considerable future potential for the diagnostics of small-scale magnetic fields.

12.10. Model Building

Early attempts to derive the temperature-density structure of magnetic field concentrations, so-called facular models, were based on the use of unpolarized spectra and the assumption of a plane-parallel stratification of the atmosphere. We now know that such an approach is entirely inadequate, since the observed Stokes I spectra are dominated by the contributions from the non-magnetic atmosphere, and the magnetic field elements have a geometry that is vastly different from that of a plane-parallel stratification. Only through the use of polarized spectra can the contributions from the flux elements be separated from the surroundings.

If we neglect magnetic tension forces, which are only important where the curvature of the magnetic field lines is large, then horizontal force balance requires equality between the external gas pressure P_{nm} of the non-magnetic surroundings and the total pressure inside the flux element, due to the gas pressure P_m and the magnetic pressure $B^2/(2\mu_0)$:

$$P_m + \frac{B^2}{2\mu_0} = P_{nm} \, . \tag{12.26}$$

Although this means that $P_m < P_{nm}$, it does not necessarily imply that the density $\rho_m < \rho_{nm}$, since this depends on the temperature difference $\Delta T = T_m - T_{nm}$, which is determined by the energy transport mechanisms. Nevertheless it turns out that the flux tubes are in fact highly evacuated, as we have seen from the low empirical values of the plasma β in the preceding section (Fig. 12.16). $\beta \approx 0.3$ means that P_m is about 30 % of the value of $B^2/(2\mu_0)$ and thus only 20–25 % of P_{nm} according to Eq. (12.26). Such a large reduction of the internal gas pressure can only be achieved if the density is also substantially reduced, which implies lower opacities in the flux tubes. The surfaces of equal optical depth are thus depressed in the flux elements, which constitutes the so-called Wilson depression that had first been discovered for sunspots.

Eq. (12.26) implies that the directly observed magnetic field structure of a flux element is coupled to its thermodynamic and opacity structure. A stronger magnetic field generally requires a deeper Wilson depression, which means that the height of line formation is lower. The diagnostic information then emanates from a region with larger field strength (since the field strength increases with depth). A change of the internal temperature modifies the opacity and thus shifts the formation heights, with the result that the *apparent*, observed field strength changes, although the field strength at a given *geometrical* height may not change. What we see is related to the optical depth scale, and it is the opacity that dictates the relation between the optical and geometrical depth scales.

Another consequence of Eq. (12.26) is that B decreases with height, since P_{nm} drops off almost exponentially with height. Due to flux conservation a decreasing value of B implies an increasing cross section of the flux element. The rate of divergence is determined by $P_{nm} - P_m$. In the lower layers the expansion rate is moderate and the curvature of the field lines is small. In the upper photosphere and lower chromosphere, however, the field lines flare out to become almost horizontal. In this region with large field-line curvature the tension forces become important and need to be accounted for in the force balance equation. The assumption (12.26) that the tension forces can be neglected is called the *thin tube approximation* (cf. Roberts and Webb, 1978). It is a good approximation throughout most of the line-forming regions in the photosphere.

To model the flux tube interior we first have to adopt a model for the non-magnetic external atmosphere. A natural first choice is to use some standard, reference atmosphere model that is representative of the average quiet sun, like the Harvard-Smithsonian Reference Atmosphere HSRA (Gingerich et al., 1971). Due to the Wilson depression the flux tube atmosphere that is accessible to observations however extends to larger depths than those covered by such reference atmospheres, which therefore need to be extended into the convection zone. Such an extension has been done for HSRA by combining it with the convection zone model of Spruit (1974). The resulting extended reference model atmosphere is called HSRASP (Chapman, 1979).

The initial flux tube models may be stepwise improved to increasing levels of sophistication. Thus the thin-tube approximation may be abandoned and replaced by the solution of the complete magnetohydrostatic equations for an axially symmetric flux tube with an arbitrarily prescribed temperature structure, as done by Steiner et al. (1986). The flux tube temperature at different heights may then be varied as free parameters, the values of which get determined by fitting synthetic Stokes spectra to the observed ones. Each choice of internal temperature stratification and field strength at a reference level automatically determines the magnetic field structure via the constraints imposed by the magnetohydrostatic equations.

The next step in the hierarchy of models is to include convective flow fields in the surrounding, non-magnetic atmosphere, as illustrated in Fig. 12.18, from Bünte et al. (1993b). In these models the type of flow and the flow amplitudes are again specified as free parameters to be determined by the observations, but the flow pattern and its modifying influence on the field structure are calculated self-consistently from these free parameters.

Models such as that of Fig. 12.18 are used to calculate synthetic polarized spectra through numerical solutions of the LTE polarized radiative transfer equations, e.g. using the DELO method of Sect. 11.11. When the flux tubes are being viewed obliquely, which is the case when we approach the solar limb, we need to take into account the contributions from the neighboring flux tubes as well. Since the flux tubes are optically rather thin in the horizontal direction, the contribution function along the line of sight may sample more than one flux tube. Therefore the structure of Fig. 12.18 has been used as part of the periodic arrangement of flux tubes that was illustrated in Fig. 11.5. The right panel of Fig. 12.18 shows an example of 20 oblique lines of sight with the grid points for the radiative transfer computations marked. The synthetic spectra from all the lines of sight have to be averaged before they may be compared with the observations.

The model of Figs. 12.18 and 11.5 was specially designed to explain the observed center-to-limb variation of the Stokes V area asymmetry that was shown in Fig. 12.13. The sign reversal at a center-to-limb distance of $\mu \approx 0.4$ could not be explained by an outside downdraft alone. A complete convection cell was required,

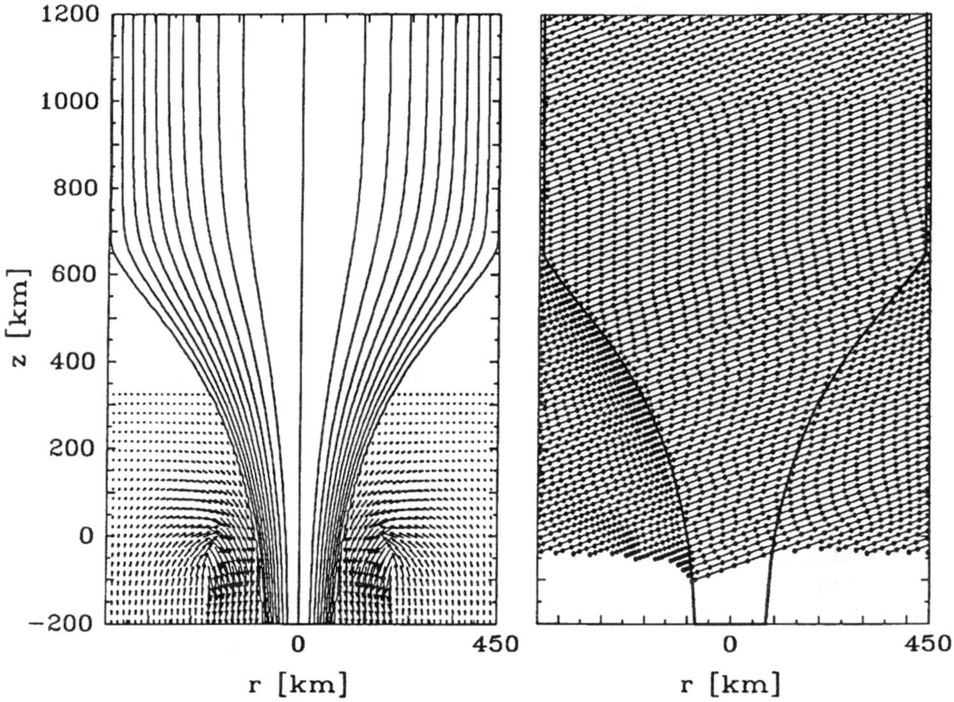

Fig. 12.18. Example of an axially symmetric model of a magnetic flux tube with a surrounding convective velocity field. It has been used to successfully reproduce the observed center-to-limb variation of the Stokes V area asymmetry of Fig. 12.13. Left panel: Magnetic field lines and flow lines of the velocity field. Right panel: 20 lines of sight at an angle of 70° with the vertical direction, used for the radiative transfer calculations of the synthetic Stokes V spectra. The grid points for the computations (filled circles) have been selected with an adaptive step size method. The flux tube is part of the array of flux tubes illustrated in Fig. 11.5. A line of sight that exits the left boundary enters the model again at the same height at the right boundary, to simulate the effect of neighboring flux tubes. From Bünte et al. (1993b).

which included an upflow and a temperature-velocity correlation of the kind that is observed in the solar granulation.

The calculation of synthetic Stokes spectra with such sophisticated models requires massive computations on supercomputers and have therefore so far only been used for *direct* modelling, i.e., for the computation of synthetic spectra for a few hand-picked choices of free model parameters with subsequent visual comparison with the observations. The desired, ultimate approach is however Stokes *inversion*, which was discussed in Sect. 11.13. The free parameters of the physical model are then determined by an iterative least-squares fit to a carefully selected set of observables that respond to the model parameters in ways that are mutually as "orthogonal" as possible. As this approach requires the calculation of synthetic observables for a rather large number of models, it is not yet computationally feasible to use for models as sophisticated as that of Fig. 12.18.

For the application of Stokes inversion to determine the temperature-density stratification of the flux tube interiors it has been found adequate to use the thin tube approximation without any convective mass flows (apart from the line-broadening turbulent motions). The free parameters of the model are then the field strength B at a given height, the temperature at several different vertical grid points, and the macroturbulent broadening velocity, parametrized with a regression equation such that its value for any line strength and excitation potential can be obtained. The temperature between the selected grid points is obtained by cubic spline interpolation. The small set of free parameters in combination with the thin tube approximation, the requirement of hydrostatic balance, etc., fully specifies the model and allows synthetic spectra to be computed. A suitable set of observables is a combination of the magnetic and thermal line ratios, Stokes V area ratios between weak and strong Fe I and II lines, Stokes V line widths, and the separation between the red and blue Stokes V peaks.

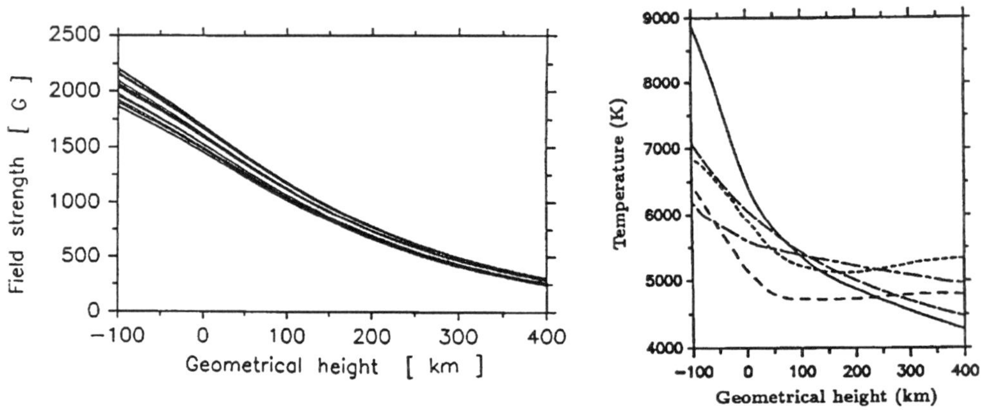

Fig. 12.19. Left panel: Variation of the field strength with geometrical height, determined from Stokes inversion by Zayer et al. (1990). The various curves correspond to different filling factors, with an order of magnitude difference between the smallest and largest filling factor. Right panel: Variation of the temperature with geometrical height. The solid curve represents the external atmosphere. The two dashed curves that run almost parallel to each other have been obtained through Stokes inversion by Keller et al. (1990). The upper curve is representative of a network element with a small filling factor, while the lower curve corresponds to a facular element with a larger filling factor. The two remaining curves represent purely theoretical models. The dashed-dotted curve is due to Grossmann-Doerth et al. (1989b), while the steeper, long-dashed curve is due to Steiner and Stenflo (1990).

Figure 12.19 illustrates some flux tube properties derived from such Stokes inversions. The left diagram, from Zayer et al. (1990), shows the variation of the field strength with geometrical height z. The various curves in the diagram refer to different magnetic filling factors. Although the filling factors vary by an order of magnitude, the different $B(z)$ curves are practically identical, within the instrumental noise of the data and the uncertainties of the inversion procedure. When the field strength is plotted as a function of optical depth τ, however, the spread of the various $B(\tau)$ curves is very large, since the flux tube temperature and thus also the opacities vary significantly with the filling factor. As however the flux tubes are so highly evacuated, with a plasma $\beta \approx 0.3$ as shown by Fig. 12.16, the $B(z)$ curves are primarily determined by the gas pressure $P_{nm}(z)$ of the external atmosphere. All $B(z)$ curves therefore resemble the asymptotic case of completely evacuated flux tubes (with $P_m = 0$) and are thus insensitive to the thermodynamic state of the interior.

In the diagram to the right of Fig. 12.19 four flux tube temperature curves are compared with $T_{nm}(z)$ of the surroundings (solid curve). The two dashed curves that run almost parallel to each other have been obtained via Stokes inversion by Keller et al. (1990). While the upper, hotter curve represents a network element with a small filling factor, the lower, cooler curve represents a facular element with a larger filling factor. The two remaining curves represent purely theoretical models that we will discuss later. Common to all these flux tube curves is a temperature deficit in the lower layers with respect to the surroundings and a temperature excess in the upper layers.

What matters for the optical *appearance* of the flux tubes is not their $T(z)$ curves but their $T(\tau)$ curves. Although the network elements are cooler in the deeper layers at equal geometrical height, they are hotter than their surroundings *at equal optical depth.* They therefore appear as bright features in continuum radiation. This could be demonstrated directly with speckle polarimetry by Keller (1992), who spatially resolved a flux tube with a diameter of about 200 km and a continuum brightness exceeding the surroundings by 30 % near the center of the solar disk. The facular elements with larger filling factors on the other hand are cooler and appear as dark features in disk-center continuum radiation. Using Stokes V profiles of C I lines to diagnose the conditions at the level of continuum formation, Solanki and Brigljević (1992) were able to clarify this filling-factor dependence of the continuum brightness. Since the continuum radiation has no polarized Zeeman component that can be used to isolate the radiation from flux tubes with small filling factors, one has to use spectral diagnostics like deeply forming C I lines to diagnose the continuum layers.

In all the MHD models of flux tubes that we have discussed so far the thermodynamic properties have been left as free parameters to be determined by fitting synthetic Stokes spectra to observed ones. The thermal structure however no longer would remain free if an energy equation were to be included in the MHD problem. In its full generality the energy balance is a formidable problem, which has so far only been approached with incomplete and approximate energy equations.

The main energy transport processes are due to radiation, convection, and waves. While the 2-D radiative transfer problem for flux tube geometries has been taken care of in previous work, convection has been treated in an approximate and idealized way, and wave heating has hardly been included at all. Spruit (1976) did the pioneering work, while treating the radiative transport in the diffusion approximation and the convective transport with a mixing-length theory. The most comprehensive 2-D models of the flux concentrations have been developed for slab geometry by Deinzer et al. (1984a,b), Knölker et al. (1988), Knölker and Schüssler (1988), and Grossmann-Doerth et al. (1989b). Instead of seeking stationary solutions, the time-dependent MHD equations in a compressible medium are solved. One lets the solution relax towards a stationary end state that describes the properties of the flux concentration. Convection is treated with the mixing-length theory and the radiative transfer with a grey atmosphere (wavelength-independent opacity).

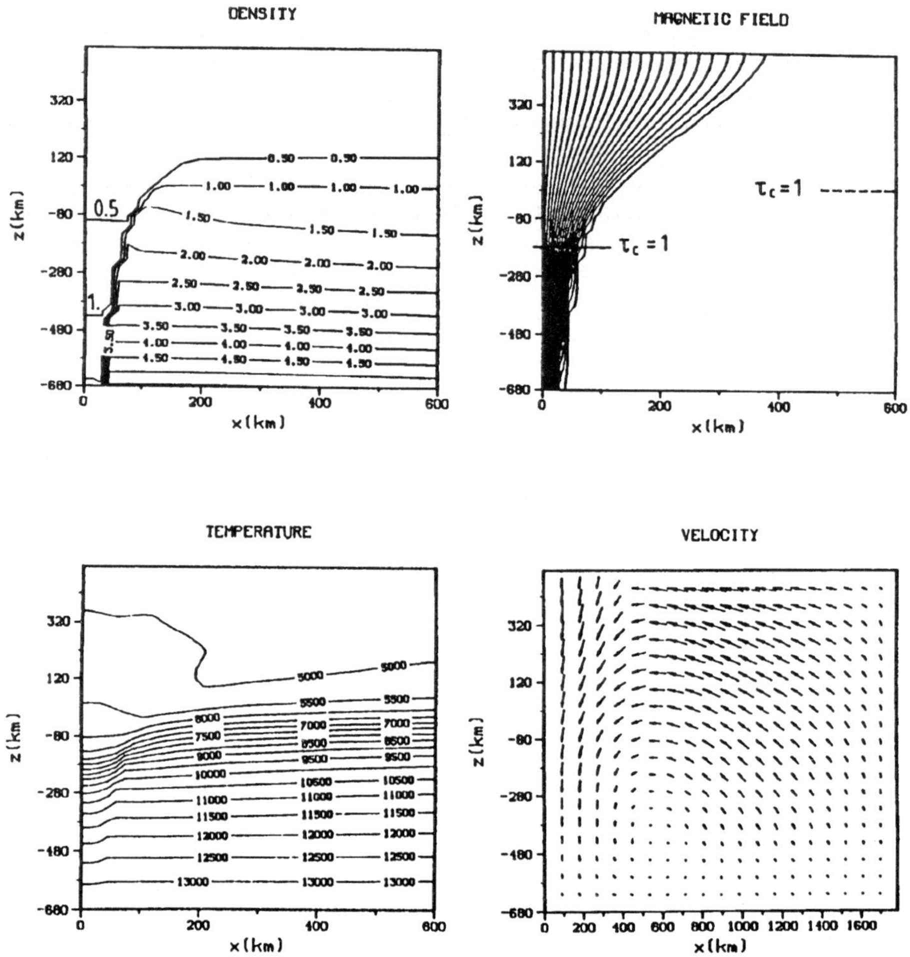

Fig. 12.20. Theoretical model of a magnetic element in slab geometry, from Grossmann-Doerth et al. (1989b). The horizontal coordinate scale has been stretched to enhance the visibility of the structure. The densities are in units of 3×10^{-4} kg m^{-3}, the temperatures in K.

Figure 12.20 illustrates the result of such calculations for a slab of thickness 150 km at optical depth $\tau_{5000} = 1$ inside the slab. The final end state has been approached but not yet reached for the model results that are shown in Fig. 12.20. The Wilson depression is about 180 km. Cooling of the lower layers, seen in the figure as depressions of the temperature contours, is due to inhibition by the magnetic field of lateral convective energy transport. The heating of the upper, observable layers is due to radiation from the hot bottom and walls of the fairly transparent flux slab.

The most complete treatment of the 2-D radiative transfer with proper opacities has been done by Steiner and Stenflo (1990) for stationary, axially symmetric flux tubes. Their calculations also lead to radiative heating of the upper flux tube atmosphere. The $T(z)$ temperature curves that result from the models of Grossmann-Doerth et al. (1989b) and Steiner and Stenflo (1990) at the flux tube axis have been given in the right panel of Fig. 12.19 (as the dashed-dotted and the long-dashed curves, respectively), where they may be directly compared with the empirically determined curves from Stokes inversion. The theoretical and empirical curves share the same qualitative properties, namely a temperature deficit with respect to the surroundings in the lower layers, which changes to become a temperature excess higher up. The detailed shapes of the curves however differ, indicating that all the physics has not yet been properly modelled, or that some of the assumptions underlying the empirical models are inadequate (e.g. LTE, simple geometry, no cool surroundings).

The theoretical models may be made more realistic by going to 3-D, as done in the extensive numerical simulations of granular convection by Nordlund (1983, 1986). Although these models can also include magnetic fields, they have not yet been able to treat the detailed physics of flux tubes due to the inadequate spatial resolution of the computational grid. Much progress can be expected in the near future from the application of new numerical techniques with grid refinement procedures on ever more powerful computer systems.

CHAPTER 13

INSTRUMENTATION FOR SOLAR POLARIMETRY

13.1. Overview of System Components

The aim of solar polarimetry is to measure the Stokes vector $\boldsymbol{I}$ (the four Stokes parameters I, Q, U, and V) with the highest possible polarimetric accuracy, spectral, spatial, and time resolution, and spectral and spatial coverage. As it is not possible to optimize all these parameters simultaneously, various combinations of trade-offs are needed. The type of trade-off that one selects depends on the scientific objective of the observations.

As the sun appears so bright to us due to its proximity, it may come as a surprise that solar observations are photon starved, i.e., the trade-offs mentioned become necessary because of the limited photon statistics. This can be understood better if we recall that the surface brightness (received radiation energy per unit solid angle) is independent of distance (assuming that the stellar disk is resolved). Thus for a given angular resolution the number of photons received per resolution element on the solar disk is the same as the photon number from a barely resolved disk of a star with the same effective temperature as the sun. The difference caused by the proximity of the sun is that the solar disk contains a large number (more than 10^8 for a diffraction-limited 1-m telescope) of resolution elements, while only some nearby giant stars (like α Orionis) can be resolved at all (although optical interferometry is now improving this situation).

The main components of a polarimetric observing system are:

- The earth's atmosphere
- The image-forming optics
- The optics for polarization analysis
- The wavelength-selecting optics
- The detector and data handling system

Turbulence in the earth's atmosphere, both outside and inside the telescope, causes fluctuating wavefront distortions that may spatially smear the image and produce spurious polarization signals. Seeing effects can also be produced inside a solar telescope due to the high concentration of heat (like in a solar oven), but they may be eliminated by evacuating the telescope. Evacuation is however only feasible for apertures $\lesssim 1\,\mathrm{m}$, since larger entrance windows that can sustain a pressure difference of one atmosphere cannot be manufactured with the required optical and polarimetric quality. This problem can be solved by filling the telescope with helium to equalize the pressures (see Sect. 13.10 below).

The removal of the seeing-induced effects of the external atmosphere will be discussed in Sect. 13.6. Spatial smearing of the polarization images may be overcome either by image reconstruction techniques (like "speckle imaging" and "phase diversity") or by adaptive optics. Spurious polarization signals can be avoided by using polarization modulation schemes with modulation frequencies that are much higher than the frequencies prevailing in the atmospheric turbulence. An alternative to such temporal modulation is to record orthogonal polarization states strictly simultaneously (so that the seeing effects in the different polarization states are identical), either via the use of beam splitters or by spatial modulation, but the polarimetric accuracy is then limited by the accuracy with which the flat field may be determined and by the effects of differential optical aberrations.

The image-forming optics generally corrupts the polarization information by producing instrumental polarization, which will be discussed in Sect. 13.5. Ideally, to entirely avoid this problem the polarization analyzer should be the first optical element of the system, placed in front of the whole telescope. This is not feasible in practice, since while the apertures of solar telescopes are large, those of high-precision polarization optics are small.

The largest instrumental polarization effects arise from oblique reflections, e.g. in the commonly used telescopes that have a heliostat (like the NSO McMath-Pierce facility) or a coelostat mirror in front (like the tower telescopes at Mt Wilson, NSO/Sacramento Peak and Kitt Peak, La Palma, and Tenerife). The best solution is to build "polarization-free" telescopes (like THEMIS and LEST), which are rotationally symmetric with respect to the optical axis until the beam reaches the secondary focus, where the polarization analysis takes place. Still, even for such telescopes, small, residual polarization effects are unavoidable, e.g. due to stresses in the entrance window or other optical imperfections.

The polarization optics needs to be placed early in the optical train to minimize the instrumental polarization, since it is only the optics in front of the polarization analysis package that is of concern. The task of the optics for the polarization analysis is to encode the incoming polarization information into intensity differences, since the detector systems used are only sensitive to photon numbers (intensity), not to polarization. After the optical encoding of the polarization has taken place and the beam leaves the optical analysis package, any oblique reflections or instrumental polarization may occur without being of any consequence for the polarization analysis.

For the optical polarization analysis a linear polarizer is needed for the linear polarization (Stokes Q and U), a retarder plus linear polarizer for the circular polarization (Stokes V). The linear polarizer may be replaced by a polarizing beam splitter, which gives two output beams of orthogonal polarization states.

Modulation of the polarization state (see Sect. 13.7) may be done mechanically, e.g. with a rotating wave plate, or electrooptically, without any moving parts. An electrooptical modulator serves as a variable retarder and can easily modulate at frequencies in the kHz range, much above the seeing frequencies. To record all four Stokes parameters simultaneously, two such non-moving variable retarders having their optical axes at different position angles are needed.

Wavelength selection can be made with a spectrograph, a narrow-band filter, or a Fourier Transform Spectrometer (FTS). The 2-D image at the end focus of a spectrograph represents (x, λ)-space, i.e., one spatial dimension (x) and the wavelength dimension (λ) are covered for each given value of the other spatial coordinate (y). With a narrow-band filter one covers (x, y)-space for a given value of λ. To cover a certain wavelength range the filter has to be tuned stepwise. Since the intrinsic half width of photospheric spectral lines is typically 0.1 Å only, it is hard to obtain well resolved line profiles this way.

The ultimate performance in terms of spectral resolution, spectral coverage, and absence of spectral stray light, is provided by the FTS. The solar spectrum can be fully resolved, and all wavelengths within the prefilter band pass, typically 1000 Å, are covered simultaneously. Interferometric scanning of the path length difference is however needed before the spectrum can be retrieved by Fourier transformation. So far the FTS has only been used for interferometric recordings at single (x, y) points (zero-D spatial information) on the sun, but as the 2-D entrance aperture of the FTS is properly imaged on the exit focal plane, 2-D spatial (x, y) information can in principle be obtained with array detectors. The high data rate however imposes unusual demands on the data handling system of such an imaging FTS, and its realization for solar applications still lies in the future.

The detector system can be placed in the exit focus of any wavelength-selecting system, be it a spectrograph, narrow-band filter, or an imaging FTS. If polarization modulation has been used to encode the polarization information, then the intensity sensed by the detector contains not only the usual dc component (representing Stokes I) but also ac components, out of which Stokes Q, U, or V have to be retrieved by demodulation. For single-channel detectors, like photomultipliers, the demodulation has usually been done with lock-in amplifiers. This is not possible with imaging detectors like CCDs (charge coupled devices), which contain a large number of pixels (on the order of 10^6). The simplest solution would be to gate the read-out in synchrony with the modulation, but since the read-out time for large array detectors is $\gtrsim 0.1$ s, it is then not possible to modulate faster than the seeing fluctuations.

The problem of speed may be overcome if instead of frequent read-outs one may create fast buffers within the CCD chip, to which the charges may be quickly shifted for temporary storage while a new image is being exposed. In Sect. 13.8 we will describe how four separate image planes can be created in this way within a single detector chip. This allows all four Stokes parameters to be demodulated within a single CCD. As a byproduct the flat-field or gain table problem is eliminated, since all four image planes use the identical pixels for exposure. A cylindrical microlens array prevents any light losses at the pixels used for buffer storage.

As an alternative to temporal modulation with its complex demodulation schemes we will also discuss spatial modulation, which for instance can be done with a Babinet compensator. Demodulation is then accomplished simply by spatial Fourier filtering of the final image off-line.

13.2. General Treatment of an Optical Train

Let $\boldsymbol{I}$ be the Stokes vector that enters an optical system composed of n separate components, whose polarization properties are described by the Mueller matrices $\boldsymbol{M}_i$, where the index numbers $i = 1, 2, \ldots, n$ increase in the same direction as the light travels (element 1 being encountered first, element n last). Let $\boldsymbol{I}'$ be the

Stokes vector of the beam that exits the optical train. Then the relation between $\boldsymbol{I}$ and $\boldsymbol{I}'$ is given by

$$\boldsymbol{I}' = \boldsymbol{M}\boldsymbol{I}\,, \tag{13.1}$$

where

$$\boldsymbol{M} = \boldsymbol{M}_n \boldsymbol{M}_{n-1} \ldots \boldsymbol{M}_2 \boldsymbol{M}_1\,. \tag{13.2}$$

The various optical elements may be of different kinds, like partial polarizers, beam splitters, retarders, modulators, and may also represent reflections at metallic or dielectric surfaces.

In the next section we will give the Mueller matrices for the main categories of optical elements. These matrix expressions are most conveniently written down in the special coordinate system where the polarizing direction, the fast axis of a retarder, etc., are parallel to the Stokes Q direction. We call this Mueller matrix $\boldsymbol{M}_{i,0}$. To obtain the corresponding Mueller matrix $\boldsymbol{M}_i$ for an arbitrary position angle α (measured in the counter-clockwise direction as seen from the observer), we have to apply the rotation matrix

$$\boldsymbol{R}(\alpha) = \begin{pmatrix} 1 & 0 & 0 & 0 \\ 0 & \cos 2\alpha & \sin 2\alpha & 0 \\ 0 & -\sin 2\alpha & \cos 2\alpha & 0 \\ 0 & 0 & 0 & 1 \end{pmatrix} \tag{13.3}$$

and its inverse, $\boldsymbol{R}(-\alpha)$:

$$\boldsymbol{M}_i = \boldsymbol{R}(-\alpha)\boldsymbol{M}_{i,0}\boldsymbol{R}(\alpha)\,. \tag{13.4}$$

With this general formalism it is straightforward to treat any linear optical train of any complexity that does not involve coherence phenomena (coherent superposition of wave trains from different subapertures). Matrix multiplication is easy to do on the computer, both numerically and analytically (using algebraic languages like MAPLE or Mathematica).

Coherent superposition of wave trains from different subapertures is only of interest if one wants to know the subtle instrumental polarization patterns within the diffraction element (Airy disk) of the telescope or within the individual speckles in speckle interferometry. The calculation of these effects requires the use of ray tracing with a Jones formalism rather than a Stokes formalism. These tiny polarization effects however vanish when spatially averaging over the Airy disk (Sánchez-Almeida and Martínez Pillet, 1992), which is approximately done in solar observations, since no solar information is contained at scales beyond the diffraction limit. It has been claimed that a significant net effect remains after such averaging, even on the optical axis of perfectly rotationally symmetric telescopes (McGuire and Chipman, 1988). However, it immediately follows from simple symmetry arguments (when the constraints of rotational and left-right symmetries are applied) that the instrumental polarization on the optical axis of such an ideal telescope is exactly zero, regardless of the focal ratios or mirror coatings that are assumed. Experiments on real telescopes also verify that the instrumental polarization is orders of magnitude smaller than the claimed values. Thus, for all practical purposes, the Mueller calculus of Eqs. (13.1)–(13.4) is sufficient.

13.3. Mueller Matrices for Different Types of Optics

We will now give the Mueller matrices $\boldsymbol{M}_{i,0}$ in the preferred Stokes coordinate system for the main types of optical components that will be encountered.

13.3.1. Partial linear polarizers

If the complex electrical vector components E_j, $j = 1,2$, of Eq. (2.27) are differentially attenuated, such that $E'_j = k_j E_j$, where k_j are real numbers, then one can easily derive (e.g. using the expressions of Sect. 2.6.3) that the Mueller matrix for such an optical element is

$$\boldsymbol{M} = \frac{1}{2} \begin{pmatrix} k_1^2 + k_2^2 & k_1^2 - k_2^2 & 0 & 0 \\ k_1^2 - k_2^2 & k_1^2 + k_2^2 & 0 & 0 \\ 0 & 0 & 2k_1 k_2 & 0 \\ 0 & 0 & 0 & 2k_1 k_2 \end{pmatrix} . \tag{13.5}$$

The $j = 1$ direction corresponds to the positive Stokes Q direction. When $k_1 = k_2 = 1$ we retrieve the identity matrix $\boldsymbol{E}$ as expected. When $k_1 = 1$ while $k_2 = 0$, we obtain the Mueller matrix for a 100 % linear polarizer that has its polarizing axis along the Stokes Q direction.

13.3.2. Polarizing beam splitter

For a polarizing beam splitter (like a Wollaston prism or a calcite beam splitter) that produces two output beams of orthogonal linear polarization, parallel and perpendicular to the positive Stokes Q direction (at position angles $\alpha = 0$ and $\pi/2$, respectively), we immediately obtain from Eq. (13.5) that

$$\boldsymbol{M} = \frac{1}{2} \begin{pmatrix} 1 & \pm 1 & 0 & 0 \\ \pm 1 & 1 & 0 & 0 \\ 0 & 0 & 0 & 0 \\ 0 & 0 & 0 & 0 \end{pmatrix} . \tag{13.6}$$

13.3.3. Retarder

For a retarder that has its *fast* axis along the positive Stokes Q direction it is straightforward to derive (cf. Sect. 2.6.3) that

$$\boldsymbol{M} = \begin{pmatrix} 1 & 0 & 0 & 0 \\ 0 & 1 & 0 & 0 \\ 0 & 0 & \cos\delta & \sin\delta \\ 0 & 0 & -\sin\delta & \cos\delta \end{pmatrix} , \tag{13.7}$$

where δ is the phase retardation in radians.

13.3.4. MODULATOR

The polarization modulators that have been used in solar work are variable retarders. Mechanical modulation is usually done with a rotating wave plate (retarder). Its Mueller matrix is obtained if we insert the matrix of Eq. (13.7) in Eq. (13.4) and let the position angle $\alpha = \Omega(t - t_0)$ be a linear function of time t, with Ω being the angular frequency of rotation.

Among the electrooptical modulators the main types in use are Pockels cells, piezoelastic modulators, and recently also liquid crystals. The Pockels cells have the advantage that the frequency and wave form (e.g. a square wave) can be freely chosen, but their acceptance angle is small (about 2°), they are alignment and temperature sensitive, require a high modulation voltage (several kV), suffer from electrode degradation, and have field of view errors associated with their natural birefringence. Piezoelastic modulators on the other hand are stable and very easy to handle, have an acceptance angle of about 45°, require a small driving power (generally < 1 W), and have very small residual natural birefringence. Any glass material can be used, e.g. $MgFl_2$ for the vacuum ultraviolet, or $CaFl_2$ for the infrared.

In contrast to Pockels cells the wave form for the piezoelastic modulators cannot be chosen, it is always sinusoidal. Because of the superior properties of the piezoelastic modulators, and since the polarimeter systems that we will later discuss make use of them, we will now describe them in some detail including their Mueller matrices.

The principles of a piezoelastic modulator were first given by Kemp (1969). Various applications have been discussed by Stenflo (1984) in terms of a Mueller matrix formalism. A piezoelastic transducer sets up a standing acoustic wave at the fundamental-mode frequency. This standing pressure wave induces a variable strain, which generates a birefringence that oscillates with the acoustic wave.

Since the natural resonant frequency of the glass plate is used, the frequency of the oscillation is very stable, and the driving electric power needed is very small. The resonant frequency is determined by the dimensions of the glass plate, by the ratio between the sound speed and twice the length of the plate. A typical value is 50 kHz. As the oscillation is sinusoidal, the retardation δ_m will also vary sinusoidally. Thus

$$\delta_m = A \sin \omega t \,, \tag{13.8}$$

where A is the retardation amplitude in radians, and ω is the modulation frequency. The Mueller matrix for a piezoelastic modulator is obtained simply by replacing δ in Eq. (13.7) by δ_m.

The $\cos\delta_m$ and $\sin\delta_m$ terms can be expanded in Bessel functions. This results in the appearance of harmonics of the modulation frequency ω as well as the dc (zero) frequency. Retaining only the dominating terms of the expansion,

$$\begin{aligned} \sin\delta_m &\approx 2J_1(A)\sin\omega t \,, \\ \cos\delta_m &\approx J_0(A) + 2J_2(A)\cos 2\omega t \,, \end{aligned} \tag{13.9}$$

where $J_{0,1,2}$ are the Bessel functions of orders $0, 1, 2$. It is convenient to separate the contributions to the Mueller matrix $\boldsymbol{M}$ from the three different frequencies and write

$$\boldsymbol{M}(A) \approx \boldsymbol{M}_0(A) + \boldsymbol{M}_\omega(A) \sin \omega t + \boldsymbol{M}_{2\omega}(A) \cos 2\omega t \,. \tag{13.10}$$

Inserting Eq. (13.9) in Eq. (13.7) we find

$$\begin{aligned} \boldsymbol{M}_0(A) &= \begin{pmatrix} 1 & 0 & 0 & 0 \\ 0 & 1 & 0 & 0 \\ 0 & 0 & J_0(A) & 0 \\ 0 & 0 & 0 & J_0(A) \end{pmatrix}, \\ \boldsymbol{M}_\omega(A) &= 2J_1(A) \begin{pmatrix} 0 & 0 & 0 & 0 \\ 0 & 0 & 0 & 0 \\ 0 & 0 & 0 & 1 \\ 0 & 0 & -1 & 0 \end{pmatrix}, \\ \boldsymbol{M}_{2\omega}(A) &= 2J_2(A) \begin{pmatrix} 0 & 0 & 0 & 0 \\ 0 & 0 & 0 & 0 \\ 0 & 0 & 1 & 0 \\ 0 & 0 & 0 & 1 \end{pmatrix}. \end{aligned} \tag{13.11}$$

13.3.5. Reflection and Transmission at a Dielectric Surface

For light impinging on a dielectric surface Snell's law of refraction states that

$$\sin i = n \sin b \,, \tag{13.12}$$

where i is the angle of incidence (in air or vacuum), b the angle of refraction, and n the refractive index of the dielectric medium. Let us decompose the electric vector into its components $E_\parallel$ and $E_\perp$ parallel and perpendicular to the plane of incidence. Then, according to Fresnel's equations for reflection (Born and Wolf, 1970; Collett, 1993), the reflected vector $\boldsymbol{E}'$ is determined by the relations

$$\begin{aligned} \frac{E'_\parallel}{E_\parallel} &= \frac{n \cos i - \cos b}{n \cos i + \cos b} \,, \\ \frac{E'_\perp}{E_\perp} &= \frac{\cos i - n \cos b}{\cos i + n \cos b} \,. \end{aligned} \tag{13.13}$$

With Snell's law (13.12) this becomes

$$\begin{aligned} \frac{E'_\parallel}{E_\parallel} &= \frac{\tan(i-b)}{\tan(i+b)} \,, \\ \frac{E'_\perp}{E_\perp} &= -\frac{\sin(i-b)}{\sin(i+b)} \,. \end{aligned} \tag{13.14}$$

Let us choose the Stokes coordinate system such that the positive Stokes Q direction is perpendicular to the plane of incidence (along the $E_\perp$ direction). Then the expressions (13.14) may be used for k_1 and k_2 in Eq. (13.5), which gives us the Mueller matrix $\boldsymbol{M}_r$ for reflection:

$$M_r = \frac{1}{2}\frac{1-c_-^2}{c_-^2(1-c_+^2)}\begin{pmatrix} c_-^2+c_+^2 & c_-^2-c_+^2 & 0 & 0 \\ c_-^2-c_+^2 & c_-^2+c_+^2 & 0 & 0 \\ 0 & 0 & -2c_-c_+ & 0 \\ 0 & 0 & 0 & -2c_-c_+ \end{pmatrix}, \tag{13.15}$$

where

$$c_\pm = \cos(i \pm b)\,. \tag{13.16}$$

When comparing with the matrix (13.5) for a partial polarizer we notice the minus signs in Eq. (13.15) in the last two diagonal terms. The reason for this sign change is that a reflection reverses left and right, so that Stokes U becomes $-U$ ($+45°$ becomes $-45°$), and Stokes V becomes $-V$ (counterclockwise becomes clockwise).

Similarly the Mueller matrix M_t for transmission at a dielectric surface can be obtained. It also has the form (13.5) of a partial linear polarizer:

$$M_t = \frac{1}{2}\frac{c_-^2-c_+^2}{c_-^2(1-c_+^2)}\begin{pmatrix} c_-^2+1 & c_-^2-1 & 0 & 0 \\ c_-^2-1 & c_-^2+1 & 0 & 0 \\ 0 & 0 & 2c_- & 0 \\ 0 & 0 & 0 & 2c_- \end{pmatrix}. \tag{13.17}$$

It is easy to verify that

$$\begin{aligned} M_{r,11} + M_{t,11} &= 1\,, \\ M_{r,12} + M_{t,12} &= 0 \end{aligned} \tag{13.18}$$

as required by energy conservation ($I'_r + I'_t = I$ regardless of the polarization state of the incident radiation).

The above expressions are valid for the transition air $\rightarrow$ dielectric. The expressions for the reverse direction remain the same if we let angle b refer to the incident, angle i to the refracted beam.

For small angles, when 2nd and higher orders are neglected in an expansion in terms of i and b,

$$\begin{aligned} M_r &\approx \left(\frac{n-1}{n+1}\right)^2 \begin{pmatrix} 1 & 0 & 0 & 0 \\ 0 & 1 & 0 & 0 \\ 0 & 0 & -1 & 0 \\ 0 & 0 & 0 & -1 \end{pmatrix}, \\ M_t &\approx \frac{4n}{(n+1)^2}E\,, \end{aligned} \tag{13.19}$$

where as before E is the identity matrix (diagonal elements unity, off-diagonal elements zero).

At the Brewster angle $i = i_B$, which is determined by

$$\tan i_B = n\,, \tag{13.20}$$

the Mueller reflection matrix M_r of Eq. (13.15) represents a 100 % linear polarizer:

$$M_{r,B} = \tfrac{1}{2}\cos^2 2i_B \begin{pmatrix} 1 & 1 & 0 & 0 \\ 1 & 1 & 0 & 0 \\ 0 & 0 & 0 & 0 \\ 0 & 0 & 0 & 0 \end{pmatrix}. \tag{13.21}$$

Brewster-angle reflection can be used for polarization analysis in cases when transmission optics will not work, e.g. in the EUV region of the spectrum ($\lesssim$ 1000 Å), but it has the disadvantage that the reflectivity $\cos^2 2i_B$ is rather low. A better solution is to use special metallic coatings, like gold or platinum, which exhibit partial but strong Brewster-like polarization properties while having high reflectivity. This method was used with a gold-coated mirror to explore the solar linear polarization in the hydrogen Lyman α line at 1216 Å from the satellite Intercosmos 16 (Stenflo et al., 1976, 1980b).

13.3.6. Reflection by a mirror

A metallic surface differs from a dielectric by its electrical conductivity. This leads to absorption of the impinging radiation due to induced electric currents in the conducting metallic medium. In spite of this difference we can still retain the Fresnel expressions (13.13) if we replace the ordinary refractive index n by a complex refractive index $\mathbf{n}$ defined by

$$\mathbf{n} = n - ik\,, \tag{13.22}$$

where k is the extinction coefficient. We can then proceed from Eq. (13.13) and the expressions of Sect. 2.6.3 to derive the Mueller matrix for reflection. It can be regarded as a generalization of Eq. (13.15) to the case of a non-zero k.

The presence of an imaginary term in the refractive index gives rise to a phase shift and makes the surface behave like a retarder in addition to a partial linear polarizer. The resulting reflection Mueller matrix is in normalized form (not accounting for the common reflectivity factor)

$$M = \frac{1}{2}\begin{pmatrix} 1+\rho^2 & 1-\rho^2 & 0 & 0 \\ 1-\rho^2 & 1+\rho^2 & 0 & 0 \\ 0 & 0 & -2\rho\cos\delta & -2\rho\sin\delta \\ 0 & 0 & 2\rho\sin\delta & -2\rho\cos\delta \end{pmatrix}. \tag{13.23}$$

A comparison with Eqs. (13.5) and (13.7) shows that Eq. (13.23) represents a combination of a partial linear polarizer and a retarder.

The expressions for ρ and δ can most conveniently be given in terms of the following parameters:

$$\begin{aligned} p &= n^2 - k^2 - \sin^2 i\,, \\ q &= 2nk\,, \\ r_\pm &= \frac{1}{\sqrt{2}}(\,\pm p + \sqrt{p^2+q^2}\,)^{1/2}\,, \\ s &= \sin i\,\tan i\,. \end{aligned} \tag{13.24}$$

We then get

$$\begin{aligned} \rho^2 &= \frac{\sqrt{p^2+q^2}\ + s^2 - 2\,s\,r_+}{\sqrt{p^2+q^2}\ + s^2 + 2\,s\,r_+}\,, \\ \tan\delta &= \frac{2\,s\,r_-}{\sqrt{p^2+q^2}\ - s^2}\,. \end{aligned} \tag{13.25}$$

Note that in contrast to the expressions usually given in previous literature, the above expressions involve no approximations and are valid for all angles of incidence.

It is straightforward to verify that in the special case when $k = 0$ the Mueller matrix $\boldsymbol{M}_r$ of Eq. (13.15) for a dielectric is retrieved (in normalized form) from the general matrix (13.23).

13.4. Weakly Polarizing Optical Train

The Mueller matrices of different optical components generally do not commute. Thus for the optical train described by Eq. (13.2) the order in which the components occur is essential. If the beam direction is reversed, the resulting polarization will usually be quite different. For instance, a linear polarizer followed by a quarter-wave retarder ($\lambda/4$ plate) with its fast optical axis at 45° to the polarizing direction produces a circularly polarized output beam. If the beam direction is reversed, linearly polarized output is obtained.

In the case when the optical components are only weakly polarizing, however, symmetry is restored, and the Mueller matrices commute. This greatly simplifies the analysis and calibration of almost polarization-free telescopes. Such telescopes will be increasingly used for precision solar polarimetry in the future. Although "polarization-free" in a first approximation, their small residual polarization (e.g. due to stresses or imperfections) needs to be accounted for.

Let us thus assume that all components of the optical train are weakly polarizing, which means that their Mueller matrices $\boldsymbol{M}_i$ may be written in the form

$$\boldsymbol{M}_i = \boldsymbol{E} + \boldsymbol{m}_i\,, \tag{13.26}$$

where the components $m_{i,jk} \ll 1$ for all i, j, k. Let us next consider the two components with $i = 1$ and 2. With Eq. (13.26) we get

$$\boldsymbol{M}_2\boldsymbol{M}_1 = \boldsymbol{E} + \boldsymbol{m}_1 + \boldsymbol{m}_2 + \boldsymbol{m}_2\boldsymbol{m}_1\,. \tag{13.27}$$

To first order in $\boldsymbol{m}_i$ we have

$$\boldsymbol{M}_2\boldsymbol{M}_1 \approx \boldsymbol{M}_1\boldsymbol{M}_2 \approx \boldsymbol{E} + \boldsymbol{m}_1 + \boldsymbol{m}_2\,. \tag{13.28}$$

Commutation has been achieved, because in the weakly polarizing limit matrix multiplication can be converted to matrix summation.

Repeating the above procedure we obtain for the full optical train of Eq. (13.2):

$$\boldsymbol{M} \approx \boldsymbol{E} + \sum_{i=1}^{n} \boldsymbol{m}_i\,. \tag{13.29}$$

For a weakly polarizing optical system we may thus simply add all the individual matrices $\boldsymbol{M}_i - \boldsymbol{E}$ of all the components. In addition we may also average the Mueller matrices of the separate subapertures over the entire aperture. Optical components and subapertures may be mixed in any arbitrary order (Stenflo, 1991d).

If we rotate the matrix (13.5) for a partial linear polarizer, non-zero matrix components M_{31} and M_{13} connecting Stokes I and U with each other will appear. The most general form of a weakly polarizing partial linear polarizer is therefore

$$\boldsymbol{M} = \boldsymbol{E} + \begin{pmatrix} 0 & p & q & 0 \\ p & 0 & 0 & 0 \\ q & 0 & 0 & 0 \\ 0 & 0 & 0 & 0 \end{pmatrix} . \tag{13.30}$$

Similarly, rotation of a retarder matrix (13.7) introduces non-zero M_{42} and M_{24} components. The most general form for a weak retarder is thus

$$\boldsymbol{M} = \boldsymbol{E} + \begin{pmatrix} 0 & 0 & 0 & 0 \\ 0 & 0 & 0 & -s \\ 0 & 0 & 0 & r \\ 0 & s & -r & 0 \end{pmatrix} . \tag{13.31}$$

We notice that while the matrix in Eq. (13.30) is symmetric, the one in Eq. (13.31) is anti-symmetric.

As we have seen from Eqs. (13.15), (13.17), and (13.23), the polarizing effects of different optical components can generally be described in terms of a partial linear polarizer, a retarder, or a combination of both. Stresses in an entrance window may for instance generate phase retardations, but the orientation of the stress axis may vary randomly over the window. Since however the symmetry properties are not changed by averaging, the form (13.31) will be preserved when calculating the average Mueller matrix of the entire entrance window. Similarly, as the polarization effects of weakly polarizing optical components may be added in any arbitrary order, and since the operation of addition does not alter the symmetry properties of Eqs. (13.30) and (13.31), the general Mueller matrix for an optical train of weakly polarizing optical components with arbitrary orientations of their polarizing or retarding axes is given by

$$\begin{aligned} \boldsymbol{M} &= \boldsymbol{E} + \boldsymbol{m} , \\ \boldsymbol{m} &= \begin{pmatrix} 0 & p & q & 0 \\ p & 0 & 0 & -s \\ q & 0 & 0 & r \\ 0 & s & -r & 0 \end{pmatrix} . \end{aligned} \tag{13.32}$$

This is a general result, valid as long as $|p|$, $|q|$, $|r|$, $|s| \ll 1$. It simplifies the calibration problem tremendously, since instead of 16 free parameters we need only the four p, q, r, and s to completely determine the polarization properties of the optical train.

For a general optical system that is not limited to a combination of retarders and partial linear polarizers it can be shown (Sánchez Almeida and Vela Villahoz, 1993) that the weakly polarizing condition leads to the same form as Eq. (13.32) but with the additional non-zero components $m_{14} = m_{41}$ and $m_{23} = -m_{32}$. This form is identical to that of the Mueller matrix $\boldsymbol{\eta}$ of Eq. (11.3) for differential absorption of polarized light in a magnetized atmosphere (including anomalous dispersion).

13.5. Instrumental Polarization

An observing system for solar polarimetry, like a magnetograph or a Stokesmeter, is calibrated by inserting polarizing plates via a filter wheel in front of the optics used for polarization analysis (which may include polarization modulators). In principle four filter wheel positions are sufficient, representing the filters of Fig. 2.1 that were used for an operational definition of the Stokes parameters: empty position, linear polarizers at position angles 0° and 45°, and a circular polarizer. This calibration however provides no information on the instrumental polarization *in front of* the filter wheel.

With instrumental polarization we thus mean any polarization that arises in the telescope optics that precedes the location where the polarimetry system is calibrated. The Mueller matrix that characterizes this part of the telescope will be denoted $\boldsymbol{M}_T$. Any polarization produced after this place in the telescope is of no consequence for the polarimetric accuracy (assuming that the last element of the optics used for polarization analysis is a perfect linear polarizer, which is important to prevent polarization effects later in the system to affect the measurements). The calibration should thus be made as early as possible in the system, and the front end of the telescope should be designed to have minimal polarization.

The corrupting effects of the Mueller matrix $\boldsymbol{M}_T$ may be eliminated either by measuring $\boldsymbol{M}_T$, by optically compensating for it, or by correcting for it in the off-line analysis.

13.5.1. Measurement of the Instrumental Polarization

The polarimetry system measures an apparent Stokes vector $\boldsymbol{I}'$ that has been corrupted by the instrumental polarization. If the instrumental polarization matrix $\boldsymbol{M}_T$ can be determined, the true Stokes vector $\boldsymbol{I}$ of the light that enters the telescope can be retrieved by simple matrix inversion and multiplication:

$$\boldsymbol{I} = \boldsymbol{M}_T^{-1}\,\boldsymbol{I}' . \tag{13.33}$$

Direct measurement of $\boldsymbol{M}_T$ is problematic for telescopes with large apertures, since polarizing filters that cover the entire aperture would have to be inserted in front of the telescope. Such huge polarizers or mosaics of smaller polarizers cannot be manufactured to be very homogeneous, which means that the polarimetric accuracy that can be reached is rather limited.

For heliostat or coelostat systems with an oblique reflection at the first optical element, the instrumental polarization varies strongly with hour angle and declination. To account for these large variations one may develop an analytical model of the telescope, in which the oblique reflections at the coated surfaces are described via Eqs. (13.23)–(13.25). Such a model is characterized only by the two free parameters n and k, since all the angles of incidence and position angles are known from the geometry of the beam path (using the tools of spherical trigonometry). Furthermore, since n and k are known as a function of wavelength for different metals, they are approximately known for the actual coatings used, so the measurement of $\boldsymbol{M}_T$ will only consist of determining the corrections to the nominal values of n and k that best fit the measurements. For aluminum the nominal values at a wavelength of 5000 Å are $n = 0.77$ and $k = 6.1$ (Palik, 1985). If n and k are empirically determined in this way for a few representative wavelengths and hour angles, the model should give a reasonably good representation of $\boldsymbol{M}_T$ for all times and wavelengths of interest.

If entrance and exit windows are part of $\boldsymbol{M}_T$, they may be analytically modeled as retarders to first order. Each window is then characterized by the two free parameters δ (retardation) and α (azimuth angle). All the free parameters of the analytical model may be determined by an iterative least squares fit to suitable observables, as indicated in Sect. (13.5.3).

When high-precision polarimetry of the full Stokes vector is needed, heliostat or coelostat systems are not adequate, but rotationally symmetric optics in front of the location where the polarization calibration is carried out should be used. Such rotationally symmetric systems are usually called "polarization free", and they may be described in terms of the weakly polarizing matrix (13.32) of the previous section. A strictly polarization free telescope would have $\boldsymbol{m} = \boldsymbol{0}$, but in practice mechanically and thermally induced stresses and optical imperfections may produce non-zero elements of $\boldsymbol{m}$. Since the mechanical stresses depend on the pointing direction of the telescope and the thermal stresses on the heat load, precision calibration cannot be made by pointing the telescope at some external (non-solar) light source. It has to be done when the telescope is pointed towards the sun.

As will be shown in Sect. 13.5.3, much, although not all, of the instrumental polarization effects may be determined and corrected for in the off-line data analysis. One way to directly measure the telescope Mueller matrix of a weakly polarizing telescope has been described in Stenflo (1991d). A laser near the final telescope focus (in the observing room) is used to send a beam backwards through the optical system. Most of the beam, which has been spatially filtered and expanded to fit the telescope f-ratio, will exit the telescope towards the sun, but a fraction of it will be reflected at the two plane surfaces of the entrance window at the front end of the telescope. The two reflected beams will after their second pass through the telescope form two images, which will not overlap if the entrance window has been manufactured with a small angle between the front and back surfaces to avoid fringing. With this double-pass method the modulator package will have the dual function of calibrating and analysing polarization optics.

It can be shown that the components of $\boldsymbol{M}_T$ may be determined this way. It is the double-pass polarization matrix $\boldsymbol{E} + 2\boldsymbol{m}$ that is measured, from which the wanted single-pass matrix $\boldsymbol{M}_T = \boldsymbol{E} + \boldsymbol{m}$ is immediately obtained. The method is only applicable for nearly polarization-free systems, since only for such systems the order in which the polarizing components are encountered is immaterial (cf. Eq. (13.29)).

13.5.2. Compensation of the Instrumental Polarization

An approximate method that has often been used to compensate the instrumental polarization is to place in front of the optics for polarization analysis a tilting plane-parallel glass plate to generate partial linear polarization and a Soleil compensator to produce any desired phase retardation. The settings of these compensating elements are chosen such that the observed linear and circular polarization in the continuous spectrum at the center of the solar disk vanish.

Such a method is however of limited value since it will only eliminate the polarization effects in the first column of $\boldsymbol{M}_T$, which may be easily eliminated in the off-line analysis as well without any compensation. Furthermore, this type of compensation of the first column of $\boldsymbol{M}_T$ only works for one wavelength at a time. Nearly complete and achromatic compensation of *all* elements of $\boldsymbol{M}_T$ is however possible by designing the compensator as a model "anti-telescope", as has been done by Harvey (1985) for the NSO McMath-Pierce facility. This is achieved by letting the compensator consist of the same type of optical components as the telescope itself, with identical angles of incidence, but with the difference that the elements of the compensator optics are rotated in azimuth by 90° with respect to the telescope, and the order in which the optical elements are encountered by the beam is reversed.

Let $\boldsymbol{M}_i(\alpha)$ be the Mueller matrix of an optical component at position angle α. Then, if $\boldsymbol{M}_i$ is a partial linear polarizer or a retarder, one easily finds that

$$\boldsymbol{M}_i(\alpha)\boldsymbol{M}_i(\alpha \pm 90^\circ) = \boldsymbol{E}\,, \tag{13.34}$$

which shows that the polarization effect of $\boldsymbol{M}_i(\alpha)$ is compensated by $\boldsymbol{M}_i(\alpha \pm 90^\circ)$. Hence, if $\boldsymbol{M}_T$ is composed of a train of partial linear polarizers and/or retarders (which includes mirror reflections as shown by the matrix (13.23)) such that

$$\boldsymbol{M}_T = \boldsymbol{M}_n(\alpha_n)\boldsymbol{M}_{n-1}(\alpha_{n-1})\ldots\boldsymbol{M}_2(\alpha_2)\boldsymbol{M}_1(\alpha_1)\,, \tag{13.35}$$

then the "anti-telescope" compensator matrix $\boldsymbol{M}_C$ of an ideal, achromatic compensator that makes

$$\boldsymbol{M}_C\boldsymbol{M}_T = \boldsymbol{E} \tag{13.36}$$

is

$$\boldsymbol{M}_C = \boldsymbol{M}_1(\alpha_1 \pm 90^\circ)\boldsymbol{M}_2(\alpha_2 \pm 90^\circ)\ldots\boldsymbol{M}_{n-1}(\alpha_{n-1} \pm 90^\circ)\boldsymbol{M}_n(\alpha_n \pm 90^\circ)\,. \tag{13.37}$$

Note that $\boldsymbol{M}_C$ cannot in general be represented by a tilting glass plate plus a Soleil compensator, since when $\boldsymbol{M}_T$ is composed of a train of not weakly polarizing retarders and partial linear polarizers, the resulting matrix product does not have the form (13.23) or (13.32) of a partial polarizer – retarder combination.

13.5.3. OFF-LINE CORRECTION FOR THE INSTRUMENTAL POLARIZATION

The various components of the instrumental polarization matrix $\boldsymbol{M}_T$ represent various forms of cross talk between the Stokes parameters. Since the intrinsic degree of polarization $(Q^2 + U^2 + V^2)^{1/2}/I$ is usually $\ll 1$ on the sun except in sunspots and spatially resolved magnetic flux tubes, it is the cross talk from Stokes I into the other Stokes parameters that is the dominating effect. This cross talk, which is represented by the first column of $\boldsymbol{M}_T$, can however normally be calibrated for or be eliminated in the off-line analysis. In the uncorrected fractional polarization images Q'/I', U'/I', and V'/I' when $I' \approx M_{T,11}I$ (insignificant distortion of Stokes I by cross talk *from* Q, U, and V) the non-zero off-diagonal elements of the first column of $\boldsymbol{M}_T$ cause an unstructured zero-line offset that is equal to $M_{T,i1}/M_{T,11}$, $i = 2, 3, 4$.

In a spectral recording this offset has no spectral features (the wavelength dependence of $M_{T,i1}$ is very slow in comparison with the widths of spectral lines), so the shapes of the polarized line profiles are unaffected. In monochromatic polarized images obtained with a narrow-band filter this offset is spatially flat and does not distort the structures or the morphology. As it is safe to assume that the spatially averaged continuum polarization at disk center is intrinsically zero, the absolute value of the offset can easily be determined from spectral recordings. In narrow-band filter images the absolute zero level can often be found from the circumstance that the solar magnetic flux occurs in discrete elements separated by larger areas of nearly zero net flux, which may thus be used for the offset calibration.

The first column of $\boldsymbol{M}_T$ thus does not cause problems. It is the other cross talk terms that are the difficult ones. For certain types of observing programs their effects can however be neglected. When one for instance explores the center-to-limb variation of the polarization due to coherent scattering, symmetry arguments demand that the resulting, spatially averaged polarization is linear, with the polarization plane either parallel or perpendicular to the nearest solar limb. If Stokes Q is defined to be along this direction, then the intrinsic Stokes U and V can be assumed to be zero. Similarly, when observing Stokes V near disk center away from sunspots, the $Q, U \to V$ cross talk may usually be neglected since the polarization from the transverse Zeeman effect is generally much smaller than that of the longitudinal Zeeman effect as long as the magnetic sensitivity is not too large. The flux tubes are expected to be more vertical than horizontal and thus more longitudinal than transverse. The validity of the assumption that Stokes Q and U are smaller than V can easily be checked if the apparent values of Q' and U' are also recorded simultaneously. Whenever possible the full Stokes vector should be recorded to keep the instrumental polarization effects under control, also when one is only interested in the determination of one of the three parameters Q, U, and V.

In the general case when Q, U, and V are of the same orders of magnitude, a more sophisticated analysis is needed. November (1991) has suggested that one may use the symmetry properties of the Zeeman-effect line profiles to sort out the cross-talk terms. Thus, for an atmosphere without velocity gradients, the Stokes Q and U line profiles are symmetric with respect to the line center, while the Stokes V profiles are anti-symmetric. It could then be possible to determine the various cross-talk terms by measuring the symmetric component of V' and the anti-symmetric components of Q' and U'. The accuracy of this method is however limited by the circumstance that the symmetry properties are broken in the real solar atmosphere due to the correlated magnetic and velocity field gradients of magnetic flux concentrations, as we have seen in Chapter 12. In principle these symmetry-breaking effects may be calibrated, since they depend on the line strengths and excitation potentials of the various spectral lines, but such a calibration would be of a statistical nature and would thus be of limited accuracy when applied to a smaller data set. Nevertheless such symmetry considerations may still provide valuable first approximations to the various cross-talk terms. Of particular interest in this respect is the 1.5649 μm line ($g = 3$), since its Stokes V profile is practically anti-symmetric in kG flux tubes, as has been pointed out by Solanki et al. (1992).

The situation is helped by the circumstance that $\boldsymbol{M}_T$ itself exhibits symmetry properties. The structure represented by Eq. (13.32) for a nearly polarization-free system is often a good approximation also for a rather strongly polarizing system. Thus the $\boldsymbol{M}_T$ determined by Harvey (1985) for the strongly polarizing NSO McMath-Pierce facility (a heliostat system) is approximately of this form. When this is the case, $\boldsymbol{M}_T$ may be characterized by only the four parameters p, q, r, and s. As p and q are immediately obtained from the observed offset in the fractional Q'/I' and U'/I' images, there are actually only *two* unknown parameters remaining, i.e., r and s. They represent cross talk $Q, U \leftrightarrow V$ (between linear and circular polarization).

A much more precise determination of $\boldsymbol{M}_T$ in the off-line analysis without relying on symmetry assumptions is possible by modelling $\boldsymbol{M}_T$ analytically as indicated in Sect. (13.5.1) and determining the free parameters of the model by an iterative least squares fit to the measured $I \rightarrow Q, U, V$ and $V \rightarrow Q, U$ cross talks. The $I \rightarrow Q, U, V$ cross talk terms are readily obtained from the off-sets that we have discussed. The $V \rightarrow Q, U$ cross talk will be considered next.

One way to determine the $V \rightarrow Q, U$ cross talk through solar observations anywhere in the spectrum is to locate regions near the center of the solar disk with nearly vertical magnetic fields and record the Q', U', and V' signals in lines with moderately small Landé factors. In such lines the transverse Zeeman-effect polarization is much smaller than the longitudinal Zeeman-effect polarization, even for kG fields that are inclined as much as 45° with respect to the vertical direction. Without the need for any subtle discussion of the symmetry properties of the Zeeman effect on the sun, the observed Q' and U' line profiles should arise almost entirely from the cross talk $V \rightarrow Q, U$. This expectation can be immediately checked, since the Q' and U' line profiles should then have the same shape as the observed V' profile (after the $I \rightarrow V$ cross talk has been eliminated via the off-set determinations). If we thus fit the observed and off-set corrected Q'/I' and U'/I' line profiles with a function $\epsilon_{Q,U}\, V'/I'$, then the values of the two free parameters ϵ_Q and ϵ_U determined from the fit give us the wanted components $M_{T,24}$ and $M_{T,34}$. This procedure should be able to give quite accurate results.

An even more accurate method that however only works for certain wavelengths is to use a spectral line for which the Zeeman displacements of the π and σ components are identical, with the consequence that all linear polarization effects exactly cancel out. The intrinsic values of Q and U are therefore zero. Such spectral lines are fairly rare, but one useful representative is the Fe I 6149 Å line (Lites, 1993). Measurements of the apparent Q' and U' signals in such a line directly gives the $I \rightarrow Q, U$ and $V \rightarrow Q, U$ cross talk terms. This technique has been implemented with the Advanced Stokes Polarimeter (Lites et al., 1991). Other useful spectral lines of this kind have been listed by Vela Villahoz et al. (1993).

13.6. Removal of Atmospheric Disturbances

Turbulence in the earth's atmosphere may corrupt the polarimetric information in two fundamentally different ways: (1) Spurious or artificial polarization features may be produced if the images that represent different polarization states have not been affected by the identical seeing disturbance. (2) Seeing causes a spatial redistribution of the rays from the sun (image motions, smearing, stray light), which affects polarized and unpolarized images in the same way. While the first problem has to be eliminated in the design of the system for polarization analysis, as will be explained in the next subsection, the second problem can be addressed in terms of methods introduced for unpolarized radiation, like speckle interferometry, phase diversity, and adaptive optics. The application of the image reconstruction techniques (speckle interferometry and phase diversity) to polarimetry however requires a special "trick" to overcome the signal-to-noise "hurdle", as we will see.

13.6.1. Elimination of seeing-induced polarization signals

Stokes Q, U, and V images can only be formed from linear combinations of two or more images recorded in different polarization states. With an ideal linear polarizer oriented at 0° and 90°, for instance, the two recorded images have intensities $\frac{1}{2}(I \pm Q)$, which means that the sum image represents Stokes I, the difference Stokes Q. An image of the full Stokes vector requires recording and linear combination of at least four images representing different polarization states. If the detailed distribution of the seeing distortions and stray light is not fully identical in these four images, spurious polarization features in Q, U, and V will be produced when such images are combined, due to the differential seeing effects.

The atmospheric seeing fluctuations vary over a broad range of frequencies extending upwards beyond 100 Hz at visible wavelengths, while being slower in the infrared. To avoid spurious seeing-induced polarization effects the images that represent the different polarization states need to be recorded simultaneously within about 1 ms or less in the visible. The state of the atmosphere can be regarded as "frozen" on such short time scales. This requirement implies that we either have to record the different polarization states simultaneously using polarizing beam splitters, or we have to modulate the polarization state at frequencies in the kHz range.

The static beam-splitter method however does not enable precision polarimetry. As two images of orthogonal polarization states fall on two different detector areas, drifts and small non-linearities in the detector will contribute to errors in the calibrated differential pixel gain table or flat field that will limit the accuracy of such polarimetry (Harvey, 1985). In the absence of seeing the gain-table effects could in principle be eliminated by using an optical "phase switch" in front of the beam-splitting analyser, such that the sign of the polarization information is reversed, while the spurious gain-table features keep their sign. Combination of two pairs of such images would then eliminate the flat-field problems. However, since the two image pairs corresponding to the two phase-switch settings have been recorded under different seeing conditions, gain-table errors will still be a source of fictitious polarization features, now due to the coupled seeing-flat field effects.

There are several additional drawbacks of beam-splitter systems. To record the full Stokes vector a complex system generating a minimum of four output beams is needed. In practice a six-beam system representing $I \pm Q$, $I \pm U$, and $I \pm V$ would be used (Wittmann, 1973; Harvey, 1985). This requires many optical surfaces. The optical polarization analysis has to be performed as early in the telescope system as possible to minimize the instrumental polarization, and the six beams would have to be transmitted together all the way from the polarimeter package to the end focus of the spectrograph or filter system, which is likely to introduce some differential aberrations between the six images.

All these problems may be avoided if we can instead modulate the polarization state in the kHz range. While mechanical modulation is generally too slow and introduces image motions due to residual wobbling or other imperfections, electrooptical modulation has the required speed and avoids all moving parts. As will be seen in Sect. 13.7 below one needs two electrooptical modulators to encode all four Stokes parameters simultaneously. One modulator followed by a linear polarizer is sufficient for the recording of the three Stokes parameters I, Q, and V, but a second modulator with a different position angle of its optical axis is needed to take care of Stokes U.

The big problem with such modulation schemes, however, is how they can be combined with the use of large array detectors (in particular charge-coupled devices or CCDs) with a slow read-out (0.2 s for a 1000×1000 pixel array with a read-out pixel rate of 5 MHz). Various approaches to this problem will be discussed in Sect. 13.8. As we will see the optimum solution appears to be the creation of several fast image buffers for temporary storage within the CCD chip itself.

13.6.2. Speckle Polarimetry

The degradation of the spatial resolution by the earth's atmosphere may be removed off-line by image reconstruction if the observations contain information on the instantaneous optical transfer function (the Fourier transform $\tilde{s}$ of the instantaneous point spread function s). The method of speckle interferometry, pioneered by Labeyrie (1970), has been successfully applied to solar observations (cf. von der Lühe, 1989) to accomplish such image restoration. Another even more promising method that is currently being tested is that of "phase-diversity restoration" (Gonsalves and Chidlaw, 1979; Löfdahl and Scharmer, 1993), but it is still in an experimental stage for solar observations.

Since the atmospheric turbulence varies on a short time scale, one needs short exposures ($\lesssim$ 20 ms in the visible) to "freeze" the seeing effects such that the image will contain information out to the diffraction limit of the telescope. The recorded image i will then also contain information on the instantaneous point spread function s, since it represents a convolution between s and the true image or "object" o:

$$i = o * s , \tag{13.38}$$

where the operator $*$ stands for convolution. In the Fourier domain convolution becomes multiplication, so that

$$\tilde{i} = \tilde{o}\,\tilde{s} , \tag{13.39}$$

where $\tilde{}$ means Fourier transformation.

With ordinary speckle interferometry about 100 short exposures are made within a time interval that is shorter than the evolutionary time scale of the solar structures (typically $\frac{1}{2}$ min), so that the object o can be regarded as identical for all the exposures (while s varies). The images i are then evaluated with special numerical techniques to recover the amplitudes and phases of $\tilde{o}$ and thus to reconstruct the true image o. It is of advantage to record a much larger number of frames within the 30 s interval and apply frame selection to pick the 100 best images to be used for the reconstruction. Such frame selection is very important to provide a data set of rather homogeneous quality, needed for the numerical reconstruction to work well.

Speckle interferometry in this form cannot be applied directly to observations with the high spectral resolution that is needed for most polarimetric work, since the photon flux is too small. The direct applications to resolve the fine structure of solar granulation have been made with broad-band filters (with a typical band width of 100 Å). In polarimetry, however, we want a spectral resolution of $\lesssim$ 1/8 Å with filters and $\lesssim$ 0.02 Å with spectrographs. With such narrow spectral bands the signal-to-noise ratio that can be achieved in a 20 ms exposure is far too low to allow the amplitude and phase information of $\tilde{o}$ to be retrieved. This situation is not improved by using telescopes with a larger photon-collecting area, since for such telescopes the diffraction-limited resolution element is correspondingly smaller. It is the number of collected photons per diffraction element that counts, and this quantity does not change with aperture size. An additional, complicating factor is that the evolutionary time scale of the smallest resolved solar structures decreases linearly with the diameter of the telescope aperture.

This problem has been solved (Keller and von der Lühe, 1992; Keller, 1992) by using two simultaneous channels fed by a beam splitter, one narrow-band channel for the polarimetry, and one broad-band channel for retrieving the information on the instantaneous optical transfer function. The short exposures are strictly simultaneous and of identical duration in the two channels. Since the instantaneous s is a slow function of wavelength, it can be considered to be identical in the two channels. We thus get

$$\begin{aligned} \tilde{i}_n &= \tilde{o}_n \tilde{s} , \\ \tilde{i}_b &= \tilde{o}_b \tilde{s} , \end{aligned} \tag{13.40}$$

where indices n and b refer to the narrow and broad-band channels.

With standard reconstruction algorithms the object transform $\tilde{o}_b$ of the broad-band channel can be retrieved. This then allows us to derive the instantaneous optical transfer function $\tilde{s}$ of each individual frame from

$$\tilde{s} = \tilde{i}_b / \tilde{o}_b \,. \tag{13.41}$$

Inserting this in Eq. (13.40) for the narrow-band channel and averaging over the about 100 images, we obtain the true narrow-band image transform as

$$\tilde{o}_n = \left\langle \frac{\tilde{i}_n}{\tilde{i}_b} \right\rangle \tilde{o}_b \,, \tag{13.42}$$

where $\tilde{o}_b$ appears outside the averaging bracket (ensemble average), since it is invariant.

The form (13.42) is however not suitable for calculations, since $\tilde{s}$ and $\tilde{i}_b$ have zeros at certain spatial frequencies. This problem can be avoided by introducing weighting with $|\tilde{i}_b|^2$:

$$\begin{aligned} \tilde{o}_n &= \frac{\langle\, |\tilde{i}_b|^2 \, \tilde{i}_n / \tilde{i}_b \,\rangle}{\langle\, |\tilde{i}_b|^2 \,\rangle} \, \tilde{o}_b \\ &= \frac{\langle\, \tilde{i}_n \tilde{i}_b^* \,\rangle}{\langle\, |\tilde{i}_b|^2 \,\rangle} \, \tilde{o}_b \,. \end{aligned} \tag{13.43}$$

Here $\tilde{i}_b^*$ denotes the complex conjugate of $\tilde{i}_b$. The reason why $\langle\, |\tilde{i}_b|^2 \,\rangle$ does not have the zero problem is that the frequency positions of the zeros of $\tilde{s}$ and $\tilde{i}_b$ fluctuate from image to image, with the consequence that the ensemble average $\langle\, |\tilde{i}_b|^2 \,\rangle$ becomes finite over the whole spatial frequency range of relevance.

Application of such speckle polarimetry, with the Swedish 50 cm La Palma telescope, has for the first time given us diffraction-limited magnetograms (Stokes V/I images) with an angular resolution of about 0.2 arcsec, revealing individual magnetic flux tubes with a diameter of about 200 km and a continuum brightness that is 30 % higher than the surroundings (Keller, 1992). A quarter-wave retarder and a calcite beam splitter were used to produce simultaneous images of $I \pm V$ in the narrow-band channel, out of which the reconstructed Stokes I and V/I images could be derived. Some results are displayed in Fig. 13.1.

The method of using simultaneous narrow and broad-band channels would also work to overcome the "signal-to-noise hurdle" with other image reconstruction methods, for instance the promising new technique of phase-diversity restoration (Gonsalves and Chidlaw, 1979; Högbom, 1989; Löfdahl and Scharmer, 1993). This technique has the great advantage that the amplitude and phase information may be retrieved from a single exposure instead of the 100 or so needed in speckle interferometry. Two cameras record simultaneously one focused and one defocused image, from which the phase errors of the wave front can be derived.

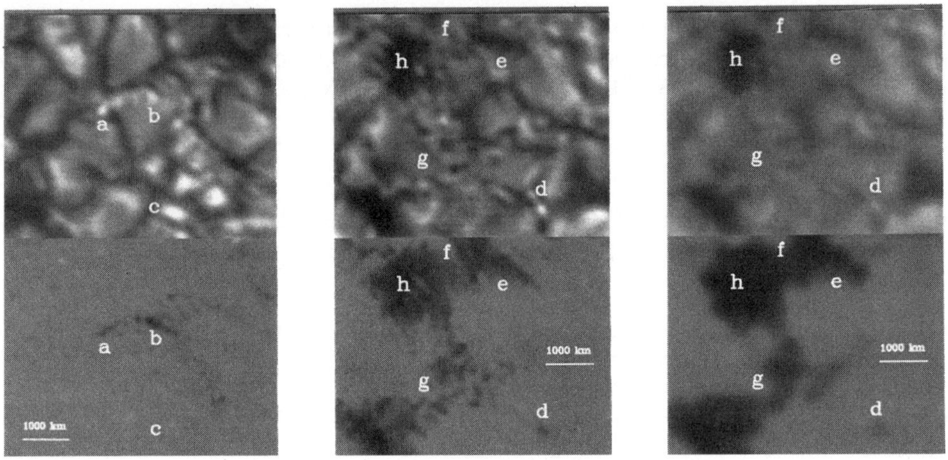

Fig. 13.1. Examples of first results from speckle polarimetry, from Keller (1992). The upper images show Stokes I of the broad-band channel, the lower images Stokes V/I of the narrow-band channel (in the blue wing of the Fe I 5250.22 Å line). The left and middle panels show reconstructed images of a weak plage region and a pore (small sunspot) region, respectively. The right panel represents unreconstructed images of the same region as in the middle panel. The sharpest single white-light image is shown in the upper image of the right panel, while the lower image represents the average Stokes V/I after correction for image motion of the individual frames. Note the image scale, indicated by the 1000 km bar (the typical best resolution in previous magnetograph work has been about 1000 km). The various features marked are described in the original publication. The feature at 'd' represents a resolved magnetic flux tube with a diameter of about 200 km, a continuum brightness contrast of about +30 % , and an observed V/I amplitude of about 8 % .

13.6.3. Adaptive optics

Adaptive optics allow the errors in the incoming wave front to be corrected for in real time by measuring them with a wave front sensor and converting the recorded information to control signals that govern the deformation of an adaptive mirror. The delay between error detection and correction has to be $\lesssim 10$ ms, so that the atmosphere does not have time to change. As the entrance aperture is imaged on the deformable mirror, the deformation is optically equivalent to a deformation of the much larger main mirror on a time scale of ms. The number of actuators to be used for the deformation should be at least as large as the number of coherence areas over the entrance aperture, or $(D/r_0)^2$, where D is the diameter of the entrance aperture, and r_0 is the so-called Fried parameter or size of the coherence area that is characteristic for the optical effects of the atmospheric turbulence. The wave front sensor is fed by a beam splitter before the end focus of the telescope. An overview of adaptive optics for solar applications has been given in Merkle et al. (1987).

Adaptive optics works the same way for polarimetry as for unpolarized applications. Some additional considerations need to be made, however. Thus the same constraints as those discussed in Sect. 13.6.1 for the elimination of seeing-induced polarization effects also apply for polarimetry with adaptive optics. The frequencies used for polarization modulation have to be well above the frequencies involved in the adaptive optics correction, otherwise the modulation may interfere with the operation of the adaptive optics system. A non-modulating double or multiple beam system early in the telescope (before the adaptive mirror and the beam splitter for the wave front sensor) would hardly be compatible with the use of adaptive optics. Thus a fast polarization modulation scheme is required.

13.7. Modulation Schemes

Detectors are sensitive to intensity (the first component of the Stokes vector), not to polarization (the other components of the Stokes vector). The task of the optical polarization analyser in the telescope is therefore to encode the polarization information into intensity information. In the following we will refer to this polarization analyser or encoder as the modulator package. Let $\boldsymbol{I}$ and $\boldsymbol{I}'$ be the Stokes vectors entering and leaving the modulator package, which is characterized by the Mueller matrix $\boldsymbol{M}$, i.e., $\boldsymbol{I}' = \boldsymbol{M}\boldsymbol{I}$. As the optical polarization analysis is completed when the beam leaves the modulator package (assuming that the last component of the modulator package is a perfect polarizer, otherwise instrumental polarization later in the telescope will affect the analysis), it is only the intensity part of $\boldsymbol{I}'$, i.e., I', not the other Stokes parameters, that is of relevance for the detection system. In other words, only the first row of the matrix $\boldsymbol{M}$ is of interest. For convenience we call the four first-row components $\frac{1}{2}x_{i,q,u,v}$ and normalize the matrix such that $x_i = 1$. Thus

$$I' = \tfrac{1}{2}(I + x_q Q + x_u U + x_v V)\,. \tag{13.44}$$

For an analysis package consisting of only a polarizing beam splitter and no retarder we would for instance have $x_q = \pm 1$ and $x_{u,v} = 0$ according to Eq. (13.6). To make a determination of all four of I, Q, U, and V, we need to have at least four different readings of I' that correspond to different, linearly independent combinations of the coefficients $x_{q,u,v}$. This can be done by having a polarizing beam splitter system with four simultaneous output beams, or better, by modulating the coefficients $x_{q,u,v}$. In the present section we will describe the main types of modulation schemes, including mechanical, electrooptical, and spatial modulation. Afterwards (in Sect. 13.8) we will address the question of demodulation, which is done by the detector and data handling system.

13.7.1. Rotating retarder

The main method to encode all four Stokes parameters via mechanical modulation is through the use of a rotating wave plate (retarder) followed by a stationary linear polarizer (e.g. Lites et al., 1991). If the wave plate retardation is δ and the angular velocity of rotation is Ω in the counter-clockwise direction, then

$$\begin{aligned} x_q &= \tfrac{1}{2}[(1+\cos\delta)+(1-\cos\delta)\cos 4\Omega(t-t_0)]\,, \\ x_u &= \tfrac{1}{2}(1-\cos\delta)\sin 4\Omega(t-t_0)\,, \\ x_v &= -\sin\delta \sin 2\Omega(t-t_0)\,, \end{aligned} \tag{13.45}$$

where the temporal zero point t_0 determines the phase of the modulation.

We see that Stokes Q and U are both modulated at frequency 4Ω (with a relative phase shift of $\pi/2$ between them), while Stokes V is modulated at 2Ω. The choice of retardation δ determines the modulation efficiences for the various Stokes parameters.

The Stokes parameters can be retrieved by making a number of read-outs of I' during each modulation cycle. The sampling interval for I' should not be longer than the $\pi/2$ phase shift between x_q and x_u for optimum sampling of *both* Q and U. This interval is 1/8 of the modulation period for x_v. A minimum of 8 samplings of I' (and thus 8 image buffers) per modulation cycle is thus needed with such a modulation scheme.

To avoid seeing-induced polarization features the full modulation cycle (8 samplings of I') should be $\ll$ 10 ms, i.e., the time between two successive images representing different polarization states should be $\ll$ 1 ms. It follows that the rotation rate of the plate should be $\gg$ 50 rotations per sec. As these requirements are hard to satisfy, compromises allowing for some seeing effects are made. Thus the wave plate in the Advanced Stokes Polarimeter (ASP) of Lites et al. (1991) turns at a rate of 3.75 rotations per sec.

Another significant problem with mechanically moving parts is a wobbling of the beam during modulation due to optical and alignment imperfections. In the ASP this wobbling is continually compensated for by an image motion stabilizer.

13.7.2. Uncoupled electrooptical modulators

Electrooptical modulation schemes have the great advantage that moving parts can be avoided altogether (which for instance eliminates beam wobbling), and there is no problem of making the modulation frequency sufficiently high to be beyond the range of the seeing and adaptive optics frequencies.

As was noted in Sect. 13.6.1, a single, non-moving modulator cannot encode both Stokes Q and U. Two modulators with different position angles of their retardation axes are needed for the encoding of all four Stokes parameters. The simplest (and at the same time most efficient) modulation scheme for the full Stokes vector consists of two electrooptical modulators (variable retarders) followed by a stationary linear polarizer.

Although Pockels cell modulators (in particular KD*P crystals) have been widely used, the piezoelastic modulators have superior properties, as explained in Sect. 13.3.4. Our explicit treatment here will be for a system based on such modulators. The optimization of such a system has been described in Stenflo (1991c).

Let the modulation frequencies of the two piezoelastic modulators be ω_1 and ω_2. Then, according to Eq. (13.10), the Mueller matrix of the modulator package will contain $\sin \omega_{1,2}t$ and $\cos 2\omega_{1,2}t$ terms, cross products between these trigonometric functions (leading to beat frequencies), and dc terms. We may however choose the modulation amplitudes $A_{1,2}$ and position angles $\alpha_{1,2}$ of the modulators with respect to the polarizing direction of the linear polarizer (the third and last component of the modulator package) such that the information on the four Stokes parameters I, Q, U, and V gets optimized at the four frequencies dc (zero), $2\omega_1$, $2\omega_2$, and ω_1, respectively. Then it is convenient to express each of the $x_{q,u,v}$ coefficients as a sum of these four frequency contributions:

$$x = x_{\mathrm{DC}} + x_{2\omega_1} \cos 2\omega_1 t + x_{2\omega_2} \cos 2\omega_2 t + x_{\omega_1} \sin \omega_1 t\,, \tag{13.46}$$

where we have omitted indices q, u, v for clarity.

We will in the present subsection assume that the two modulators are not coupled to each other, so that there are no fixed phase relations between the ω_1 and ω_2 modulations. The case of phase-coupled modulators will be treated in the following subsection.

Let us assume that the signals at each of the three ac frequencies ($2\omega_1$, $2\omega_2$, and ω_1) are demodulated by reading out the images I'_+ and I'_- that have been averaged over each of the two half periods of the sinusoidal modulation. We define a *signal vector*

$$\boldsymbol{S} = \begin{pmatrix} S_1 \\ S_2 \\ S_3 \\ S_4 \end{pmatrix} \tag{13.47}$$

with the components

$$\begin{aligned} S_1 &= (I'_+ + I'_-)_\omega\,, \\ S_2 &= (I'_+ - I'_-)_{2\omega_1}\,, \\ S_3 &= (I'_+ - I'_-)_{2\omega_2}\,, \\ S_4 &= (I'_+ - I'_-)_{\omega_1}\,. \end{aligned} \tag{13.48}$$

Index ω in the equation for S_1 can be any one of the three frequencies $2\omega_1$, $2\omega_2$, or ω_1. We then get (in normalized form with the coefficient for Stokes I being unity)

$$\begin{aligned} S_1 &= I + x_{q,\mathrm{DC}} Q + x_{u,\mathrm{DC}} U + x_{v,\mathrm{DC}} V\,, \\ S_2 &= \frac{2}{\pi}\,(x_{q,2\omega_1} Q + x_{u,2\omega_1} U + x_{v,2\omega_1} V)\,, \\ S_3 &= \frac{2}{\pi}\,(x_{q,2\omega_2} Q + x_{u,2\omega_2} U + x_{v,2\omega_2} V)\,, \\ S_4 &= \frac{2}{\pi}\,(x_{q,\omega_1} Q + x_{u,\omega_1} U + x_{v,\omega_1} V)\,. \end{aligned} \tag{13.49}$$

The reason for the factor $2/\pi$ is that the average of $|\sin|$ and $|\cos|$ is $2/\pi$.

In analogy with Mueller calculus it is convenient to express the relation between the signal vector $\boldsymbol{S}$ and the incident Stokes vector $\boldsymbol{I}$ as a matrix equation

$$\boldsymbol{S} = \boldsymbol{X} \boldsymbol{I}\,, \tag{13.50}$$

where the "response matrix" $\boldsymbol{X}$ is defined by Eq. (13.49). The wanted Stokes vector is then obtained by inversion of $\boldsymbol{X}$:

$$\boldsymbol{I} = \boldsymbol{X}^{-1}\boldsymbol{S}\,. \tag{13.51}$$

Let $\alpha_{1,2}$ be the position angles of the modulators measured in the counter-clockwise direction relative to the position angle of the linear polarizer, which is assumed to be α_0 with respect to the positive Stokes Q direction (the modulators are thus at angles $\alpha_{1,2} + \alpha_0$ with respect to the Stokes Q direction). When these angles satisfy the relations

$$\begin{aligned} \alpha_0 + \alpha_1 &= \pm 45^\circ\,, \\ \alpha_2 - \alpha_1 &= \pm 45^\circ\,, \end{aligned} \tag{13.52}$$

then $\boldsymbol{X}$ becomes diagonal below the first row:

$$\boldsymbol{X} = \begin{pmatrix} 1 & X_{12} & X_{13} & X_{14} \\ 0 & X_{22} & 0 & 0 \\ 0 & 0 & X_{33} & 0 \\ 0 & 0 & 0 & X_{44} \end{pmatrix}. \tag{13.53}$$

Equal response to Stokes Q and U, i.e., $X_{22} = X_{33}$, is obtained when $\alpha_2 = \pm 22\overset{\circ}{.}5$. We may thus choose

$$\begin{aligned} \alpha_1 &= +22\overset{\circ}{.}5\,, \\ \alpha_2 &= -22\overset{\circ}{.}5\,, \\ \alpha_0 &= +22\overset{\circ}{.}5\,. \end{aligned} \tag{13.54}$$

Then

$$\begin{aligned} X_{12} &= \frac{1}{\sqrt{2}}\,J_0(A_1)\,, \\ X_{13} &= \frac{1}{\sqrt{2}}\,J_0(A_2)\,, \\ X_{14} &= 0\,, \\ X_{22} &= \;\;\frac{2}{\pi}\,\sqrt{2}\,J_2(A_1)\,, \\ X_{33} &= \;\;\frac{2}{\pi}\,\sqrt{2}\,J_2(A_2)\,, \\ X_{44} &= -\frac{2}{\pi}\,\sqrt{2}\,J_1(A_1)\,, \end{aligned} \tag{13.55}$$

where $J_{0,1,2}$ are the Bessel functions of order 0, 1, and 2, which appear as a result of relations (13.9) and (13.11).

An optimum choice of modulation amplitudes is $A_1 = A_2 = 2.40$ (corresponding to $\lambda/2.62$ modulation, since $2\pi/2.40 = 2.62$). With this choice the off-diagonal terms in the first row of the response matrix vanish (since J_0 has a zero crossing), while the diagonal components (determined by J_1 and J_2) are fairly close to their maximum values. We then get

$$X = \begin{pmatrix} 1 & 0 & 0 & 0 \\ 0 & 0.39 & 0 & 0 \\ 0 & 0 & 0.39 & 0 \\ 0 & 0 & 0 & -0.47 \end{pmatrix} . \tag{13.56}$$

A search of parameter space (Stenflo, 1991c) has shown that the settings of $\alpha_{0,1,2}$ and $A_{1,2}$ used for Eq. (13.56) are optimum in terms of efficiency or signal-to-noise ratio for simultaneous recording of all four Stokes parameters.

13.7.3. Coupled modulators

With uncoupled modulators the signal at frequency $2\omega_2$ has an arbitrary phase with respect to the signals at frequencies ω_1 and $2\omega_1$ and needs to be demodulated independently of these other frequencies. This requires the use of beam splitters sending the light to different exit ports, where each of the ac frequencies can be demodulated independently of the others. As the light level at each port is correspondingly reduced due to the split beam, the photon efficiency is diminished. Furthermore, the demodulated signals at the three output ports need to be recombined to form the Stokes vector.

A much more elegant solution that allows demodulation at a single exit port is to use two mutually phase-locked piezoelastic modulators (Stenflo et al., 1992). Then Stokes Q, U, and V can be modulated in a common cycle similar to the rotating wave plate modulation. The drawback of needing as many as 8 samplings of I' in the case of the mechanically rotating retarder can be overcome with the coupled electrooptical modulators, such that the minimum of 4 samplings per cycle becomes both sufficient and optimum in terms of efficiency.

If the instantaneous retardations of the two piezoelastic modulators are δ_1 and δ_2, then Mueller matrix multiplication for the optical train of two retarders and a linear polarizer for the particular position angle settings of Eq. (13.54) gives for the $x_{q,u,v}$ coefficients of Eq. (13.44):

$$\begin{aligned} x_q &= \frac{1}{\sqrt{2}}(\cos\delta_1 + \sin\delta_1 \sin\delta_2)\,, \\ x_u &= \frac{1}{\sqrt{2}}\cos\delta_2\,, \\ x_v &= \frac{1}{\sqrt{2}}(-\sin\delta_1 + \cos\delta_1 \sin\delta_2)\,, \end{aligned} \tag{13.57}$$

where according to Eq. (13.9)

$$\begin{aligned} \sin\delta_k &= 2J_1(A_k)\sin\omega_k(t - t_{0k})\,, \\ \cos\delta_k &= J_0(A_k) + 2J_2(A_k)\cos 2\omega_k(t - t_{0k}) \end{aligned} \tag{13.58}$$

($k = 1, 2$). The reference times t_{01} and t_{02} define the relative phases of the two modulations.

The $\sin\delta_1 \sin\delta_2$ and $\cos\delta_1 \sin\delta_2$ terms give rise to signals at the sum and beat frequencies $|\omega_1 \pm \omega_2|$ and $|2\omega_1 \pm \omega_2|$, which could be ignored in the case of the uncoupled modulators, but which have small, significant contributions in the phase-locked case. Thus, if we choose the modulators such that $\omega_1 = \omega_2 = \omega$, then the beat frequency $2\omega_1 - \omega_2 = \omega$, while $\omega_1 + \omega_2 = 2\omega$.

With the choice $\omega_1 = \omega_2 = \omega$, $t_{01} = 0$, and $t_{02} = \pi/(4\omega)$, Stokes Q is modulated as $\cos 2\omega t$, Stokes U as $\sin 2\omega t$, and Stokes V as $\sin \omega t$. This modulation scheme is therefore equivalent to that of a mechanically rotating wave plate (if we make the identification $\omega = 2\Omega$) and is subject to the same sampling conditions: 8 samplings of I' per modulation cycle are needed for a determination of all four Stokes parameters.

To reduce the number of required samplings from 8 to the ideal minimum of 4, we need to change the modulation frequency for Stokes U from 2ω to ω and have the U modulation phase shifted by $\pi/2$ with respect to the Stokes V modulation. In this case the sampling interval can be $(\pi/2)/\omega$. The full cycle then consists of 4 such intervals, which makes the cycle length $2\pi/\omega$. One way to accomplish this is to choose the modulation frequency of the second modulator, which is responsible for the modulation of Stokes U, to be $\omega/2$, half that of the first modulator. This would however require the *size* of the second modulator to be twice as large as that of the first, since the size of the modulating glass plate scales in inverse proportion to the resonant acoustic frequency (cf. Sect.13.3.4).

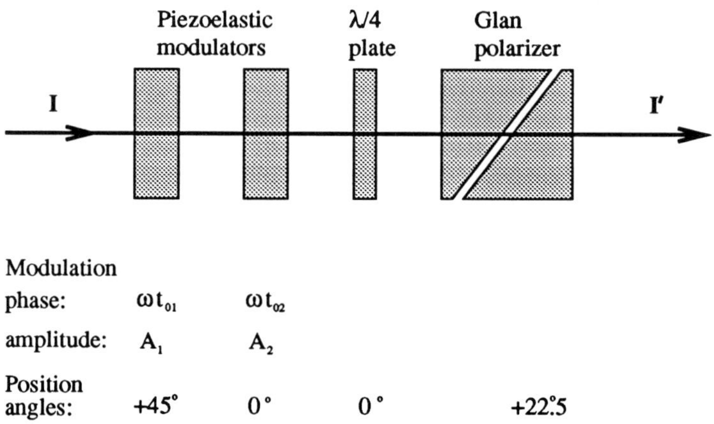

Fig. 13.2. Optical scheme of a modulator package with two coupled (phase-locked) piezoelastic modulators. The position angles are defined with respect to the positive Stokes Q direction. The $\lambda/4$ plate makes Stokes U be modulated at frequency ω instead of 2ω, with the consequence that four instead of eight samplings per modulation cycle becomes sufficient for a determination of all four Stokes parameters. From Stenflo et al. (1992).

A more elegant way to solve this problem is to keep $\omega_1 = \omega_2 = \omega$ but add a fixed retardance of $\pi/2$ to δ_2. Then $\cos \delta_2$ in Eq. (13.57) gets converted to $\sin \delta_2$, which according to Eq. (13.58) is modulated at the fundamental frequency ω_2 $(= \omega)$. This is accomplished by adding a superachromatic $\lambda/4$ plate with its fast optical axis parallel to the axis of the second modulator, as shown in Fig. 13.2. Eq. (13.57) then gets transformed into

$$\begin{aligned} x_q &= \frac{1}{\sqrt{2}}(\cos\delta_1 + \sin\delta_1 \cos\delta_2)\,, \\ x_u &= -\frac{1}{\sqrt{2}}\sin\delta_2\,, \\ x_v &= \frac{1}{\sqrt{2}}(-\sin\delta_1 + \cos\delta_1\cos\delta_2)\,. \end{aligned} \tag{13.59}$$

The relative phase shifts that allow optimum sampling efficiency are defined by

$$\begin{aligned} \omega t_{01} &= \pi/4\,, \\ \omega t_{02} &= -\pi/4\,. \end{aligned} \tag{13.60}$$

The three curves $\cos 2(\omega t - \pi/4)$, $\sin(\omega t + \pi/4)$, and $\sin(\omega t - \pi/4)$ that are shown in Fig. 13.3 represent the modulations of the three Stokes parameters Q, U, and V if we disregard the small beating contributions. The four sampling intervals over which the signal I' is integrated are bounded by the vertical lines in the figure.

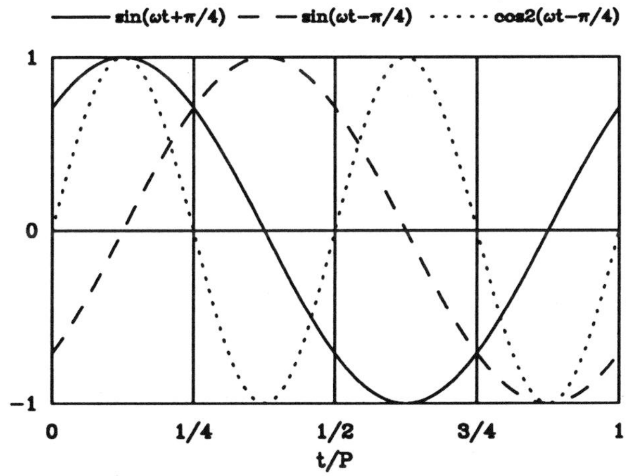

Fig. 13.3. Modulation cycle with the system of Fig. 13.2. The solid, dashed, and dotted curves indicate the modulations of Stokes U, V, and Q, respectively. The four sampling periods are bounded by the vertical lines in the figure. From Stenflo et al. (1992).

Let us denote the average value of I' over each of the four sampling intervals by B_n, $n = 1, 2, 3, 4$ ($B_n = \langle I' \rangle_n$). Similarly we denote the averages of the $x_{q,u,v}$ coefficients of Eq. (13.44) by $q_n = \langle x_q \rangle_n$, $u_n = \langle x_u \rangle_n$, and $v_n = \langle x_v \rangle_n$. This allows us to define the 4-vectors

$$\boldsymbol{B} = \begin{pmatrix} B_1 \\ B_2 \\ B_3 \\ B_4 \end{pmatrix}, \quad \boldsymbol{i} = \begin{pmatrix} 1 \\ 1 \\ 1 \\ 1 \end{pmatrix}, \quad \boldsymbol{q} = \begin{pmatrix} q_1 \\ q_2 \\ q_3 \\ q_4 \end{pmatrix}, \quad \boldsymbol{u} = \begin{pmatrix} u_1 \\ u_2 \\ u_3 \\ u_4 \end{pmatrix}, \quad \boldsymbol{v} = \begin{pmatrix} v_1 \\ v_2 \\ v_3 \\ v_4 \end{pmatrix}. \tag{13.61}$$

Eq. (13.44) then becomes the vector relation

$$\boldsymbol{B} = \tfrac{1}{2}(I\boldsymbol{i} + Q\boldsymbol{q} + U\boldsymbol{u} + V\boldsymbol{v})\,. \tag{13.62}$$

It is convenient to define the linearly independent unit 4-vectors $\boldsymbol{e}_{I,Q,U,V}$ as

$$\begin{aligned} \boldsymbol{e}_I &= \begin{pmatrix} 1 \\ 1 \\ 1 \\ 1 \end{pmatrix}, \quad \boldsymbol{e}_Q = \begin{pmatrix} 1 \\ -1 \\ 1 \\ -1 \end{pmatrix}, \\ \boldsymbol{e}_U &= \begin{pmatrix} 1 \\ 0 \\ -1 \\ 0 \end{pmatrix}, \quad \boldsymbol{e}_V = \begin{pmatrix} 0 \\ 1 \\ 0 \\ -1 \end{pmatrix}. \end{aligned} \tag{13.63}$$

We then obtain for the modulation scheme of Fig. 13.3 using Eqs. (13.59) and (13.60)

$$\begin{aligned} \boldsymbol{q} &= \frac{1}{\sqrt{2}} J_0(A_1)\,\boldsymbol{e}_I + \frac{2\sqrt{2}}{\pi} J_2(A_1)\,\boldsymbol{e}_Q \\ &\quad + \frac{4}{\pi} J_1(A_1) J_2(A_2)\,\boldsymbol{e}_U + \frac{4}{\pi} J_1(A_1)[\,J_0(A_2) + \tfrac{1}{3} J_2(A_2)\,]\,\boldsymbol{e}_V\,, \\ \boldsymbol{u} &= -\frac{4}{\pi} J_1(A_2)\,\boldsymbol{e}_U\,, \\ \boldsymbol{v} &= [\,\frac{1}{\sqrt{2}} J_0(A_1) J_0(A_2) - \sqrt{2} J_2(A_1) J_2(A_2)\,]\,\boldsymbol{e}_I \\ &\quad + \frac{2\sqrt{2}}{\pi}[\,J_0(A_2) J_2(A_1) - J_0(A_1) J_2(A_2)\,]\,\boldsymbol{e}_Q - \frac{4}{\pi} J_1(A_1)\,\boldsymbol{e}_V\,. \end{aligned} \tag{13.64}$$

Let us now define the signal vector $\boldsymbol{S}$ by

$$\begin{aligned} S_1 &= \tfrac{1}{2}(B_1 + B_2 + B_3 + B_4)\,, \\ S_2 &= \tfrac{1}{2}(B_1 - B_2 + B_3 - B_4)\,, \\ S_3 &= \tfrac{1}{2}(B_1 - B_3)\,, \\ S_4 &= \tfrac{1}{2}(B_2 - B_4)\,. \end{aligned} \tag{13.65}$$

The "response matrix" $\boldsymbol{X}$ defined by Eq. (13.50) can then be derived from Eq. (13.64), with the result

$$\boldsymbol{X} = \begin{pmatrix} 1 & X_{12} & 0 & X_{14} \\ 0 & X_{22} & 0 & X_{24} \\ 0 & X_{32} & X_{33} & 0 \\ 0 & X_{42} & 0 & X_{44} \end{pmatrix}, \tag{13.66}$$

where

$$\begin{aligned}
X_{22} &= \frac{2\sqrt{2}}{\pi} J_2(A_1)\,,\\
X_{33} &= -\frac{2}{\pi} J_1(A_2)\,,\\
X_{44} &= -\frac{2}{\pi} J_1(A_1)\,,\\
X_{12} &= \frac{1}{\sqrt{2}} J_0(A_1)\,,\\
X_{14} &= \frac{1}{\sqrt{2}} J_0(A_1) J_0(A_2) - \sqrt{2}\, J_2(A_1) J_2(A_2)\,,\\
X_{24} &= \frac{2\sqrt{2}}{\pi} [J_0(A_2) J_2(A_1) - J_0(A_1) J_2(A_2)]\,,\\
X_{32} &= \frac{2}{\pi} J_1(A_1) J_2(A_2)\,,\\
X_{42} &= \frac{2}{\pi} J_1(A_1) [J_0(A_2) + \tfrac{1}{3} J_2(A_2)]\,.
\end{aligned} \tag{13.67}$$

All the off-diagonal terms except for X_{12} are due to beating caused by the mutual phase locking. For the nominal, recommended choice of modulation amplitudes $A_{1,2} = 2.40$ ($\lambda/2.62$ modulation), matrix elements X_{12} and X_{24} vanish, and we get

$$\boldsymbol{X} = \begin{pmatrix} 1 & 0 & 0 & -0.26 \\ 0 & 0.39 & 0 & 0 \\ 0 & 0.14 & -0.33 & 0 \\ 0 & 0.05 & 0 & -0.33 \end{pmatrix}. \tag{13.68}$$

The Stokes vector $\boldsymbol{I}$ can then be retrieved through Eq. (13.51).

The modest off-diagonal terms, which are due to the beating, will not introduce significant noise. They are automatically and accurately determined together with the diagonal terms by the standard calibration procedure (inserting with the calibration filter wheel in front of the modulator package known linear and circular polarizers).

13.7.4. Spatial modulation

Instead of letting the retardance of the wave plate vary in time, we may let it vary in space across the field of view (Stenflo, 1969; Beckers and Stenflo, 1969; Michard and Rayrole, 1970). This causes a spatial modulation of the intensity I' wherever there is non-zero polarization in the image. The amplitude of the modulation is proportional to the amount of polarization.

A convenient way to produce a spatial modulation is to use a Babinet compensator in a focal plane of the telescope (Beckers and Stenflo, 1969). A Babinet compensator is a retarder for which the amount of retardation δ varies linearly across the plate. Since the Mueller matrix (13.7) becomes periodic when δ increases linearly, the polarization state gets modulated sinusoidally across the field.

This method is constrained by the requirement that for a quantitative analysis the modulation needs to be done at a spatial frequency that is higher than any spatial frequency caused by structures in the solar image. Thus the period of the retardance variation needs to be smaller than the spatial resolution element (which is normally determined by the seeing fluctuations, but which may with image reconstruction or adaptive optics techniques be as small as the diffraction disk of the telescope aperture). At the same time the pixels of the detector need to be sufficiently small to resolve these high spatial frequencies. This constraint limits the accessible field of view for a given number of pixels of the detector.

In the case of temporal modulation, two non-moving modulators were needed to record the full Stokes vector. Similarly one may use a combination of two Babinet compensators with their optical axes at 45° relative to each other to simultaneously record all four Stokes parameters. The retrieval (demodulation) of the polarization information is done in the Fourier domain $\tilde{I}'$ of the I' image. The polarization information is located at discrete positions in $\tilde{I}'$ and may be extracted by applying numerical spatial filters in the off-line analysis.

Although temporal modulation is usually preferrable, spatial modulation may be of interest when rapid read-out (faster than the seeing frequencies) or temporal demodulation of 2-D images is hard to accomplish. Such temporal demodulation is the subject of the next section.

13.8. Demodulation Schemes

Before the advent of 2-D array detectors demodulation was usually done by a lock-in amplifier for each channel that needed to be demodulated. With a 1000×1000 pixel array we have a million channels instead of a few. Other ways of doing the demodulation therefore have to be invented, which will be discussed next.

13.8.1. Synchronized Read-Out

The most direct solution is to read out the image from the detector after each sampling interval of a modulation cycle. The read-out pixel rate is however limited to about 5 MHz. Thus a 1000×1000 image requires at least about 0.2 s to be read. To determine all four Stokes parameters a minimum of four images is needed during one modulation cycle, as we have seen, which implies a total read-out time of about 1 s. To avoid spurious polarization patterns due to seeing fluctuations this time would however need to be shortened by more than two orders of magnitude. This is only possible by limiting the read-out array area to 100×100 pixels or less. It would thus be impossible to use the advantages of large-array detectors for precision polarimetry in ground-based observations with temporal modulation, so one would be limited to the use of stationary beam-splitter systems or spatial modulation. Fortunately solutions to the demodulation problem compatible with large-array detectors exist, as will be described next.

13.8.2. OPTICAL DEMODULATION

The encoding of the polarization done by the modulation package is in the form of sinusoidal fluctuations of the output beam intensity I'. The encoded information would get lost if straight temporal integration of the detector signal were done, since the average of a sine or cosine function is zero. The purpose of demodulation is to convert the sinusoidal variations to something that does not vanish when the signal is integrated. This is done by multiplying the sine (or cosine) function by another sine (or cosine) function of the same frequency and phase. Since $\langle \sin^2 \rangle = \langle \cos^2 \rangle = \frac{1}{2}$, the modulated part now contributes to the temporally integrated signal in proportion to its amplitude.

This phase-locked multiplication is effectively what is done electronically in a lock-in amplifier, but it can also be done optically before the light is received by the detector. In this case the demodulation can be accomplished by a single demodulator simultaneously for all the pixels in the detector array, regardless of the number of pixels.

The optical demodulation is achieved by using at the exit focus, immediately in front of the detector, an "optical chopper" that is phase locked to the particular modulation frequency that we want to demodulate. Such an optical demodulator has been described by Stenflo and Povel (1985) and Stenflo (1991c). Its principle components are a variable retarder (modulator) followed by a polarizing beam splitter. The modulation of the variable retarder is synchronized (phase locked) with the corresponding encoding modulation done by the modulator package. The beam splitter produces two output beams, the intensities of which are chopped in anti-phase relative to each other. This has the consequence that the demodulated signal gets superposed on the ordinary intensity, with opposite signs in the two beams. The dc difference between the two images on the detector will therefore represent an image of the polarized component (Stokes Q, U, or V, depending on whether the frequency $2\omega_1$, $2\omega_2$, or ω_1 of Eq. (13.48) is being demodulated). To simultaneously demodulate all three ac frequencies in the case of a modulation scheme based on two uncoupled modulators, a beam splitter system with three simultaneous exit ports, each equipped with a demodulator and an array detector, would be needed.

Since the difference between pairs of images has to be determined, it may seem that the accuracy of the method will be limited by the accuracy with which the gain table or flat field of the detector can be determined. It is important to note, however, that what is needed is not the absolute gain table used for ordinary photometry, but the ***polarization gain table***, which can be determined to much higher levels of precision. Let us denote the gain table values for the corresponding pixels in the two images (refered to by indices + and $-$) by e_+ and e_-, respectively. In contrast to ordinary photometry it is not e_+ and e_- by themselves that we need, but only the relative difference $p_0 = (e_+ - e_-)/(e_+ + e_-)$, which appears as a background to be subtracted out in the polarization images. To find p_0 we do not need to use illumination that is flat in intensity (the flatness of which is a limiting factor in the accuracy of photometric gain tables), but only one that is flat in the degree of polarization. This is accomplished simply by switching off the modulation. In this case the two images are identical, and the relative difference between them directly represents the polarization gain table p_0 (Stenflo, 1991c).

13.8.3. Hidden fast buffers with synchronized charge shifting

As we have seen demodulation of one ac frequency requires the recording of two images, which we may call I'_+ and I'_-. The encoded polarization information is contained in the intensity difference between these two images. This was explicitly expressed by the signal vector of Eq. (13.48), where the two images represent the odd and even half periods of the modulation. The image cannot be read out at the fast modulation rate, but demodulation may still be achieved electronically if we could rapidly store an exposed image in a temporary fast buffer and bring it in and out of the buffer in synchrony with the modulation.

It has been found possible to create hidden, fast buffers by using the fast, bi-directional charge transfer allowed by certain types of CCD detector arrays (Povel et al., 1990, 1991). A mask is deposited onto the CCD chip when it is manufactured, such that every second pixel row is covered and can be used for buffer storage. Thus two simultaneous image planes can be handled within the same CCD chip. When image plane no. 1 resides in the exposed area of the CCD, during one half period of the modulation, image plane no. 2 resides in the hidden area. During the next half period the charges have been shifted by one pixel row, such that the image planes have traded places. A half period later the charges have been shifted back, etc., with the consequence that image plane no. 1 exclusively represents I'_+ (the odd half periods), while image plane no. 2 represents I'_- (the even half periods). As the charge transfer losses are generally smaller than 10^{-6}, we may integrate over a large number of modulation cycles until the CCD has been filled. Only then is it read out, with its two image planes.

As the identical pixels have been used to expose both image planes, all gain table effects divide out when forming the fractional polarization image $(I'_+ - I'_-)/(I'_+ + I'_-)$. It is a huge advantage not to be limited in any way by the flat-field determination.

The principle of a polarimetry system based on this demodulation idea is illustrated in Fig. 13.4, from Keller et al. (1992). It has been built and implemented as ZIMPOL I, the first generation of the **Z**urich **Im**aging Stokes **Pol**arimeter. Two piezoelastic modulators, PEM1 and PEM2, are used to modulate all four Stokes parameters as described in Sect. 13.7.2, with Stokes Q at frequency $2\omega_1$, U at $2\omega_2$, and V at ω_1. As each frequency is demodulated by one CCD in this scheme, a beam splitter system (BS) is used to send the light to three CCDs, such that all three frequencies can be demodulated simultaneously. In the figure the demodulation of Stokes U is shown. The reference signal at frequency ω_2 ($= f_2$ in Fig. 13.4) from the oscillator driver (OSC2) for the modulator (PEM2) is passed through a phase shifter (PH) that also doubles the frequency, after which it is fed into the CCD electronics block (EL), where it is used by the clock sequencer for demodulation. At read-out the analog signal from the CCD is amplified and processed (in EL) and is subsequently digitized by the fast analog-to-digital converter (ADC).

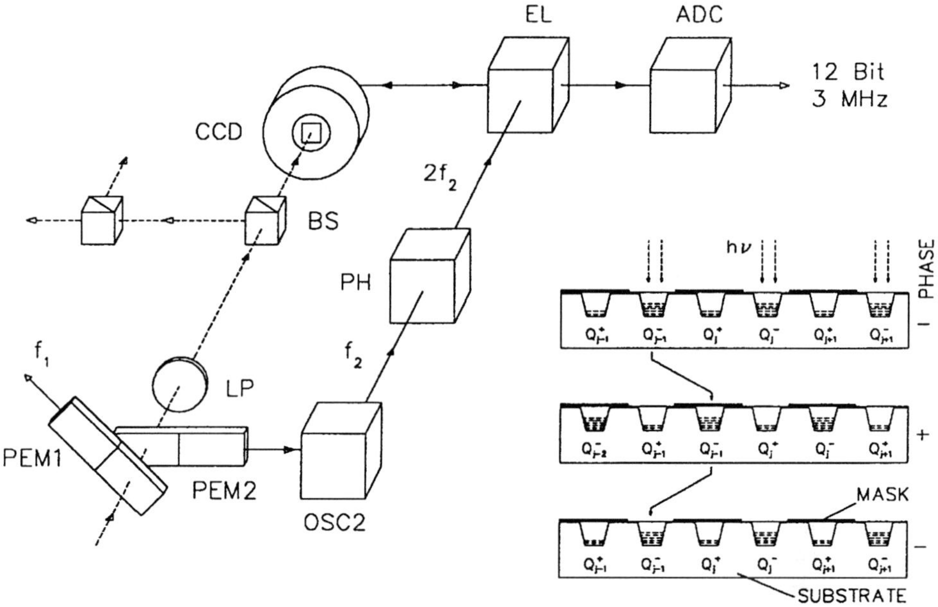

Fig. 13.4. Illustration of demodulation by synchronized charge shifting, as done by ZIMPOL I (Keller et al., 1992). See text for details.

Although this scheme may achieve simultaneous electronic demodulation of all four Stokes parameters, it has several important drawbacks. A factor of two in photon efficiency is lost by the mask, while another factor of three is lost by having to split up the beam to three exit ports. Furthermore, the separate Stokes Q, U, and V images have to be exactly matched to allow the formation of a correct, full Stokes vector at each image point.

All these drawbacks will be eliminated in the second generation of this polarimeter, ZIMPOL II (Stenflo et al., 1992). By creating three hidden, fast image buffers in the detector, four image planes can be handled simultaneously within a single CCD chip. This is sufficient to retrieve the full Stokes vector. Since the identical pixels are being used for exposure of all four image planes, the gain table factors divide out in the fractional polarization images (Q/I, U/I, V/I), and the four images are geometrically perfectly matched. Light losses at the mask are eliminated by using a cylindrical microlens array that concentrates all the light on the unmasked pixels. No beam splitter is needed, since there is only a single exit port for the full Stokes vector.

The concept is illustrated in Fig. 13.5. For each group of four pixel rows, one row is used for exposure, the other three for buffer storage. A periodic array of cylindrical microlenses is placed immediately in front of the CCD. Each microlens has a width of four pixels and focuses the incoming light over the four pixel rows such that the photons fall on a single row. The mask, which may be deposited on the back of the lens array, has here the purpose of eliminating stray light that would otherwise contaminate the buffers. The polarization modulation is done with two *coupled* piezoelastic modulators according to the scheme described by Figs. 13.2 and 13.3 in Sect. 13.7.3. The modulation of Stokes Q, U, and V indicated to the right in Fig. 13.5 is the same as was shown in Fig. 13.3.

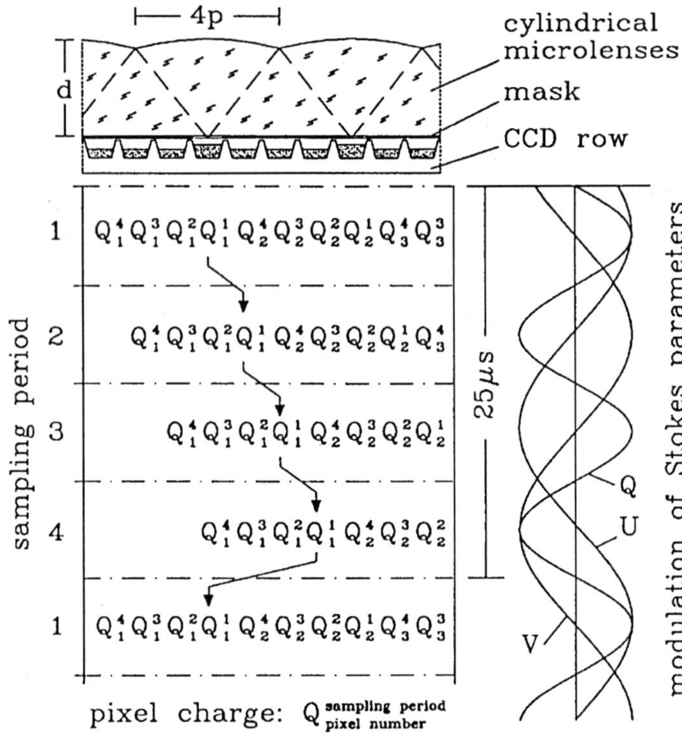

Fig. 13.5. Principle of demodulation with ZIMPOL II. All four Stokes parameters are demodulated in a single CCD. A cylindrical microlens array focuses all the light on the exposed pixel rows. The charges Q_k^j are shifted between the four image buffers in synchrony with the four sampling intervals of the modulation cycle (which has a total length of 25 μs, corresponding to a modulation frequency of 40 kHz for the two coupled, piezoelastic modulators). See text for details. Courtesy H.P. Povel.

Let lower index k refer to the pixel group number (for each group of four pixel rows), while upper index j refers to the sampling interval during a modulation cycle (out of the four sampling intervals marked in Fig. 13.3). Then Q_k^j in Fig. 13.5 denotes the corresponding accumulated charges. Figure 13.5 illustrates how these charges are shifted in synchrony with the polarization modulation, such that the four images $B_{1,2,3,4}$ of Eq. (13.61) will be obtained. If we define the signal vector $\boldsymbol{S}$ as in Eq. (13.65), the Stokes images can be retrieved with Eq. (13.51) through multiplication of $\boldsymbol{S}$ by the inverted response matrix $\boldsymbol{X}$ of Eqs. (13.66)–(13.68).

13.9. Imaging FTS Polarimeter

Much of our knowledge of the physics of solar magnetic flux tubes has been based on observations with the Fourier Transform Spectrometer (FTS) at the NSO/McMath-Pierce facility as a polarimeter. Examples of such data were given in the preceding chapter. FTS polarimetry has great future potential but is technologically difficult, so very few of its possibilities have been realized so far. Thus it has not yet been used with array detectors, the spatial resolution has been low ($\gtrsim$ 5 arcsec), the time resolution has also been low ($\gtrsim$ 7 min, more typically 30 min), and only three of the four Stokes parameters have been recorded simultaneously (Brault, 1978; Stenflo et al., 1983b, 1984, 1987a). These severe trade-offs have however been outweighed (for the quantitative analysis of flux tube physics) by the superior spectral resolution of the FTS (fully resolved solar spectrum without significant spectral smearing), large simultaneous spectral coverage, absence of spectral stray light, and excellent signal-to-noise ratio.

The above-mentioned trade-offs however do not represent any inherent limitations of the possibilities with an FTS polarimeter. In principle an FTS can be used for imaging with array detectors, all four Stokes parameters may be imaged simultaneously, and high spatial and temporal resolutions may be achieved. Since the 2-D entrance aperture of the FTS is properly imaged onto the exit pupil, the FTS delivers at the same time 2-D spatial (x, y) and spectral (λ) information. The spectral information is collected by scanning in the Fourier domain (interferometer path length difference). The multiplexing advantage of 2-D array detectors can thus be used to cover an area of the sun with good spatial resolution, while obtaining a high-resolution spectrum at each spatial location (pixel). We will now outline the concept of such an imaging FTS polarimeter (cf. Stenflo and Solanki, 1992).

An FTS is a modified Michelson interferometer, in which the path-length difference Δ is being scanned. A beam splitter divides the input beam into two beams, which are later recombined to interfere with each other. Due to this interference the output intensity $I_{\rm out}$ is a function of the path-length difference Δ between the two beams. For a monochromatic input beam we have

$$I_{\rm out}(\Delta) \sim [\,1 \pm \cos(2\pi\sigma\Delta)\,]\, I_{\rm inp}(\sigma)\,, \tag{13.69}$$

where $I_{\rm inp}$ is the intensity of the input beam, and $\sigma = 1/\lambda$ is the wave number. The $\pm$ refers to the two possible output channels of the FTS. For a continuous spectral distribution one has to sum over the individual contributions from the various wave number intervals. If we omit the constant term, or take the difference between the two output channels, we get

$$I_{\rm out}(\Delta) \sim \int_0^\infty I_{\rm inp}(\sigma) \cos(2\pi\sigma\Delta)\, {\rm d}\sigma\,. \tag{13.70}$$

The interferogram $I_{\rm out}(\Delta)$ is thus the real part of the Fourier transform of the input spectrum $I_{\rm inp}(\sigma)$ with respect to wave number (cf. Brault, 1985). This spectrum can be retrieved by calculating the inverse transform in the off-line data analysis.

The path-length difference is scanned with a constant speed $v = \Delta/t$ (if we choose the phase such that $\Delta = 0$ at $t = 0$). We can therefore write

$$\sigma\Delta = \nu t \tag{13.71}$$

to make it manifest that the output signal I_{out} is modulated at the frequency

$$\nu = v\sigma \tag{13.72}$$

as a result of the scanning. As the scanning speed v is constant, the modulation frequency ν is proportional to the wave number σ. This implies that the wave numbers are linearly mapped on temporal frequencies. In the case of the NSO FTS the scanning speed is such that the visible spectrum is mapped on the audio range, $\sim$ 0.8–1.0 kHz. To sample the visible spectrum without aliasing, the sampling frequency for the interferogram (I_{out}) has to be at least twice as high, 2.5 kHz in the case of the NSO FTS.

Let us now assume that we modulate the full polarization state with two piezoelastic modulators, as described in Sects. 13.7.2 or 13.7.3. The following discussion will be in terms of the more advanced concept of Sect. 13.7.3 of coupled modulators and the demodulation scheme of ZIMPOL II (Sect. 13.8.3) with four simultaneous image planes in a single CCD. An interferogram is produced at each of the pixels of the detector array. The sampling of all these simultaneous interferograms is done by making read-outs of the detector array (including all four image planes) after a given, integer number of polarization modulation cycles. The number of interferograms read out this way thus equals the number of exposed pixels times four. Fourier conversion of each interferogram and linear combination of the four image planes gives the spectrum of the full Stokes vector at each pixel.

To enhance the temporal resolution for a given ceiling on the sampling rate one may use under-sampling in higher-order aliases (cf. Brault, 1985) in combination with a faster scanning speed v. The trade-off here is that the free spectral range is decreased, so the spectral band width has to be reduced accordingly with a prefilter. Temporal resolution of say 10 s may thus be achieved with useful free spectral ranges. The smaller spectral band width limits the amount of simultaneous spectral coverage but has the advantage of also reducing the noise.

The probably most severe limitation in FTS imaging applications is in the array size that can be used. The sampling rate of the interferograms has to be rather high (of the order of kHz) to allow good time resolution (like 10 s) without too severe a trade-off with the free spectral range. The read-out rate of CCD-type detectors is limited to about 10 MHz. With a sampling rate of a few kHz one can then hardly use arrays much larger than 32×32 pixels. However, even with such small arrays it will be a great challenge to handle the massive data flow from the detector. With the rapid advances in detector and computer technologies, practical solutions to these problems should become available in the not too distant future.

13.10. LEST

All the existing major solar telescopes in the world have large instrumental polarization due to heliostat or coelostat reflections at the beginning of the optical system. The entrance apertures are smaller than 1 m with the exception of the NSO McMath-Pierce facility (1.5 m). An international effort to create a next-generation type facility that is optimized for solar polarimetry and magnetic field work is under way, with the current participation of 8 countries (Germany, Israel, Italy, Norway, Spain, Sweden, Switzerland, USA). This facility, called LEST (**L**arge **E**arth-based **S**olar **T**elescope), is supposed to be located on the island of La Palma (Canary Islands, Spain), 2400 m above sea level near the existing Swedish 50 cm vacuum tower telescope there. This is the best known site in the world for solar observations. As the design of LEST has been largely driven by the needs of solar polarimetry, we will outline the main design features here and refer to Engvold and Andersen (1990) for further details.

Figure 13.6 shows a cross section of the tower structure (left) and the optical scheme (right). The telescope, with a free aperture of 2.4 m, sits on top of a 30 m high tower. It has a "polarization-free" design, which means that the part of the optical system that precedes the modulator package (where the optical polarization analysis takes place) is rotationally symmetric. Since modulators and other polarization optics have rather small free apertures, they have to sit in a concentrated part of the beam, near a focus. In the case of LEST this is at the secondary focus, at F2 in Fig. 13.6. Any oblique reflections or other polarization effects that occur after this place in the optical system are of no consequence for the polarization analysis. Presently the system of coupled piezoelastic modulators described by Fig. 13.2, which is part of ZIMPOL II (cf. Sect. 13.8.3), is foreseen to be used as the modulator package at F2.

The large, 2.4 m aperture is needed for achieving high-precision polarimetry combined with high spatial, spectral, and time resolution. Even a telescope like LEST gets "photon starved" when one tries to minimize the trade-offs between the various resolutions and the polarimetric accuracy.

The thermal design of solar telescopes is of crucial importance due to the severe heat load, which degrades the optics and induces seeing effects. The heat load in the LEST f/2.3 prime focus F1 that is generated by the parabolic main mirror M1 is about 3.7 kW. An efficiently cooled field stop at F1 lets through a 2 arcmin field of view and thus removes 99.6 % of the heat energy. An elliptical, concave mirror M2 images the prime focus on the secondary focus F2, where the polarization analysis takes place. This type of design with a field stop at the prime focus and a concave secondary mirror is called a Gregorian design. The beam next passes through a hole in the flat mirror M4 before it hits the elliptical, concave mirror M3, which images F2 via the plane mirrors M4, M5, and M6 on the final focus F3 in the observing room at the base of the tower. The instrument table in the observing room will have space for about twenty different auxiliary instruments.

In many existing modern solar telescopes the internal seeing is suppressed by evacuation of the telescope tube. This solution is not possible with LEST, since a 2.4 m telescope entrance window that supports a pressure difference of one atmosphere cannot be manufactured to the optical and polarimetric quality

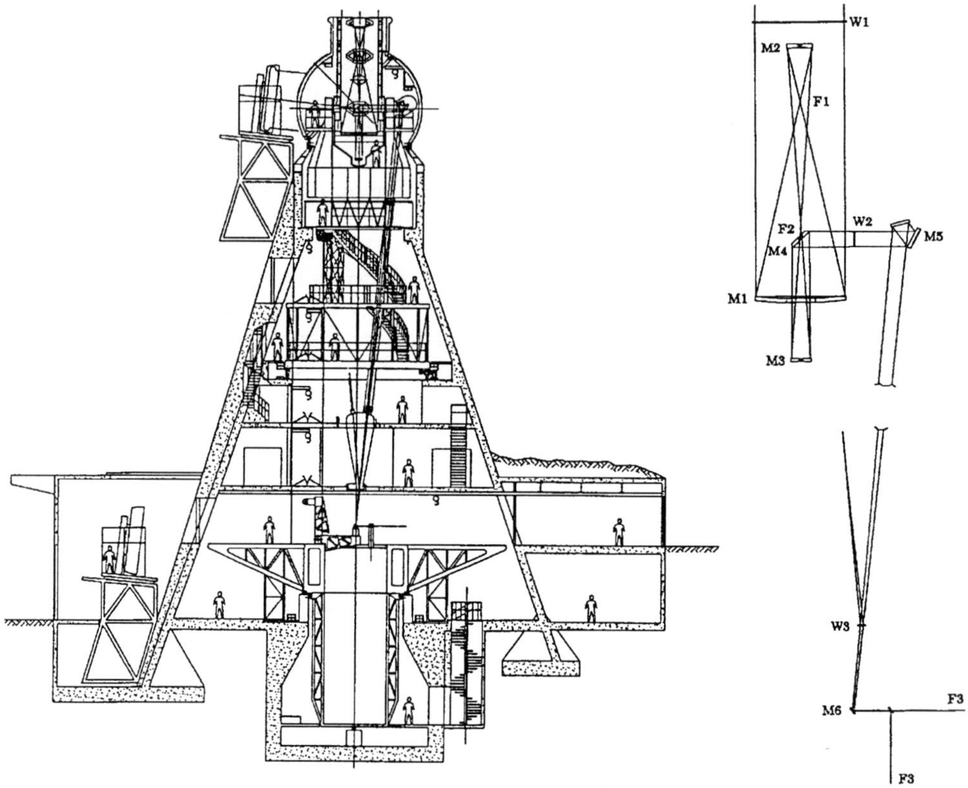

Fig. 13.6. LEST design, with the cross section of the tower structure (left) and the optical scheme (right). From Andersen (1993).

required. Instead the tube will be filled with helium, which allows equal pressures on the front and back sides of the window W1. As helium has much smaller refractive effects and much larger thermal conductivity than air, it will greatly reduce the seeing effects and smooth out the temperature gradients, such that an optical performance similar to that of an evacuated telescope may be reached.

In the narrower portion of the beam, between windows W2 and W3, the optical path is evacuated. A small portion of the beam is reflected by W3 to the wave front sensor, which measures the deformations of the incoming wave front so that they may be corrected for in real time by the live optics system. The low-frequency components of the wave front errors are compensated for by slow adjustments of the shape and tilt of the primary, meniscus-shaped mirror M1 and axial translation (focusing) of mirror M3. The high-frequency components are compensated for by the two flat mirrors at M5, which are located at an image of the entrance pupil of the telescope. The first of these two mirrors is a fast tilting, beam-steering mirror that takes out coherent shifts of the image (including shifts due to vibrations of the telescope). The second mirror is the segmented, "adaptive" mirror that in real time corrects for the wave front errors due to turbulence in the earth's atmosphere.

REFERENCES

Abramowitz, M., Stegun, I.A.: 1970, *Handbook of Mathematical Functions*, Dover.

Alfvén, H., Fälthammar, C.-G.: 1963, *Cosmical Electrodynamics — Fundamental Principles*, 2nd edition, Oxford Univ. Press, p. 137.

Andersen, T.: 1993, *LEST Foundation Annual Report 1992*, p. 7.

Auer, L.H., Heasley, J.N.: 1978, *Astron. Astrophys.* **64**, 67.

Auer, L.H., Heasley, J.N., House, L.L.: 1977, *Solar Phys.* **55**, 47.

Babcock, H.W.: 1953, *Astrophys. J.* **118**, 387.

Balthasar, H.: 1984, *Solar Phys.* **93**, 219.

Beckers, J.M.: 1969, *A Table of Zeeman Multiplets*, AFCRL-69-0115, US Air Force.

Beckers, J.M., Stenflo, J.O.: 1969, *Solar Phys.* **6**, 480.

Berman, P.R., Lamb, W.E. Jr.: 1969, *Phys. Rev.* **187**, 221.

Bethe, H.: 1933, *Handbuch der Physik* **24/1**, 438.

Bevington, P.R.: 1969, *Data Reduction and Error Analysis for the Physical Sciences*, McGraw-Hill, New York.

Bommier, V: 1980, *Astron. Astrophys.* **87**, 109.

Bommier, V., Sahal-Bréchot, S.: 1978, *Astron. Astrophys.* **69**, 57.

Bommier, V., Sahal-Bréchot, S.: 1982, *Solar Phys.* **78**, 157.

Bommier, V., Leroy, J.-L., Sahal-Bréchot, S.: 1986, *Astron. Astrophys.* **156**, 90.

Born, M., Wolf, E.: 1970, *Principles of Optics*, 4th edition, Pergamon Press.

Brault, J.W.: 1978, *Osserv. Mem. Oss. Astrofis. Arcetri* **106**, 33.

Brault, J.W.: 1985, in A.O. Benz, M.C.E. Huber, M. Mayor (eds.), *High Resolution in Astronomy*, 15th Advanced Course, Saas-Fee, Geneva Obs., p. 1.

Brault, J.W., Noyes, R.: 1983, *Astrophys. J.* **269**, L61.

Brink, D.M., Satchler, G.R.: 1968, *Angular Momentum*, 2nd ed., Clarendon Press.

Bruls, J.H.M.J., Solanki, S.K.: 1994, *Astron. Astrophys.*, in press.

Bünte, M.: 1993, *Astron. Astrophys.* **276**, 236.

Bünte, M., Steiner, O., Pizzo, V.J.: 1993a, *Astron. Astrophys.* **268**, 299.

Bünte, M., Solanki, S.K., Steiner, O.: 1993b, *Astron. Astrophys.* **268**, 736.

Bünte, M., Hasan, S., Kalkofen, W.: 1993c, *Astron. Astrophys.* **273**, 287.

Carlsson, M., Rutten, R.J., Shchukina, N.G.: 1992, *Astron. Astrophys.* **253**, 567.

Chandrasekhar, S.: 1950, *Radiative Transfer*, Clarendon Press, Oxford.

Chapman, G.A.: 1979, *Astrophys. J.* **232**, 923.

Collett, E.: 1993, *Polarized Light — Fundamentals and Applications*, Marcel Dekker, Inc., New York/Basel/Hong Kong.

Condon, E.U., Shortley, G.H.: 1970, *The Theory of Atomic Spectra*, Cambridge Univ. Press.

Corney, A.: 1977, *Atomic and Laser Spectroscopy*, Clarendon Press, Oxford.

Deinzer, W., Hensler, G., Schüssler, M., Weisshaar, E.: 1984a, *Astron. Astrophys.* **139**, 426.

Deinzer, W., Hensler, G., Schüssler, M., Weisshaar, E.: 1984b, *Astron. Astrophys.* **139**, 435.
Dolginov, A.Z., Pavlov, G.G.: 1973, *Astron. Zhurnal* **50**, 762.
Domke, H., Hubeny, I.: 1988, *Astrophys. J.* **334**, 527.
Domke, H., Staude, J.: 1973, *Solar Phys.* **31**, 279.
Durney, B.R., De Young, D.S., Roxburgh, I.W.: 1993, *Solar Phys.* **145**, 207.
Emonet Th.: 1992, Diplomarbeit, ETH Zürich.
Engvold, O., Andersen, T. (eds.): 1990, *LEST Design*, LEST Foundation.
Faurobert-Scholl, M.: 1992, *Astron. Astrophys.* **258**, 521.
Faurobert-Scholl, M.: 1993, *Astron. Astrophys.* **268**, 765.
Frazier, E.N.: 1970, *Solar Phys.* **14**, 89.
Frazier, E.N., Stenflo, J.O.: 1972, *Solar Phys.* **27**, 330.
Frazier, E.N., Stenflo, J.O.: 1978, *Astron. Astrophys.* **70**, 789.
Gilman, P.A.: 1983, in J.O. Stenflo (ed.), *Solar and Stellar Magnetic Fields: Origins and Coronal Effects*, *IAU Symp.* **102**, 247.
Gingerich, O., Noyes, R.W., Kalkofen, W., Cuny, Y.: 1971, *Solar Phys.* **18**, 347.
Giovanelli, R.G.: 1980, *Solar Phys.* **68**, 49.
Gokhale, M.H., Zwaan, C.: 1972, *Solar Phys.* **26**, 52.
Gonsalves, R.A., Chidlaw, R.: 1979, in A.G. Tescher (ed.), *Applications of Digital Image Processing III*, Proc. Soc. Photo-Opt. Instrum. Eng., p. 32.
Grossmann-Doerth, U., Larsson, B., Solanki, S.K.: 1988a, *Astron. Astrophys.* **204**, 266.
Grossmann-Doerth, U., Schüssler, M., Solanki, S.K.: 1988b, *Astron. Astrophys.* **206**, L37.
Grossmann-Doerth, U., Schüssler, M., Solanki, S.K.: 1989a, *Astron. Astrophys.* **221**, 338.
Grossmann-Doerth, U., Knölker, M., Schüssler, M., Weisshaar, E.: 1989b, in R.J. Rutten, G. Severino (eds.), *Solar and Stellar Granulation*, Kluwer, p. 481.
Grossmann-Doerth, U., Schüssler, M., Solanki, S.K.: 1991, *Astron. Astrophys.* **249**, 239.
Hale, G.E.: 1908, *Astrophys. J.* **28**, 100.
Harvey, J.W.: 1977, in E.A. Müller (ed.), *Highlights of Astronomy* **4**, 223.
Harvey, J.W.: 1985, in M.J. Hagyard, *Measurements of Solar Vector Magnetic Fields*, NASA Conf. Publ. 2374, p. 109.
Harvey, J.W., Hall, D.: 1975, *Bull. American Astron. Soc.* **7**, 459.
Harvey, J.W., Livingston, W., Slaughter, C.: 1972, in *Line Formation in the Presence of Magnetic Fields*, High Altitude Obs., NCAR, Boulder, CO, p. 227.
Harvey, K.L. (ed.): 1992, *The Solar Cycle*, PASP Conf. Ser. **27**.
Harvey, K.L., Martin, S.F.: 1973, *Solar Phys.* **32**, 389.
Harvey, K.L., Harvey, J.W., Martin, S.F.: 1975, *Solar Phys.* **40**, 87.
Henze, W. Jr., Stenflo, J.O.: 1987, *Solar Phys.* **111**, 243.
Henze, W. Jr., Tandberg-Hanssen, E., Hagyard, M.J., Woodgate, B.E., Shine, R.A., Beckers, J.M., Bruner, M., Gurman, J.B., Hyder, C.L., West, E.A.: 1982, *Solar Phys.* **81**, 231.
Hewagama, T., Deming, D., Jennings, D.E., Osherovich, V., Wiedemann, G., Zipoy, D., Mickey, D.L., Garcia, H.: 1993, *Astrophys. J. Suppl. Ser.* **86**, 313.

Högbom, J.: 1989, in O. von der Lühe (ed.), *High Spatial Resolution Solar Observations*, National Solar Obs., Sunspot, NM, p. 166.
House, L.L.: 1970a, *J. Quant. Spectrosc. Radiat. Transfer* **10**, 909.
House, L.L.: 1970b, *J. Quant. Spectrosc. Radiat. Transfer* **10**, 1171.
House, L.L.: 1971, *J. Quant. Spectrosc. Radiat. Transfer* **11**, 367.
House, L.L., Steinitz, R.: 1975, *Astrophys. J.* **195**, 235.
Howard, R., Stenflo, J.O.: 1972, *Solar Phys.* **22**, 402.
Illing, R.M.E., Landman, D.A., Mickey, D.L.: 1975, *Astron. Astrophys.* **41**, 183.
Jackson, J.D.: 1975, *Classical Electrodynamics*, Wiley, New York.
Jefferies, J.T., Mickey, D.L.: 1991, *Astrophys. J.* **372**, 694.
Jefferies, J.T., White, O.R.: 1960, *Astrophys. J.* **132**, 767.
Jefferies, J.T., Lites, B.W., Skumanich, A.: 1989, *Astrophys. J.* **343**, 920.
Jones, H.P., Giovanelli, R.G.: 1983, *Solar Phys.* **87**, 37.
Keller, C.U.: 1992, *Nature* **359**, 307.
Keller, C.U., von der Lühe, O.: 1992, *Astron. Astrophys.* **261**, 321.
Keller, C.U., Solanki, S.K., Steiner, O., Stenflo, J.O.: 1990, *Astron. Astrophys.* **233**, 583.
Keller, C.U., Aebersold, F., Egger, U., Povel, H.P., Steiner, P., Stenflo, J.O.: 1992, *LEST Foundation Technical Report* No. **53**.
Kemp, J.C.: 1969, *J. Opt. Soc. Am.* **59**, 950.
Kiepenheuer, K.O.: 1953, *Astrophys. J.* **117**, 447.
Klein, K.-L.: 1992, in M. Faurobert-Scholl, H. Frisch, N. Mein (eds.), *Methodes de determination des champs magnetiques solaires et stellaires*, Observatoire de Paris, p. 113.
Knölker, M., Schüssler, M.: 1988, *Astron. Astrophys.* **202**, 275.
Knölker, M., Schüssler, M., Weisshaar, E.: 1988, *Astron. Astrophys.* **194**, 257.
Krause, F., Rädler, K.-H.: 1980, *Mean-Field Magnetohydrodynamics*, Pergamon Press, Oxford.
Labeyrie, A.: 1970, *Astron. Astrophys.* **6**, 85.
Lamb, F., ter Haar, D.: 1971, *Phys. Rep.* **2C**, 253.
Landi Degl'Innocenti, E.: 1982, *Solar Phys.* **79**, 291.
Landi Degl'Innocenti, E.: 1983, *Solar Phys.* **85**, 3.
Landi Degl'Innocenti, E.: 1984, *Solar Phys.* **91**, 1.
Landi Degl'Innocenti, E.: 1985a, *Solar Phys.* **99**, 1.
Landi Degl'Innocenti, E.: 1985b, *Solar Phys.* **102**, 1.
Landi Degl'Innocenti, E.: 1988, *Astron. Astrophys.* **192**, 374.
Landi Degl'Innocenti, E.: 1992, in F. Sánchez, M. Collados, M. Vázquez (eds.), *Solar Observations: Techniques and Interpretation*, Cambridge Univ. Press, p. 73.
Landi Degl'Innocenti, E., Landi Degl'Innocenti, M.: 1972, *Solar Phys.* **27**, 319.
Landi Degl'Innocenti, E., Landi Degl'Innocenti, M.: 1977, *Astron. Astrophys.* **56**, 111.
Landi Degl'Innocenti, E., Landi Degl'Innocenti, M.: 1985, *Solar Phys.* **97**, 239.
Landi Degl'Innocenti, E., Landolfi, M.: 1982, *Solar Phys.* **77**, 13.
Landi Degl'Innocenti, E., Bommier, V., Sahal-Bréchot, S.: 1991a, *Astron. Astrophys.* **244**, 391.

Landi Degl'Innocenti, E., Bommier, V., Sahal-Bréchot, S.: 1991b, *Astron. Astrophys.* **244**, 401.
Leroy, J.-L.: 1985, in M.J. Hagyard (ed.), *Measurements of Solar Vector Magnetic Fields*, NASA Conf. Publ. 2374, p. 121.
Leroy, J.-L., Ratier, G., Bommier, V.: 1977, *Astron. Astrophys.* **54**, 811.
Lites, B.W.: 1993, *Solar Phys.* **143**, 229.
Lites, B.W., Elmore, D., Murphy, G., Skumanich, A., Tomczyk, S., Dunn, R.B.: 1991, in L.J. November (ed.), *Solar Polarimetry*, NSO/SP Summer Workshop Series No. **11**, 3.
Löfdahl, M.G., Scharmer, G.B.: 1993, in R.R. Radick, *Real Time and Post-Facto Solar Image Correction*, NSO/SP Summer Workshop Series, in press.
Loudon, R.: 1983, *The Quantum Theory of Light*, 2nd ed., Clarendon Press.
Marquardt, D.W.: 1963, *J. Soc. Ind. Appl. Math.* **11**, 431.
Martin, S.F.: 1990, in J.O. Stenflo (ed.), *Solar Photosphere: Structure, Convection and Magnetic Fields, IAU Symp.* **138**, 129.
Martin, S.F., Harvey, K.L.: 1979, *Solar Phys.* **64**, 93.
Martínez Pillet, V., Moreno-Insertis, F., Vázquez, M.: 1993, *Astron. Astrophys.* **274**, 521.
Mathys, G., Stenflo, J.O.: 1987a, *Astron. Astrophys.* **171**, 368.
Mathys, G., Stenflo, J.O.: 1987b, *Astron. Astrophys. Suppl. Ser.* **67**, 557.
McGuire, J.P., Chipman, R.A.: 1988, *LEST Foundation Technical Report* No. **36**.
Merkle, F., Engvold, O., Falomo, R. (eds.): 1987, *LEST Foundation Technical Report* No. **28**.
Meyer, E., Schmidt, H.U., Weiss, N.O.: 1977, *Monthly Notices Roy. Astron. Soc.* **179**, 741.
Michard, R., Rayrole, J.: 1970, *Astron. Astrophys.* **4**, 36.
Mihalas, D.: 1978, *Stellar Atmospheres*, Freeman and Company, San Francisco.
Moruzzi, G., Strumia, F. (eds.): 1991, *The Hanle Effect and Level-Crossing Spectroscopy*, Plenum Press.
Murphy, G.A.: 1990, NCAR Cooperative Thesis No. 124.
Nordlund, Å.: 1983, in J.O. Stenflo (ed.), *Solar and Stellar Magnetic Fields: Origins and Coronal Effects, IAU Symp.* **102**, 79.
Nordlund, Å.: 1986, in W. Deinzer, M. Knölker, H.H. Voigt (eds.), *Small Scale Magnetic Flux Concentrations in the Solar Photosphere*, Vandenhoeck & Ruprecht, Göttingen, p. 83.
Nordlund, Å, Stein, R.F.: 1990, in J.O. Stenflo (ed.), *Solar Photosphere: Structure, Convection and Magnetic Fields, IAU Symp.* **138**, 191.
November, L.J.: 1991, in L.J. November (ed.), *Solar Polarimetry*, NSO/SP Summer Workshop Series No. **11**, p. 149.
Ogawara, Y., Takano, T., Kato, T., Kosugi, T., Tsuneta, S., Watanabe, T., Kondo, I., Uchida, Y.: 1991, *Solar Phys.* **136**, 1.
Omont, A., Smith, E.W., Cooper, J.: 1972, *Astrophys. J.* **175**, 185.
Omont, A., Smith, E.W., Cooper, J.: 1973, *Astrophys. J.* **182**, 283.
Palik, E.D.: 1985, *Handbook of Optical Constants of Solids*, Academic Press, Inc.
Parker, E.N.: 1955, *Astrophys. J.* **122**, 293.
Parker, E.N.: 1978, *Astrophys. J.* **221**, 368.
Petrovay, K., Szakály, G.: 1993, *Astron. Astrophys.* **274**, 543.

Povel, H.P., Aebersold, H., Stenflo, J.O.: 1990, *Appl. Optics* **29**, 1186.
Povel, H.P., Keller, C.U., Stenflo, J.O.: 1991, in L.J. November (ed.), *Solar Polarimetry*, NSO/SP Summer Workshop Series No. **11**, p. 102.
Priest, E.R.: 1982, *Solar Magnetohydrodynamics*, Kluwer.
Querfeld, C.W., Smartt, R.N., Bommier, V., Landi Degl'Innocenti, E., House, L.L.: 1985, *Solar Phys.* **96**, 277.
Rabin, D.: 1992a, *Astrophys. J.* **390**, L103.
Rabin, D.: 1992b, *Astrophys. J.* **391**, 832.
Rachkovsky, D.N.: 1962a, *Izv. Krymsk. Astrofiz. Obs.* **27**, 148.
Rachkovsky, D.N.: 1962b, *Izv. Krymsk. Astrofiz. Obs.* **28**, 259.
Rachkovsky, D.N.: 1963, *Izv. Krymsk. Astrofiz. Obs.* **29**, 97.
Roberts, B., Webb, A.R.: 1978, *Solar Phys.* **56**, 5.
Robson, B.A.: 1974, *The Theory of Polarization Phenomena*, Clarendon Press.
Rüedi, I., Solanki, S.K., Livingston, W., Stenflo, J.O.: 1992, *Astron. Astrophys.* **263**, 323.
Ruiz Cobo, B., del Toro Iniesta, J.C.: 1992, *Astrophys. J.* **398**, 375.
Sahal-Bréchot, S., Bommier, V., Leroy, J.-L.: 1977, *Astron. Astrophys.* **59**, 223.
Sahal-Bréchot, S., Malinovsky, M., Bommier, V.: 1986, *Astron. Astrophys.* **168**, 284.
Saliba, G.J.: 1985, *Solar Phys.* **98**, 1.
Sánchez Almeida, J.: 1992, *Solar Phys.* **137**, 1.
Sánchez Almeida, J., Martínez Pillet, V.: 1992, *Astron. Astrophys.* **260**, 543.
Sánchez Almeida, J., Vela Villahoz, E.: 1993, *Astron. Astrophys.* **280**, 688.
Schüssler, M.: 1984, *Astron. Astrophys.* **140**, 453.
Schüssler, M.: 1987, in E.H. Schröter, M. Vázquez, A.A. Wyller (eds.), *The Role of Fine-scale Magnetic Fields on the Structure of the Solar Atmosphere*, Cambridge Univ. Press, p. 223.
Sheeley, N.R. Jr., Nash, A.G., Wang, Y.-M.: 1987, *Astrophys. J.* **319**, 481.
Shore, B.W., Menzel, D.H.: 1968, *Principles of Atomic Spectra*, Wiley, New York.
Shurcliff, W.A.: 1962, *Polarized Light: Production and Use*, Harvard Univ. Press.
Šidlichovský, M.: 1974, *Bull. Astron. Inst. Czech.* **25**, 198.
Šidlichovský, M.: 1976, *Bull. Astron. Inst. Czech.* **27**, 71.
Simon, G.W., Leighton, R.B.: 1964, *Astrophys. J.* **140**, 1120.
Skumanich, A., Lites, B.W.: 1987, *Astrophys. J.* **322**, 473.
Snodgrass, H.B.: 1983, *Astrophys. J.* **270**, 288.
Solanki, S.K.: 1984, in T.D. Guyenne, J.J. Hunt (eds.), *The Hydromagnetics of the Sun*, Proc. Fourth European Meeting on Solar Physics, ESA SP–220, p. 63.
Solanki, S.K.: 1985, in H.U. Schmidt (ed.), *Theoretical Problems in High Resolution Solar Physics*, Max Planck Inst. f. Astrophys., Munich, p. 172.
Solanki, S.K.: 1986, *Astron. Astrophys.* **168**, 311.
Solanki, S.K.: 1987, *Ph.D. Thesis*, No. 8309, ETH Zürich.
Solanki, S.K.: 1989, *Astron. Astrophys.* **224**, 225.
Solanki, S.K.: 1993, *Space Sci. Rev.* **63**, 1.
Solanki, S.K., Brigljević, V.: 1992, *Astron. Astrophys.* **262**, L29.
Solanki, S.K., Steiner, O.: 1990, *Astron. Astrophys.* **234**, 519.
Solanki, S.K., Stenflo, J.O.: 1984, *Astron. Astrophys.* **140**, 185.
Solanki, S.K., Stenflo, J.O.: 1985, *Astron. Astrophys.* **148**, 123.

Solanki, S.K., Rüedi, I., Livingston, W.: 1992, *Astron. Astrophys.* **263**, 312.
Spiegel, E.A., Weiss, N.O.: 1980, *Nature* **287**, 616.
Spruit, H.C.: 1974, *Solar Phys.* **34**, 277.
Spruit, H.C.: 1976, *Solar Phys.* **50**, 269.
Spruit, H.C.: 1979, *Solar Phys.* **61**, 363.
Spruit, H.C., Zweibel, E.G.: 1979, *Solar Phys.* **62**, 15.
Spruit, H.C., Title, A.M., van Ballegooijen, A.A.: 1987, *Solar Phys.* **110**, 115.
Steenbeck, M., Krause, F.: 1969, *Astron. Nachr.* **291**, 49.
Steiner, O., Stenflo, J.O.: 1990, in J.O. Stenflo (ed.), *Solar Photosphere: Structure, Convection and Magnetic Fields, IAU Symp.* **138**, 181.
Steiner, O., Pneuman, G.W., Stenflo, J.O.: 1986, *Astron. Astrophys.* **170**, 126.
Stenflo, J.O.: 1969, *Solar Phys.* **6**, 476.
Stenflo, J.O.: 1973, *Solar Phys.* **32**, 41.
Stenflo, J.O.: 1975, *Solar Phys.* **42**, 79.
Stenflo, J.O.: 1976a, in V. Bumba, J. Kleczek (eds.), *Basic Mechanisms of Solar Activity, IAU Symp.* **71**, 69.
Stenflo, J.O.: 1976b, *Astron. Astrophys.* **46**, 61.
Stenflo, J.O.: 1977, in R.M. Bonnet, P. Delache (eds.), *The Energy Balance and Hydrodynamics of the Solar Chromosphere and Corona, IAU Coll.* **36**, 143.
Stenflo, J.O.: 1978a, *Rep. Prog. Phys.* **41**, 865.
Stenflo, J.O.: 1978b, *Astron. Astrophys.* **66**, 241.
Stenflo, J.O.: 1980, *Astron. Astrophys.* **84**, 68.
Stenflo, J.O.: 1982, *Solar Phys.* **80**, 209.
Stenflo, J.O.: 1984, *Appl. Optics* **23**, 1267.
Stenflo, J.O.: 1985, in M.J. Hagyard, *Measurements of Solar Vector Magnetic Fields*, NASA Conf. Publ. 2374, p. 263.
Stenflo, J.O.: 1987, *Solar Phys.* **114**, 1.
Stenflo, J.O.: 1988, *Astrophys. Space Sci.* **144**, 321.
Stenflo, J.O.: 1989, *Astron. Astrophys. Rev.* **1**, 3.
Stenflo, J.O. (ed.): 1990, *Solar Photosphere: Structure, Convection and Magnetic Fields, IAU Symp.* **138**.
Stenflo, J.O.: 1991a, in G. Moruzzi, F. Strumia (eds.), *The Hanle Effect and Level-Crossing Spectroscopy*, Plenum Press, p. 237.
Stenflo, J.O.: 1991b, in L.J. November (ed.), *Solar Polarimetry*, NSO/SP Summer Workshop Series No. **11**, 416.
Stenflo, J.O.: 1991c, *LEST Foundation Technical Report* No. **44**.
Stenflo, J.O.: 1991d, *LEST Foundation Technical Report* No. **45**.
Stenflo, J.O.: 1992, in K.L. Harvey (ed.), *The Solar Cycle*, ASP Conf. Ser. **27**, 83.
Stenflo, J.O., Güdel, M.: 1988, *Astron. Astrophys.* **191**, 137.
Stenflo, J.O., Harvey, J.W.: 1985, *Solar Phys.* **95**, 99.
Stenflo, J.O., Lindegren, L.: 1977, *Astron. Astrophys.* **59**, 367.
Stenflo, J.O., Povel, H.P.: 1985, *Appl. Optics* **24**, 3893.
Stenflo, J.O., Solanki, S.K.: 1992, in L. Damé, T.D. Guyenne (eds.), *Solar Physics and Astrophysics at Interferometric Resolution*, ESA SP-344, p. 197.
Stenflo, J.O., Stenholm, L.: 1976, *Astron. Astrophys.* **46**, 69.
Stenflo, J.O., Vogel, M.: 1986, *Nature* **319**, 285.
Stenflo, J.O., Biverot, H., Stenmark, L.: 1976, *Appl. Optics* **15**, 1188.

Stenflo, J.O., Baur, T.G., Elmore, D.F.: 1980a, *Astron. Astrophys.* **84**, 60.
Stenflo, J.O., Dravins, D., Wihlborg, N., Öhman, Y., Bruns, A., Prokof'ev, V.K., Zhitnik, I.A., Biverot, H., Stenmark, L.: 1980b, *Solar Phys.* **66**, 13.
Stenflo, J.O., Twerenbold, D., Harvey, J.W.: 1983a, *Astron. Astrophys. Suppl. Ser.* **52**, 161.
Stenflo, J.O., Twerenbold, D., Harvey, J.W., Brault, J.W.: 1983b, *Astron. Astrophys. Suppl. Ser.* **54**, 505.
Stenflo, J.O., Harvey, J.W., Brault, J.W., Solanki, S.K: 1984, *Astron. Astrophys.* **131**, 33.
Stenflo, J.O., Solanki, S.K., Harvey, J.W.: 1987a, *Astron. Astrophys.* **171**, 305.
Stenflo, J.O., Solanki, S.K., Harvey, J.W.: 1987b, *Astron. Astrophys.* **173**, 167.
Stenflo, J.O., Keller, C.U., Povel, H.P.: 1992, *LEST Foundation Technical Report* No. **54**.
Stenholm, L.G., Stenflo, J.O.: 1978, *Astron. Astrophys.* **67**, 33.
Stepanov, V.E.: 1958a, *Izv. Krymsk. Astrofiz. Obs.* **18**, 136.
Stepanov, V.E.: 1958b, *Izv. Krymsk. Astrofiz. Obs.* **19**, 20.
Stix, M.: 1989, *The Sun — An Introduction*, Springer.
Tanenbaum, A.S., Wilcox, J.M., Frazier, E.N., Howard, R.: 1969, *Solar Phys.* **9**, 328.
Tuominen, I., Moss, D., Rüdiger, G. (eds.): 1991, *The Sun and Cool Stars: Activity, Magnetism, Dynamos. Lecture Notes in Physics* **380**.
Uchida, Y.: 1993, in M. Pick, M. Machado (eds.), *Fundamental Problems of Solar Activity*, Pergamon Press.
Unno, W.: 1956, *Publ. Astron. Soc. Japan* **8**, 108.
Unno, W.: 1959, *Astrophys. J.* **129**, 375.
Unno, W., Ando, H.: 1979, *Geophys. Astrophys. Fluid Dyn.* **12**, 107.
Unsöld, A.: 1955, *Physik der Sternatmosphären*, 2nd ed., Springer.
van Ballegooijen, A.A.: 1985, in M.J. Hagyard (ed.), *Measurements of Solar Vector Magnetic Fields*, NASA Conf. Publ. 2374, p. 322.
Veeder, G.J., Zirin, H.: 1970, *Solar Phys.* **12**, 391.
Vela Villahoz, E., Sánchez Almeida, J., Wittmann, A.: 1993, *Astron. Astrophys.*, in press.
von der Lühe, O.: 1989, in O. von der Lühe (ed.), *High Spatial Resolution Solar Observations*, National Solar Obs., Sunspot, NM, p. 147.
Wittmann, A.: 1973, *Solar Phys.* **33**, 107.
Zayer, I., Solanki, S.K., Stenflo, J.O.: 1989, *Astron. Astrophys.* **211**, 463.
Zayer, I., Solanki, S.K., Stenflo, J.O., Keller, C.U.: 1990, *Astron. Astrophys.* **239**, 356.
Zeeman, P.: 1897a, *Phil. Mag.* **43**, 226.
Zeeman, P.: 1897b, *Phil. Mag.* **44**, 55.
Zeeman, P.: 1897c, *Phil. Mag.* **44**, 255.
Zeldovich, Ya.B., Ruzmaikin, A.A., Sokoloff, D.D.: 1983, *Magnetic Fields in Astrophysics*, Gordon and Breach.
Zirin, H.: 1987, *Solar Phys.* **110**, 101.
Zwaan, C.: 1987, *Ann. Rev. Astron. Astrophys.* **25**, 83.

SYMBOL INDEX

The page number indicates where in the book the physical symbol is first defined or used. The same symbol is listed more than once if it appears with more than one meaning. The main letter of a symbol determines the alphabetical order, with capitals appearing before lower-case symbols. The second-rank order is only partly determined by the index letters, since it has been found more useful to group together physically related symbols in the list. When alphabetically ordered, latin letters come before greek, with numbers being last.

LATIN ALPHABET

GREEK ALPHABET

SUBJECT INDEX

Among the many occurrences of the index words in the book, only the most pertinent ones have been selected for the subject index. To further aid the reader in finding an explanation of a concept, some page numbers have been given in italics to suggest where to look first. An f after a page number means occurrence on the following page, ff means occurrence on two or more following pages.